THE USSR OLYMPIAD PROBLEM BOOK

Selected Problems and Theorems
of Elementary Mathematics

D. O. SHKLARSKY
N. N. CHENTZOV
I. M. YAGLOM

REVISED AND EDITED BY
IRVING SUSSMAN,
University of Santa Clara

TRANSLATED BY
JOHN MAYKOVICH,
University of Santa Clara

DOVER PUBLICATIONS
Garden City, New York

Bibliographical Note

This Dover edition, first published in 1993, is an unabridged and unaltered republication of the work first published by W.H. Freeman and Company, San Francisco, in 1962.

Library of Congress Cataloging-in-Publication Data

Shkliarskiĭ, D. O. (David Oskarovich), 1918–1942.
    [Izbrannye zadachi i teoremy elementarnoĭ matematiki, ch 1. English]
    The USSR Olympiad problem book : selected problems and theorems of elementary mathematics / D.O. Shklarsky, N.N. Chentzov, I.M. Yaglom ; translated by John Maykovich.—3rd ed / rev. and edited by Irving Sussman.
        p.   cm.
    ISBN-13: 978-0-486-27709-7
    ISBN-10: 0-486-27709-7
    1. Mathematics—Problems, exercises, etc.   I. Chentsov, N. N. (Nikolaĭ Nikolaevich) II. IAglom, I. M. (Isaak Moiseevich), 1921–.   III. Sussman, Irving.   IV. Title.
QA43.S5813   1994
510′.76—dc20   93-11553
                                                            CIP

Manufactured in the United States of America
27709721   2023
www.doverpublications.com

FOREWORD TO THE

THIRD (Russian) EDITION

THIS BOOK CONTAINS 320 unconventional problems in algebra, arithmetic, elementary number theory, and trigonometry. Most of these problems first appeared in competitive examinations sponsored by the School Mathematical Society of the Moscow State University and in the Mathematical Olympiads held in Moscow. The book is designed for students having a mathematical background at the high school level;[†] very many of the problems are within reach of seventh and eighth grade students of outstanding ability. Solutions are given for all the problems. The solutions for the more difficult problems are especially detailed.

The third (Russian) edition differs from the second chiefly in the elimination of errors detected in the second edition. Therefore, the preface to the second edition is retained.

---

† The level of academic attainment referred to as "high school level" is the American ninth to twelfth grades. The USSR equivalent is seventh to tenth grades. This means that this material is introduced about two years earlier in the Russian schools. Since Russian children begin their first grade studies about a year later than do American children, the actual age disparity is not as much as two years [*Editor*].

PREFACE TO THE

SECOND (Russian) EDITION

THE PRESENT VOLUME, which constitutes the first part of a collection, contains 320 problems involving principally algebra and arithmetic, although several of the problems are of a type meant only to encourage the development of logical thought (see, for example, problems 1-8).

The problems are grouped into twelve separate sections. The last four sections (Complex Numbers, Some Problems from Number Theory, Inequalities, Numerical Sequences and Series) contain important theoretical material, and they may well serve as study topics for school mathematical societies or for the Society on Elementary Mathematics at the pedagogical institutes. In this respect the supplementary references given in various sections will also prove useful. All the other sections [especially Alterations of Digits in Integers and Solutions of Equations in Integers (Diophantine equations)] should yield material profitable for use in mathematics clubs and societies.

Of the twelve sections, only four (Miscellaneous Problems in Algebra, Polynomial Algebra, Complex Numbers, Inequalities) concern algebra; the remaining sections deal with arithmetic and number theory. A special effort has been made to play down problems (particularly those in algebra) involving detailed manipulative matter. This was done to avoid duplicating material in the excellent *Problem Book in Algebra,* by V. A. Kretchmar (Government Technical Publishing House, Moscow 1950). On the other hand, an effort has been made to render much of the book attainable to eighth grade, and even seventh grade, students.

More than three years have passed since the appearance of the first edition of this book. During this period the original authors received a great many written and oral communications with respect to it, and these have been seriously considered in the reworking of the material and in deciding which features were worth retaining and emphasizing and which aspects were weak. As a result, the book has undergone considerable revision. About sixty problems that were in the first edition have been omitted—some appeared to be too difficult, or were insufficiently interesting, and others did not fit into the new structure of the book. Approximately 120 new problems have been added. The placing of each problem into a suitable section has been restudied; the sections have been repositioned; all the solutions have been reworked (several were replaced by simplified or better solutions); and alternative solutions have been provided for some of the problems. Hints have been given for every problem, and those problems which to the authors appear of greater difficulty have been starred(*). Sections 3, 5, 6, 9, and 10 have undergone such significant changes that they may be considered as having been completely rewritten. Sections 1, 2, 4, 7, and 11 have been revised radically, and only Sections 8 and 12 have had relatively minor alterations.

The first edition of the book was prepared by I. M. Yaglom in collaboration with G. M. Adelson-Vel'sky (who contributed the section on alteration of digits in integers and also a number of problems to other sections, particularly to the section on Diophantine equations). An important contribution was made to the first edition by E. E. Balash (who contributed the section on numerical sequences and series) and Y. I. Khorgin (who made the principal contribution to the section on inequalities). Solutions for other problems were written by various directors of the School Mathematical Society of the Moscow State University. About 20 problems were taken from manuscripts of the late D. O. Shklarsky.

The rewriting of the book for the second edition was done by I. M. Yaglom, who made extensive use of the material of the first edition.

In conclusion, the author wishes to thank A. M. Yaglom, whose advice was of invaluable assistance while the book was being written and who initiated the rewriting of the section on complex numbers.

The author is also indebted to the editor, A. Z. Rivkin, whose inde-fatigable labors on the first and second editions made possible many improvements, and to all the readers who made valuable suggestions, especially I. V. Volkova, L. I. Golovina, R. S. Guter, G. Lozanovsky, I. A. Laurya, Y. B. Rutitsky, A. S. Sokolin, and I. Y. Tanatar.

I. M. Yaglom

EDITOR'S FOREWORD TO THE

ENGLISH EDITION

One of the important facets of science education in the USSR has been their series of mathematical competitive examinations held for students of high ability in the secondary schools. Those contests, which are being emulated increasingly in our own educational system, culminate each year in the Soviet Union in their Mathematical Olympiads held at Moscow University, preliminary qualifying and elimination examinations having been held nationwide throughout the academic year.

This book, compiled over a twenty-year period, is a collection of the most interesting and instructive problems posed at these competitions and in other examination centers of the USSR, plus additional problems and material developed for use by the School Mathematics Study Societies. Perhaps the greatest compliment which can be paid to the problems created for this purpose by leading Soviet mathematicians (or taken and adapted from the literature) has been the extent to which the problems have been used in our own contests and examinations.

Soviet students and teachers have had available in published form the problems, and their solutions, given in such examinations, but this material has not generally been available in the United States.

A few of these problems have been translated and published in such American journals as *The American Mathematical Monthly* of The Mathematical Association of America, and problems of similar scope appear as regular features of Several American journals. Except for some compilations from these sources, little exists by way of problems which deal with real and active mathematics instead of the fringe and recreational aspects of the science or with conventional textbook exercises.

This translation and revision of the Third Revised and Augmented Edition of the Olympiad Problem Book should therefore fill a very definite need in American schools and colleges. It contains 320 problems—a few of them merely recreational and thought-provoking, but most of them seriously engaged with solid and important mathematical theory, albeit the preparational background is assumed to be elementary. The problems are from algebra, arithmetic, trigonometry, and number theory, and all of them emphasize the creative aspects of these subjects. The material coordinates beautifully with the new concepts which are being emphasized in American schools, since the "unconventional" designation attributed to the problems by the original authors means that they stress originality of thought rather than mere manipulative ability and introduce the necessity for finding new methods of attack.

In this respect I am reminded of the observation made by some forgotten character in some forgotten novel who opined that the ultimate test of an educative effort lay not nearly so much in what sort of questions the students could finally answer as in what sort of questions they could finally be asked!

Complete solutions to all problems are given; in many cases, alternate solutions are detailed from different points of view. Although most of the problems presuppose only high school mathematics, they are not in any sense easy: some are of uncommon difficulty and will challenge the ingenuity of any research mathematician. On the other hand, many of the problems will yield readily to a normally bright high school student willing to use his head. Where more advanced concepts are employed, the concepts are discussed in the section preceding the problems, which gives the volume the aspect of a textbook as well as a problem book. The solutions to more advanced problems are given in considerable detail.

Hence this book can be put to use in a variety of ways for students of ability in high schools and colleges. In particular, it lends itself exceptionally well to use in the various Institutes for high school mathematics teachers. It is certainly required reading for teachers dealing with the gifted student and advanced placement classes. It will furnish them with an invaluable fund for supplementary teaching material, for self-study, and for acquiring depth in elementary mathematics.

Except for the elimination of the few misprints and errors found in the original, and some recasting of a few proofs which did not appear to jell when translated literally, the translation is a faithful one: it was felt that the volume would lose something by too much tampering. (For this reason the original foreword and preface have also been retained). Thus the temptation to radically alter or simplify any understandable solution was resisted (as, for example, in the sections on number theory and inequalities, where congruence arithmetic would certainly have supplied some neater and more direct proofs). Some notations which differ in minor respects from the standard American notations have been retained (as, for example, C_n^k instead of C_k^n). These will cause no difficulty.

All references made in the text to books not available in English translation have been retained; no one can know when translations of some of those volumes will appear. Whenever an English translation was known to exist, the translated edition is referred to.

The translation was made from the Third (Russian) Edition of *Selected Problems and Theorems of Elementary Mathematics*, which is the title under which the original volume appeared in the Soviet Union. Mr. John Maykovich, instructor at the University of Santa Clara, was the translator, and he was assisted by Mrs. Alvin (Myra) White, who translated fifty pages. The writing out, revising, editing, annotating, and checking against the original Russian were by my own hand.

Thanks are due the following persons for their assistance in reading portions of the translation, pointing out errors, and making valuable suggestions: Professor George Polya of Stanford University,[†] Professor Abraham Hillman of the University of Santa Clara, and Professor Robert Rosenbaum of Wesleyan University.

I shall be very grateful to readers who are kind enough to point out errors, misprints, misleading statements of problems, and incorrect or obscure proofs found in this edition.

January 1962   Irving Sussman

[†] I would also like to call attention to Professor Polya's new book *Mathematical Discovery* (Wiley) which contains elementary problems and valuable textual discussion of approaches to, and techniques of, problem solving.

CONTENTS

FROM THE AUTHORS

THE THREE VOLUMES that make up the present collection of problems
are the commencement of a series of books based on material
gathered by the School Mathematics Society of the Moscow State
University over a twenty-year period. The text consists of problems
and theorems, most of which have been presented during meetings
of the various sections of the School Mathematical Society of the
M.S.U. as well as in the Mathematical Olympiads held in Moscow.
(The numbers of the problems given in the Olympiads are listed on
p. 5).

These volumes are directed to students, teachers, and directors of
school mathematical societies and societies on elementary mathematics
of the pedagogical institutes. The first volume (Part I) contains
problems in arithmetic, algebra, and number theory. The second
volume is devoted to problems in plane geometry, and the third to
problems in solid geometry.

In contrast to the majority of problem books intended for high
school students, these books are designed not only to reinforce the
student's formal knowledge, but also to acquaint him with methods
and ideas new to him and to develop his predilection for, and ability
in, original thinking. Here, there are few problems whose solutions

1

require mere formal mastery of school mathematics. Also, there are few problems intended for the superficially "clever" or adroit student —that is, problems involving artificial methods for solving equations or systems of equations of higher degree. On the other hand, these books do contain many problems demanding originality and non-standardized formulations.

In the selection of problems, emphasis has been given to those aspects of elementary mathematics which are pertinent to contemporary mathematical developments and new directions. Several groups of problems are worked out in detail in the section on answers and hints, especially when individual problems involve more mature mathematics (for instance, elementary number theory and inequalities). Some of the problems have been taken from classics of the mathematical literature and from articles published in recent mathematical journals.

In view of the unconventional nature of the problems, they may prove difficult for students accustomed to conventional high school exercises. Nevertheless, the School Mathematical Society of the M.S.U. and the directors of the Moscow Mathematical Olympiads believe that such problems are not beyond the persevering student.

It is recommended that the suggestions for using this book be read before the problems are undertaken.

Parts I and II of this work were compiled by I. M. Yaglom, and Part III was done principally by N. N. Chentzov. In addition to the listed authors, many directors of the School Mathematical Society of the M.S.U. contributed; their names are listed in the Preface to each volume. Some forty problems were taken from the manuscripts of D. O. Shklarsky, who worked with the School Mathematical Society from 1936 to 1941, and who was killed in action on the military front in 1942. In view of the very great influence which D. O. Shklarsky exerted through his work with the Society, and in particular upon the contents of this volume, it is appropriate to place his name first in authorship of it.

The authors will be grateful to readers who send them new, and possibly better, solutions to the problems or new problems suitable for inclusion in such a book as this.

SUGGESTIONS FOR USING THIS BOOK

THIS BOOK CONTAINS (1) statements of problems, (2) solutions, (3) answers and hints for solving the problems. For more effective use of the book, the answers and hints appear at the end.

The starred problems are more difficult, in the opinion of the authors, than the others; the few double-starred problems are the most difficult. (Naturally, there will be differences of opinion as to which problems are more difficult than others.)

For most of the problems the authors recommend that the reader first attempt a solution without recourse to the hints. If this attempt is unsuccesful, the hint can be referred to, which should aid in arriving at a solution. If, then, the reader cannot solve a problem, he can (of course) read the solution; but if he appears to be successful in finding a solution, he should compare his answer with that given in the answers and hints section. If his answer disagrees with that given, he should try to determine his possible mistake and correct it. If the answers agree, he should compare his solution with that given in the solutions section. If several solutions are given in the answers section, the reader will profit by comparing the various solutions.

These suggestions are perhaps not as pertinent to the starred problems as they are to the others. For the starred problems it might prove advisable for the reader to read the hint before attempting the problem. For the double-starred problems it is recommended that the hints be consulted first. These problems may profitably be considered as theoretical developments and their solutions read as textual material. Also, each of the double-starred problems might be considered as a topic for a special report, or paper, to be given before a mathematics club. Before attempting one of the more difficult problems, the reader should solve and analyze the simpler neighboring problems.

Some solutions involve techniques not ordinarily found in the high school curriculum. With each such problem this information is given in small type.

The problems are, in general, independent of each other; only rarely does the solution of one problem involve the results of another. Some exception is made in the final four sections, where the problems are more closely allied.

NUMERICAL REFERENCE TO THE

PROBLEMS GIVEN IN THE

MOSCOW MATHEMATICAL OLYMPIADS

THE OLYMPIAD MATHEMATICAL COMPETITION for seventh to tenth grade students consists of two examinations. The first (Type I question) is for elimination purposes; the second (Type II question) is for the finalists.

Olympiads	Type I	Type II
For 7th-8th Grade Students		
VI (1940)	48	110(a)
VII (1941)	75	68, 208
VIII (1945)	64(a), 110(b), 152(a)	78(b), 83
IX (1946)	76, 198	30, 125
X (1947)	71(a), 201(a)	5, 91(a), 140
XI (1948)	122	—
XII (1949)	39	9, 11, 92(b), 117(a)
XIII (1950)	—	141(a)
XIV (1951)	7(b)[1]	203
XV (1952)	8, 54[2]	75
For 9th-10th Grade Students		
I (1935)	—	134(d), 176
II (1936)	—	56
V (1939)	168	43. 165, 217
VI (1940)	80, 113	81, 144, 269(b)
VII (1941)	75, 172, 177(a), 214	209
VIII (1945)	33, 64(b), 173	195
IX (1946)	29, 131, 192(a)	95, 126
X (1947)	71(b), 197, 200	10, 91(c)
XI (1948)	190(a)	124(a)
XII (1949)	169	9, 11, 88, 117(b)
XIII (1950)	82, 171	90(b)
XIV (1951)	7(b)	98
XV (1952)	193[3]	194[3]

[1] For sixty teams.
[2] Problems given to eighth to ninth grade students.
[3] Problems given only to tenth grade students.

1

INTRODUCTORY PROBLEMS

1. Every living person has shaken hands with a certain number of other persons. Prove that a count of the number of people who have shaken hands an odd number of times must yield an even number.

2. In chess, is it possible for the knight to go (by allowable moves) from the lower left-hand corner of the board to the upper right-hand corner and in the process to light exactly once on each square?

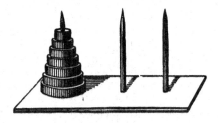

Figure 1

3. (a) N rings having different outer diameters are slipped onto an upright peg, the largest ring on the bottom, to form a pyramid (Figure 1). We wish to transfer all the rings, one at a time, to a second peg, but we have a third (auxiliary) peg at our disposal. During the transfers it is not permitted to place a larger ring on a smaller one. What is the smallest number, k, of moves necessary to complete the transfer to peg number 2?†

(b)* A brain-teaser called the game of Chinese Rings is constructed as follows: n rings of the same size are each connected to a plate by a series of wires, all of which are the same length (see Figure 2). A thin, doubled rod is slipped through the rings in such a way that all the wires are inside the U-opening of the rod. (The wires are free to slide in holes in the plate, as shown.) The problem consists of removing all the rings from the rod. What is the least number of moves necessary to do this?

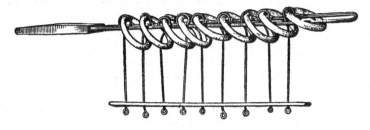

Figure 2

4. (a) We are given 80 coins of the same denomination; we know that one of them is counterfeit and that it is lighter than the others. Locate the counterfeit coin by using four weighings on a pan balance.

(b) It is known that there is one counterfeit coin in a collection of n similar coins. What is the least number of weight trials necessary to identify the counterfeit?

5. Twenty metal blocks are of the same size and external appearance; some are aluminum, and the rest are duraluminum, which is heavier. Using at most eleven weighings on a pan balance, how can we determine how many blocks are aluminum?

6. (a)* Among twelve similar coins there is one counterfeit. It is not known whether the counterfeit coin is lighter or heavier than a genuine one (all genuine coins weigh the same). Using three weighings on a pan balance, how can the counterfeit be identified and in the process determined to be lighter or heavier than a genuine coin?

---

† This is sometimes referred to as the Tower of Hanoi problem [*Editor*].

(b)** There is one counterfeit coin among 1000 similar coins. It is not known whether the counterfeit coin is lighter or heavier than a genuine one. What is the least number of weighings, on a pan balance, necessary to locate the counterfeit and to determine whether it is light or heavy?

Remark: Using the conditions of problem (a) it is possible to locate, in three weighings, one counterfeit out of thirteen coins, but we cannnot determine whether it is light or heavy. For fourteen coins, four weighings are necessary.

It would be interesting to determine the least number of weighings necessary to locate one counterfeit out of 1000 coins if we are relieved of the necessity of determining whether it is light or heavy.

7. (a) A traveler having no money, but owning a gold chain having seven links, is accepted at an inn on the condition that he pay one link per day for his stay. If the traveler is to pay daily, but may take change in the form of links previously paid, and if he remains seven days, what is the least number of links that must be cut out of the chain? (*Note*: A link may be taken from any part of the chain.)

(b) A chain consists of 2000 links. What is the least number of links that must be disengaged from the chain in order that any specified number of links, from 1 to 2000, may be gathered together from the parts of the chain thus formed?

8. Two-hundred students are positioned in 10 rows, each containing 20 students. From each of the 20 columns thus formed the shortest student is selected, and the tallest of these 20 (short) students is tagged A. These students now return to their initial places. Next the tallest student in each row is selected, and from these 10 (tall) students the shortest is tagged B. Which of the two tagged students is the taller (if they are different people)?

9. Given thirteen gears, each weighing an integral number of grams. It is known that any twelve of them may be placed on a pan balance, six on each pan, in such a way that the scale will be in equilibrium. Prove that all the gears must be of equal weight.

10. Refer to the following number triangle.

$$1$$
$$1 \quad 1 \quad 1$$
$$1 \quad 2 \quad 3 \quad 2 \quad 1$$
$$1 \quad 3 \quad 6 \quad 7 \quad 6 \quad 3 \quad 1$$
$$\cdot \quad \cdot \quad \cdot \quad \cdot \quad \cdot \quad \cdot \quad \cdot \quad \cdot$$

Each number is the sum of three numbers of the previous row: the number immediately above it and the numbers immediately to the right and left of that one. If no number appears in one or more of these locations, the number zero is used. Prove that every row, beginning with the third row, contains at least one even number.

11. Twelve squares are laid out in a circular pattern [as on the circumference of a circle]. Four different colored chips, red, yellow, green, blue, are placed on four consecutive squares. A chip may be moved in either a clockwise or a counterclockwise direction over four other squares to a fifth square, provided that the fifth square is not occupied by a chip. After a certain number of moves the same four squares will again be occupied by chips. How many permutations (rearrangements) of the four chips are possible as a result of this process?

12. An island is inhabited by five men and a pet monkey. One afternoon the men gathered a large pile of coconuts, which they proposed to divide equally among themselves the next morning. During the night one of the men awoke and decided to help himself to his share of the nuts. In dividing them into five equal parts he found that there was one nut left over. This one he gave to the monkey. He then hid his one-fifth share, leaving the rest in a single pile. Later during the night another man awoke with the same idea in mind. He went to the pile, divided it into five equal parts, and found that there was one coconut left over. This he gave to the monkey, and then he hid his one-fifth share, restoring the rest to one pile. During the same night each of the other three men arose, one at a time, and in ignorance of what had happened previously, went to the pile, and followed the same procedure. Each time one coconut was left over, and it was given to the monkey. The next morning all five men went to the diminished nut pile and divided it into five equal parts, finding that one nut remained over. What is the least number of coconuts the original pile could have contained?

13. Two brothers sold a herd of sheep which they owned. For each sheep they received as many rubles as the number of sheep originally in the herd. The money was then divided in the following manner. First, the older brother took ten rubles, then the younger brother took ten rubles, after which the older brother took another ten rubles, and so on. At the end of the division the younger brother, whose turn it was, found that there were fewer than ten rubles left, so he took what remained. To make the division just, the older brother gave the younger his penknife. How much was the penknife worth?

14.* (a) On which of the two days of the week, Saturday or Sunday, does New Year's Day fall more often?

(b) On which day of the week does the thirtieth of the month most often fall?

2

ALTERATIONS OF DIGITS IN INTEGERS

15. Which integers have the following property? If the final digit is deleted, the integer is divisible by the new number.

16. (a) Find all integers with initial digit 6 which have the following property, that if this initial digit is deleted, the resulting number is reduced to $\frac{1}{25}$ its original value.

(b) Prove that there does not exist any integer with the property that if its first digit is deleted, the resulting number is $\frac{1}{35}$ the original number.

17.* An integer is reduced to $\frac{1}{6}$ its value when a certain one of its digits is deleted, and the resulting number is again divisible by 9.

(a) Prove that division of this resulting integer by 9 results in deleting an additional digit.

(b) Find all integers satisfying the conditions of the problem.

18. (a) Find all integers having the property that when the third digit is deleted the resulting number divides the original one.

(b)* Find all integers with the property that when the second digit is deleted the resulting number divides the original one.

19. (a) Find the smallest integer whose first digit is 1 and which

11

has the property that if this digit is transferred to the end of the number the number is tripled. Find all such integers.

(b) With what digits is it possible to begin a (nonzero) integer such that the integer will be tripled upon the transfer of the initial digit to the end? Find all such integers.

20. Prove that there does not exist a natural number which, upon transfer of its initial digit to the end, is increased five, six, or eight times.

21. Prove that there does not exist an integer which is doubled when the initial digit is transferred to the end.

22. (a) Prove that there does not exist an integer which becomes either seven times or nine times as great when the initial digit is transferred to the end.

(b) Prove that no integer becomes four times as great when its initial digit is transferred to the end.

23. Find the least integer whose first digit is seven and which is reduced to $\frac{1}{3}$ its original value when its first digit is tranferred to the end. Find all such integers.

24. (a) We say one integer is the "inversion" of another if it consists of the same digits written in reverse order. Prove that there exists no natural number whose inversion is two, three, five, seven, or eight times that number.

(b) Find all integers whose inversions are four or nine times the original number.

25. (a) Find a six-digit number which is multiplied by a factor of 6 if the final three digits are removed and placed (without changing their order) at the beginning.

(b) Prove that there cannot exist an eight-digit number which is increased by a factor of 6 when the final four digits are removed and placed (without changing their order) at the beginning.

26. Find a six-digit number whose product by 2, 3, 4, 5, or 6 contains the same digits as did the original number (in different order, of course).

3

THE DIVISIBILITY OF INTEGERS

27. Prove that for every integer n:
   (a) $n^3 - n$ is divisible by 3;
   (b) $n^5 - n$ is divisible by 5;
   (c) $n^7 - n$ is divisible by 7;
   (d) $n^{11} - n$ is divisible by 11;
   (e) $n^{13} - n$ is divisible by 13.

Note: Observe that $n^9 - n$ is not necessarily divisible by 9 (for example, $2^9 - 2 = 510$ is not divisible by 9).
Problems (a-e) are special cases of a general theorem; see problem 240.

28. Prove the following:
   (a) $3^{6n} - 2^{6n}$ is divisible by 35, for every positive integer n;
   (b) $n^5 - 5n^3 + 4n$ is divisible by 120, for every integer n;
   (c)* for all integers m and n, $mn(m^{60} - n^{60})$ is divisible by the number 56,786,730.

29. Prove that $n^2 + 3n + 5$ is never divisible by 121 for any posi-

---

† For a discussion of the general concepts involved in the solution of the majority of the problems in this section, see the book by B. B. Dynkin and V. A. Uspensky, *Mathematical Conversations*, Issue 6, Section 2, "Problems in Number Theory," Library of the USSR Mathematical Society.

tive integer n.

30. Prove that the expression

$$m^5 + 3m^4n - 5m^3n^2 - 15m^2n^3 + 4mn^4 + 12n^5$$

cannot have the value 33, regardless of what integers are substituted for m and n.

31. What remainders can result when the 100th power of an integer is divided by 125?

32. Prove that if an integer n is relatively prime to 10, the 101st power of n ends with the same three digits as does n. (For example, 1233^{101} ends with the digits 233, and 37^{101} ends with the digits 037.)

33. Find a three-digit number all of whose integral powers end with the same three digits as does the original number.

34. Let N be an even number not divisible by 10. What digit will be in the tens place of the number N^{20}, and what digit will be in the hundreds place of N^{200}?

35. Prove that the sum

$$1^k + 2^k + 3^k + \cdots + n^k ,$$

where n is an arbitrary integer and k is odd, is divisible by $1 + 2 + 3 + \cdots + n$.

36. Give a criterion that a number be divisible by 11.

37. The number 123456789(10)(11)(12)(13)(14) is written in the base 15—that is, the number is equal (in the base 10) to

$$14 + (13)\cdot 15 + (12)\cdot 15^2 + (11)\cdot 15^3 + \cdots + 2\cdot 15^{12} + 15^{13} .$$

What is the remainder upon dividing the number by 7?

38. Prove that 1, 3, and 9 are the only numbers K having the property that if K divides a number N, it also divides every number obtained by permuting the digits of N. (For $K = 1$, the condition given is trivial; for $K = 3$, or 9, the condition follows from the well-known fact that a number is divisible by 3, or 9, if and only if the sum of its digits is divisible by 3, or 9.)

39. Prove that $27{,}195^8 - 10{,}887^8 + 10{,}152^8$ is exactly divisible by 26,460.

40. Prove that $11^{10} - 1$ is divisible by 100.

41. Prove that $2222^{5555} + 5555^{2222}$ is divisible by 7.

42. Prove thas a number consisting of 3^n identical digits is divisible by 3^n. (For example, the number 222 is divisible by 3, the number 777,777,777, is divisible by 9, and so on).

43. Find the remainder upon dividing the following number by 7:

$$10^{10} + 10^{(10^2)} + \cdots + 10^{(10^{10})} .$$

44. (a) Find the final digit of the numbers $9^{(9^9)}$ and $2^{(3^4)}$.
(b) Find the final two digits of the numbers 2^{999} and 3^{999}.
(c)* Find the final two digits of the number $14^{(14^{14})}$.

45. (a) What is the final digit of the number

$$(\ldots(((7^7)^7)^7)\cdots^7)$$

(where the 7th power is taken 1000 times)? What are the final two digits?

(b) What is the final digit of the number

$$7^{\left(\cdot^{\cdot^{7^{\left(7^{(7^7)}\right)}}}\cdots\right)} ,$$

which contains 1001 sevens, as does the number given in problem (a), but with the exponents used differently? What are the final two digits of this number?

46.* Determine the final five digits of the number

$$N = 9^{\left(\cdot^{\cdot^{9^{\left(9^{(9^9)}\right)}}}\cdots\right)} ,$$

which contains 1001 nines, positioned as shown.

47.* Find the last 1000 digits of the number

$$N = 1 + 50 + 50^2 + 50^3 + \cdots + 50^{999} .$$

48. How many zeros terminate the number which is the product of all the integers from 1 to 100, inclusive?

Here we may use the following well-known notation:

$$1 \cdot 2 \cdot 3 \cdot 4 \cdots (n - 1) \cdot n = n!$$

(called factorial n). The problem can then be stated more succinctly: How many zeros are at the end of 100!?

49. (a) Prove that the product of n consecutive integers is divisible by $n!$.

(b) Prove that if $a + b + \cdots + k \leqq n$, then the fraction

$$\frac{n!}{a!\,b!\cdots k!}$$

is an integer.

(c) Prove that $(n!)!$ is divisible by $n!^{(n-1)!}$.

(d)* Prove that the product of the n integers of an arithmetic progression of n terms, where the common difference is relatively prime to $n!$, is divisible by $n!$.

Note: Problem 49 (d) is a generalization of 49 (a).

50. Is the number, C_{1000}^{500}, of combinations of 1000 elements, taken 500 at a time, divisible by 7?†

51. (a) Find all numbers n between 1 and 100 having the property that $(n - 1)!$ is not divisible by n.

(b) Find all numbers n between 1 and 100 having the property that $(n - 1)!$ is not divisible by n^2.

52.* Find all integers n which are divisible by all integers not exceeding $\sqrt{n}$.

53. (a) Prove that the sum of the squares of five consecutive integers cannot be the square of any integer.

(b) Prove that the sum of even powers of three consecutive numbers cannot be an even power of any integer.

(c) Prove that the sum of the same even power of nine consecutive integers, the first of which exceeds 1, cannot be any integral power of any integer.

54. (a) Let A and B be two distinct seven-digit numbers, each of which contains all the digits from 1 to 7. Prove that A is not divisible by B.

(b) Using all the digits from 1 to 9, make up three, three-digit numbers which are related in the ratio $1 : 2 : 3$.

55. Which integers can have squares that end with four identical digits?

56. Prove that if two adjacent sides of a rectangle and its diagonal can be expressed in integers, then the area of the rectangle is divi-

---

† More "standard" notations for this are $C(1000, 500)$ or C_{500}^{1000} or $\binom{1000}{500}$. However, retention of the notation used in the original will cause no difficulty [*Editor*].

sible by 12.

57. Prove that if all the coefficients of the quadratic equation

$$ax^2 + bx + c = 0$$

are odd integers, then the roots of the equation cannot be rational.

58. Prove that if the sum of the fractions

$$\frac{1}{n} + \frac{1}{n+1} + \frac{1}{n+2}$$

(where n is a positive integer) is put in decimal form, it forms a nonterminating decimal of deferred periodicity.[‡]

59. Prove that the following numbers (where m and n are natural numbers) cannot be integers:

(a) $$M = \frac{1}{2} + \frac{1}{3} + \cdots + \frac{1}{n} \; ;$$

(b) $$N = \frac{1}{n} + \frac{1}{n+1} + \cdots + \frac{1}{n+m} \; ;$$

(c) $$K = \frac{1}{3} + \frac{1}{5} + \cdots + \frac{1}{2n+1} \; .$$

60.** (a) Prove that if p is a prime number greater than 3, then the numerator of the (reduced) fraction

$$1 + \frac{1}{2} + \frac{1}{3} + \cdots + \frac{1}{p-1}$$

is divisible by p^2. For example,

$$1 + \frac{1}{2} + \frac{1}{3} + \frac{1}{4} = \frac{25}{12} \; ,$$

the numerator of which is 5^2.

(b) Prove that if p is a prime number exceeding 3, then the numerator of the (reduced) fraction which is the sum

$$1 + \frac{1}{2^2} + \frac{1}{3^2} + \cdots + \frac{1}{(p-1)^2}$$

is divisible by p. For example,

[‡] Deferred periodicity means that the periodic portion is preceded by one or more nonrepeating digits. The criterion is whether the denominator of the (reduced) fraction has a common factor with 10 [*Editor*].

$$1 + \frac{1}{2^2} + \frac{1}{3^2} + \frac{1}{4^2} = \frac{205}{144}$$

has a numerator which is divisible by 5.

61. Prove that the expression

$$\frac{a^3 + 2a}{a^4 + 3a^2 + 1} ,$$

where a is any positive integer, is a fraction in lowest terms.

62.* Let $a_1, a_2, \cdots, a_n$ be n distinct integers. Show that the product of all the fractions of form $\frac{a_k - a_l}{k - l}$, where $n \geq k > l$, is an integer.

63. Prove that all numbers made up as follows,

$$10001, \ 100010001, \ 1000100010001, \ \cdots$$

(three zeros between the ones), are composite numbers.

64. (a) Divide $a^{128} - b^{128}$ by

$$(a + b)(a^2 + b^2)(a^4 + b^4)(a^8 + b^8)(a^{16} + b^{16})(a^{32} + b^{32})(a^{64} + b^{64}) .$$

   (b) Divide $a^{2^{k+1}} - b^{2^{k+1}}$ by

$$(a + b)(a^2 + b^2)(a^4 + b^4)(a^8 + b^8) \cdots (a^{2^{k-1}} + b^{2^{k+1}})(a^{2^k} + b^{2^k}) .$$

65. Prove that any two numbers of the following sequence are relatively prime:

$$2 + 1, \ 2^2 + 1, \ 2^4 + 1, \ 2^8 + 1, \ 2^{16} + 1, \ \cdots, \ 2^{2^n} + 1, \ \cdots .$$

Remark: The result obtained here proves that there is an infinite number of primes (see also problems 159 and 253).

66. Prove that if one of the numbers $2^n - 1$ and $2^n + 1$ is prime, where $n > 2$, then the other number is composite.

67. (a) Prove that if p and $8p - 1$ are both prime, then $8p + 1$ is composite.
   (b) Prove that if p and $8p^2 + 1$ are both prime, then $8p^2 - 1$ is also prime.

68. Prove that the square of every prime number greater than 3 yields a remainder of 1 when divided by 12.

69. Prove that if three prime numbers, all greater than 3, form an arithmetic progression, then the common difference of the progression is divisible by 6.

70.* (a) Ten primes, each less than 3000, form an arithmetic progression. Find these prime numbers.

(b) Prove that there do not exist eleven primes, all less than 20,000, which can form an arithmetic progression.

71. (a) Prove that, given five consecutive positive integers, it is always possible to find one which is relatively prime to all the rest.

(b) Prove that among sixteen consecutive integers it is always possible to find one which is relatively prime to all the rest.

4

SOME PROBLEMS FROM ARITHMETIC

72. The integer A consists of 666 threes, and the integer B has 666 sixes. What digits appear in the product $A \cdot B$?

73. What quotient and what remainder are obtained when the number consisting of 1001 sevens is divided by the number 1001?

74. Find the least square which commences with six twos.

75. Prove that if the number α is given by the decimal $0.999\ldots$, where there are at least 100 nines, then $\sqrt{\alpha}$ also has 100 nines at the beginning.

76. Adjoin to the digits 523... three more digits such that the resulting six-digit number is divisible by 7, 8, and 9.

77. Find a four-digit number which, on division by 131, yields a remainder of 112, and on division by 132 yields a remainder of 98.

78. (a) Prove that the sum of all the n-digit integers ($n > 2$) is equal to

$$49499\cdots95500\cdots0 \, .$$
$$\underbrace{\qquad}_{(n-3) \text{ nines}} \underbrace{\qquad}_{(n-2) \text{ zeros}}$$

(For example, the sum of all three-digit numbers is equal to 494,550, and the sum of all six-digit numbers is 494,999,550,000.)

(b) Find the sum of all the four-digit even numbers which can be written using 0, 1, 2, 3, 4, 5 (and where digits can be repeated in a number).

79. How many of each of the ten digits are needed in order to write out all the integers from 1 to 100,000,000 inclusive?

80. All the integers beginning with 1 are written successively (that is, 1234567891011121314···). What digit occupies the 206,788th position?

81. Does the number 0.1234567891011121314···, which is obtained by writing successively all the integers, represent a rational number (that is, is it a periodic decimal)?

82. We are given 27 weights which weigh, respectively, $1^2, 2^2, 3^2,$ ···, 27^2 units. Group these weights into three sets of equal weight.

83. A regular polygon is cut from a piece of cardboard. A pin is put through the center to serve as an axis about which the polygon can revolve. Find the least number of sides which the polygon can have in order that revolution through an angle of $25\frac{1}{2}$ degrees will put it into coincidence with its original position.

84. Using all the digits from 1 to 9, make up three, three-digit numbers such that their product will be:

(a) least;   (b) greatest.

85. The sum of a certain number of consecutive positive integers is 1000. Find these integers.

86. (a) Prove that any number which is not a power of 2 can be represented as the sum of at least two consecutive positive integers, but that such a representation is impossible for powers of 2.

(b) Prove that any composite odd number can be represented as a sum of some number of consecutive odd numbers, but that no prime number can be represented in this form. Which even numbers can be represented as the sum of consecutive odd numbers?

(c) Prove that every power of a natural number n ($n > 1$) can be represented as the sum of n positive odd numbers.

87. Prove that the product of four consecutive integers is one less than a perfect square.

88. Given $4n$ positive integers such that if any four distinct integers are taken, it is possible to form a proportion from them. Prove that at least n of the given numbers are identical.

89.* Take four arbitrary natural numbers, A, B, C, and D. Prove that if we use them to find the four numbers A_1, B_1, C_1, and D_1, which are equal, respectively, to the differences between A and B, B and C, C and D, D and A (taking the positive difference each time), and then we repeat this process with A_1, B_1, C_1 and D_1 to obtain four other numbers A_2, B_2, C_2, and D_2, and so on, we eventually must obtain four zeros.

For example, if we begin with the numbers 32, 1, 110, 7, we obtain the following pattern:

32,	1,	110,	7,
31,	109,	103,	25,
78,	6,	78,	6,
72,	72,	72,	72,
0,	0,	0,	0.

90.* (a) Rearrange the integers from 1 to 100 in such an order that no eleven of them appear in the rearrangement (adjacently or otherwise) in either ascending or descending order.

(b) Prove that no matter what rearrangement is made with the integers from 1 to 101 it will always be possible to choose eleven of them which appear (adjacently or otherwise) in the arrangement in either an ascending or a descending order.

91. (a) From the first 200 natural numbers, 101 of them are arbitrarily chosen. Prove that among the numbers chosen there exists a pair of numbers such that one of them is divisible by the other.

(b) From the first 200 natural numbers select a set of 100 numbers such that no one of them is divisible by any other.

(c) Prove that if one of 100 numbers taken from the first 200 natural numbers is less than 16, then one of those 100 numbers is divisible by another.

92. (a) Prove that, given any 52 integers, there exist two of them whose sum, or else whose difference, is divisible by 100.

(b) Prove that out of any 100 integers, none divisible by 100, it is always possible to find two or more integers whose sum is divisible by 100.

93.* A chess master who has eleven weeks to prepare for a tournament decides to play at least one game every day, but in order not to tire himself he agrees to play not more than twelve games during any one week. Prove that there exists a succession of days during which the master will have played exactly twenty games.

94. Let N be an arbitrary natural number. Prove that there exists a multiple of N which contains only the digits 0 and 1. Moreover, if N is relatively prime to 10 (that is, is not divisible by 2 or 5), then some multiple of N consists entirely of ones. (If N is not relatively prime to 10, then, of course, there exists no number of form $11 \cdots 1$ which is divisible by N.)

95.* Given the sequence of numbers

$$0, 1, 1, 2, 3, 5, 8, 13, 21, 34, 55, 89, \cdots,$$

where each number, beginning with the third, is the sum of the two preceding numbers (this is called a Fibonacci sequence). Does there exist, among the first 100,000,001 numbers of this sequence, a number terminating with four zeros?

96.* Let α be an arbitrary irrational number. Clearly, no matter which integer n is chosen, the fraction taken from the sequence $\dfrac{0}{n} = 0, \dfrac{1}{n}, \dfrac{2}{n}, \dfrac{3}{n}, \cdots$, and which is closest to α, differs from α by no more than half of $1/n$. Prove that there exist n's such that the fraction closest to α differs from α by not more than $0.001\left(\dfrac{1}{n}\right)$.

97. Let m and n be two relatively prime natural numbers. Prove that if the $m + n - 2$ fractions

$$\frac{m+n}{m}, \quad \frac{2(m+n)}{m}, \quad \frac{3(m+n)}{m}, \quad \cdots, \quad \frac{(m-1)(m+n)}{m},$$

$$\frac{m+n}{n}, \quad \frac{2(m+n)}{n}, \quad \frac{3(m+n)}{n}, \quad \cdots, \quad \frac{(n-1)(m+n)}{n}$$

are points on the real-number axis, then precisely one of these fractions lies inside each one of the intervals $(1, 2), (2, 3), (3, 4), \cdots,$ $(m + n - 2, m + n - 1)$ (see Figure 3, in which $m = 3$, $n = 4$).

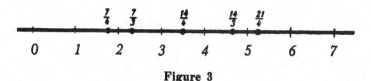

Figure 3

98.* Let $a_1, a_2, a_3, \cdots, a_n$ be n natural numbers, each less than

1000, but where the least common multiple of any two of the numbers exceeds 1000. Prove that the sum of the reciprocals of these numbers is less than 2.

99.* The fraction q/p, where $p \neq 5$ is an odd prime, is expanded as a (periodic) decimal fraction. Prove that if the number of digits appearing in the period of the decimal is even, then the arithmetic mean of these digits is 9/2 (that is, coincides with the arithmetic mean of the digits $0, 1, 2, \cdots, 9$ (this shows that the "greater" and the "lesser" digits of the period appear "equally often"). If the number of digits in the period is odd, then the arithmetic mean of these digits is different from 9/2.

100.* Prove that if the numbers of the following sequence are written as decimals,

$$\frac{a_1}{p}, \frac{a_2}{p^2}, \frac{a_3}{p^3}, \cdots, \frac{a_n}{p^n}, \cdots,$$

(where p is a prime different from 2 or 5, and where $a_1, a_2, \cdots, a_n$ are all relatively prime to p), then some (perhaps only one) of the first few decimal fractions may contain the same number of digits in their periods, but the subsequent decimal fractions of the sequence will all have p times as many digits in their periods as has the preceding term.

For example: $\frac{1}{3} = 0.\overline{3}$; $\frac{4}{9} = 0.\overline{4}$, $\frac{10}{27} = 0.\overline{370}$; $\frac{80}{81} = 0.\overline{987654320}$; $\frac{116}{143}$ has 27 digits in its period; $\frac{953}{128}$ has 81 digits in its period; and so on.

Remark: By "the greatest integer in x" we shall mean the greatest integer not exceeding x (that is, to the left of x on the number axis if x is not a whole number). This concept will be designated by the use of brackets, that is, by writing $[x]$. For example: $[2.5] = 2$, $[2] = 2$, $[-2.5] = -3$.

101. Prove the following properties of the greatest integer in a number.

    (1) $[x + y] \geq [x] + [y]$.

    (2) $\left[\dfrac{[x]}{n}\right] = \left[\dfrac{x}{n}\right]$, where n is an integer.

    (3) $[x] + \left[x + \dfrac{1}{n}\right] + \cdots + \left[x + \dfrac{n-1}{n}\right] = [nx]$.

102.* Prove that if p and q are relatively prime natural numbers, then

$$\left[\frac{p}{q}\right]+\left[\frac{2p}{q}\right]+\left[\frac{3p}{q}\right]+\cdots+\left[\frac{(q-1)p}{q}\right]$$
$$=\left[\frac{q}{p}\right]+\left[\frac{2q}{p}\right]+\left[\frac{3q}{p}\right]+\cdots+\left[\frac{(p-1)q}{p}\right]=\frac{(p-1)(q-1)}{2}.$$

103. (a) Prove that

$$t_1+t_2+t_3+\cdots+t_n=\left[\frac{n}{1}\right]+\left[\frac{n}{2}\right]+\left[\frac{n}{3}\right]+\cdots+\left[\frac{n}{n}\right],$$

where t_n is the number of divisors of the natural number n. [*Note:* 1 and n are always counted as divisors.]

(b) Prove that

$$s_1+s_2+s_3+\cdots+s_n=\left[\frac{n}{1}\right]+2\left[\frac{n}{2}\right]+3\left[\frac{n}{3}\right]+\cdots+n\left[\frac{n}{n}\right],$$

where s_n is the sum of the divisors of the integer n.

104. Does there exist a natural number n such that the fractional part of the number $(2+\sqrt{2})^n$, that is, the difference

$$(2+\sqrt{2})^n-[(2+\sqrt{2})^n],$$

exceeds 0.999999?

105.* (a) Prove that for any natural number n, the integer $[(2+\sqrt{3})^n]$ is odd.

(b) Find the highest power of 2 which divides the integer $[(1+\sqrt{3})^n]$.

106. Prove that if p is an odd prime, it divides the difference

$$[(2+\sqrt{5})^p]-2^{p+1}.$$

107.* Prove that if p is a prime number, the difference

$$C_n^p-\left[\frac{n}{p}\right]$$

is divisible by p. (C_n^p is the number of combinations of n elements taken p at a time, where n is a natural number not less than p.)

For example,

$$C_{11}^5=\frac{11\cdot10\cdot9\cdot8\cdot7}{1\cdot2\cdot3\cdot4\cdot5}=462;$$
$$C_{11}^5-\left[\frac{11}{5}\right]=462-2,$$

which is divisible by 5.

108.* Prove that if the positive numbers α and β have the property that among the numbers

$$[\alpha], [2\alpha], [3\alpha], \cdots; \qquad [\beta], [2\beta], [3\beta], \cdots$$

every natural number appears exactly once, then α and β are irrational numbers such that $1/\alpha + 1/\beta = 1$. Conversely, if α and β are irrational numbers with the property that $1/\alpha + 1/\beta = 1$, then every natural number N appears precisely once in the sequence

$$[\alpha], [2\alpha], [3\alpha], \cdots; \qquad [\beta], [2\beta], [3\beta], \cdots.$$

We shall designate by (a) the whole number nearest a. If a lies exactly between two integers, then (a) will be defined to be the larger integer. For example: $(2.8) = 3$; $(4) = 4$; $(3.5) = 4$.

109.* Prove that in the equality

$$N = \frac{N}{2} + \frac{N}{4} + \frac{N}{8} + \cdots + \frac{N}{2^n} + \cdots$$

(where N is an arbitrary natural number) every fraction may be replaced by the nearest whole number:

$$N = \left(\frac{N}{2}\right) + \left(\frac{N}{4}\right) + \left(\frac{N}{8}\right) + \cdots + \left(\frac{N}{2^n}\right) + \cdots.$$

5

EQUATIONS HAVING INTEGER SOLUTIONS

110. (a) Find a four-digit number which is an exact square, and such that its first two digits are the same and also its last two digits are the same.

 (b) When a certain two-digit number is added to the two-digit number having the same digits in reverse order, the sum is a perfect square. Find all such two-digit numbers.

111. Find a four-digit number equal to the square of the sum of the two two-digit numbers formed by taking the first two digits and the last two digits of the original number.

112. Find all four-digit numbers which are perfect squares and are written:

 (a) with four even integers;

 (b) with four odd integers.

113. (a) Find all three-digit numbers equal to the sum of the factorials of their digits.

 (b) Find all integers equal to the sum of the squares of their digits.

114. Find all integers equal to:

 (a) the square of the sum of the digits of the number;

(b) the sum of the digits of the cube of the number.

115. Solve, in whole numbers, the following equations.
  (a) $1! + 2! + 3! + \cdots + x! = y^2$.
  (b) $1! + 2! + 3! + \cdots + x! = y^z$.

116. In how many ways can 2^n be expressed as the sum of four squares of natural numbers?

117. (a) Prove that the only solution in integers of the equation

$$x^2 + y^2 + z^2 = 2xyz$$

is $x = y = z = 0$.
  (b) Find integers x, y, z, v such that

$$x^2 + y^2 + z^2 + v^2 = 2xyzv .$$

118.* (a) For what integral values of k is the following equation possible (where x, y, z are natural numbers)?

$$x^2 + y^2 + z^2 = kxyz.$$

  (b) Find (up to numbers less than 1000) all possible triples of integers the sum of whose squares is divisible by their product.

119.* Find (within the first thousand) all possible pairs of relatively prime numbers such that the square of one of the integers when increased by 125 is divisible by the other.

120.* Find four natural numbers such that the square of each of them, when added to the sum of the remaining numbers, again yields a perfect square.

121. Find all integer pairs having the property that the sum of the two integers is equal to their product.

122. The sum of the reciprocals of three natural numbers is equal to one. What are the numbers?

123. (a) Solve, in integers (positive and negative),

$$\frac{1}{x} + \frac{1}{y} = \frac{1}{14} .$$

  (b)* Solve, in integers,

$$\frac{1}{x} + \frac{1}{y} = \frac{1}{z}$$

(write a formula which gives all solutions.)

124. (a) Find all distinct pairs of natural numbers which satisfy the equation

$$x^y = y^x \, .$$

(b) Find all positive rational number pairs, not equal, which satisfy the equation

$$x^y = y^x$$

(write a formula which gives all solutions).

125. Two seventh-grade students were allowed to enter a chess tournament otherwise composed of eighth-grade students. Each contestant played once against each other contestant. The two seventh graders together amassed a total of 8 points, and each eighth grader scored the same number of points as his classmates. (In the tournament, a contestant received 1 point for a win and $\frac{1}{2}$ point for a tie.) How many eighth graders participated?

126. Ninth- and tenth-grade students participated in a tournament. Each contestant played each other contestant once. There were ten times as many tenth-grade students, but they were able to win only four-and-a-half times as many points as ninth graders. How many ninth-grade students participated, and how many points did they collect?

127.* An *integral triangle* is defined as a triangle whose sides are measurable in whole numbers. Find all integral triangles whose perimeter equals their area.

128.* What sides are possible in:
    (a) a right-angled integral triangle;
    (b) an integral triangle containing a 60° angle;
    (c) an integral triangle containing a 120° angle?
(Write a formula giving all solutions.)

Remark: It can be shown that an integral triangle cannot have a rational angle (that is, an angle whose degree measure is a rational number) other than one of 90°, 60°, or 120°.

129.* Find the lengths of the sides of the smallest integral triangle for which:
    (a) one of the angles is twice another;
    (b) one of the angles is five times another;
    (c) one angle is six times another.

130.* Prove that if the legs of right-angle triangle are expressible as the squares of integers, the hypotenuse cannot be an integer.

6

EVALUATING SUMS AND PRODUCTS

131. Prove that

$$(n + 1)(n + 2)(n + 3)\cdots(2n - 1)2n = 2^n \cdot 1 \cdot 3 \cdot 5 \cdots (2n - 3)(2n - 1) .$$

132. Calculate the following sums.

(a) $\dfrac{1}{1 \cdot 2} + \dfrac{1}{2 \cdot 3} + \dfrac{1}{3 \cdot 4} + \cdots + \dfrac{1}{(n - 1)n}$;

(b) $\dfrac{1}{1 \cdot 2 \cdot 3} + \dfrac{1}{2 \cdot 3 \cdot 4} + \dfrac{1}{3 \cdot 4 \cdot 5} + \cdots + \dfrac{1}{(n - 2)(n - 1)n}$;

(c) $\dfrac{1}{1 \cdot 2 \cdot 3 \cdot 4} + \dfrac{1}{2 \cdot 3 \cdot 4 \cdot 5} + \dfrac{1}{3 \cdot 4 \cdot 5 \cdot 6}$

$$+ \cdots + \dfrac{1}{(n - 3)(n - 2)(n - 1)n} .$$

133. Prove that

(a) $1 \cdot 2 + 2 \cdot 3 + 3 \cdot 4 + \cdots + n(n + 1) = \dfrac{n(n + 1)(n + 2)}{3}$;

(b) $1 \cdot 2 \cdot 3 + 2 \cdot 3 \cdot 4 + 3 \cdot 4 \cdot 5 + \cdots + n(n + 1)(n + 2)$

$$= \dfrac{n(n + 1)(n + 2)(n + 3)}{4} ;$$

(c) $1 \cdot 2 \cdot 3 \cdots p + 2 \cdot 3 \cdots p(p + 1) + \cdots + n(n + 1)$

$$+ \cdots + (n + p - 1) = \frac{n(n + 1)(n + 2) \cdots (n + p)}{p + 1}$$

for any p.

134. Calculate the following sums.

(a) $1^2 + 2^2 + 3^2 + \cdots + n^2$;

(b) $1^3 + 2^3 + 3^3 + \cdots + n^3$;

(c) $1^4 + 2^4 + 3^4 + \cdots + n^4$;

(d) $1^3 + 3^3 + 5^3 + \cdots + (2n - 3)^3$.

135. Prove the identity

$a + b(1 + a) + c(1 + a)(1 + b) + d(1 + a)(1 + b)(1 + c)$

$$+ \cdots + l(1 + a)(1 + b) \cdots (1 + k)$$
$$= (1 + a)(1 + b)(1 + c) \cdots (1 + l) - 1 .$$

Investigate the case in which $a = b = c = \cdots = l$.

136. Calculate the following.

(a) $1 \cdot 1! + 2 \cdot 2! + 3 \cdot 3! + \cdots + n \cdot n!$;

(b) $C_{n+1}^1 + C_{n+2}^2 + C_{n+3}^3 + \cdots + C_{n+k}^k$.

137. Prove that

$$\frac{1}{\log_2 N} + \frac{1}{\log_3 N} + \frac{1}{\log_4 N} + \cdots + \frac{1}{\log_{100} N} = \frac{1}{\log_{100!} N} ,$$

where 100! is the product $1 \cdot 2 \cdot 3 \cdots 100$.

138. Given n positive numbers $a_1, a_2, \cdots, a_n$. Find the sum of all the fractions

$$\frac{1}{a_{k_1}(a_{k_1} + a_{k_2})(a_{k_1} + a_{k_2} + a_{k_3}) \cdots (a_{k_1} + a_{k_2} + \cdots + a_{k_n})} ,$$

where the set $k_1, k_2, \cdots, k_n$ of indices runs through all possible permutations of $1, 2, \cdots, n$ (of which there are $n!$).

139. Simplify the following expressions.

(a) $\left(1 + \dfrac{1}{3}\right)\left(1 + \dfrac{1}{9}\right)\left(1 + \dfrac{1}{81}\right)\left(1 + \dfrac{1}{3^8}\right) \cdots \left(1 + \dfrac{1}{3^{2^n}}\right)$;

(b) $\cos \alpha \cos 2\alpha \cos 4\alpha \cdots \cos 2^n \alpha$.

140. How many digits are there in the integer 2^{100} after it has been "multiplied out"?

141. (a) Prove that

$$\frac{1}{15} < \frac{1}{10\sqrt{2}} < \frac{1}{2} \cdot \frac{3}{4} \cdot \frac{5}{6} \cdots \frac{99}{100} < \frac{1}{10}.$$

(b) Prove that

$$\frac{1}{2} \cdot \frac{3}{4} \cdot \frac{5}{6} \cdots \frac{99}{100} < \frac{1}{12}.$$

Remark: The result of problem (b) is evidently a refinement of that of problem (a).

142. Prove that

$$\frac{2^{100}}{10\sqrt{2}} < C_{100}^{50} < \frac{2^{100}}{10}.$$

(C_{100}^{50} is the number of combinations of one-hundred elements taken fifty at a time.)

143. Which is larger, $99^n + 100^n$ or 101^n (where n is a natural number)?

144. Which is larger, 100^{300} or $300!$?

145. Prove that, for any natural number n, the following is true:

$$2 < \left(1 + \frac{1}{n}\right)^n < 3.$$

146. Which is larger, $(1.000001)^{1,000,000}$ or 2?

147. Which is larger, 1000^{1000} or 1001^{999}?

148. Prove that for any integer $n > 6$

$$\left(\frac{n}{2}\right)^n > n! > \left(\frac{n}{3}\right)^n.$$

149.* Prove that if $m > n$ (where m, n are natural numbers):

(a) $\left(1 + \frac{1}{m}\right)^m > \left(1 + \frac{1}{n}\right)^n.$

For example,

$$\left(1 + \frac{1}{2}\right)^2 = \frac{9}{4} = 2\frac{1}{4}, \text{ and } \left(1 + \frac{1}{3}\right)^3 = \frac{64}{27} = 2\frac{10}{27} > 2\frac{1}{4}.$$

(b) $\left(1 + \dfrac{1}{m}\right)^{m+1} < \left(1 + \dfrac{1}{n}\right)^{n+1}$ $(n \geqq 2)$.

For example,

$$\left(1 + \frac{1}{2}\right)^3 = \frac{27}{8} = 3\frac{3}{8}, \text{ and } \left(1 + \frac{1}{3}\right)^4 = \frac{256}{81} = 3\frac{13}{81} < 3\frac{3}{8} .$$

From problem (a) it follows that in the sequence of numbers $(1 + 1/2)^2$, $(1 + 1/3)^3, \cdots, (1 + 1/n)^n, \cdots$, each is greater than that preceding. Since, on the other hand, no member of the sequence exceeds 3 (see problem 145), it follows that if $n \to \infty$, the magnitude of $(1 + 1/n)^n$ approaches some definite limit (which is evidently a number between 2 and 3). This limiting number is designated by e. It is equal, approximately, to $2.718281828459045\cdots$.

Analogously, problem 149 (b) shows that in the sequence $(1 + 1/2)^3, (1 + 1/3)^4$, $(1 + 1/4)^5, \cdots, (1 + 1/n)^{n+1}, \cdots$ every number is less than that preceding. Since every number of the sequence exceeds 1, the magnitude $(1 + 1/n)^{n+1}$, where n increases without bound, tends toward some limiting number. The numbers of the second sequence then become successively closer and closer to the numbers of the first [that is, the ratios $(1 + 1/n)^{n+1} : (1 + 1/n)^n = 1 + 1/n$ become closer and closer to 1]. Hence, the limiting number must, in the second case, also be equal to e. This number, e, plays a very important role in higher mathematics, and is encountered in a wide variety of problems (see, for example, problems 156 and 159).

150. Prove that, for any integer n, the following inequality holds,

$$\left(\frac{n}{e}\right)^n < n! < n\left(\frac{n}{e}\right)^n ,$$

where $e = 2.71828\cdots$ is the limit of $(1 + 1/n)^n$ as $n \to \infty$.

This result is an extension of the result of problem 148. It follows, in particular, that for any two numbers, a_1 and a_2, such that $a_1 < e < a_2$ (for example, for $a_1 = 2.7$ and $a_2 = 2.8$; for $a_1 = 2.71$ and $a_2 = 2.72$; for $a_1 = 2.718$ and $a_2 = 2.719$, and so on) for all integers n which are "large enough" (greater than some integer N, where the magnitude of N depends on what a_1 we consider), the following inequality holds:

$$\left(\frac{n}{a_1}\right)^n > n! > \left(\frac{n}{a_2}\right)^n .$$

Thus, the number e is that limitig number which separates the numbers a for which $(n/a)^n$ exceeds, or "dominates," $n!$ from those numbers a for which the $(n/a)^n$ are "dominated" by $n!$. (The existence of such a limiting number follows from problem 148.)

Actually, $(n/a_2)^n < n!$ for every n exceeding 6 [if $a_2 > e$, and if $n > 6$, in view of problem 150, $n! > (n/e)^n$]. Further, from the results of problems 145 and 149, it follows that, for $n \geqq 3$, the following inequalities hold:

$$n > e > \left(1 + \frac{1}{n}\right)^n = \frac{(n+1)^n}{n^n},$$

$$n^{n+1} > (n+1)^n,$$

$$\sqrt[n]{n} > \sqrt[n+1]{n+1};$$

consequently, for $n \geq 3$, $\sqrt[n]{n}$ diminishes as n increases. It is readily seen that if n becomes very large, $\sqrt[n]{n}$ approaches as close to unity as we wish. It follows, for example, that $\log \sqrt[10^k]{10^k} - k/10^k$, for sufficiently large k, can be made as small as we wish. Let us now select an N such that the inequality $\sqrt[N]{N} < e/a_1$ holds. Then for $n > N$ the approximation $\sqrt[n]{n} < e/a_1$ is still more improved, and from problem 150 it follows that

$$n! < \left(\frac{n/e}{\sqrt[n]{n}}\right)^n < \left(\frac{n}{a_1}\right)^n.$$

The inequality of problem 150 admits a great deal of precision. It is possible to show that for sufficiently large n the number $n!$ is approximated by $C\sqrt{n}\,(n/e)^n$, where C is a constant equal to $\sqrt{2\pi}$:

$$n! \approx \sqrt{2\pi n}\left(\frac{n}{e}\right)^n \dagger$$

[more precisely, it is possible to prove that if n increases without bound, ratio

$$\frac{n!}{\sqrt{2\pi n}\,(n/e)^n}$$

tends to unity. (See the book by A. M. Yaglom and E. M. Yaglom, *Non-elementary Problems Treated by Elementary Means*, Library of the Mathematical Society, Volume 5)].

151. Prove that

$$\frac{1}{k+1}n^{k+1} < 1^k + 2^k + 3^k + \cdots + n^k$$

$$< \left(1 + \frac{1}{n}\right)^{k+1}\frac{1}{k+1}n^{k+1}$$

(n and k are arbitrary integers).

Remark: A particular consequence of problem 151 is the following:

$$\lim_{n\to\infty}\frac{1^k + 2^k + 3^k + \cdots + n^k}{n^{k+1}} = \frac{1}{k+1}.$$

(See also problem 316.)

152. Prove that for all integers $n > 1$:

---

† The approximation given for $n!$ is usually referred to as *Stirling's formula* [*Editor*].

(a) $\dfrac{1}{2} < \dfrac{1}{n+1} + \dfrac{1}{n+2} + \cdots + \dfrac{1}{2n} < \dfrac{3}{4}$;

(b) $1 < \dfrac{1}{n+1} + \dfrac{1}{n+2} + \cdots + \dfrac{1}{3n+1} < 2$.

153.* (a) Calculate the whole part of the number

$$1 + \frac{1}{\sqrt{2}} + \frac{1}{\sqrt{3}} + \frac{1}{\sqrt{4}} + \cdots + \frac{1}{\sqrt{1,000,000}} \ .$$

(b) Calculate the sum

$$\frac{1}{\sqrt{10,000}} + \frac{1}{\sqrt{10,001}} + \frac{1}{\sqrt{10,002}} + \cdots + \frac{1}{\sqrt{1,000,000}}$$

to within a tolerance (allowable error) of 1/50.

154.* Find the whole part of the number

$$\frac{1}{\sqrt[3]{4}} + \frac{1}{\sqrt[3]{5}} \ \frac{1}{\sqrt[3]{6}} + \cdots + \frac{1}{\sqrt[3]{1,000,000}} \ .$$

155. (a) Determine the sum

$$\frac{1}{10^2} + \frac{1}{11^2} + \frac{1}{12^2} + \cdots + \frac{1}{1000^2}$$

to a tolerance of 0.006.

(b) Determine the sum

$$\frac{1}{10!} + \frac{1}{11!} + \frac{1}{12!} + \cdots + \frac{1}{1000!}$$

to a tolerance of 0.000000015.

156. Prove that the sum

$$1 + \frac{1}{2} + \frac{1}{3} + \frac{1}{4} + \cdots + \frac{1}{n}$$

is greater than any previously selected number N, if n is taken sufficiently great.

Remark: The calculation of this sum can be made very precise. It is possible to show that the sum

$$1 + \frac{1}{2} + \frac{1}{3} + \frac{1}{4} + \cdots + \frac{1}{n} \ ,$$

for large n, is very close to the value of $\log n$ (this logarithm taken to the base $e = 2.718\cdots$). In every case, it can be shown that for any n the difference

$$1 + \frac{1}{2} + \frac{1}{3} + \cdots + \frac{1}{n} - \log n$$

does not exceed unity (see the reference following problem 150 to the book by A. M. Yaglom and E. M. Yaglom).

157. Prove that if in the summation

$$1 + \frac{1}{2} + \frac{1}{3} + \frac{1}{4} + \cdots + \frac{1}{n}$$

we throw out every term which contains the digit 9 in its denominator, then the sum of the remaining terms, for any n, will be less than 80.

158. (a) Prove that, for any n, the following holds:

$$1 + \frac{1}{4} + \frac{1}{9} + \frac{1}{16} + \frac{1}{25} + \cdots + \frac{1}{n^2} < 2 .$$

(b) Prove that for all n

$$1 + \frac{1}{4} + \frac{1}{9} + \frac{1}{16} + \cdots + \frac{1}{n^2} < 1\frac{3}{4} .$$

It is evident that the inequality of problem (b) is a refinement of problem (a). An even more precise bound in given by problem 233. That problem shows that the sum

$$1 + \frac{1}{4} + \frac{1}{9} + \cdots + \frac{1}{n^2}$$

is less than $\pi^2/6 = 1.6449340668\cdots$ (but for any number less than $\pi^2/6$, for instance for $N = 1.64$ or for $N = 1.644934$, it is possible to find an n such that the sum

$$1 + \frac{1}{4} + \frac{1}{9} + \cdots + \frac{1}{n^2}$$

will exceed N).

159.* Consider the sum

$$1 + \frac{1}{2} + \frac{1}{3} + \frac{1}{5} + \frac{1}{7} + \frac{1}{11} + \frac{1}{13} + \frac{1}{17} + \frac{1}{19} + \cdots + \frac{1}{p} ,$$

in which the denominators run through the prime numbers from 2 to some prime number p. Prove that this sum becomes greater than any preassigned number N, provided the prime p is taken sufficiently great.

Remark: The summation of the series in this problem can be found with great accuracy. For large p, the sum

$$1 + \frac{1}{2} + \frac{1}{3} + \frac{1}{5} + \frac{1}{7} + \cdots + \frac{1}{p}$$

differs relatively little from $\log \log p$ (where the logarithms are taken to the base $e = 2.718 \cdots$), and the differences

$$1 + \frac{1}{2} + \frac{1}{3} + \frac{1}{5} + \frac{1}{7} + \cdots + \frac{1}{p} - \log \log p$$

never exceed 15 (refer to the book by A. M. Yaglom and E. M. Yaglom).

Comparison of the results of this problem with those of problems 157 and 158 emphasizes that among the prime numbers may be found arbitrarily large integers (this problem reaffirms that there are infinitely many). It is possible, for example, to say that the primes are "more numerous" in the sequence of natural numbers than either squares or numbers failing to contain the digit 9, inasmuch as the sum of the reciprocals of all the squares, as well as the sum of all those reciprocals of whole numbers not containing the digit 9, are bounded (by $1\frac{3}{4}$ and by 80, respectively), whereas the sum of the reciprocals of all the primes becomes arbitrarily great.

7

MISCELLANEOUS PROBLEMS

FROM ALGEBRA

---

160. If $a + b + c = 0$, what does the following expression equal?

$$\left(\frac{b-c}{a} + \frac{c-a}{b} + \frac{a-b}{c}\right)\left(\frac{a}{b-c} + \frac{b}{c-a} + \frac{c}{a-b}\right).$$

161. Prove that if $a + b + c = 0$, then

$$a^3 + b^3 + c^3 = 3abc.$$

162. Factor the following:

(a) $a^3 + b^3 + c^3 - 3abc$;

(b) $(a + b + c)^3 - a^3 - b^3 - c^3$.

163. Rationalize the denominator:

$$\frac{1}{\sqrt[3]{a} + \sqrt[3]{b} + \sqrt[3]{c}}.$$

164. Prove that

$$(a + b + c)^{333} - a^{333} - b^{333} - c^{333}$$

is divisible by

$$(a + b + c)^3 - a^3 - b^3 - c^3.$$

165. Factor the following expression:

$$a^{10} + a^5 + 1 .$$

166. Prove that the polynomial

$$x^{9999} + x^{8888} + x^{7777} + \cdots + x^{2222} + x^{1111} + 1$$

is divisible by

$$x^9 + x^8 + x^7 + \cdots + x^2 + x + 1 .$$

167. Using the result of problem 162 (a), find the general formula for the solution of the cubic equation

$$x^3 + px + q = 0 .$$

Remark: This result enables us to solve any equation of the third degree. Let

$$x^3 + Ax^2 + Bx + C = 0$$

be any cubic equation (the coefficient of x^3 is taken as 1, since in any other case we can divide through by the coefficient of x^3). We make the substitution

$$x = y + c ,$$

and we obtain

$$y^3 + 3cy^2 + 3c^2y + c^3 + A(y^2 + 2cy + c^2) + B(y + c) + C = 0 ,$$

or

$$y^3 + (3c + A)y^2 + (3c^2 + 2Ac + B)y + (c^3 + Ac^2 + Bc + C) = 0 .$$

From this, if we set $c = -A/3$ (that is, $x = y - A/3$), we arrive at

$$y^3 + \left(\frac{3A^2}{9} - \frac{2A^2}{9} + B\right)y + \left(-\frac{A^3}{27} + \frac{A^3}{9} - \frac{AB}{3} + C\right) = 0 ,$$

which has the same forms as that of the given problem:

$$y^3 + py + q = 0 ,$$

where

$$p = -\frac{A^2}{3} + B \quad \text{and} \quad q = \frac{2A^3}{27} - \frac{AB}{3} + C .$$

168. Solve the equation

$$\sqrt{a - \sqrt{a + x}} = x .$$

169.* Find the real roots of the equation

$$x^3 + 2ax + \frac{1}{16} = -a + \sqrt{a^2 + x - \frac{1}{16}} \quad \left(0 < a < \frac{1}{4}\right).$$

170. (a) Find the real roots of the equation

$$\sqrt{x + 2\sqrt{x + 2\sqrt{x + \cdots + 2\sqrt{x + 2\sqrt{3x}}}}} = x$$
$$\underbrace{\qquad\qquad\qquad\qquad\qquad\qquad\qquad}_{\{n \text{ radicals}\}}$$

(all roots are considered positive).

(b) Solve the equation

$$\cfrac{1}{1 + \cfrac{1}{1 + \cfrac{1}{1 + \cdot \atop \qquad \ddots \atop \qquad\quad 1 + \cfrac{1}{x}}}} = x.$$

(In the expression on the left the fraction designation is repeated *n* times.)

171. Find the real roots of the equation

$$\sqrt{x + 3 - 4\sqrt{x - 1}} + \sqrt{x + 8 - 6\sqrt{x - 1}} = 1.$$

(All square roots are to be taken as positive.)

172. Solve the equation

$$|x + 1| - |x| + 3|x - 1| - 2|x - 2| = x + 2.$$

173. A system of two second-degree equations

$$\begin{cases} x^2 - y^2 = 0, \\ (x - a)^2 + y^2 = 1 \end{cases}$$

has, in general, four solutions. For what values of *a* is the number of solutions of this system decreased to three or to two?

174. (a) Solve the system of equations

$$\begin{cases} ax + y = a^2, \\ x + ay = 1. \end{cases}$$

For what values of *a* does this system fail to have solutions, and for what values of *a* are there infinitely many solutions?

(b) Answer the above question for the system

$$\begin{cases} ax + y = a^3, \\ x + ay = 1. \end{cases}$$

(c) Answer the above question for the system

$$\begin{cases} ax + y + z = 1 \,, \\ x + ay + z = a \,, \\ x + z + az = a^2 \,. \end{cases}$$

175. Find the conditions which must be satisfied by the numbers $\alpha_1, \alpha_2, \alpha_3, \alpha_4$ such that the following system of six equations in four unknowns has a solution:

$$\begin{cases} x_1 + x_2 = \alpha_1\alpha_2 \,, \\ x_1 + x_3 = \alpha_1\alpha_3 \,, \\ x_1 + x_4 = \alpha_1\alpha_4 \,, \\ x_2 + x_3 = \alpha_2\alpha_3 \,, \\ x_2 + x_4 = \alpha_2\alpha_4 \,, \\ x_3 + x_4 = \alpha_3\alpha_4 \,. \end{cases}$$

Find the values of the unknowns x_1, x_2, x_3, x_4.

176. How many real solutions has the following system?

$$\begin{cases} x + y = 2 \,, \\ xy - z^2 = 1 \,. \end{cases}$$

177. (a) How many roots has the following equation?

$$\sin x = \frac{x}{100} \,.$$

(b) How many roots has the following equation?

$$\sin x = \log x \,.$$

(*Note*: $\log x \equiv \log_{10} x$.)

178.* Prove that if x_1 and x_2 are roots of the equation $x^2 - 6x + 1 = 0$, then $x_1^n + x_2^n$ is, for any natural number n, an integer not divisible by 5.

179. Is it possible for the expression

$$(a_1 + a_2 + \cdots + a_{999} + a_{1000})^2 = a_1^2 + a_2^2 + \cdots + a_{999}^2 + a_{1000}^2$$
$$+ 2a_1a_2 + 2a_1a_3 + \cdots + 2a_{999}a_{1000}$$

(where some of the numbers $a_1, a_2, \cdots, a_{999}, a_{1000}$ are positive and the test are negative) to contain the same number of positive and negative terms in a_ia_j?

Investigate the analogous problem for the expression

$$(a_1 + a_2 + \cdots + a_{9999} + a_{10,000})^2 \,.$$

180. Prove that any integral power of the number $\sqrt{2} - 1$ can be expressed in the form $\sqrt{N} - \sqrt{N-1}$, where N is an integer (for example: $(\sqrt{2} - 1)^2 = 3 - 2\sqrt{2} = \sqrt{9} - \sqrt{8}$, and $(\sqrt{2} - 1)^3 = 5\sqrt{2} - 7 = \sqrt{50} - \sqrt{49}$.

181. Prove that the number $99{,}999 + 111{,}111\sqrt{3}$ cannot be written in the form $(A + B\sqrt{3})^2$, where A and B are integers.

182. Prove that $\sqrt[3]{2}$ cannot be represented in the form $p + q\sqrt{r}$, where p, q, r are rational numbers.

183. (a) Which of the following two expressions is greater?

$$\frac{2.00000000004}{(1.00000000004)^2 + 2.00000000004} \; ;$$

$$\frac{2.00000000002}{(1.00000000002)^2 + 2.00000000002} \; .$$

(b) Let $a > b > 0$. Which of the following is greater?

$$\frac{1 + a + a^2 + \cdots + a^{n-1}}{1 + a + a^2 + \cdots + a^n} \; ;$$

$$\frac{1 + b + b^2 + \cdots + b^{n-1}}{1 + b + b^2 + \cdots + b^n} \; .$$

184. Given n numbers $a_1, a_2, \cdots, a_n$. Find the number x such that the sum

$$(x - a_1)^2 + (x - a_2)^2 + \cdots + (x - a_n)^2$$

has the least possible value.

185. (a) Given four distinct numbers $a_1 < a_2 < a_3 < a_4$. Put these numbers in such an order, $a_{i_1}, a_{i_2}, a_{i_3}, a_{i_4}$ (i_1, i_2, i_3, i_4 being some rearrangement of $1, 2, 3, 4$) that the sum

$$\emptyset = (a_{i_1} - a_{i_2})^2 + (a_{i_2} - a_{i_3})^2 + (a_{i_3} - a_{i_4})^2 + (a_{i_4} - a_{i_1})^2$$

has the least possible value.

(b)* Given n real distinct numbers $a_1, a_2, \cdots, a_n$. Put these numbers in such an order $a_{i_1}, a_{i_2}, \cdots, a_{i_n}$ that the sum

$$\emptyset = (a_{i_1} - a_{i_2})^2 + (a_{i_2} - a_{i_3})^2 + \cdots + (a_{i_{n-1}} - a_{i_n})^2 + (a_{i_n} - a_{i_1})^2$$

has the least possible value.

186. (a) Prove that, regardless of what numbers $a_1, a_2, \cdots, a_n, b_1, b_2, \cdots, b_n$ are taken, the following relation always holds:

$$\sqrt{a_1^2 + b_1^2} + \sqrt{a_2^2 + b_2^2} + \cdots + \sqrt{a_n^2 + b_n^2}$$
$$\geq \sqrt{(a_1 + a_2 + \cdots + a_n)^2 + (b_1 + b_2 + \cdots + b_n)^2} .$$

Under what conditions does the equality hold?

(b) A pyramid is called *a right pyramid* if, when a circle is inscribed in its base, the altitude of the pyramid falls on the center of the circle. Prove that a right pyramid has less lateral surface area than any other pyramid of the same altitude and base area and having the same perimeter.

Remark: The inequality of problem (a) is a special case of what is called the Inequality of Minkowski (see problem 308).

187.* Prove that for any real numbers $a_1, a_2, \cdots, a_n$ the following inequality holds:

$$\sqrt{a_1^2 + (1 - a_2)^2} + \sqrt{a_2^2 + (1 - a_3)^2}$$
$$+ \cdots + \sqrt{a_{n-1}^2 + (1 - a_n)^2} + \sqrt{a_n^2 + (1 - a_1)^2} \geq \frac{n\sqrt{2}}{2} .$$

For what values of the numbers is the left member equal to the right member?

188. Prove that if the numbers x_1 and x_2 do not exceed 1 in absolute value, then

$$\sqrt{1 - x_1^2} + \sqrt{1 - x_2^2} \leq 2\sqrt{1 - \left(\frac{x_1 + x_2}{2}\right)^2} .$$

For what numbers x_1 and x_2 does the equality hold?

189. Which is greater, $\cos(\sin x)$ or $\sin(\cos x)$?

190. Prove, without using logarithm tables, that:

(a) $\dfrac{1}{\log_2 \pi} + \dfrac{1}{\log_5 \pi} > 2$;

(b) $\dfrac{1}{\log_2 \pi} + \dfrac{1}{\log_\pi 2} > 2$.

191. Prove that if α and β are acute angles, with $\alpha < \beta$, then

(a) $\alpha - \sin \alpha < \beta - \sin \beta$;

(b) $\tan \alpha - \alpha < \tan \beta - \beta$.

192.* Prove that if α and β are acute angles and $a < \beta$, then

$$\frac{\tan \alpha}{\alpha} < \frac{\tan \beta}{\beta} \, .$$

193. Find the relationship between $\arcsin [\cos (\arcsin x)]$ and $\arccos [\sin (\arccos x)]$.

194. Prove that for arbitrary coefficients $a_{31}, a_{30}, \cdots, a_2, a_1$, the sum
$$\cos 32x + a_{31} \cos 31x + a_{30} \cos 30x + \cdots + a_2 \cos 2x + a_1 \cos x$$
cannot take on only positive values for all x.

195. Let some of the numbers $a_1, a_2, \cdots, a_n$ be $+1$ and the rest be -1. Prove that

$$2 \sin \left(a_1 + \frac{a_1 a_2}{2} + \frac{a_1 a_2 a_3}{4} + \cdots + \frac{a_1 a_2 \cdots a_n}{2^{n-1}} \right) 45°$$
$$= a \sqrt{2 + a_2 \sqrt{2 + a_3 \sqrt{2 + \cdots + a_n \sqrt{2}}}} \, .$$

For example, let $a_1 = a_2 = \cdots = a_n = 1$:

$$2 \sin \left(1 + \frac{1}{2} + \frac{1}{4} + \cdots + \frac{1}{2^{n-1}} \right) 45° = 2 \cos \frac{45°}{2^{n-1}}$$
$$= \sqrt{2 + \sqrt{2 + \cdots + \sqrt{2}}} \, .$$

8

THE ALGEBRA OF POLYNOMIALS

196. Find the sum of the coefficients of the polynomial obtained after expanding and collecting the terms of the product

$$(1 - 3x + 3x^2)^{743}(1 + 3x - 3x^2)^{744} .$$

197. Which of the expressions,

$$(1 + x^2 - x^3)^{1000} \quad \text{or} \quad (1 - x^2 + x^3)^{1000} ,$$

will have the larger coefficient for x^{20} after expansion and collecting of terms?

198. Prove that in the product

$$(1 - x + x^2 - x^3 + \cdots - x^{99} + x^{100})(1 + x + x^2 + \cdots + x^{99} + x^{100}) ,$$

after multiplying and collecting terms, there does not appear a term in x of odd degree.

199. Find the coefficient of x^{50} in the following polynomials:

(a) $(1 + x)^{1000} + x(1 + x)^{999} + x^2(1 + x)^{998} + \cdots + x^{1000}$;

(b) $(1 + x) + 2(1 + x)^2 + 3(1 + x)^3 + \cdots + 1000(1 + x)^{1000}$.

200.* Find the coefficient of x^2 upon the expansion and collecting of terms in the expression

$$\underbrace{((((x - 2)^2 - 2)^2 - 2)^2 - \cdots - 2)^2}_{n \text{ times}}.$$

201. Find the remainders upon dividing the polynomial $x + x^3 + x^9 + x^{27} + x^{81} + x^{243}$

     (a) by $x - 1$;

     (b) by $x^2 - 1$.

202. An unknown polynomial yields a remainder of 2 upon division by $x - 1$, and a remainder of 1 upon division by $x - 2$. What remainder is obtained if this polynomial is divided by $(x - 1)(x - 2)$?

203. If the polynomial $x^{1951} - 1$ is divided by $x^4 + x^3 + 2x^2 + x + 1$, a quotient and remainder are obtained. Find the coefficient of x^{14} in the quotient.

204. Find an equation with integral coefficients whose roots include the numbers

     (a) $\sqrt{2} + \sqrt{3}$,

     (b) $\sqrt{2} + \sqrt[3]{3}$.

205. Prove that if α and β are the roots of the equation

$$x^2 + px + 1 = 0,$$

and if γ and δ are the roots of the equation

$$x^2 + qx + 1 = 0,$$

then

$$(\alpha - \gamma)(\beta - \gamma)(\alpha + \delta)(\beta + \delta) = q^2 - p^2.$$

206. Let α and β be the roots of the equation

$$x^2 + px + q = 0,$$

and γ and δ be the roots of the equation

$$x^2 + Px + Q = 0.$$

Express the product

$$(\alpha - \gamma)(\beta - \gamma)(\alpha - \delta)(\beta - \delta)$$

in terms of the coefficients of the given equations.

207. Given the two polynomials

$$x^2 + ax + 1 = 0,$$
$$x^2 + x + a = 0.$$

Determine all values of the coefficient a for which these equations have at least one common root.

208. Find an integer a such that $(x - a)(x - 10) + 1$ can be written as a product $(x + b)(x + c)$ of two factors with integers b and c.

209. Find (nonzero) distinct integers a, b, and c such that the following fourth-degree polynomial with integral coefficients, can be written as the product of two other polynomials with integral coefficients:

$$x(x - a)(x - b)(x - c) + 1$$

210. For what integers $a_1, a_2, \cdots, a_n$, where these are all distinct, are the following polynomials with integral coefficients expressible as the product of two polynomials with integral coefficients?

   (a) $(x - a_1)(x - a_2)(x - a_3) \cdots (x - a_n) - 1$;

   (b) $(x - a_1)(x - a_2)(x - a_3) \cdots (x - a_n) + 1$.

211.* Prove that if the integers $a_1, a_2, \cdots, a_n$ are all distinct, then the polynomial

$$(x - a_1)^2(x - a_2)^2 \cdots (x - a_n)^2 + 1$$

cannot be written as a product of two other polynomials with integral coefficients.

212. Prove that if the polynomial

$$P(x) = a_0 x^n + a_1 x^{n-1} + \cdots + a_{n-1} x + a_n ,$$

with integral coefficients, takes on the value 7 for four integral values of x, then it cannot have the value 14 for any integral value of x.

213. Prove that if the polynomial

$$a_0 x^7 + a_1 x^6 + a_2 x^5 + a_3 x^4 + a_4 x^3 + a_5 x^2 + a_6 x + a_7 ,$$

of seventh degree, with integral coefficients, has for seven integral values of x the value $+1$ or -1, then it cannot be factored as the product of two polynomials with integral coefficients.

214. Prove that if the polynomial

$$P(x) = a_0 x^n + a_1 x^{n-1} + \cdots + a_{n-1} x + a_n ,$$

with integral coefficients, has odd values for $x = 0$ and $x = 1$, then the equation $P(x) = 0$ cannot have integral roots.

215.* Prove that if the polynomial

$$P(x) = a_0 x^n + a_1 x^{n-1} + \cdots + a_{n-1} x + a_n ,$$

with integral coefficients, is equal in absolute value to 1 for two integers $x = p$ and $x = q$ ($p > q$), and if the equation $P(x) = 0$ has rational roots a, then $p - q$ is equal to 1 or 2, and $a = \dfrac{p+q}{2}$.

216.* Prove that neither of the following polynomials can be written as a product of two polynomials with integral coefficients:

(a) $x^{2222} + 2x^{2220} + 4x^{2218} + 6x^{2216} + 8x^{2214}$
$$+ \cdots + 2218x^4 + 2220x^2 + 2222 ;$$

(b) $x^{250} + x^{249} + x^{248} + \cdots + x^2 + x + 1$.

217. Prove that if the product of two polynomials with integral coefficients is a polynomial with even coefficients, not all of which are divisible by 4, then in one of the polynomials all coefficients must be even, and in the other not all coefficients will be even.

218. Prove that all the rational roots of the polynomial

$$P(x) = x^n + a_1 x^{n-1} + a_2 x^{n-2} + \cdots + a_{n-1} x + a_n ,$$

with integral coefficients and with leading coefficient 1, are integers.

219.* Prove that there does not exist a polynomial

$$P(x) = a_0 x^n + a_1 x^{n-1} + \cdots + a_{n-1} x + a_n$$

such that $P(0)$, $P(1)$, $P(2)$, $\cdots$ are all prime numbers.

Remark: The proposition stated in this problem was first proven by the mathematician L. Euler. Also credited to him are polynomials whose values for many consecutive integers are prime numbers. For example, for the polynomial $P(x) = x^2 - 79x + 1601$, the 80 numbers $P(0) = 1$, $P(1) = 1523$, $P(2)$, $P(3), \cdots, P(79)$ are all primes.

220. Prove that if the polynomial

$$P(x) = x^n + A_1 x^{n-1} + A_2 x^{n-2} + \cdots + A_{n-1} + A_n$$

assumes integral values for all integral values for x, then it is possible to represent it as a sum of polynomials

$$P_0(x) = 1, \ P_1(x) = x, \ P_2(x) = \frac{x(x-1)}{1 \cdot 2}, \ \cdots, \ P_n(x)$$
$$= \frac{x(x-1)(x-2)\cdots(x-n+1)}{1 \cdot 2 \cdot 3 \cdots n} ,$$

having the same property [in view of problem 49 (a)] and having integral coefficients.

221. (a) Prove that if the nth degree polynomial $P(x)$ has integral values for $x = 0, 1, 2, \cdots, n$, then it has integral values for all integral values of x.

(b) Prove that if a polynomial $P(x)$ of degree n has integral values for $n + 1$ successive integers x, then it is integral valued for all integers x.

(c) Prove that if the polynomial $P(x)$ of degree n has integral values for $x = 0, 1, 4, 9, 16, \cdots, n^2$, then it has integral values for all integers x which are perfect squares (but this does not necessarily follow for all integers x).

Give an example of a polynomial which assumes integral values for all integers x which are perfect squares, but which for some other value of x yields a rational (not whole) number.

9

COMPLEX NUMBERS

In many of the problems in this section the following formulas are useful.

(1) The formula for the product of complex numbers in trigonometric form:

$$(\cos \alpha + i \sin \alpha)(\cos \beta + i \sin \beta) = \cos (\alpha + \beta) + i \sin (\alpha + \beta) .$$

(2) De Moivre's formula:

$$(\cos \alpha + i \sin \alpha)^n = \cos n\alpha + i \sin n\alpha$$

(where n is a natural number), which is an n-fold application of the previous formula.

(3) The formula for the roots of complex numbers:

$$\sqrt[n]{\cos \alpha + i \sin \alpha} = \cos \frac{\alpha + 360° \cdot k}{n} + i \sin \frac{\alpha + 360° \cdot k}{n}$$
$$(k = 0, 1, 2, \cdots, n - 1) ,$$

which is an extended form of De Moivre's theorem.

In particular, a large role is played in the following problems by the formula for the nth rooths of unity, that is, the roots of the nth-degree equation

$$x^n - 1 = 0 ,$$

which are given by the following formulas:

$$\sqrt[n]{1} = \sqrt[n]{\cos 0 + i \sin 0} = \cos \frac{360° \cdot k}{n} + i \sin \frac{360° \cdot k}{n}$$
$$(k = 0, 1, 2, \cdots, n - 1) ,$$

The following observation will often be useful in solving the problems of this section. Let the equation of degree n,

$$x^n + a_1 x^{n-1} + a_2 x^{n-2} + \cdots + a_{n-1} x + a_n = 0 ,$$

have the n roots $x_1, x_2, \cdots, x_{n-1}, x_n$. Then the left member of the equation is divisible by $(x - x_1)(x - x_2)\cdots(x - x_n)$; that is,

$$x^n + a_1 x^{n-1} + \cdots + a_{n-1} x + a_n = (x - x_1)(x - x_2)\cdots(x - x_{n-1})(x - x_n) .$$

If we multiply out the second member of this equation and equate coefficients of like powers of x from both members, we obtain the following formulas giving relationships between the coefficients on the left and the roots of the equation (Vieta's formulas).

$$a_1 = -(x_1 + x_2 + \cdots + x_{n-1} + x_n) ,$$
$$a_2 = x_1 x_2 + x_1 x_3 + \cdots + x_{n-1} x_n ,$$
$$a_3 = -(x_1 x_2 x_3 + \cdots + x_{n-2} x_{n-1} x_n) ,$$
$$\cdots\cdots\cdots\cdots\cdots\cdots\cdots\cdots\cdots\cdots ,$$
$$a_{n-1} = (-1)^{n-1}(x_1 x_2 \cdots x_{n-1} + x_1 x_2 \cdots x_{n-2} x_n + \cdots + x_2 x_3 \cdots x_n) ,$$
$$a_n = (-1)^n x_1 x_2 x_3 \cdots x_n .$$

222. (a) Prove that

$$\cos 5\alpha = \cos^5 \alpha - 10 \cos^3 \alpha \sin^2 \alpha + 5 \cos \alpha \sin^4 \alpha ,$$
$$\sin 5\alpha = \sin^5 \alpha - 10 \sin^3 \alpha \cos^2 \alpha + 5 \sin \alpha \cos^4 \alpha .$$

(b) Prove that for integers n

$$\cos n\alpha = \cos^n \alpha - C_n^2 \cos^{n-2} \alpha \sin^2 \alpha + C_n^4 \cos^{n-4} \alpha \sin^4 \alpha$$
$$- C_n^6 \cos^{n-6} \alpha \sin^6 \alpha + \cdots ,$$
$$\sin n\alpha = C_n^1 \cos^{n-1} \alpha \sin \alpha - C_n^3 \cos^{n-3} \alpha \sin^3 \alpha$$
$$+ C_n^5 \cos^{n-5} \alpha \sin^5 \alpha - \cdots ,$$

where the terms designated by $\cdots$, which are readily identified from the given terms, are continued while they preserve the sense of the binomial coefficients.

Remark: Problem (b) is, of course, a generalization of problem (a).

223. Express $\tan 6\alpha$ in terms of $\tan \alpha$.

224. Prove that if $x + \dfrac{1}{x} = 2 \cos \alpha$, then

$$x^n + \frac{1}{x^n} = 2 \cos n\alpha .$$

225. Prove that

$$\sin \varphi + \sin (\varphi + \alpha) + \sin (\varphi + 2\alpha) + \cdots + \sin (\varphi + n\alpha)$$
$$= \frac{\sin \dfrac{(n + 1)\alpha}{2} \sin \left(\varphi + \dfrac{n\alpha}{2}\right)}{\sin \dfrac{\alpha}{2}} ,$$

and that

$$\cos \varphi + \cos (\varphi + \alpha) + \cos (\varphi + 2\alpha) + \cdots + \cos (\varphi + n\alpha)$$
$$= \frac{\sin \dfrac{(n + 1)\alpha}{2} \cos \left(\varphi + \dfrac{n\alpha}{2} \right)}{\sin \dfrac{\alpha}{2}}.$$

226. Find the value of

$$\cos^2 \alpha + \cos^2 2\alpha + \cdots + \cos^2 n\alpha ,$$

and of

$$\sin^2 \alpha + \sin^2 2\alpha + \cdots + \sin^2 n\alpha .$$

227. Evaluate

$$\cos \alpha + C_n^1 \cos 2\alpha + C_n^2 \cos 3\alpha + \cdots + C_n^{n-1} \cos n\alpha + \cos (n + 1)\alpha$$

and

$$\sin \alpha + C_n^1 \sin 2\alpha + C_n^2 \sin 3\alpha + \cdots + C_n^{n-1} \sin n\alpha + \sin (n + 1)\alpha .$$

228. Prove that if m, n, and p are arbitrary integers, then

$$\sin \frac{m\pi}{p} \sin \frac{n\pi}{p} + \sin \frac{2m\pi}{p} \sin \frac{2n\pi}{p} + \sin \frac{3m\pi}{p} \sin \frac{3n\pi}{p}$$
$$+ \cdots + \sin \frac{(p - 1)m\pi}{p} \sin \frac{(p - 1)n\pi}{p} \Big]$$

$$= \begin{cases} -\dfrac{p}{2}, \text{ if } m + n \text{ is divisible by } 2p \text{ and } m - n \text{ is not divisible by } 2p; \\ \dfrac{p}{2}, \text{ if } m - n \text{ is divisible by } 2p \text{ and } m + n \text{ is not divisible by } 2p; \\ 0, \text{ if } m + n \text{ and } m - n \text{ are both divisible by } 2p, \text{ or if neither} \\ \quad \text{ is divisible by } 2p. \end{cases}$$

229. Prove that

$$\cos \frac{2\pi}{2n + 1} + \cos \frac{4\pi}{2n + 1} + \cos \frac{6\pi}{2n + 1} + \cdots + \cos \frac{2n\pi}{2n + 1} = -\frac{1}{2} .$$

230. Construct an equation whose roots are the numbers:

(a) $\sin^2 \dfrac{\pi}{2n + 1}$, $\sin^2 \dfrac{2\pi}{2n + 1}$, $\sin^2 \dfrac{3\pi}{2n + 1}$, $\cdots$, $\sin^2 \dfrac{n\pi}{2n + 1}$;

(b) $\cot^2 \dfrac{\pi}{2n + 1}$, $\cot^2 \dfrac{2\pi}{2n + 1}$, $\cot^2 \dfrac{3\pi}{2n + 1}$, $\cdots$, $\cot^2 \dfrac{n\pi}{2n + 1}$.

231. Find the following sums.

(a) $\cot^2 \dfrac{\Pi}{2n+1} + \cot^2 \dfrac{2\Pi}{2n+1} + \cot^2 \dfrac{3\Pi}{2n+1}$

$$+ \cdots + \cot^2 \dfrac{n\Pi}{2n+1} \, .$$

(b) $\csc^2 \dfrac{\Pi}{2n+1} + \csc^2 \dfrac{2\Pi}{2n+1} + \csc^2 \dfrac{3\Pi}{2n+1}$

$$+ \cdots + \csc^2 \dfrac{n\Pi}{2n+1} \, .$$

232. Calculate the following products.

(a) $\sin \dfrac{\Pi}{2n+1} \; \sin \dfrac{2\Pi}{2n+1} \; \sin \dfrac{3\Pi}{2n+1} \; \cdots \; \sin \dfrac{n\Pi}{2n+1} \, ,$

and

$$\sin \dfrac{\Pi}{2n} \; \sin \dfrac{2\Pi}{2n} \; \sin \dfrac{3\Pi}{2n} \; \cdots \; \sin \dfrac{(n-1)\Pi}{n} \, .$$

(b) $\cos \dfrac{\Pi}{2n+1} \; \cos \dfrac{2\Pi}{2n+1} \; \cos \dfrac{3\Pi}{2n+1} \; \cdots \; \cos \dfrac{n\Pi}{2n+1} \, ,$

and

$$\cos \dfrac{\Pi}{2n} \; \cos \dfrac{2\Pi}{2n} \; \cos \dfrac{3\Pi}{2n} \; \cdots \; \cos \dfrac{(n-1)\Pi}{2n} \, .$$

233. Using the results of problems 231 (a) and (b), show that for any natural number n the sum

$$1 + \frac{1}{2^2} + \frac{1}{3^2} + \cdots + \frac{1}{n^2}$$

lies between the values

$$\left(1 - \frac{2}{n+1}\right)\left(1 - \frac{2}{2n+1}\right)\frac{\Pi^2}{6}$$

and

$$\left(1 - \frac{1}{2n+1}\right)\left(1 + \frac{1}{2n+1}\right)\frac{\Pi^2}{6} \, .$$

Remark: A particular result which follows from problem 233 is

$$1 + \frac{1}{2^2} + \frac{1}{3^2} + \frac{1}{4^2} + \cdots = \frac{\pi^2}{6} \, ,$$

where the summation on the left is the limit to which $1 + 1/2^2 + \cdots + 1/n^2$ tends as $n \to \infty$.

234. (a) On a circle which circumscribes an n-sided polygon $A_1A_2\cdots A_n$, a point M is taken. Prove that the sum of the squares of the distances from this point to all the vertices of the polygon is a number independent of the position of the point M on the circle, and that this sum is equal to $2nR^2$, where R is the radius of the circle.

(b) Prove that the sum of the squares of the distances from an arbitrary point M, taken in the plane of a regular n-sided polygon $A_1A_2\cdots A_n$ to all the vertices of the polygon, depends only upon the distance l of the point M from the center O of the polygon, and is equal to $n(R^2 + l^2)$, where R is the radius of the circle circumscribing the regular n-sided polygon.

(c) Prove that statement (b) remains correct even when point M does not lie in the plane of the n-sided polygon $A_1A_2\cdots A_n$.

235. Let M be a point on the circle circumscribing a regular n-sided polygon $A_1A_2\cdots A_n$. Prove the following.

(a) If n is even, then the sum of the squares of the distances from M to the vertices indicated by even-numbered subscripts (for example, A_2, A_4, and so on) is equal to the sum of the squared distances to the vertices having odd subscripts.

(b) If n is odd, then the sum of the distances from the point M to the vertices of the polygon which are even-numbered is equal to the sum of the distances to those which are odd-numbered.

236. The radius of a circle which circumscribes a regular n-sided polygon $A_1A_2\cdots A_n$ is equal to R. Prove the following.

(a) The sum of the squares of all the sides and all the diagonals is equal to n^2R^2.

(b) The sum of all the sides and all the diagonals of the polygon is equal to $n \cot \dfrac{\pi}{2n} R$.

(c) The product of all the sides and all the diagonals of the polygon is equal to $n^{n/2} R^{[n(n-1)]/2}$.

237.* Find the sum of the 50th powers of all the sides and all the diagonals of the regular 100-sided polygon inscribed in a circle of radius R.

238.* Prove that in a triangle whose sides have integral length it is not possible to find angles differing from 60°, 90°, and 120°, and commensurable with a right angle.

239. * (a) Prove that for any odd integer $p > 1$ the angle arc $\cos \dfrac{1}{p}$ cannot contain a rational number of degrees.

(b) Prove that an angle arc $\tan \dfrac{p}{q}$, where p and q are distinct positive integers, cannot contain a rational number of degrees.

10

SOME PROBLEMS OF

NUMBER THEORY

These problems are concerned with that division of mathematics treating properties of integers, Elementary Number Theory. Many of the problems in other sections of this book also deal with number theory—particularly Sections 3, 4, and 5. Several of the following theorems, stated here as problems, play an important role in number theory (see, for example, problems 240, 241, 245–247, 249, 253). Clearly, these problems do not pretend to explore with any completeness the rich variety of methods and ideas that have permeated this discipline, which is at once one of the most fruitful and one of the most difficult of all mathematical endeavors. A good systematic account of some parts of number theory is given in the book by B. B. Dynkin and V. A. Uspensky, *Mathematical Conversations*, Issue 6, Library of the USSR Mathematical Society. There the reader will find alternate solutions to some of the problems of this section. An excellent condensed treatment is the article by A. Y. Khinchin, "Elementary Number Theory," appearing in the *Encyclopedia of Elementary Mathematics*, Government Technical Publishing House, Moscow, 1951, which contains, as an appendix, an extensive bibliography covering the topics touched on in the article.

240. *Fermat's Theorem.* Prove that if p is a prime number, then the difference $a^p - a$ is, for any integer a, divisible by p.

Remark: Problems 27 (a)-(d) are special cases of this theorem.

241. *Euler's Theorem.* Let N be any natural number and let r

be the number of integers in the sequence $1, 2, 3, \cdots, N - 1$ which are relatively prime to N. Prove that if a is any integer which is relatively prime to N, then $a^r - 1$ is divisible by N.

Remark: If the number N is prime, then all the integers of the sequence are, of course, relatively prime to N; that is, $r = N - 1$. In this case, Euler's theorem assumes the form $a^{N-1} - 1$ is divisible by N, if N is prime. It is clear that Fermat's theorem (problem 240) can be considered a special case of Euler's theorem.

If $N = p^n$, where p is a prime number, then of the first $N - 1 = p^n - 1$ positive integers, those not relatively prime to $N = p^n$ will be $p, 2p, 3p, \cdots,$ $N - p = (p^{n-1} - 1)p$. Therefore, we have $r = (p^n - 1) - (p^{n-1} - 1) = p^n - p^{n-1}$, and Euler's theorem provides the following corollary: The difference $a^{p^n - p^{n-1}} - 1$, where p is prime and a is not divisible by p, is divisible by p^n.

If $N = p_1^{\alpha_1} p_2^{\alpha_2} \cdots p_k^{\alpha_k}$, where $p_1, p_2, \cdots, p_k$ are distinct primes, then the number r of prime numbers less than N and relatively prime to N is given by the formula

$$r = N\left(1 - \frac{1}{p_1}\right)\left(1 - \frac{1}{p_2}\right) x \cdots \left(1 - \frac{1}{p_k}\right).$$

(See, for example, the article by A. Y. Khinchin, referred to above.) If $N = p^n$ is a power of the prime p, this formula yields

$$r = p^n\left(1 - \frac{1}{p}\right) = p^n - p^{n-1},$$

which is the result obtained previously.

242.* According to Euler's theorem, the difference $2^k - 1$, where $k = 5^n - 5^{n-1}$, is divisible by 5^n (see problem 241, and the remark following it). Prove that there exists no k less than $5^n - 5^{n-1}$ such that $2^k - 1$ is divisible by 5^n.

243. Let us write, in order, the consecutive powers of the number 2: 2, 4, 8, 16, 32, 64, 128, 256, 512, 1024, 2048, 4096, $\cdots$. Note that in this sequence the final digits periodically repeat with a period of 4:

$$2, 4, 8, 6, 2, 4, 8, 6, 2, 4, 8, 6, \cdots.$$

Prove that, if we begin at a suitable point of the sequence, the last ten digits of the numbers of the sequence will also repeat periodically. Find the length of the period and the number of integers in the sequence for which this observed periodicity occurs.

244.* Prove that there exists some power of 2 whose final 1000 digits are all ones and twos.

245. *Wilson's Theorem.* Prove that: if the integer p is prime, then the number $(p-1)! + 1$ is divisible by p; if p is composite, then $(p-1)! + 1$ is not divisible by p.

246.* Let p be a prime number which yields the remainder 1 upon division by 4. Prove that there exists an integer x such that $x^2 + 1$ is divisible by p.

247.** Prove the following.

   (a) If each of the two integers A and B can be represented as the sum of two squares, then their product $A \cdot B$ can also be represented in this manner.

   (b) All prime numbers of form $4n + 1$ can be written as the sum of two squares, and no number of form $4n + 3$ can be so expressed.

   (c) A composite number N can be written as the sum of two squares if and only if all its prime factors of form $4n + 3$ occur an even number of times.

For example, the numbers $10,000 = 2^4 \cdot 5^4$ and $2430 = 2^2 \cdot 3^2 \cdot 5 \cdot 13$ can be represented as the sum of the squares of two integers (in the first number there are no factors of form $4n + 3$, and in the second number there is one such factor, 3, which occurs twice); the number $2002 = 2 \cdot 7 \cdot 11 \cdot 13$ cannot be represented as the sum of two squares (the factors 7 and 11, of form $4n + 3$, appear once).

248. Prove that, for any prime p, it is possible to find integers x and y such that $x^2 + y^2 + 1$ is divisible by p.

249.** Prove the following.

   (a) If each of two numbers A and B can be written as the sum of the squares of four integers, then their product $A \cdot B$ can also be represented in this manner.

   (b) Every natural number can be written as the sum of not more than four squares.

example, $35 = 25 + 9 + 1 = 5^2 + 3^2 + 1^2$; $60 = 49 + 9 + 1 + 1 = 7^2 + 3^2 + 1^2 + 1^2$; $1000 = 900 + 100 = 30^2 + 10^2$, and so on.

250. Prove that no number of the form $4^n(8k - 1)$, where n and k are integers, that is, no number belonging to the geometric progressions

$$7, \quad 28, \quad 112, \quad 448, \quad \cdots,$$
$$15, \quad 60, \quad 240, \quad 960, \quad \cdots,$$
$$23, \quad 92, \quad 368, \quad 1472, \quad \cdots,$$
$$31, \quad 124, \quad 496, \quad 1984, \quad \cdots.$$

can be a square or the sum of two squares or three squares of integers.

Remark: It has been shown that every integer which *cannot* be written in form $4^n(8k - 1)$ is representable as the sum of three or fewer squares. However, the proof is very complicated.

251.** Prove that every positive integer can be written as the sum of not more than 53 fourth powers of integers.

Remark: Experimental trials indicate that integers of moderate size are representable as the sum of far fewer fourth powers than 53. To the present time, no integer has been produced which cannot be given as the sum of 19, or fewer, fourth powers. (Of the numbers less than 100, only one—the number 79—requires as many as 19 fourth powers; that is four terms of 2^4 and 15 units). It has been conjectured that 19 fourth powers suffice for every integer, but no proof of this has as yet appeared. The best result in this direction has been the proof that every natural number can be written as the sum of not more than 21 fourth powers. This is a substantial improvement over the proposition given as problem 251, but the proof of it involves considerable higher mathematics.

In problem 239 (b) it was stated that every integer can be written as the sum of not more than four squares. It has also been shown that every integer can be written as the sum of not more than nine cubes.

All these propositions are embraced by the following remarkable theorem: *For every positive integer k there exists a positive integer N (depending, of course, on k) such that every integer may be written as the sum of not more than N kth powers of positive integers.* This theorem has been provided with several different proofs, but only recently has a proof been given which does not require considerable higher mathematics. In 1942 the Soviet mathematician U. V. Linnik gave the elementary proof. This proof is presented in the popular little book by A. Y. Khinchin, *Three Pearls of Number Theory*, Government Technical Publishing House, Moscow, 1949.[†] Although Linnik's proof is elementary, it is not easy reading. Khinchin himself remarks that almost anybody can understand it with "only two or three weeks work with pencil and paper."

252.** Prove that every positive rational number (in particular, every positive integer) can be written as the sum of three cubes of positive rational numbers.

Remark: Not all positive rational numbers can be represented as the sum of two cubes of positive rational numbers. Consider, for example, the number 1. The equation

[†] An English translation has been published by Graylock Press, Rochester. N. Y., 1952, 64 pp., $2.00 [*Editor*].

$$1 = \left(\frac{m}{n}\right)^3 + \left(\frac{p}{q}\right)^3$$

can be written

$$(nq)^3 = (mq)^3 + (np)^3 ,$$

where m, n, p, and q are integers. But it is known that no solution in integers exists for the equation

$$x^3 + y^3 = z^3$$

(a proof of this may be found in most standard texts on number theory).

253. Prove that there exists an infinite number of prime numbers.

254. (a) Prove that among the numbers of the arithmetic progressions 3, 7, 11, 15, 19, 23, $\cdots$ and 5, 11, 17, 23, 29, 35, $\cdots$ there are an infinite number of primes.

(b)* Prove that there are an infinite number of primes in the arithmetic progression

$$5, 9, 13, 17, 21, 25, \cdots .$$

(c)* Prove that there are an infinite number of primes in the arithmetic progression

$$11, 21, 31, 41, 51, 61, \cdots .$$

Remark: The following more general theorem holds: *If the first term of an infinite arithmetic progression of integers is relatively prime to the common difference, the progression contains an infinite number of primes.* However, the proof of this theorem is quite complicated. (It is interesting that an elementary, albeit very difficult, proof of this classical theorem of number theory was published for the first time only in 1952 by the Danish mathematician Selberg. Prior to this the only known proofs involved higher mathematics).

11

SOME DISTINCTIVE INEQUALITIES

This section presents several problems relating to inequalities stemming from two important inequalities which play a major role in mathematical analysis and in geometry. These are the theorem relating arithmetic and geometric means (problem 268), and the so-called Cauchy-Buniakowski inequality (problem 289). Many problems on inequalities, not related to these two but of importance in other applications, appear in other sections of this book (see, in particular, Sections 6 and 7).

A great many interesting inequalities may be found in the *Problem Book in Algebra*, by V. A. Kretchmer, Government Technical Publishing House, Moscow, 1950, where an entire chapter is devoted to inequalities. That book offers alternative proofs of several of the inequalities presented here. There is also much interesting material in the books by P. P. Korovkin, *Inequalities* (Government Technical Publishing House, Moscow, 1951), by G. L. Nevyashy, *Inequalities* (Pedagogical Publishing House, Moscow, 1947), and particularly that by Hardy, Littlewood, and Polya, *Inequalities*, (Government Technical Publishing House, Moscow, 1949).[†]

The initial chapters of the last book may be read by persons not acquainted with higher mathematics·

The following problems are not presented in order of increasing difficulty.

[†] The last book was originally written in English. It is published by Cambridge University Press, revised edition, [*Editor*].

The ordering is such that in some instances the result of one problem will be useful in solving the next; in other instances problems conceptually related are grouped together. The simplest properties of inequalities are assumed known.

In all the problems of this section, small English letters designate real numbers.

Theorems on Arithmetic and Geometric Means and Their Applications

We know, from formal mathematics courses, that the geometric mean of two positive numbers a and b is less than, or equal to, their arithmetic mean,

$$\sqrt{ab} \leqq \frac{a+b}{2} , \tag{1}$$

and the equality holds only if $a = b$. This is proved as follows.

If we square both members of the inequality and clear of fractions, we arrive at

$$4ab \leqq (a+b)^2 .$$

Expanding the right member, transposing $4ab$ to the right side, and so on, we obtain

$$0 \leqq a^2 - 2ab + b^2 = (a-b)^2 ,$$

which clearly is true for all numbers a and b, since the square of any real number is nonnegative.

Hence, inequality (1) holds for every real number. Moreover, it is evident that $(a-b)^2$ can be zero only if $a = b$; that is, the last inequality reduces to the equality only for $a = b$. Therefore, this criterion must hold also for inequality (1).

Inequality (1) may be rewritten in the following equivalent form, which we shall use hereafter:

$$\left(\frac{a+b}{2}\right)^2 \leqq \frac{a^2 + b^2}{2} . \tag{1'}$$

If we expand the left member of (1'), clear of fractions and put all terms in the right member, we obtain

$$0 \leqq 2a^2 + 2b^2 - (a^2 + 2ab + b^2) = (a-b)^2 .$$

Use of inequalities (1) and (1') simplifies the solution of the first of the problems which follow. These two forms of the inequality are useful in the derivation of many generalizations, the most important of which are the propositions of problems 268 and 283.

The arithmetic mean of n positive numbers $a_1, a_2, \cdots, a_n$ is defined by the following expression:

$$A_n(a) = \frac{a_1 + a_2 + \cdots + a_n}{n} .$$

The geometric mean of n positive numbers $a_1, a_2, \cdots, a_n$ is defined as the nth root of their product:

$$\Gamma_n(a) = \sqrt[n]{a_1 a_2 \cdots a_n} .$$

Finally, *the harmonic mean* of n positive numbers is the number $H(a)$ such that

$$\frac{1}{H(a)} = \frac{1/a + 1/a_2 + \cdots + 1/a_n}{n}$$

(the reciprocal of the harmonic mean of n numbers is the arithmetic mean of the numbers inverse to the given ones). In particular, the harmonic mean of two numbers a and b is determined by the equation

$$\frac{1}{c} = \frac{1}{2}\left(\frac{1}{a} + \frac{1}{b}\right),$$

from which $c = 2ab/(a + b)$.

255. (a) Prove that, of all rectangles having the same given perimeter P, the square encloses the greatest area.

(b) Prove that, of all rectangles having the same given area S, that of smallest perimeter is the square.

256. Prove that the sum of the legs of a right triangle never exceeds $\sqrt{2}$ times the hypotenuse of the triangle.

257. Prove that for every acute angle α

$$\tan \alpha + \cot \alpha \geq 2 .$$

258. Prove that if $a + b = 1$, where a and b are positive numbers, then

$$\left(a + \frac{1}{a}\right)^2 + \left(b + \frac{1}{b}\right)^2 \geq \frac{25}{2} .$$

Determine for what values of a and b the equality holds.

259. Prove, given any three positive numbers a, b, and c, the following inequality holds:

$$(a + b)(b + c)(c + a) \geq 8abc .$$

Show that the equality holds only for $a = b = c$.

260. For what values of x does the following fraction have the least value?

$$\frac{a + bx^4}{x^2} \quad (a \text{ and } b \text{ positive}).$$

261. A butcher has an inaccurate balance scale (its beams are of unequal length). Knowing that it is inaccurate, and being an honest merchant, he weighs his meat as follows. He takes half of it and places it on one pan, and he places the weights on the other pan;

then he weighs the other half of the meat by reversing this procedure, that is, by removing the weights and placing the meat on that pan. Thus, the butcher believes he is giving honest weight. Is his assumption correct?

262. (a) Prove that the geometric mean of two positive numbers is equal to the geometric mean of their arithmetic and harmonic means.

(b) Prove that the harmonic mean of two positive numbers a and b does not exceed the geometric mean, and that the equality holds only if $a = b$.

263.* Prove that the arithmetic mean of three positive numbers is not less than their geometric mean, that is,

$$\frac{a + b + c}{3} \geqq \sqrt[3]{abc} \,,$$

and that the equality holds only if $a = b = c$.

264. Prove that, of all triangles with the same given perimeter, the greatest area is enclosed by the equilateral triangle.

265. Given a three-faced pyramid having a right trihedral angle at the vertex. Designate the edges from the vertex by x, y, and z. For what x, y, and z is the volume of the pyramid a maximum if it is known that

$$x + y + z = a?$$

266. Given six positive numbers $a_1, a_2, a_3, b_1, b_2, b_3$. Prove that the following inequality holds:

$$\sqrt[3]{(a_1 + b_1)(a_2 + b_2)(a_3 + b_3)} \geqq \sqrt[3]{a_1 a_2 a_3} + \sqrt[3]{b_1 b_2 b_3} \,.$$

267. *A Special Case of the Theorem Concerning the Arithmetic and Geometric Means.* Given 2^m positive numbers $a_1, a_2, \cdots, a_{2^m}$. Prove the inequality

$$\Gamma_{2^m} \leqq A_{2^m}(a) \,,$$

that is,

$$\sqrt[2^m]{a_1 a_2 \cdots a_{2^m}} \leqq \frac{a_1 + a_2 + \cdots + a_{2^m}}{2^m} \,,$$

and that the equality holds only if all the numbers $a_1, a_2, \cdots, a_{2^m}$ are equal.

268.* *Theorem of the Arithmetic and Geometric Means for n Num-*

bers. Prove that for any n positive numbers $a_1, a_2, \cdots, a_n$

$$\Gamma_n(a) \leqq A_n(a) \, ,$$

that is,

$$\sqrt[n]{a_1 a_2 \cdots a_n} \leqq \frac{a_1 + a_2 + \cdots + a_n}{n} \, ,$$

and that the equality holds only if $a_1 = a_2 = \cdots = a_n$.

269. (a) Consider all sets of n positive numbers whose sum is a given number k. Prove that the maximum product of the numbers of any such set is attained when all the numbers are equal.

(b) Given n positive numbers $a_1, a_2, \cdots, a_n$. Prove that

$$\frac{a_1}{a_2} + \frac{a_2}{a_3} + \cdots + \frac{a_n}{a_1} \geqq n \, .$$

270. Prove that for n positive numbers $a_1, a_2, \cdots, a_n$ the following inequality holds,

$$H(a) \leqq \Gamma(a) \, ,$$

that is,

$$\frac{n}{\left(\dfrac{1}{a_1} + \dfrac{1}{a_2} + \cdots + \dfrac{1}{a_n} \right)} \leqq \sqrt[n]{a_1 a_2 \cdots a_n} \, ,$$

and that the equality is obtained only if $a_1 = a_2 = \cdots = a_n$.

271. Prove that for two positive numbers a and b

$$\sqrt[n+1]{ab^n} \leqq \frac{a + nb}{n + 1} \, ,$$

and that equality can hold only if $a = b$.

272. Prove that for any set of positive numbers $a_1, a_2, \cdots, a_n$

$$(a_1 + a_2 + \cdots + a_n)\left(\frac{1}{a_1} + \frac{1}{a_2} + \cdots + \frac{1}{a_n} \right) \geqq n^2 \, .$$

When does the equality hold?

273. Prove that for any integer $n > 1$

$$n! < \left(\frac{n + 1}{2} \right)^n \, .$$

274. Prove that the following inequality holds for any four positive numbers a_1, a_2, a_3, a_4:

$$a_1 a_2^2 a_3^3 a_4^4 \leq \left(\frac{a_1 + 2a_2 + 3a_3 + 4a_4}{10}\right)^{10}.$$

275. Prove the following.

(a) $1 \cdot \dfrac{1}{2^2} \cdot \dfrac{1}{3^3} \cdot \dfrac{1}{4^4} \cdots \dfrac{1}{n^n} < \left[\dfrac{2}{n+1}\right]^{[n(n+1)]/2}$;

(b) $1 \cdot 2^2 \cdot 3^3 \cdot 4^4 \cdots n^n < \left[\dfrac{2n+1}{3}\right]^{[n(n+1)]/2}$

($[a]$ means "the largest integer in a").

276. Let $a_1, a_2, \cdots, a_n$ be positive numbers, and let

$$s = a_1 + a_2 + \cdots + a_n.$$

Prove that

$$(1 + a_1)(1 + a_2)\cdots(1 + a_n) \leq 1 + s + \frac{s^2}{2!} + \frac{s^3}{3!} + \cdots + \frac{s^n}{n!}.$$

277. Prove that for every integer n

$$\sqrt{2} \sqrt[4]{4} \sqrt[8]{8} \cdots \sqrt[2^n]{2^n} \leq n + 1.$$

278. For which value of x is the product

$$(1 - x)^5(1 + x)(1 + 2x)^2$$

a maximum, and what is this value?

279.* Inscribe between a given segment of a circle and the arc of the circle the rectangle of greatest area.

280. From a square piece of cardboard measuring $2a$ on each side

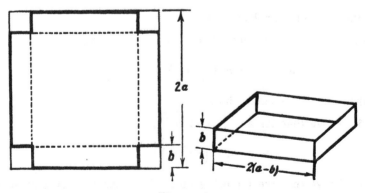

Figure 4

a box with no top is to be formed by cutting out from each corner
a square with sides b and bending up the flaps, as shown in Figure
4. For what value of b will the box contain the greatest volume?

Two Generalizations of the Theorem Concerning
Arithmetic and Geometric Means

The *power mean* of order α of n positive numbers $a_1, a_2, \cdots, a_n$ is defined
to be the number

$$S_\alpha(a) = \left(\frac{a_1^\alpha + a_2^\alpha + \cdots + a_n^\alpha}{n} \right)^{1/\alpha} ;$$

in particular, if $\alpha = k$ is a whole number, we obtain

$$S_k(a) = \sqrt[k]{\frac{a_1^k + a_2^k + \cdots + a_n^k}{n}} .$$

It is easy to see that $S_1(a) = A(a)$ and $S_{-1}(a) = H(a)$.

If $\alpha = 0$, the expression for S_α is meaningless. On the other hand, it can
be proved that if $\alpha \to 0$, then $S_\alpha(a)$ tends to the geometric mean $\Gamma(a)$.† Therefore, it is convenient to define $S_0(a) = \Gamma(a)$. (An additional justification for
this definition is given in problem 282.) The power mean of order 2 is referred
to as the *quadratic mean*.

Inequality (1′) (see the remark at the beginning of this section) can now be
stated as follows: *The arithmetic mean of two numbers does not exceed their
quadratic mean* (and the equality holds only if the numbers are equal).

281.* (a) Prove that the arithmetic mean of n positive numbers
does not exceed the quadratic mean:

$$\left(\frac{a_1 + a_2 + \cdots + a_n}{n} \right)^2 \leqq \frac{a_1^2 + a_2^2 + \cdots + a_n^2}{n} .$$

The equality holds only if the numbers are all equal.

(b) Let k be any integer greater than 1. Prove that the
arithmetic mean of n positive numbers does not exceed their power
mean of order k:

$$\left(\frac{a_1 + a_2 + \cdots + a_n}{n} \right)^k \leqq \frac{a_1^k + a_2^k + \cdots + a_n^k}{n} ,$$

The equality holds only if all the numbers are equal.

---

† That is,

$$\lim_{\alpha \to 0} \left(\frac{a_1^\alpha + a_2^\alpha + \cdots + a_n^\alpha}{n} \right)^{1/\alpha} = \sqrt[n]{a_1 a_2 \cdots a_n} .$$

See V. E. Levine, "Elementary proof of one theorem of the theory of means,"
Math. Educa., Issue 3, pp. 177–181, Moscow, 1958.

282. Prove that the power mean of order α of n positive numbers, for $\alpha > 0$, is not less than the geometric mean, and, for $\alpha < 0$, is not greater than the geometric mean (equality holds only if all n numbers are equal.)

Remark: The theorems of problems 268 and 270 are particular cases of this proposition.

283.* *Theorem of Power Means.* Prove that if $\alpha < \beta$, then the power mean of order α does not exceed the power mean of order β:

$$\left(\frac{a_1^\alpha + a_2^\alpha + \cdots + a_n^\alpha}{n}\right)^{1/\alpha} \leqq \left(\frac{a_1^\beta + a_2^\beta + \cdots + a_n^\beta}{n}\right)^{1/\beta}.$$

The equality holds only if $a_1 = a_2 = \cdots = a_n$.

284. (a) The sum of three positive numbers is equal to 6. What is the smallest value which the sum of their squares can have? What is the smallest value which the sum of their cubes can have?

(b) The sum of the squares of three positive numbers is equal to 18. What is the smallest possible value for the sum of the cubes of these numbers? What is the smallest possible value for the sum of these numbers?

The symmetric mean of order k of n positive numbers $a_1, a_2, \cdots, a_n$ (where k is a natural number not exceeding n) is defined to be the kth root of the sum of all possible products of these n numbers taken k at a time:

$$\Sigma_k(a) = \sqrt[k]{\frac{a_1 a_2 \cdots a_k + a_1 a_2 \cdots a_{k-1} a_{k+1} + \cdots + a_{n-k+1} a_{n-k+2} \cdots a_n}{C_n^k}}.$$

It is clear that $\Sigma_1(a) = A(a)$, $\Sigma_n(a) = \Gamma(a)$.

285. Prove that

$$(\Sigma_k)^{2k} \geqq (\Sigma_{k+1})^{k+1} \cdot (\Sigma_{k-1})^{k-1}.$$

286. *Theorem of the Symmetric Mean.* Prove that if $k > l$, then

$$\Sigma_k(a) \leqq \Sigma_l(a).$$

The equality holds only if $a_1 = a_2 = \cdots = a_n$.

287. Given that the sum of all six possible pairwise products of four numbers is equal to 24. What is the smallest value possible for the sum of the four numbers? What is the greatest possible value for the product of the numbers?

288. Let $\alpha + \beta + \gamma = \pi$.

(a) Find the smallest possible value for

$$\tan \frac{\alpha}{2} + \tan \frac{\beta}{2} + \tan \frac{\gamma}{2} \, .$$

(b) Find the largest possible value for

$$\tan \frac{\alpha}{2} \cdot \tan \frac{\beta}{2} \cdot \tan \frac{\gamma}{2} \, .$$

The Cauchy-Buniakowski Inequality

The following elementary inequality is readily verified:

$$a_1 b_1 + a_2 b_2 \leqq \sqrt{a_1^2 + a_2^2} \sqrt{b_1^2 + b_2^2}$$

or,

$$(a_1 b_1 + a_2 b_2)^2 \leqq (a_1^2 + a_2^2)(b_1^2 + b_2^2) \, . \qquad (1)$$

Expanding both sides and collecting all terms on one side we obtain

$$(a_1 b_2 - a_2 b_1)^2 \geqq 0 \, .$$

It follows that inequality (1) becomes an equality if

$$a_1 b_2 = a_2 b_1 \, ,$$

that is, if

$$\frac{a_1}{b_1} = \frac{a_2}{b_2} \, .$$

Inequality (1) yields a significant generalization which is important in inequality theory and has useful applications in mathematics and physics.[†]

289. *The Cauchy-Buniakowski Inequality.* Prove that for any $2n$ real numbers $a_1, a_2, \cdots, a_n$ and $b_1, b_2, \cdots, b_n$ the following inequality holds:

$$(a_1 b_1 + a_2 b_2 + \cdots + a_n b_n)^2 \leqq (a_1^2 + a_2^2 + \cdots + a_n^2)(b_1^2 + b_2^2 + \cdots + b_n^2) \, .$$

The equality holds only if

$$\frac{a_1}{b_1} = \frac{a_2}{b_2} = \cdots = \frac{a_n}{b_n} \, .$$

290. Use the Cauchy-Buniakowski inequality to derive the results of problem 272.

291. Use the Cauchy-Buniakowski inequality to obtain the theorem of problem 281 (a).

292. Prove that if $\alpha + \beta + \gamma = \Pi$, then

[†] This inequality is sometimes referred to, in other texts, as the Cauchy-Schwarz inequality [*Editor*].

$$\tan^2 \frac{\alpha}{2} + \tan^2 \frac{\beta}{2} + \tan^2 \frac{\gamma}{2} \geqq 1 .$$

293. Prove for any positive numbers $x_1, x_2, \cdots, x_n; y_1, y_2, \cdots, y_n$:

$$\sqrt{(x_1 + y_1)^2 + (x_2 + y_2)^2 + \cdots + (x_n + y_n)^2}$$
$$\leqq \sqrt{x_1^2 + x_2^2 + \cdots + x_n^2} + \sqrt{y_1^2 + y_2^2 + \cdots + y_n^2} .$$

294. Let Q be the sum of all the possible pairwise products of the n positive numbers $a_1, a_2, \cdots, a_n$, and let P be the sum of their squares. Prove that

$$Q \leqq \frac{n-1}{2} P .$$

295. Prove that, given $2n$ positive numbers $p_1, p_2, \cdots, p_n; x_1, x_2, \cdots, x_n$, the following inequality holds:

$$(p_1 x_1 + p_2 x_2 + \cdots + p_n x_n)^2$$
$$\leqq (p_1 + p_2 + \cdots + p_n)(p_1 x_1^2 + p_2 x_2^2 + \cdots + p_n x_n^2) .$$

296. Verify that for any three arbitrary numbers x_1, x_2, x_3 the following inequality holds:

$$\left(\frac{1}{2} x_1 + \frac{1}{3} x_2 + \frac{1}{6} x_3 \right)^2 \leqq \frac{1}{2} x_1^2 + \frac{1}{3} x_2^2 + \frac{1}{6} x_3^2 .$$

297. Prove that if $x_1, x_2, \cdots, x_n; y_1, y_2, \cdots, y_n$ are positive numbers, then

$$\sqrt{x_1 y_1} + \sqrt{x_2 y_2} + \cdots + \sqrt{x_n y_n}$$
$$\leqq \sqrt{x_1 + x_2 + \cdots + x_n} \cdot \sqrt{y_1 + y_2 + \cdots + y_n} .$$

298. Let $a_1, a_2, \cdots, a_n; b_1, b_2, \cdots, b_n; c_1, c_2, \cdots, c_n; d_1, d_2, \cdots, d_n$ be four sequences of positive numbers. Prove the inequality

$$(a_1 b_1 c_1 d_1 + a_2 b_2 c_2 d_2 + \cdots + a_n b_n c_n d_n)^4$$
$$\leqq (a_1^4 + a_2^4 + a_3^4 + \cdots + a_n^4)(b_1^4 + b_2^4 + b_3^4 + \cdots + b_n^4)$$
$$\times (c_1^4 + c_2^4 + c_3^4 + \cdots + c_n^4)(d_1^4 + d_2^4 + d_3^4 + \cdots + d_n^4) .$$

299.* The Cauchy-Buniakowski inequality (problem 289) verifies that the relationship

$$\frac{(a_1^2 + a_2^2 + \cdots + a_n^2)(b_1^2 + b_2^2 + \cdots + b_n^2)}{(a_1 b_1 + a_2 b_2 + \cdots + a_n b_n)^2} ,$$

where $a_1, a_2, \cdots, a_n; b_1, b_2, \cdots, b_n$ are two sequences of positive num-

bers, is greater than or equal to 1 $\left(\text{and is equal to 1 only if } \dfrac{a_1}{b_1} = \right.$
$\left.\dfrac{a_2}{b_2} = \cdots = \dfrac{a_n}{b_n}\right)$. Prove that this value is always included between
1 and the quantity

$$1 + \left(\frac{\sqrt{M_1 M_2/m_1 m_2} - \sqrt{m_1 m_2/M_1 M_2}}{2}\right)^2$$

$$= \left.\left(\frac{\sqrt{M_1 M_2/m_1 m_2} + \sqrt{m_1 m_2/M_1 M_2}}{2}\right)^2, \right]$$

where M_1 and m_1 are, respectively, the greatest and the least of the
numbers $a_1, a_2, \cdots, a_n$, and M_2 and m_2 are, respectively, the greatest
and least of the numbers $b_1, b_2, \cdots, b_n$. In which case does the value
exactly equal the following?

$$1 + \left(\frac{\sqrt{M_1 M_2/m_1 m_2} - \sqrt{m_1 m_2/M_1 M_2}}{2}\right)^2.$$

Some Additional Inequalities

300. *Chebycheff's Inequality.* Let $a_1, a_2, \cdots, a_n$ and $b_1, b_2, \cdots, b_n$ be
two nonincreasing sequences of numbers. The following inequality
holds:

$$\frac{a_1 + a_2 + \cdots + a_n}{n} \cdot \frac{b_1 + b_2 + \cdots + b_n}{n} \leq \frac{a_1 b_1 + a_2 b_2 + \cdots + a_n b_n}{n},$$

the equality holding only if $a_1 = a_2 = \cdots = a_n$ and $b_1 = b_2 = \cdots = b_n$.

Remark: It is possible to show that if $a_1, a_2, \cdots, a_n$ is a nonincreasing
sequence of numbers, and if $b_1, b_2, \cdots, b_n$ is nondecreasing, then

$$\frac{a_1 + a_2 + \cdots + a_n}{n} \cdot \frac{b_1 + b_2 + \cdots + b_n}{n} \geq \frac{a_1 b_1 + a_2 b_2 + \cdots + a_n b_n}{n}.$$

The proof of this proposition is left to the reader.

301. Let p and q be positive rational numbers for which

$$\frac{1}{p} + \frac{1}{q} = 1.$$

Prove that for any positive numbers x and y the following inequality
holds:

$$xy \leq \frac{1}{p} x^p + \frac{1}{q} y^q.$$

Remark: For $p = q = 2$ we obtain the Elementary Theorem of the Arithmetic and Geometric Means.

302. Let α and β be positive rational numbers, where $\alpha + \beta = 1$. Prove that for any positive numbers $a_1, a_2, \cdots, a_n;\ b_1, b_2, \cdots, b_n$ the following inequality holds:

$$a_1^\alpha b_1^\beta + a_2^\alpha b_2^\beta + \cdots + a_n^\alpha b_n^\beta \leqq (a_1 + a_2 + \cdots + a_n)^\alpha (b_1 + b_2 + \cdots + b_n)^\beta .$$

Remark: If $\alpha = \beta = \frac{1}{2}$, it is readily seen that we obtain the inequality of problem 297, which is equivalent to the Cauchy-Buniakowski inequality.

303. *Holder's Inequality.* Let p and q be positive rational numbers such that

$$\frac{1}{p} + \frac{1}{q} = 1 .$$

Prove that for any positive numbers $x_1, x_2, \cdots, x_n;\ y_1, y_2, \cdots, y_n$ the following inequality holds:

$$x_1 y_1 + x_2 y_2 + \cdots + x_n y_n$$
$$\leqq (x_1^p + x_2^p + \cdots + x_n^p)^{1/p} (y_1^q + y_2^q + \cdots + y_n^q)^{1/q} .$$

Remark: If $p = q = 2$, this inequality becomes that of Cauchy-Buniakowski (problem 289), which, in turn, is a special case of Hölder's inequality.

304. Let $a_1, a_2, \cdots, a_n;\ b_1, b_2, \cdots, b_n;\ \cdots;\ l_1, l_2, \cdots, l_n$ be k sequences of positive numbers, and $\alpha, \beta, \cdots, \lambda$ be k positive number such that

$$\alpha + \beta + \cdots + \lambda = 1 .$$

Prove that

$$a_1^\alpha b_1^\beta \cdots l_1^\lambda + a_2^\alpha b_2^\beta \cdots l_2^\lambda + \cdots + a_n^\alpha b_n^\beta \cdots l_n^\lambda$$
$$\leqq (a_1 + a_2 + \cdots + a_n)^\alpha (b_1 + b_2 + \cdots + b_n)^\beta \cdots (l_1 + l_2 + \cdots + l_n)^\lambda .$$

305. Let $a_1, a_2, \cdots, a_n$ be n positive numbers, and let g be their geometric mean ($g = \sqrt[n]{a_1 a_2 \cdots a_n}$). Prove that

$$(1 + a_1)(1 + a_2) \cdots (1 + a_n) \geqq (1 + g)^n .$$

306. Prove that if $a_1, a_2, \cdots, a_n;\ b_1, b_2, \cdots, b_n;\ \cdots;\ l_1, l_2, \cdots, l_n$ are k sequences of positive numbers, then the following holds:

$$\sqrt[n]{a_1 a_2 \cdots a_n} + \sqrt[n]{b_1 b_2 \cdots b_n} + \cdots + \sqrt[n]{l_1 l_2 \cdots l_n}$$
$$\leqq \sqrt[n]{(a_1 + b_1 + \cdots + l_1)(a_2 + b_2 + \cdots + l_2)(a_n + b_n + \cdots + l_n)} .$$

307. Let $x, y,$ and z be positive numbers for which

$$x + y + z = 1 .$$

Prove that

$$\left(1 + \frac{1}{x}\right)\left(1 + \frac{1}{y}\right)\left(1 + \frac{1}{z}\right) \geqq 64 .$$

308. *Minkowski's Inequality.* Let $a_1, a_2, \cdots, a_n$; $b_1, b_2, \cdots, b_n$; $\cdots$; $l_1, l_2, \cdots, l_n$, be k sequences of positive numbers. Prove that

$$\sqrt{a_1^2 + a_2^2 + \cdots + a_n^2} + \sqrt{b_1^2 + b_2^2 + \cdots + b_n^2}$$
$$+ \cdots + \sqrt{l_1^2 + l_2^2 + \cdots + l_n^2}$$
$$\geqq \sqrt{(a_1 + b_1 + \cdots + l_1)^2 + (a_2 + b_2 + \cdots + l_2)^2 + \cdots + (a_n + b_n + \cdots + l_n)^2} .$$

Remark: The inequality of problem 308 [a generalization of the result of problem 186 (a)] can also be written in the form

$$S_2(a) + S_2(b) + \cdots + S_2(l) \geqq S_2(a + b + \cdots + l) ,$$

where S_2 is the quadratic mean of n numbers (see p. 67).

A more general formulation of Minkowski's inequality is as follows. If $a_1, a_2, \cdots, a_n$; $b_1, b_2, \cdots, b_n$; $\cdots$; $l_1, l_2, \cdots, l_n$, are k sequences of positive numbers, then

$$S_2(a) + S_2(b) + \cdots + S_2(l)\begin{cases} \geqq S_\alpha(a + b + \cdots + l) & \text{if } \alpha > 1 ; \\ \leqq S_\alpha(a + b + \cdots + l) & \text{if } \alpha < 1 . \end{cases}$$

In particular, the inequality of problem 306, which may be written

$$\Gamma(a) + \Gamma(b) + \cdots + \Gamma(l) \leqq \Gamma(a + b + \cdots + l)$$

or

$$S_0(a) + S_0(b) + \cdots + S_0(l) \leqq S_0(a + b + \cdots + l)$$

is a special case of Minkowski's inequality for $\alpha = 0$.

12

DIFFERENCE SEQUENCES AND SUMS

Consider the sequence of numbers

$$u_0, u_1, u_2, \cdots, u_n, \cdots .$$

The *first difference sequence* of this sequence is the set of numbers

$$u_0^{(1)} = u_1 - u_0 \; ;$$
$$u_1^{(1)} = u_2 - u_1 \; ;$$
$$u_2^{(1)} = u_3 - u_2 \; ;$$
$$\cdots\cdots\cdots \; ;$$
$$u_n^{(1)} = u_{n+1} - u_n \; , \; \cdots .$$

The *second difference sequence* is the difference sequence of the previous sequence:

$$u_0^{(2)} = u_1^{(1)} - u_0^{(1)} \; ;$$
$$u_1^{(2)} = u_2^{(1)} - u_1^{(1)} \; ;$$
$$u_2^{(2)} = u_3^{(1)} - u_2^{(1)} \; ;$$
$$\cdots\cdots\cdots \; ;$$
$$u_n^{(2)} = u_{n+1}^{(1)} - u_n^{(1)} \; .$$

Analogously, the sequence of *differences of kth order*, $u_0^{(k)}, u_1^{(k)}, u_2^{(k)}, \cdots, u_n^{(k)}$, is the sequence obtained by working on the $(k-1)$st sequence of differences, $u_0^{(k-1)}, u_1^{(k-2)}, u_2^{(k-3)}, \cdots$. For example, if the initial sequence of numbers is the arithmetic progression $1, 5, 9, 13, 17, \cdots$, then the first row of differences consists of the numbers $4, 4, 4, 4, \cdots$, and the differences of second order form a sequence of zeros: $0, 0, 0, 0, \cdots$. If the initial sequence is the set of squares of integers, $1, 4, 9, 16, 25, 36, 49, \cdots$, then the differences of first order form the sequence of odd numbers: $3, 5, 7, 9, 11, 13, \cdots$; the differences of second order form the sequence $2, 2, 2, 2, \cdots$, and the third sequence (differences of third order) consists of zeros. [In the examples investigated we quickly arrived at a sequence of zeros, and this is related to the general proposition of 309 (b).]

The sequences of differences of a finite sequence of numbers can be conveniently written in *triangular array*:

$$
\begin{array}{cccccc}
u_0 & & u_1 & & u_2 & & \cdots & & u_n \\
& u_0^{(1)} & & u_1^{(1)} & & u_2^{(1)} & \cdots & u_{n-1}^{(1)} \\
& & u_0^{(2)} & & u_1^{(2)} & & \cdots u_{n-2}^{(2)} \\
& & & \cdots\cdots\cdots\cdots\cdots \\
& & & & u_0^{(n)}
\end{array}
$$

Here it is apparent that each number is the difference of the two adjacent numbers of the row above. For an infinite sequence of numbers the triangular (infinite) array has the form

$$
\begin{array}{ccccccc}
u_0 & & u_1 & & u_2 & & \cdots & & u_n & & \cdots \\
& u_0^{(1)} & & u_1^{(1)} & & u_2^{(1)} & & \cdots & & u_n^{(1)} & & \cdots \\
& & u_0^{(2)} & & u_1^{(2)} & & u_2^{(2)} & & \cdots & & u_n^{(2)} & & \cdots \\
& & & \cdots\cdots\cdots\cdots\cdots\cdots\cdots\cdots\cdots\cdots\cdots\cdots \\
& & & \cdots\cdots\cdots\cdots\cdots\cdots\cdots\cdots\cdots\cdots\cdots
\end{array}
$$

In a fashion analogous to finding the successive sequences of differences of a set of numbers we can also define sequences of sums. The sequence of *sums of first order* of the set of numbers $u_0, u_1, u_2, \cdots, u_n, \cdots$, which we shall designate by writing $\bar{u}_0^{(1)}, \bar{u}_1^{(1)}, \bar{u}_2^{(1)}, \cdots, \bar{u}_n^{(1)}, \cdots$, is defined by

$$\bar{u}_0^{(1)} = u_0 + u_1 ;$$
$$\bar{u}_1^{(1)} = u_1 + u_2 ;$$
$$\cdots\cdots\cdots\cdots ;$$
$$\bar{u}_n^{(1)} = u_n + u_{n+1} ;$$
$$\cdots\cdots\cdots\cdots ;$$

The sequence of *sums of* kth *order* of the numbers $u_0, u_1, \cdots, u_n, \cdots$ is obtained from the $(k-1)$st row of such sums. We shall designate the sums of kth order by $\bar{u}_0^{(k)}, \bar{u}_1^{(k)}, \cdots, \bar{u}_n^{(k)}, \cdots$.

For example, if the initial set is the sequence of ones, $1, 1, 1, 1, \cdots$, then the sequence of sums of first order consists of two's: $2, 2, 2, 2, \cdots$; the sequence of sums of second order is $4, 4, 4, 4, \cdots$; the third sequence is $8, 8, 8, 8, \cdots$; and so on. If the initial set is the sequence of natural numbers, $1, 2, 3, 4, 5, \cdots$, then each sequence of sums will form an arithmetic progression:

$$3, \quad 5, \quad 7, \quad 9, \quad 11, \quad 13, \quad \cdots$$
$$8, \quad 12, \quad 16, \quad 20, \quad 24, \quad \cdots$$
$$20, \quad 28, \quad 36, \quad 44, \quad \cdots,$$

and so on.

Sequences of successive sums of a finite set $u_0, u_1, \cdots, u_n$ can be conveniently displayed in triangular array:

$$u_0 \quad u_1 \quad u_2 \quad \cdots \quad u_{n-1} \quad u_n$$
$$\bar{u}_0^{(1)} \quad \bar{u}_1^{(1)} \quad \bar{u}_2^{(1)} \quad \cdots \quad \bar{u}_{n-1}^{(1)}$$
$$\bar{u}_0^{(2)} \quad \bar{u}_1^{(2)} \quad \cdots \quad \bar{u}_{n-2}^{(2)}$$
$$\cdots\cdots\cdots\cdots\cdots$$
$$\bar{u}_0^{(n-1)} \quad \bar{u}_1^{(n-1)}$$
$$\bar{u}_0^{(n)}$$

Here, each number is the sum of the two adjacent numbers in the row above. If we consider an infinite sequence of numbers $u_0, u_1, u_2, \cdots, u_n$, then the triangular array continues indefinitely.

We now consider a related concept, *Pascal's Triangle* (or the *Arithmetic Triangle*):

$$1$$
$$1 \quad 1$$
$$1 \quad 2 \quad 1$$
$$1 \quad 3 \quad 3 \quad 1$$
$$1 \quad 4 \quad 6 \quad 4 \quad 1$$
$$\cdots\cdots\cdots\cdots\cdots\cdots\cdots$$

Here, the rows are bordered on each end by ones, and the interior integers are obtained as the sum of the two adjacent numbers of the previous row.

For convenience we shall start the row enumeration of the Pascal triangle with the number zero; that is, the number one at the apex of the triangle will be thought of as the 0th row; the sequence 1, 1 constitutes the first row, and so on. We shall designate the $(k+1)$st element of the nth row as C_n^k (that is, in each row, too, we shall start counting from zero). Using this terminology, we have the following format for Pascal's triangle:

$$C_0^0$$
$$C_1^0 \quad C_1^1$$
$$C_2^0 \quad C_2^1 \quad C_2^2$$
$$C_3^0 \quad C_3^1 \quad C_3^2 \quad C_3^3$$
$$C_4^0 \quad C_4^1 \quad C_4^2 \quad C_4^3 \quad C_4^4$$
$$\cdots\cdots\cdots\cdots\cdots\cdots\cdots$$

A number of properties of the members in the Pascal triangle have been developed in the book by B. B. Dynkin and V. A. Uspensky, *Mathematical Conversations*, Issue 6, Section 2, Chapter III, Library of the Mathematical Society. The material contained in the problems of this section are closely related to the material in the interesting popular book by A. E. Markuskevich, *Reflexive Series*, Government Technical Publishing House, Moscow, 1950.

The sequence of numbers obtained by successively substituting, for x in a polynomial $P(x) = a_0x^k + a_1x^{k-1} + \cdots + a_{k-1}x + a_k$ the sequence of integers $1, 2, 3, \cdots, n$ [that is, the sequence $P(1), P(2), \cdots, P(n)$] will be called the kth *order sequence* of $P(x)$. A special case of a kth order sequence is the sequence $1^k, 2^k, 3^k, 4^k, \cdots, n^k, \cdots$ [that is, $P(x) = x^k$].

309. Let $u_0, u_1, u_2, \cdots, u_n$ be a sequence of kth order; that is, let $u_n = a_0n^k + a_1n^{k-1} + \cdots + a_k$.

(a) Prove that $u_n^{(1)}$ forms a sequence of $(k-1)$st order.

(b) Prove that the $(k+1)$st difference sequence of this sequence consists only of zeros.

310. Prove that if $u_n = a_0n^k + a_1n^{k-1} + \cdots + a^k$, then all the numbers of the kth row of the difference sequences $u_0, u_1, u_2, \cdots, u_n, \cdots$ are equal to $a_0k!$.

311. Prove that:

(a) $\bar{u}_n^{(k)} = C_k^0 u_n + C_k^1 u_{n+1} + C_k^2 u_{n+2} + \cdots + C_k^k u_{n+k}$;

(b) $u_n^{(k)} = (-1)^k C_k^0 u_n + (-1)^{k-1}C_k^1 u_{n+1}$
$$+ (-1)^{k-2}C_k^2 u_{n+2} + \cdots + C_k^k u_{n+k}.$$

312. Prove that

$$C_n^k = \frac{n(n-1)(n-2)\cdots(n-k+1)}{k!} \qquad (k > 0),$$

where $k! = 1\cdot2\cdot3\cdots k$.

313. Prove that

$$u_n = C_n^0 u_0 + C_n^1 u_0^{(1)} + C_n^2 u_0^{(2)} + \cdots + C_n^k u_0^{(k)}.$$

314. Assume that the $(k+1)$st row of successive differences [differences of $(k+1)$st order] of some seqnence consists of zeros, but that the kth row consists of nonzero numbers. Prove that this sequence is a sequence of order k.

Remark: The theorem represented by this problem is the converse of the theorem of problem 309 (b). There we were to prove that the $(k+1)$st row of the differences of a kth order sequence consists of zeros. Here we are to prove that if the $(k+1)$st differences of some sequence consists of zeros, then the sequence is of kth order.

315. Find the formula giving the sum of the series

$$1^4 + 2^4 + 3^4 + \cdots + n^4 .$$

316. (a) Prove that the sum $1^k + 2^k + 3^k + \cdots + n^k$ is a polynomial in n of degree $k + 1$.

(b) Calculate the coefficients of n^{k+1} and of n^k of this polynomial.

317. We say that a sequence of integers is divisible by a number d if every number of this sequence is divisible by d. [For example, the sequence of numbers $n^{13} - n$ is divisible by 13; the sequence of numbers $3^{6n} - 2^{6n}$ is divisible by 35; the sequence of numbers $n^5 - 5n^3 + 4n$ is divisible by 120. See problems 27 (d), 28 (a), (b)].

Let u_n be a kth order sequence, $u_n = a_0 n^k + a_1 n^{k-1} + \cdots + a_k$, where the coefficients $a_0, a_1, a_2, \cdots, a_k$ are relatively prime integers. Prove that if the sequence u_n is divisible by an integer d, then d is a divisor of $k!$.

318. Calculate $(C_n^0)^2 + (C_n^1)^2 + (C_n^2)^2 + \cdots + (C_n^n)^2$.

319. Using the result of problem 313, prove Newton's binomial formula:

$$(a + b)^k = a^k + ka^{k-1}b + \frac{k(k-1)}{2!}a^{k-2}b^2 + \cdots + \frac{k(k-1)\cdots 2\cdot 1}{k!}b^k .$$

320. Consider the sequence $1, \dfrac{1}{2}, \dfrac{1}{3}, \cdots, \dfrac{1}{n}, \cdots$. Construct the successive-difference triangle:

$$
\begin{array}{ccccccc}
1 & \dfrac{1}{2} & \dfrac{1}{3} & \dfrac{1}{4} & \dfrac{1}{5} & \dfrac{1}{6} & \cdots \\[2mm]
 & -\dfrac{1}{2} & -\dfrac{1}{6} & -\dfrac{1}{12} & -\dfrac{1}{20} & -\dfrac{1}{30} & \cdots \\[2mm]
 & & \dfrac{1}{3} & \dfrac{1}{12} & \dfrac{1}{30} & \dfrac{1}{60} & \cdots \\[2mm]
 & & & -\dfrac{1}{4} & -\dfrac{1}{20} & -\dfrac{1}{60} & \cdots \\[2mm]
 & & & & \dfrac{1}{5} & \dfrac{1}{30} & \cdots \\[2mm]
 & & & & & -\dfrac{1}{6} &
\end{array}
$$

Turn this triangle 60° clockwise such that the apex consists of the number 1:

$$1$$

$$-\frac{1}{2} \quad \frac{1}{2}$$

$$\frac{1}{3} \quad -\frac{1}{6} \quad \frac{1}{3}$$

$$-\frac{1}{4} \quad \frac{1}{12} \quad -\frac{1}{12} \quad \frac{1}{4}$$

$$\frac{1}{5} \quad -\frac{1}{20} \quad \frac{1}{30} \quad -\frac{1}{20} \quad \frac{1}{5}$$

$$-\frac{1}{6} \quad \frac{1}{30} \quad -\frac{1}{60} \quad \frac{1}{60} \quad -\frac{1}{30} \quad \frac{1}{6}$$

. . . .

Disregard the minus signs of this triangle, and divide through every row by the number at the end of that row to obtain

$$1$$

$$1 \quad 1$$

$$1 \quad \frac{1}{2} \quad 1$$

$$1 \quad \frac{1}{3} \quad \frac{1}{3} \quad 1$$

$$1 \quad \frac{1}{4} \quad \frac{1}{6} \quad \frac{1}{4} \quad 1$$

$$1 \quad \frac{1}{5} \quad \frac{1}{10} \quad \frac{1}{10} \quad \frac{1}{5} \quad 1$$

. . . .

Finally, substitute for each number its reciprocal (that is, replace *a/b* by *b/a*).

Prove that this end result gives Pascal's triangle.

SOLUTIONS

1. Consider the total number of handshakes which have been completed at any moment. This must be an even number, since every handshake is participated in by two people, thus the total number is increased by two. The number of handshakes, however, is also the sum of the handshakes made by each individual person. Since this sum is an even number, the count of the people who have shaken hands an odd number of times must be even (otherwise, odd times odd would given an odd contribution to the total).

2. In order to traverse the chessboard, stopping precisely once on each square, the knight must move 63 times. At each move the knight goes from a square of one color to a square of another color. Thus, in an even number of moves the knight is again on a square of the same color as that of the square he started from, and in an odd number of moves he is on a square of the other color. Therefore, the knight can not arrive at the opposite end of the diagonal of the chessboard in 63 moves: the initial and final squares are the same color.

3. (a) Let us denote the minimum number of transfers necessary to construct a pyramid consisting of n rings on the second peg (under

the conditions given by the problem) by $k(n)$. Clearly, $k(1) = 1$. Further, if $n = 2$, then to transfer the second ring to the second peg, we must first transfer the first ring from the second peg to the auxilliary peg; then we place the second ring on the second peg and transfer the first (smallest) ring to the second peg. Thus, $k(2) = 3$. If $n = 3$, then to transfer the lowest (largest) ring to the second peg, in the necessary arrangement, we must first move the two rings already on the second peg to the auxilliary peg. This requires $k(2)$ moves, and $k(2)$ moves will be required again to replace the rings on the second peg after the largest ring is moved from the first peg to the second peg. Thus,

$$k(3) = 2k(2) + 1 = 7 .$$

In an analogous way find

$$k(4) = 2k(3) + 1 = 15;$$
$$k(5) = 2k(4) + 1 = 31 .$$

In general,

$$k(n) = 2k(n - 1) + 1 .$$

Noting that, for example, $k(3) = 2^3 - 1$, $k(4) = 2^4 - 1$, and so on, we assume as an induction hypothesis that $k(n - 1) = 2^{n-1} - 1$. Then

$$k(n) = 2k(n - 1) + 1 = 2(2^{n-1} - 1) + 1 = 2^n - 1 .$$

Thus, by the principle of finite mathematical induction, it follows that $k(n) = 2^n - 1$ for all n.

(b) Designated by $K(n)$ the least number of moves necessary to remove n rings from the rod. From the beginning position it is possible to remove either the first ring [see Figure 5(a)] or the second [Figure 5(b)]; consequently, $K(1) = 1$ and $K(2) = 2$ (for two rings, we first remove the second ring, then the first).

In order to remove the ith ring it is necessary to remove the $i - 2$ preceding rings; otherwise the ith ring cannot be moved to the end of the rod. On the other hand, if the $(i - 1)$st ring is already removed, then the ith ring cannot be removed [see Figure 6(a)] (it is evident that if three rings are removed, then the fourth cannot be removed). But if the $(i - 1)$st ring is removed, then the $(i + 1)$st ring can easily be removed [see Figure 6(b) and (c)].

Now it is not difficult to answer the question posed by the problem. In order to remove the last of the n rings it is first necessary to remove the first $n - 2$ rings. This can be done in $K(n - 2)$ moves,

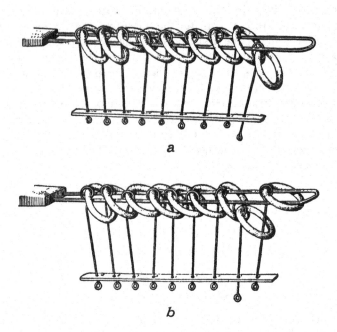

a

b

Figure 5

after this the final ring can be taken off in one more move. It then
remains to remove the $(n-1)$st ring. We will designate by $k(n)$ the
number of moves necessary to remove only the nth ring, under the
condition that all the preceding rings have already been taken off.
We obtain

$$K(n) = K(n-2) + 1 + k(n-1)\,.$$

We shall now find an expression for the number $k(n)$. Clearly, in
order to remove the nth ring from the rod we must put the $(n-1)$st
ring back on the rod. This can be done in $k(n-1)$ moves [the same
number of moves which were necessary to remove the $(n-1)$st ring
from the rod, the moves now being done in reverse order]. The nth
ring is now easily removed from the rod, requiring only one additional move. Finally, the $(n-1)$st ring must be taken off the rod,
for which $k(n-1)$ more moves are needed. Thus, we obtain

$$k(n) = 2k(n-1) + 1\,,$$

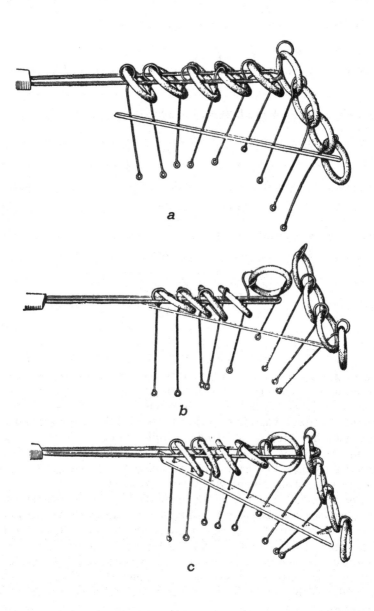

a

b

c

Figure 6

from which we easily derive

$$k(n) = 2^n - 1$$

[see the solution to problem (a)].

Now the formula which determines $K(n)$ takes the form

$$K(n) = K(n - 2) + 2^{n-1}.$$

But since $K(1) = 1$ and $K(2) = 2$, we readily find, for $n = 2m$ (an even number) that

$$\begin{aligned}
K(n) &= K(n - 2) + 2^{n-1} = K(n - 4) + 2^{n-1} + 2^{n-3} \\
&= K(n - 6) + 2^{n-1} + 2^{n-3} + 2^{n-5} = \cdots \\
&= K(2) + 2^{n-1} + 2^{n-3} + 2^{n-5} + \cdots + 2^3 \\
&= 2 + \frac{2^{n+1} - 2^3}{4 - 1} = \frac{1}{3}(2^{n+1} - 2).
\end{aligned}$$

If $n = 2m + 1$ (an odd number),

$$\begin{aligned}
K(n) &= K(n - 2) + 2^{n-1} = K(n - 4) + 2^{n-1} + 2^{n-3} \\
&= K(n - 6) + 2^{n-1} + 2^{n-3} + 2^{n-5} = \cdots \\
&= K(1) + 2^{n-1} - 2^{n-3} + 2^{n-5} + \cdots + 2^2 \\
&= 1 + \frac{2^{n+1} - 2^2}{4 - 1} = \frac{1}{3}(2^{n+1} - 1).
\end{aligned}$$

4. (a) We divide the coins into three groups: two having 27 coins each and one having 26 coins. A first weight trial is made by placing a group of 27 coins on each of the two pans. If the pans do not balance, the counterfeit coin is among those in the lighter pan. If the scale is in balance, the counterfeit coin is among the 26 unweighed coins. Therefore, it suffices to solve the following problem: To find, by three weight trials, a light counterfeit coin among 27 coins (the problem of detecting the counterfeit in the group of 26 coins can be reduced to this by simply adding to the set of 26 coins a genuine coin from the 54 which were weighed).

For a second weight trial we divide the 27 coins into three groups of nine each and place a group of nine coins on each of the two pans. This will reveal which group of nine coins contains the counterfeit. The group of nine coins containing the counterfeit is then divided into three groups of three coins each. A third weight trial will tell which group of three coins contains the counterfeit. Finally, a fourth weight trial involving two of the three doubtful coins will reveal the counterfeit.

(b) Let k be a natural number which satisfies the inequalities

$3^k \geqq n$ and $3^{k-1} < n$. We shall show that the number k satisfies the conditions of the problem.

First we show that, at most, k weight trials are enough to allow detection of the counterfeit coin. We divide the coins into three groups such that in each of two equal groups there are 3^{k-1} (or fewer) coins and the number of coins in the third group does not exceed 3^{k-1} (this is possible for $n \leqq 3^k$). The two groups having the same number of coins are placed on the pans of the balance; this enables us to determine which group contains the counterfeit [see the solution to problem (a)]. If the group containing the counterfeit coin contains fewer than 3^{k-1} coins, we add to it enough (genuine) coins from the other two groups to give it 3^{k-1} coins. This group of 3^{k-1} coins containing the counterfeit is again divided into three groups, as before, and a second weight trial is made, as before. The procedure is continued until after, at most, k weight trials we arrive at a group containing only one coin—that is, we have located the counterfeit coin.

It is now necessary to show that k is the *minimal* number of weight trials which will guarantee the detection of the counterfeit coin under all circumstances. (*Note*: We might, under certain circumstances, succeed with fewer weight trials. For example, if two coins are chosen at random and compared, we might be fortunate enough to have chosen the counterfeit coin as one of them. However, this does not give us a procedure by which we can be sure, under all possible circumstances, of detecting the counterfeit by the required number of weight trials.)

In each weight trial, as outlined above, the coins are divided into three groups, two of which are placed on each of the two pans, and the third being the unweighed group. If with an equal number of coins on each pan the scale balances, then, of course, the counterfeit is in the third group. If the pans are out of balance, we know that the counterfeit is on one of the pans, although a prior knowledge as to whether the counterfeit is lighter or heavier than the rest is needed to tell us on which of the two pans it lies.

Let us assume that in an arbitrary sequence of weight trials the result of each weighing is *least* favorable with respect to enabling us to detect the counterfeit coin; that is the counterfeit is always in that group of three which contains the largest number of coins. Then upon each weight trial the number of coins in the group which contains the counterfeit will not be less than $\frac{1}{3}$ the total number of coins (that is, upon division of the coins into three groups, one of the groups must contain not fewer than $\frac{1}{3}$ of them). Then after $k - 1$ weight trials the number of coins in the group containing the counterfeit will be not less than $\frac{n}{3^{k-1}}$, and, since $n > 3^{k-1}$, the counterfeit will not be found by $k - 1$ weight trials.

Remark: It can be shown that the answer to the problem can be expressed in the following form: The minimal number of weight trials necessary to detect the one counterfeit coin in a collection of n coins is $[\log_3 (n - \frac{1}{2})] + 1$, where the brackets designate "the largest integer in the number" (see the note just prior to problem 101).

5. One block is placed on each pan (first weight trial). There are two possible outcomes:

On the first weight trial one of the pans is heavier. In this event one of the blocks must be aluminum and the other duraluminum. We now place both blocks on one pan and weigh them against pairs of remaining blocks (those being divided into nine pairs arbitrarily). Any pair of blocks which outweighs the first pair must consist of two duraluminum blocks. If the first pair is the heavier, then both blocks of the second pair are alminum. If both pairs balance, the second pair contains one aluminum and one duraluminum block. Thus, for this first event the number of duraluminum blocks can be determined by at most ten weight trials.

On the first weight trial the pans balance. In this event both blocks are aluminum or both are duraluminum. As before, we now place both blocks on one pan and weigh them against pairs of remaining blocks. Assume that the first k pairs of blocks from the nine pairs have the same weight as the test blocks, and that the $(k + 1)$st pair tested have a different weight. (If $k = 9$, then all the blocks weigh the same, and so there are no duraluminum blocks. The event in which $k = 0$ falls into the general case.) Suppose, for definiteness, that the $(k + 1)$st pair is heavier than the test blocks [the reasoning which follows will be quite analogous if the $(k + 1)$st pair is lighter]. Then the original two blocks, as well as all those of the first k pairs tested, must be aluminum. Therefore, in the $1 + (k + 1) = k + 2$ weight trials already made we have found $k + 1$ pairs of aluminum blocks. Now we compare the two blocks of the $(k + 1)$st (heavier) pair. [This is the $(k + 3)$rd weight trial.] If both blocks are of the same weight, they must both be duraluminum; if they are not of the same weight, one is aluminum and the other is duraluminum. In either event we are able, after $k + 3$ weight trials, to display a pair of blocks of which one is aluminum and the other duraluminum.

By using this pair we can determine in, at most, $8 - k$ weight trials how many duraluminum blocks remain among the $20 - 2(k + 2) = 16 - 2k$ unweighed ones, using the technique employed in the first event. The number of weight trials used in this second event is then equal to $k + 3 + (8 - k) = 11$.

6. (a) Divide the coins into three sets of four coins each. For a first weight trial we place a group of four coins on each pan. There are two possibilities, which we shall investigate separately:

(1) The pans balance.

(2) One pan outweighs the other.

The pans balance. In this event the counterfeit coin is in the unweighed set, and all eight coins on the scale are genuine. We number the coins of the doubtful group 1, 2, 3, 4. We carry out second weight trial by placing coins 1, 2, and 3 on one pan and placing the other three coins now known to be genuine on the other pan. There are two possibilities:

(A) The pans are in balance. In this event coin 4 is the counterfeit. A third weighing, comparing coin 4 with a genuine one, will tell whether it is lighter or heavier.

(B) One pan is heavier. In this event the counterfeit is one of coins 1, 2, or 3. If the genuine coins are the heavier, then the counterfeit is a light coin, and vice-versa. One more weight trial will identify which of coins 1, 2, or 3 is a light coin, hence counterfeit [see the solution to problem 4(a)]. If the pan containing coins 1, 2, and 3 is the heavier, then the counterfeit coin is heavier than a genuine one. One weight trial will identify it.

One pan outweighs the other. In this event all the other coins are genuine. Let us designate the coins on the heavier pan by the numbers 1, 2, 3, 4 (if one of these coins is false, then it is heavier than the others) and the coins on the lighter pan by $1', 2', 3', 4'$ (if one of these coins is false, then it is lighter than the others). A second weight trial is made by placing coins, 1, 2, and $1'$ on one pan and coins 3, 4, and $2'$ on the other. Again, there are several possibilities:

(A) The pans balance. In this event the counterfeit coin is either $3'$ or $4'$ (and is lighter than a genuine coin). A third weight trial is made by placing coin $3'$ on one pan and coin $4'$ on the other; the lighter coin will be the counterfeit.

(B) The pan containing coins 1, 2, and $1'$ is heavier. In this event coins 3, 4, and $1'$ are genuine; were either coin 3 or coin 4 heavier than the others, or were coin $1'$ light, then in the second weight trial the pan containing coins 3, 4, and 2' would have been heavy, which was not the result for this case. Therefore, the counterfeit coin is either coin 1 or coin 2 (and it is a heavier coin), or else it is coin 2' (and it is a lighter coin). A third weight trial is made, placing coin 1 on one pan and coin 2 on the other. If the pans balance, then the counterfeit coin is 2'. If the pans fail to balance, then the counterfeit coin is on the heavier pan.

(C) The pan containing coins 3, 4, and 2' is heavier. Reasoning as before, we conclude that coins 1, 2, and 2' are genuine and that if one of the coins 3 or 4 is counterfeit, then it is a heavier coin than the others, and if coin 1' is the counterfeit, then it is lighter. A third weighing is made by placing coin 3 on one pan and coin 4 on the other. If the pans balance, then the counterfeit is 1'; if, on the other hand, one pan is heavier, then it contains the counterfeit coin.

(b) We shall prove in three stages, (A), (B), (C), that if the number of coins is $N \leqq \dfrac{3^n - 3}{2}$, then the counterfeit coin can be detected (and determined to be heavy or light) by n weight trials, and if $N > \dfrac{3^n - 3}{2}$, then n weighings do not necessarily suffice (see the hint to this problem). We shall solve first the following related problem, applying relaxed conditions.

(A) Suppose we are given N coins, divided into two groups, which we shall designate X and Y (we do not exclude the possibility that one of the groups contains no coins). We assume that one of the N coins is counterfeit, and also that if the counterfeit is in the X group, then it is lighter than all the others, but if it is in group Y then it is heavier than all the others. We must show that *if $N \leqq 3^n$, then the counterfeit coin can be detected by means of n weight trials on a pan balance, and if $N > 3^n$, then this is not always possible.*[†] *If group Y contains no coins, the problem becomes as follows. There is one counterfeit coin among N coins, and it is lighter than the others; to prove: that if $N \leqq 3^n$, then the counterfeit can be found by n weight trials on a pan balance, and if $N > 3^n$, then this is not always possible; see problem 4(b).]*

The proof will be given by mathematical induction. First we show that *if $N \leqq 3^n$, the counterfeit can always be detected by n weight trials.* If $n = 1$, that is if $N = 1, 2,$ or 3, the proposition is quite

---

[†] This statement has one obvious exception: If $N = 2$ and groups X and Y each contains one coin, then, of course, the counterfeit coin cannot be detected.

obvious (except as noted in the preceding footnote). For example, if $N = 3$, then it suffices to compare two of the coins from one group. Assume now that it has been shown that if $N \leqq 3^{n-1}$, then the counterfeit coin can be found by $n - 1$ weight trials. Now let $N \leqq 3^n$. Place on each pan x coins from group X and y coins from group Y, where x and y are selected to satisfy the inequalities

$$x + y \leqq 3^{n-1};$$
$$N - 2(x + y) \leqq 3^{n-1}.$$

[For $N \leqq 3^n$ the inequality $N - 2(x + y) \leqq 3^n$, or $x + y \geqq \dfrac{N - 3^{n-1}}{2}$ is compatible with $x + y \leqq \dfrac{3^n - 3^{n-1}}{2} = 3^{n-1}$; that is, it is always possible to choose numbers x and y such that $3^{n-1} \geqq x + y \geqq \dfrac{N - 3^{n-1}}{2}$].

If the pans balance, the counterfeit coin is among the $N - 2(x + y)$ coins which were not placed on the scale; if one of the pans is heavier, the counterfeit is either among the x coins of group X on the lighter pan, or among the y coins of group Y on the heavier pan. But according to the induction hypothesis we can, in either event, find the counterfeit coin by conducting $n - 1$ additional weight trials [since both $N - 2(x + y)$ and $(x + y)$ fail to exceed 3^{n-1}].[†] This proves, by induction, that if $N \leqq 3^n$, then n weight trials suffice for finding the counterfeit.

We now prove that if $N > 3^n$, *then the counterfeit cannot, in general, be detected by n weight trials*. (By "in general" we mean that there simply does not exist a *sure* program which will always locate the false coin in n trials. There is always the possibility, for example, that if we select two coins at random, one of them might be the counterfeit; but we are not concerned with this sort of accidental success.) For this proof we make an additional assumption to relax our conditions. We assume that *in addition to the two sets X and Y of coins, together containing $N > 3^n$ coins, we have at our disposal a quantity of coins all known to be genuine*. We designate this last group of coins by Z (how many there are is not important). We shall show, by using these auxiliary coins, that the counterfeit cannot, in general, be found by n weight trials.

It is easily verified that if $n = 1$ (that is, if the total number N of coins in X and Y exceeds 3), then one weighing will not always suf-

---

[†] If $N > 2$, then $x = 1$, $y = 1$ is not an exception, because there are now, beyond the two coins, some number of coins known to be genuine. Comparison of one of the genuine coins with either one of the doubtful coins will enable us to find the counterfeit by one weighing.

fice to locate the counterfeit coin. Assume now that it has been proved that if the number N of coins in groups X and Y exceeds 3^{n-1}, then it is not, in general, possible to detect the counterfeit by $n-1$ weight trials. We must show, then, that in this event if $N > 3^n$, the counterfeit cannot always be detected in n weight trials. Suppose that in the first weight trial we have placed on the first pan x coins from group X, y coins from group Y, and z coins from auxiliary group Z, and that we have put on the other pan x' coins from group X and y' coins from group Y, so that $x + y + z = x' + y'$ (clearly, it would be useless to put coins from Z on both pans). Let w be the number of coins of groups X and Y which have been left off the scale ($x + x' + y + y' + w = N$). Now the scale may be in balance, the counterfeit is among those coins not placed on the scale. If w exceeds 3^{n-1}, then, by our induction hypothesis, $n - 1$ additional weighings will not, in general, suffice to locate the counterfeit there, and we will already have used up one weight trial. Thus, on the first weight trial, in order to guarantee even the possibility that n trials will suffice, it is necessary that $w \leqq 3^{n-1}$. Hence, on this first trial,

$$x + y + x' + y' = N - 3^{n-1} > 3^n - 3^{n-1} = 2 \cdot 3^{n-1} .$$

If the pans do not balance, the imbalance is caused either by a light counterfeit in the X group or by a heavy counterfeit in the Y group. Hence, we must deal with either a group of $x + y'$ coins (if the pan with $x + y + z$ coins is lighter) or a group of $x' + y$ doubtful coins. But according to the relation $x + y + x' + y' > 2 \cdot 3^{n-1}$, the larger of the numbers $x + y'$ and $x' + y$ exceeds 3^{n-1}. We must be prepared to deal with this larger number of coins, and according to the induction hypothesis $n - 1$ additional weight trials will not always suffice to identify the counterfeit. This induction completes the proof of case (A).

(B) Suppose we are given N coins containing one counterfeit which differs in weight from the others, but that we do not know whether the counterfeit is lighter or heavier. *Assume we have available an extra coin which we know to be genuine.* We shall show that *if*

$$N \leqq \frac{3^n - 1}{2} :$$

the counterfeit can always be detected by n weight trials; the counterfeit can be shown to be lighter or heavier than the genuine coins; for

$$N > \frac{3^n - 1}{2}$$

n weight trials do not, in general, suffice.

First, let $N \leq \dfrac{3^n - 1}{2}$. If $n = 1$ (that is, if $N = 1$), one weighing will obviously suffice. Assume that for

$$N \leq \frac{3^{n-1} - 1}{2}$$

$n - 1$ weight trials suffice we shall show that for

$$N \leq \frac{3^n - 1}{2}$$

n weight trials will suffice. We place on one of the pans x of the N given coins together with the extra genuine coin, and we place on the other pan $x + 1$ coins. There remain $N - 2x - 1$ unweighed coins. We choose the number x to satisfy the requirement

$$2x + 1 \leq 3^{n-1};$$
$$N - 2x - 1 \leq \frac{3^{n-1} - 1}{2}.$$

It is clear that such an x can always be chosen if $N \leq \dfrac{3^n - 1}{2}\left(\text{in this}\right.$ case, $N - \dfrac{3^{n-1} - 1}{2} \leq \dfrac{3^n - 1}{2} - \dfrac{3^{n-1} - 1}{2} = 3^{n-1}\left.\right)$. If the pans balance, then there remain

$$N - 2x - 1 \leq \frac{3^{n-1} - 1}{2}$$

unweighed coins plus a quantity of coins (those on the balance) known to be genuine. Therefore, according to case (A) we can find the counterfeit coin by making not more than $n - 1$ additional weight trials. If the pans do not balance, then the x coins on one pan and the $x + 1$ coins on the other pan form two groups containing a total of $2x + 1 \leq 3^{n-1}$ coins, to which we can apply the result of problem A just investigated. Therefore, $n - 1$ further weight trials at most, suffice here also. This completes the proof that, under the conditions of this case, if

$$N \leq \frac{3^{n-1} - 1}{2},$$

then n weight trials suffice to identify the counterfeit.

We now show that *if*

$$N \geq \frac{3^n - 1}{2},$$

then it is not always possible to solve the problem with only n weight trials. For $n = 1$ (that is, for $N > 1$) one weighing will not be enough. Once more we proceed by induction. Assume that for

$$N > \frac{3^{n-1} - 1}{2}$$

it has already been shown that $n - 1$ weight trials are not enough. We investigate the event

$$N > \frac{3^n - 1}{2} .$$

We place on one of the pans x of the N coins and, in addition, z coins known to be genuine (here we may assume that we have not just one but many genuine coins at our disposal). We place on the other pan $x + z$ coins from the given set N. Let us designate by w the number of coins from N which remain off the balance. Inasmuch as the pans may be in balance, the number w of coins left off the balance must not exceed $\frac{3^{n-1} - 1}{2}$; if the number does exceed this value, then, according to the induction hypothesis, n weight trials will not suffice to locate the counterfeit. But in this event $2x + z > 3^{n-1}$ $\left(\text{or, } N > \frac{3^n - 1}{2}\right)$. If the balance is not in equilibrium, we can apply to the coins belonging to groups X and Y, which are made up of the doubtful coins and which lie on one or the other of the pans, the results of case (A). It follows from the inequality $2x + z > 3^{n-1}$ that for the unbalanced condition of the scale, n weight trials do not suffice.

(C) Having obtained the foregoing preliminary results, we now return to the original problem. We shall prove that *given N coins, where*

$$2 < N \leqq \frac{3^n - 3}{2} ,$$

one of which is counterfeit (but it is not known whether it is lighter or heavier than the others), the counterfeit can be detected by n weight trials and simultaneously shown to be either light or heavy.[†]

If we place x coins on each pan, there remain $N - 2x$ coins not on the scale. The number x is selected such that

$$2x \leqq 3^{n-1} ,$$

$$N - 2x \leqq \frac{3^{n-1} - 1}{2} .$$

[†] Clearly, if only two coins are involved, the counterfeit cannot be found by weighing.

If the pans balance, then the false coin is among the unweighed ones, of which there are

$$N - 2x \leqq \frac{3^{n-1} - 1}{2} \, .$$

Moreover, we now have on hand $2x$ coins known to be genuine, and in view of the result obtained in case (B) we are able to find the counterfeit by $n - 1$ additional weight trials. If one of the pans outweighs the other, then we can apply the result of case A to the coins on the scale, since the total number of coins on the two pans is $2x < 3^{n-1}$, and $n - 1$ additional weight trials suffice to locate the counterfeit.

We shall now prove that *if*

$$N > \frac{3^n - 3}{2} \, ,$$

then n weight trials will not, in general, suffice. For the first weight trial we place x coins on each pan, leaving w coins off the scale. If the pans balance, then in view of case (B), $n - 1$ further weight trials will suffice to locate the counterfeit and to determine whether it is light or heavy *only* if w does not exceed $\dfrac{3^{n-1} - 1}{2}$. But in this event

$$2x > \frac{3^n - 3}{2} - \frac{3^{n-1} - 1}{2} = \frac{2 \cdot 3^{n-1} - 2}{2} = 3^{n-1} - 1 \, .$$

Since $2x$ is an even number, $2x > 3^{n-1}$, and in view of the results of case (A), we cannot determine the counterfeit coin with n weight trials if one of the pans is heavier.

Thus, we have proved the generalized version of the given problem. Now we need note merely that

$$\frac{3^7 - 3}{2} = 1092 > 1000 > 363 = \frac{3^6 - 3}{2}$$

in order to obtain the answer to our specific problem; that is, $k = 7$.

7. (a) If the third link is disengaged from the chain, then there are three pieces: the single link, a two-link piece, and a four-link piece. On the first day the traveler gives the single link to the innkeeper. On the second day he gives the innkeeper the two-link piece, receiving the single link back in change. On the third day the traveler pays the single link; on the fourth day he presents the four-link piece, receiving as change the single link and the two-link piece. On the fifth day he gives the single link, and on the sixth day the single link is returned to him as change for the two link piece. Finally, the single link pays for the seventh day's lodging. Hence, only one link need be disengaged from the chain.

(b) It will be convenient to consider first the following problem. If k links are disengaged from an n-link chain, where k is a fixed number, how large can n be such that any number of links from 1 to n can be obtained by taking one or more of the severed pieces of the chain? To solve this problem, we first investigate which links would be most advantageous to remove. After removing k single links we can gather from these any number of links from 1 to k. But if we want $k + 1$ links, we must consider the remaining pieces of chain and in particular those pieces having $k + 1$ or fewer links.

Clearly, the most convenient arrangement would be to have a piece of chain with exactly $k + 1$ links. Then we could gather any number of links from 1 to $2k + 1$. In order to collect $2k + 2 = 2(k + 1)$ links, we need another piece containing $2(k + 1)$, or fewer, links. The most convenient situation would be to have a piece with exactly $2(k+1)$ links; we could then gather any number of links from 1 to $(2k+1) + 2(k+1) = 4k + 3$. The next step would be to have available a piece of chain with $4k + 4 = 4(k + 1)$ links. Continuing this reasoning, the most advantageous method of removing links from the chain would be to have pieces of the following lengths (setting aside the individual severed links):

$$k + 1, 2(k + 1), 4(k + 1), 8(k + 1), \cdots, 2^k(k + 1) .$$

Thus, any number of links from 1 to

$$n = k + [(k + 1) + 2(k + 1) + 4(k + 1) + \cdots + 2^k(k + 1)]$$
$$= k + (2^{k+1} - 1)(k + 1) = 2^{k+1}(k + 1) - 1$$

can be built up by taking pieces of the chain.

Thus, if $2^k k \leqq n \leqq 2^{k+1}(k + 1) - 1$, then it is possible to make k breaks in the chain, but removal of $k - 1$ links will not suffice to solve the problem. In particular, if

$$2 \leqq n \leqq 7, \text{ then } k = 1; \quad 160 \leqq n \leqq 383, \text{ then } k = 5;$$
$$8 \leqq n \leqq 23, \text{ then } k = 2; \quad 384 \leqq n \leqq 895, \text{ then } k = 6;$$
$$24 \leqq n \leqq 63, \text{ then } k = 3; \quad 896 \leqq n \leqq 2047, \text{ then } k = 7 .$$
$$64 \leqq n \leqq 159, \text{ then } k = 4;$$

Therefore, if $n = 2000$, the least number of links which can be disengaged is 7. The conditions of the problem will be satisfied if we select those links such that the 8 pieces of chain we obtain (excluding the 7 individual severed links) have, respectively, 8, 16, 32, 64, 128, 256, 512, and 977 links.

8. Let A be the first of the selected students (that is, the tallest of the short), and let B be the second of the selected students (the shortest of the tall). If A and B stand in the same row, then B is taller than A, since B is the tallest student in that row. If A and B stand in the same column, then again B is taller than A, since A is the shortest student in that column. Finally, if A and B do not stand in either the same column or the same row, let C be that student standing in the same column as does A and in the same row as does B. Then B is taller than C (since B is the tallest in that row), and C is taller than A (since A is the shortest in that column). Hence, again B is taller than A; and so in every possible case B is taller than A.

9. First, it follows from the conditions of the problem that each gear weighs either an even number of grams or an odd number of grams. The reasoning is as follows. Since any set of twelve gears can be divided into two groups of equal weight, a set of twelve gears will weigh an even number of grams. This total weight remains an even number if one of the twelve gears is exchanged with the thirteenth gear. But this is possible only if the exchanged gear and the thirteenth gear are both of even weight or both of odd weight, and this holds for any of the twelve gears initially weighed. Hence, all the gears are of even, or all are of odd, weight.

Subtract now from the weight of each gear the weight of the lightest gear (possibly two or more gears have the same minimum weight; this is unimportant). This may be thought of as producing a "new" set of gears, and this new set clearly again satisfies the conditions of the problem. (One of the gears, and possibly more, must now be thought of as having "zero weight".) It is easily seen that each gear of the new set has even weight (counting 0 as an even number), since if all the gears were of odd weight initially, then an odd number was subtracted from each individual weight; if they were all of even weight initially, then an even number was subtracted from each individual weight.

If now we divide each weight by 2 and think of this as providing a "new" set of weights, this new set again satisfies the conditions of the problem.

Assume now that not all the gears are of the same weight. In this case, not all the weights of the second set (obtained by subtracting the weights of the lightest gear from each of the original weights) will be zero. If we continue to divide by 2, thus obtaining "new" sets satisfying the conditions of the problem, we finally arrive at a set of gears of which some are of even weight (at least one is of zero weight) and some are of odd weight (continued division of an even natural number by 2 finally produces an odd number). But such a set satisfying the conditions of the problem has been shown to be impossible. This contradiction proves the assertion of the problem.

Remark: The conditions of the problem require that all the gears be of integral weight, but the result can be extended to weights which are rational numbers. If we multiply each of the weights by the least common multiple of all the denominators, we obtain a "new" set of weights, all integers, and all of which still satisfy the conditions of the problem, and we go on from there. Moreover, if we allow the weights to be irrational numbers, the extension can again be made, since irrational numbers can be approximated by rational numbers to as close a tolerance as desired. (The reader is invited to carry out such a proof, although a rigorous demonstration is by no means a simple task.)

10. Beginning with the third row of the number triangle we write the first four numbers of each row, but putting in place of an even number the letter e, and in place of an odd number the letter θ:

$$\theta \quad e \quad \theta \quad \theta$$
$$\theta \quad \theta \quad e \quad \theta$$
$$\theta \quad e \quad e \quad e$$
$$\theta \quad \theta \quad \theta \quad e$$
$$\theta \quad e \quad \theta \quad e$$
$$\cdot \quad \cdot \quad \cdot \quad \cdot$$
$$\cdot \quad \cdot \quad \cdot \quad \cdot$$

Note that the fifth row of the array coincides with the first row. But the evenness or oddness of the first four numbers of every row of the number triangle depends only upon the evenness or oddness of the first four numbers of the preceding row. Hence, in the above array any row will periodically duplicate itself in the fourth row following. Since an even number occurs in each of the first four rows shown above, an even number must occur in every subsequent row.

11. The plan is to number the squares consecutively from 1 to 12, starting, say, from the red chip, and continuing on through the chips and around the circle, and then to rearrange the squares by putting them in an order in which it is possible to move a chip from one square to the next. That is, after square 1 we place square 6 (since by the conditions of the problem it is possible for a chip to move from square 1 to square 6, and vice-versa), then after square 6 we place square 11 (since a chip may move from square 6 to 11), and after square 11 we place square 4, and so on. After this rearrangement we have the following order of squares:

$$
\begin{array}{cccccc}
R & & B & & Y & \\
1 \longleftrightarrow 6 \longleftrightarrow 11 \longleftrightarrow 4 \longleftrightarrow 9 \longleftrightarrow 2 & & & & & \\
\updownarrow & & & & \updownarrow & \\
8 \longleftrightarrow 3 \longleftrightarrow 10 \longleftrightarrow 5 \longleftrightarrow 12 \longleftrightarrow 7 & & & & & \\
G & & & & &
\end{array}
$$

The chips (red, yellow, green, blue, designated *R*, *Y*, *G*, *B*) are shown adjacent to their original positions: *R* on square 1, *Y* on 2, *G* on 3, and *B* on 4. The rule by which the chips may now move is simple: A chip may move one square in either direction, provided that square is unoccupied. Clearly, the only way in which a chip can change places with another is for it to move around the rectangle in either direction, but now a chip can neither jump over nor occupy the position of another chip. Thus, if chip *R* moves to occupy square 4, then *B* must occupy square 2, *Y* must occupy square 3, and *G* must move to square 1. If *R* occupies square 2 then *B* must occupy square 3, *Y* must move to square 1, and *G* will occupy square 4. If *R* occupies square 3, then *B* occupies square 1, *Y* moves to 4 and *G* moves to square 2. Other than these three rearrangements, no arrangement differing from the initial one is possilbe.

12. *First solution.* Let *n* be the number of coconuts each man received when the pile of coconuts was divided the next morning. Then $5n + 1$ coconuts were in the final pile. The last man to have raided the pile the night before must have taken $\dfrac{5n + 1}{4}$ nuts, since prior to that there must have been $5 \cdot \dfrac{5n + 1}{4} + 1 = \dfrac{25n - 9}{4}$ coconuts in the pile. The next-to-last (penultimate) man to have raided the pile took $\dfrac{1}{4} \cdot \dfrac{25n + 9}{4}$ nuts, and prior to that the pile contained

$5 \cdot \dfrac{1}{4} \dfrac{25n + 9}{4} + 1 = \dfrac{125n + 61}{16}$ coconuts. The man who raided prior

to that took $\dfrac{1}{4} \cdot \dfrac{125n + 61}{4}$ nuts from a pile of $5 \cdot \dfrac{1}{4} \cdot \dfrac{125n + 61}{4} + 1 =$

$\dfrac{625n + 369}{64}$; the man before him took $\dfrac{1}{4} \cdot \dfrac{625n + 369}{64}$ nuts from a

pile of $5 \cdot \dfrac{625 + 369}{64} + 1 = \dfrac{3125n + 2101}{265}$; finally, the first man to have

arisen the night before took $\dfrac{1}{4} \cdot \dfrac{3125n + 2101}{256}$ nuts from the original

pile, which contained

$$N = 5 \cdot \frac{1}{4} \cdot \frac{3125n + 2101}{256} + 1 = \frac{15625n + 11529}{1024}$$

$$= 15n + 11 + \frac{265n + 265}{1024}$$

coconuts. Since N must be an integer, $265(n + 1)$ must be divisible by 1024. The least integral value of n which will make $265(n + 1)$ divisible by 1024 is 1023 (since 265 and 1024 are relatively prime). Thus,

$$N = 15 \cdot 1023 + 11 + 265 = 15621 \ .$$

Second solution. The problem can be solved more readily and with much less calculation if we consider the conditions imposed on the total number of coconuts. The first condition asserts that the first division of the total number of nuts by 5 yields a remainder of 1, that is, for some number l, $N = 5l + 1$. The numbers N satisfying this condition appear in the sequence of natural numbers at intervals five integers apart, and if we know one of the numbers, we can find as many more as we wish by adding or subtracting multiples of 5. The second condition asserts that $k = (4/5)(N - 1) = 4l$ (k is the number of nuts remaining for the second man's raid) gives a remainder of 1 when divided by 5; that is, $k = 5l_1 + 1$, for the suitable l_1. This is equivalent to the requirement that l yield a remainder of 4 when divided by 5, or that $N = 5l + 1$ yield a remainder of 21 when divided by 25. Numbers satisfying this condition are found in the natural number sequence at intervals of 25, and if we know one such number, we can obtain the others by adding or subtracting mutliples of 25. The third condition asserts that $k_1 = (4/5)(k - 1) = 4l_1$ yields a remainder of 1 upon division by 5; this condition determines the remainder yielded upon division of l_1 by 5, or the remainders obtained by dividing k and l by 25, or the remainder obtained by dividing N by 125. All the conditions together determine the remainder obtained by dividing N by $5^6 = 15{,}625$. These numbers appear at intervals of 15,625 in the sequence of natural numbers.[†]

We can now calculate the remainder which N yields upon division by 5^6, but this is not necessary. One number satisfying all the imposed conditions is -4, which upon division by 5 yields a quotient of -1 and a remainder of $+1$. If we subtract the number 1 from -4 and divide the difference by 4/5, then divide the result by 5, we again obtain a remainder of $+1$. Accordingly, upon all further divisions by 5 we will obtain the same remainder of $+1$. The number -4 cannot, of course, be the answer to the problem, since N has to be a positive integer. But since we do know one number which satisfies the conditions of the problem, we can find as many others as we wish by adding multiples of 5^6. The least positive number which we can add to satisfy the given conditions is 5^6 itself; we obtain $-4 + 5^6 = 15625 - 4 = 15{,}621$.

13. Let n be the number of sheep in the herd. Then the brothers received n rubles for each sheep, and so they received a total of $N = n \cdot n = n^2$ rubles. Let d represent the digit in the tens place of the number n, and let e be the digit in the units place; then $n = 10d + e$, and

$$N = (10d + e)^2 = 100d^2 + 20de + e^2 .$$

It follows, from the manner in which the money was divided, and since the older brother had one more selection from the money than did his brother, that there must have been an odd number of tens in N, plus a remaining number less than 10. But $100d^2 + 20de = 20d(5d + e)$ is divisible by 20 and so contains an even number of tens. Consequently, the number e^2 must contain an odd number of tens. Since $e < 10$ (being the remainder obtained by dividing n by 10), the only possible values for e are 1, 4, 9, 16, 25, 36, 49, 64, or 81.

Of these numbers, only 16 and 36 contain an odd number of tens, so the possibilities for e^2 are limited to 16 or 36. Both of these numbers end with the digit 6, which means that the remainder the younger brother received in place of 10 rubles was 6 rubles. Thus, the older brother received 4 rubles more than did the younger. To make the division even, the older brother would have to give the

---

[†] The argument here would be carried out in American schools by the use of congruence arithmetic [*Editor*].

younger brother 2 rubles. Therefore, the penknife was worth 2 rubles.

14. (a) Our calendar is constructed on the following scheme. Every year has 365 days except for leap-years (those years whose identifying number is divisible by 4—for example, 1960). The leap-years have 366 days (the extra day being February 29). However, there is an exception to this rule: any year whose identifying number is divisible by 100 but not by 400 has only 365 days instead of 366; that is, those are *not* leap years. For example, the years 1800 and 1900 were not leap years; also, 2100 will not be a leap year. However, 2000 (being divisible by 400) will be a leap year. The problem asks on which of the two days, Saturday or Sunday, New Years Day more frequently falls.

We shall show that any 400-year period of time presents a periodic pattern, in the sense that the next 400-year period will exactly duplicate the calendars of the previous 400-year period (for example, this month's calender is exactly the same as that of exactly 400 years ago). This interval of 400 years contains an integral number of weeks. Leap years have 52 weeks and 2 days, and other years have 52 weeks and 1 day. In the course of four years, of which one is a leap year, there are 4·52 weeks plus 5 days. In the course of 400 years there would be an additional 500 days, but because three of these years are divisible by 100 but not by 400 (and so are not leap years) the 400 years will have only 497 days in addition to 400·52 weeks; that is, there are an additional 71 weeks. Therefore, the problem can be limited to any 400-year period.

Let us investigate the 400-year period from 1901 to 2301. Every fourth year is a leap year (since there is no exception for leap years in this interval). In span of 28 years there will be one "extra" week, owing to leap years; in addition, there will be an additional 28 days, or 4 weeks, over an exact number of weeks. Thus, a 28-year interval yields an integral number of weeks: 28·52 + 5 weeks.

We select a particular New Year's Day: January 1, 1952, which fell on a Tuesday. Since 1952 was a leap year, and a leap year has 2 days over 52 weeks, January 1, 1953, fell on a Thursday, but january 1, 1954, fell on a Friday (1953 not being a leap year), January 1, 1955, fell on a Saturday, and so on. January 1, 1951, fell on a Monday, January, 1, 1950, fell on a Sunday, and so on. In the 28-year period 1929–1956, inclusive, January 1 fell exactly four times on each day of the week. This distribution was exactly the same for the 28-year period 1901–1928, inclusive. We recall that a 28-year period in which every fourth year is a leap year contains an integral number of weeks, and, consequently, we repeat the same distribution of New Year's Days over the days of the week as obtained for the previous 28-year period. The interval 1901–2096 (inclusive) contains 196 years, a number divisible by 28 (the year 2000 not providing any exception for leap year). In this interval, New Year's Day will have fallen exactly the same number of times on each day of the week.

Further, January 1, 2097, will fall on the same day of the week as did january 1, 1901, January 1, 1929, and so on—it will fall on a Tuesday. Thus, New Year's Day of 2100 will fall on a Friday, and New Year's Day of 2101 will fall on a Saturday (2100 is not a leap year). The 28-year period 2101–2128 will differ from the 28-year period 1901–1928 in that the period begins not on a Tuesday but on a Saturday. This calls for a corresponding shift in the days on which New Year's Day will fall. However, in the period 1901–1928 New Year's Day fell exactly four times on each day of the week, and this will happen also in the period 2101–2128. The 28-year intervals 2129–2156 and 2157–2184 will follow the same pattern. The year 2185 begins on the same day as did 2101—Saturday; this allows us to find the data for the period 2185 to 2201. Simple calculation shows that in the interval from 2185 to 2200 New Year's Day will fall twice each on Monday, Wednesday, Thursday, Friday, and Saturday, but three times each on Sunday and Tuesday. New Year's Day of 2201 will fall on a Thursday. In the course of the $3 \cdot 28 = 84$ years, from 2201 to 2284, New Year's Day will fall the same number of times on each day of the week. New Year's Day 2285 will fall on the same day it will in 2201—Thursday. This allows us to determine the data for the period 2285–2300. In this interval January 1 will fall twice each on Monday, Tuesday, Wednesday, Thursday, and Saturday, and three times each on Sunday and Friday. Similarly, over these periods in the course of which New Year's Day will fall the same number of times on each day we counted $2 + 2 = 4$ Mondays, $1 + 3 + 2 = 6$ Tuesdays, $1 + 2 + 2 = 5$ Wednesdays, $1 + 2 + 2 = 5$ Thursdays, 6 Fridays, 4 Saturdays, and 6 Sundays. It follows that New Year's Day falls more often on Sunday than on Saturday.

(b) Employing methods analogous to those used in problem (a), we can show that in the 400-year interval the thirtieth day of the month will fall on Sunday 687 times, on Monday 685 times, on Tuesday 685 times, on Wednesday 687 times, on Thursday 684 times, on

Friday 688 times, and on Saturday 684 times. Thus, the thirtieth day of the month falls most often on a Friday.

15. It is clear that if the final digit of an integer is deleted, the integer is reduced by a factor of at least 10. If that final digit is zero then, of course, the number is reduced by exactly the factor 10, and so all such numbers automatically satisfy the conditions of the problem.

Assume now that if the final digit of a number x is deleted, then the digit is reduced by an integral factor exceeding 10, say, by the factor $10 + a\,(a \geqq 1)$. Let y be the quotient obtained by dividing x by 10, and let z be the remainder; that is, $x = 10y + z\,(z \leqq 9)$. If the last digit of x is deleted, then the deleted digit is z, and the new number is equal to y. The condition of the problem then calls for the equation

$$x = (10 + a) \cdot y \; ,$$

or the equation

$$10y + z = (10 + a) \cdot y \; ,$$

from which we obtain

$$z = ay \; .$$

Since $z < 10$, both a and y must be less than 10. Therefore, except for the numbers divisible by 10, the only integers meeting the conditions of the problem are two-digit numbers; also, if the final digit is deleted, the original integer can be reduced by, at most, a factor of $19\,(11 \leqq 10 + a < 19)$. It is easily shown that only the following two-digit numbers are reduced by a factor of 11 when their last digit is deleted: 11, 22, 33, 44, 55, 66, 77, 88, 99. (If $10 + a = 11$, then $a = 1$; hence $z = ay = y$, and $x = 10y + z = 11y$, where $y = 1, 2, 3, \cdots, 9$.) It can be shown, in an analogous way, that the only two-digit integers reduced by a factor of 12 upon deletion of the final digit are 12, 24, 36, 48 ($z = ay = 2y$, where only $y = 1, 2, 3, 4$ are possible, since $z < 10$, and $x = 12y$). The numbers diminished by a factor of 13 are 13, 26, 39; the numbers diminished by a factor of 14 are 14 and 28. It is evident that the only numbers which can be reduced by factors of 15, 16, 17, 18, 19 are these numbers themselves.

16. (a) Let the integer sought have $k + 1$ digits. It may be written in the form $6 \cdot 10^k + y$, where y is a k-digit number (it may possibly begin with one or more zeros). The conditions of the problem are expressed by

$$6 \cdot 10^k + y = 25y ,$$

or

$$y = \frac{6 \cdot 10^k}{24} .$$

Clearly, it is necessary that $k \geq 2$, since 60 is not divisible exactly by 24 and y is an integer. For $k \geq 2$ the integer y is equal to $25 \cdot 10^{k-2}$; that is, y has the form $250\cdots0$, where there are $k - 2$ zeros. Therefore, all the numbers sought are of the form $6250\cdots0$, where there are n zeros and $n = 0, 1, 2, 3, \cdots$.

(b) Solve this problem: Find a number whose first digit is a and which is reduced by a factor of 35 when its final digit is deleted Proceeding as in problem (a), we have the equation

$$y = \frac{a \cdot 10^k}{34} ,$$

where y is an integer [see solution of problem (a)]. But the right member of this equation cannot possibly be an integer $a \leq 9$ and $k \geq 1$.

Remark: Proceeding as we did in problems 16 (a) and (b), we can show *that an integer beginning with a given digit a can be reduced by an integral factor b upon deletion of the digit a only if $a < b - 1$ and the (proper) fraction $\frac{a}{b-1}$ can be represented as a finite decimal fraction* (that is, all the prime factors of $b - 1$ other than 2 and 5 are factors of a, of sufficient multiplicity). For example, no integer can be reduced by a factor of 85 by deleting its first digit, since $85 - 1 = 84$ includes 3 and 7 among its factors and no digit is divisible by both 3 and 7. Again, any number which is reduced by a factor of 15 upon deletion of its first digit must commence with 7, since 14 has 7 as a factor. A general criterion which will give the necessary form of an integer having a known first digit a and which is reduced by a factor b upon deletion of the first digit is usually easy to find.

17. (a) First we show that no integer can be diminished to $\frac{1}{9}$ its original value by deleting a digit standing farther from its beginning than the second place. Let the digits of the number N be $a_0, a_1, a_2, \cdots, a_n (a_0 \neq 0)$; that is,

$$a_0 \cdot 10^n + a_1 \cdot 10^{n-1} + \cdots + a_n = N .$$

Assume that $\frac{N}{9}$ is an integer, and that it has been obtained by deleting a digit past the second position. We have

$$a_0 10^{n-1} + a_1 10^{n-2} + \cdots = \frac{N}{9} .$$

If we multiply this equation by 10 and subtract from it the initial equation for N, we obtain

$$\frac{N}{9} < 10^{n-1} .$$

This is a contradiction, since $\frac{N}{10}$ is a smaller number, and

$$\frac{N}{10} = a_0 \cdot 10^{n-1} + \cdots \geqq 10^{n-1} .$$

We recall the criterion for divisibility of an integer by 9 (a number is divisible by 9 if and only if the sum of its digits is divisible by 9). Now, if N is divisible by 9, as is also the number obtained from N by deleting the *first* digit of N, then this first digit must be 0 or 9 (since it must by itself be divisible by 9). Since $a_0 \neq 0$, we must have $a_0 = 9$ if N is to meet the conditions of the problem. But then the number $\frac{N}{9}$ has the same number of digits as does N, and so $\frac{N}{9}$ cannot be obtained from N by deleting its fiirst digit. This contradiction means that the only possible deletion of a digit from N in order to meet the conditions given by the problem is that of the second digit. Thus,

$$\frac{N}{9} = a_0 10^{n-1} + a_2 10^{n-2} + \cdots + a_n .$$

Since $\frac{N}{9}$ must again be divisible by 9 (that is, $a_0 + a_2 + a_3 + \cdots + a_n$ is a multiple of 9, as was $a_0 + a_1 + a_2 + \cdots + a_n$) either $a_1 = 0$ or $a_1 = 9$. If we assume that $a_1 = 9$, and if we multiply the expansion for $\frac{N}{9}$ by 10 and subtract the expansion for N, we obtain $\frac{N}{9} = (a_2 - 9) \cdot 10^{n-1} + \cdots$, and since $a_2 \leqq 9$, then

$$\frac{N}{9} < 10^{n-1} ,$$

which is impossible.

Therefore, the conditions of the problem can be met only if the second digit of N is zero and if this is the digit deleted. Thus,

$$N - \frac{N}{9} = a_0 \cdot 10^n - a_0 \cdot 10^{n-1} ,$$

or

$$\frac{N}{9} = N - a_0 \cdot 10^n + a_0 \cdot 10^{n-1} = N - a_0 \cdot 9 \cdot 10^{n-1} .$$

Finally,

$$\frac{1}{9} \cdot \frac{N}{9} = \frac{N}{9} - a_0 \cdot 10^{n-1} .$$

This equation states that to divide $\frac{N}{9}$ by 9, we merely delete the first digit a_0. This solves problem (a).

(b) In solving problem (a) we arrived at the necessary relation

$$\frac{N}{9} = N - a_0 \cdot 10^{n-1} \cdot 9 ,$$

from which it follows, upon solving for N, that

$$N = \frac{a_0 \cdot 10^{n-1} \cdot 81}{8} .$$

This is the form that N must have. It is evident that a_0 cannot be either of the digits 8 or 9, inasmuch as the second digit of N is required to be zero. If we try successively $a_0 = 1, 2, 3, 4, 5, 6, 7$, along with the least necessary value of n which will make N an integer, we have the following seven numbers, which satisfy the conditions of the problem:

10,125; 2025; 30,375; 405; 50,625; 6075; 70,875 .

These numbers, together with all products by 10^k ($k = 1, 2, 3, \cdots$) constitute all the integers which satisfy the conditions of the problem.

18. (a) The proposition is equivalent to the assertion that the integer is diminished by some factor m when the third digit is deleted. Let

$$N = a_0 \cdot 10^n + a_1 \cdot 10^{n-1} + a_2 \cdot 10^{n-2} + \cdots + a_n .$$

Then

$$10 \cdot \frac{N}{m} = a_0 \cdot 10^n + a_1 \cdot 10^{n-1} + a_3 \cdot 10^{n-2} + \cdots + a_n \cdot 10 .$$

If $m < 10$, then subtraction of the first equation from the second yields the inequality $\frac{10-m}{m} N < 10^{n-1}$, which is impossible, since

$$\frac{10-m}{m} > \frac{1}{10} ,$$

and

$$\frac{1}{10} N = a_0 \cdot 10^{n-1} + \cdots \geqq 10^{n-1} .$$

If $m > 11$, then $\dfrac{m - 10}{m} \cdot N < 10^{n-1}$, which also is impossibile for an analogous reason $\left(\dfrac{m - 10}{m} > \dfrac{1}{10}\right)$. Again, if $m = 11$, then we must have $\dfrac{1}{11} N < 10^{n-1}$; that is, $\dfrac{N}{m} = \dfrac{N}{11}$ has two digits fewer than N, which is an impossibility.

The only possibility of obtaining a number satisfying the conditions of the problem is to let $m = 10$. Therefore, the conditions require numbers all of whose digits, except the first two, are zero. Such integers satisfy the requirement and hence describe the numbers sought.

Remark: It is possible to show, by similar reasoning, that the only integers which are diminished by an integral factor when the kth digit, where $k > 3$, is deleted are those having zeros after the first $k - 1$ digits.

(b)   The requirement that a number N be diminished by an integral factor m when its second digit is deleted is expressed as follows:

$$N = a_0 \cdot 10^n + a_1 \cdot 10^{n-1} + a_2 \cdot 10^{n-2} + \cdots + a_n \,,$$

$$\frac{N}{m} = a_0 \cdot 10^{n-1} + a_2 \cdot 10^{n-2} + \cdots + a_n \,.$$

It follows that

$$\frac{N}{m} = N - a_0 \cdot 10^n - a_1 \cdot 10^{n-1} + a_0 \cdot 10^{n-1},$$

or, upon solving for N,

$$N = \frac{(9a_0 + a_2) \cdot 10^{n-1} \cdot m}{m - 1} \,. \tag{1}$$

These equations can be combined to yield

$$N = a_0 \cdot 10^n + a_1 \cdot 10^{n-1} - a_0 \cdot 10^{n-1} + \frac{(9a_0 + a_1) \cdot 10^{n-1}}{m - 1} \,.$$

But, on the other hand, we know that N is an $(n + 1)$-digit number beginning with digits a_0 and a_1:

$$N = a_0 \cdot 10^n + a_1 \cdot 10^{n-1} + a_2 \cdot 10^{n-2} + \cdots + a_n \,,$$

where we may assume that not all the digits $a_2, \cdots, a_n$ are zero (otherwise the problem leads to investigation of two-digit numbers N; see the solution to problem 15). The following inequality must hold:

$$0 < -a_0 \cdot 10^{n-1} + \frac{(9a_0 + a_1) \cdot 10^{n-1}}{m-1} < 10^{n-1} ,$$

or, equivalently,

$$a_0 < \frac{9a_0 + a_1}{m-1} < a_0 + 1 . \qquad (2)$$

As a consequence, we have the following results. The required numbers N are expressed by (1), where $0 \leqq a_0 \leqq 9, 0 \leqq a_1 \leqq 9$. Since N is an integer, and m and $m-1$ are relatively prime, the proper fraction $\frac{9a_0 + a_1}{m-1}$ can be written as a terminating decimal. When this is done the possible values of a_0, a_1 and m must satisfy inequalities (2) in addition, it is necessary to add to the possible values of N the two-digit numbers obtained in the solution of problem 15). There now remains only successive investigations of the possible values for a_0.

(1) $a_0 = 1$. Here, inequality (2) yields

$$1 < \frac{18}{m-1}, m-1 < 18 ;$$

$$\frac{9}{m-1} < 2, m-1 > 4 .$$

Applying the $m-1$ successive values $5, 6, 7, \cdots, 17$, and selecting, each time the appropriate value of a_1, we obtain

$$N = 108; \ 105; \ 10{,}125; \ 1125; \ 12{,}375; \ 135; \ 14{,}625;$$
$$1575; \ 16{,}875; \ 121; \ 132; \ 143; \ 154; \ 165; \ 176;$$
$$187; \ 198; \ 1625; \ 195; \ 192; \ 180{,}625; \ 19{,}125 ,$$

to each of which we can adjoin an arbitrary number of zeros.
Proceeding in an analogous may, we obtain the following:

(2) $a_0 = 2$:

$$N = 2025; \ 21{,}375; \ 225; \ 23{,}625; \ 2425; \ 25{,}875;$$
$$231; \ 242; \ 253; \ 264; \ 275; \ 286; \ 297; \ 2925 .$$

(3) $a_0 = 3$:

$$N = 30{,}725; \ 315; \ 32{,}625; \ 3375; \ 34{,}875;$$
$$341; \ 352; \ 363; \ 374; \ 385; \ 396.$$

(4) $a_0 = 4$:

$$N = 405; \ 41{,}625; \ 4275; \ 43{,}875; \ 451; \ 462; \ 473; \ 484; \ 495 .$$

(5) $a_0 = 5$:

$$N = 50{,}625;\ 5175;\ 52{,}875;\ 561;\ 572;\ 583;\ 594\ .$$

(6) $a_0 = 6$:

$$N = 6075;\ 61{,}875;\ 671;\ 682;\ 693.$$

(7) $a_0 = 7$:

$$N = 781;\ 792\ .$$

(8) $a_0 = 8$:

$$N = 891\ .$$

There is no N for $a_0 = 9$.

In toto, including the results of problem 15, there are 104 values for N. An arbitrary number of zeros can be adjoined to each of these.

19. (a) *First solution.* Let X be the m-digit number obtained by deleting the first digit 1 from the integer sought. That integer is then $10^m + X$, and the new number is $10X + 1$. The condition of the problem yields

$$(10^m + X)\cdot 3 = 10X + 1\ ,$$

or

$$X = \frac{3\cdot 10^m - 1}{7}\ .$$

The last equation provides a condition for finding X: $3\cdot 10^m = 3000\cdots$ must yield the remainder 1 upon division by 7. Direct division by 7 shows that the least number of zeros necessary after 3 to produce a remainder of 1 is 5; that is, $3\cdot 10^5 = 300{,}000 = 7(42{,}857) + 1$. Thus, the least possible value for the integer X is 42,857, and hence the number sought is 142,857.

To find other such numbers, we note that in dividing $3000\cdots$ by 7 we need not stop at the first remainder 1; any quotient obtained for the remainder 1 will serve as an X. It is readily seen that the other numbers will be

$$142{,}857{,}142{,}857;\ \cdots;\ \underbrace{142{,}857{,}142{,}\ 857\cdots 142{,}857}_{k\ \text{times}};\cdots\ .$$

Second solution. Let x be the second digit of the number sought, let y be the third digit, and so on; that is, the number has the form $\overline{1xy\cdots zt}$ (the over bar denotes a succession of digits rather than a product). The condition of the problem states

$$\overline{1xy\cdots zt}\cdot 3 = \overline{xy\cdots zt1}\,.$$

It is at once apparent that $t = 7$ (in no other case will the product on the left yield a final digit 1). Therefore, the tens digit t on the right is 7. But this is possible only if $z\cdot 3$ ends in $7 - 2 = 5$ ($3\cdot z$ plus a carry-over of 2 must yield 7); that is, $z = 5$. There is a carry-over of 1. The product, 3 times the hundreds digit on the left, plus the 1 carried over, must yield 5; and so on. The reader can readily make up an organized format for this process. For example,

$$\begin{array}{cccccc} 1 & 4 & 2 & 8 & 5 & 7 \qquad 42857 \\ \cdot & \cdot & \cdot & \cdot & \cdot & \quad \times\,3 = \cdots\cdot 1 \end{array}$$
$$4 - 1 = 3,\, 2 - 0 = 2,\, 8 - 2 = 6,\, 5 - 1 = 4,\, 7 - 2 = 5$$

(the calculations are made from right to left). The smallest possible number occurs when we reach the digit 1 on the left, 142,857.

If this process is continued, we obtain other numbers satisfying the conditions by stopping whenever we have adjoined another 1 on the left:

$$\cdots;\ \underbrace{142{,}857{,}142{,}857\cdots 142{,}857}_{k \text{ times}};\ \cdots\,.$$

(b) When an integer is tripled, the resulting integer can have the same number of digits only if the initial digit of the first number did not exceed 3. As we saw in problem (a), the first digit can be 1. We shall show now that it cannot be 3.

If the first digit of the integer sought is 3, we see from the required equality $\overline{3xy\cdots z}\cdot 3 = \overline{xy\cdots z3}$ that its second digit (that is, x) must be 9. But 3 times any integer beginning with 39 yields an integer having one more digit. Hence the first digit of any integer meeting the conditions of the problem cannot be 3.

It is left to the reader to show that the numbers sought may begin with the digit 2. The smallest such number is 285,714; all such numbers are of form

$$\underbrace{285{,}714{,}285{,}714\cdots 285{,}714}_{k \text{ times}}\,.$$

The proof is quite similar to that of problem (a).

20. Since after multiplication by 5 the number of digits remains the same, the initial integer must begin with the digit 1. This digit then becomes the final digit of the new number, which is not divisible by 5.

The solutions for the digits 6 and 8 are similar.

21. *First solution.* Since the number of digits is not increased after multiplication by 2, the first digit of the initial number cannot exceed 4. Since after the transfer of the first digit to the end we have an even number (twice the original number), the first digit must be even; hence it must be either 2 or 4.

Consider now the number X obtained by deleting the first digit of the original number sought. Reasoning as we did for problem 19 (a), we obtain

$$(2 \cdot 10^m + X) \cdot 2 = 10 \cdot X + 2 ,$$

that is,

$$X = \frac{4 \cdot 10^m - 2}{8} = \frac{2 \cdot 10^m - 1}{4} ,$$

or else

$$(4 \cdot 10^m + X) \cdot 2 = 10 \cdot X + 4 ,$$

that is,

$$X = \frac{8 \cdot 10^m - 4}{8} = \frac{2 \cdot 10^m - 1}{2} .$$

However, both of these formulas are impossible, since they do not yield integers (the numerator is odd for both).

Second solution. As in the first solution we conclude that the first digit of the number sought must be either 2 or 4. We use the notation employed in problem 19 (a). We have

$$\overline{2xy \cdots zt} \cdot 2 = \overline{xy \cdots zt2}$$

or else

$$\overline{4xy \cdots zt} \cdot 2 = \overline{xy \cdots zt4} .$$

In the first case, t can be only 1 or 6 (otherwise the product on the left cannot end with the digit 2). But if $t = 1$, then on the left is a number not divisible by 4, and on the right is a number divisible by 4 (an integer being divisible by 4 if and only if the number composed of its last two digits is divisible by 4). If $t = 6$, then on the left is a number divisible by 4 (product of two even numbers), but on the right is a number not divisible by 4 (ending in 62).

In the second case, t can be only 2 or 7. If $t = 2$, then [as in the second solution of problem 19(a)], necessarily, $z = 1$, and it follows that the product on the left is divisible by 8 (product of a number divisible by 4, since it ends in 12, and 2), but the number on the right is not divisible by 8 (ending in 124).

The demonstration for $t = 7$ is left to the reader. It is quite similar to the foregoing solutions.

22. (a) *First solution.* A number increased seven-fold upon transfer of its first digit to the end must commence with the digit 1 (otherwise the larger number would contain more digits than the original). As was done in problem 19 (a), we let X be the m-digit integer obtained by deleting the first digit, 1, of the number sought. Then, as in the previous problems,

$$(1 \cdot 10^m + X) \cdot 7 = 10 \cdot X + 1 ,$$

which yields

$$X = \frac{7 \cdot 10^m - 1}{3} .$$

But it is obvious that there is no m for which X can be an m-digit number $\left(\dfrac{7 \cdot 10^m - 1}{3} > 10^m \right)$.

A similar demostration will prove that there is no number which is increased nine-fold by transfer of its initial digit to the end of the number.

Second solution. We conclude, as in the first solution, that the number sought must begin with the digit 1. Using the terminology explained in problem 19 (b), we have the following statement of our problem:

$$\overline{1xy \cdots zt} \cdot 7 = \overline{xy \cdots zt1} .$$

It follows that the product $t \cdot 7$ must end with the digit 1, which requires that $t = 3$. Insertion of this digit for t yields $\overline{1y \cdots z3} \cdot 7 = \overline{y \cdots z31}$. Since $3 \cdot 7 = 21$, and the product of of $\overline{z3}$ by 7 ends in the succession of digits 31, this product has 1 as a final digit. Consequently, z is also equal to 3. A similar procedure shows that each successive digit (from this end forward) is equal to 3. However, the initial digit of the number must be 1, which is impossible.

Therefore, there does not exist any integer which increases seven-fold upon transfer of its first digit to the end.

(b) *First solution.* Inasmuch as the number to be obtained upon multiplying the sought integer by 4 must not contain a greater number of digits than before, the initial digit cannot exceed 2. Since transfer of the initial digit to the end must produce an even number, that digit must be 2. If now we designate by X the m-digit number obtained from the number sought, upon removing the first digit we obtain

$$(2 \cdot 10^m + X) \cdot 4 = 10X + 2 ,$$

or

$$X = \frac{8 \cdot 10^m - 2}{6} ,$$

which is an impossibility, since $\dfrac{8 \cdot 10^m - 2}{6} > 10^m$ [see the first solution of problem (a)].

Second solution. As in the first solution, the first digit of the number sought must be 2. Moreover,

$$\overline{2xy \cdots zt} \cdot 4 = \overline{xy \cdots zt2} ,$$

and it follows that $t = 3$ or 8 ($t \cdot 4$ ends in 2).

If $t = 8$, then the final two digits on the right form the number 82, and the integer is not divisible by 4. If $t = 3$, then

$$\overline{2xy \cdots z3} \cdot 4 = \overline{xy \cdots z32} ,$$

hence

$$\overline{2xy \cdots z0} \cdot 4 = \overline{xy \cdots z20} ,$$

and

$$\overline{2xy \cdots z} \cdot 4 = \overline{xy \cdots z2} .$$

In the same manner we find that the number $\overline{2xy \cdots z}$ has the same property as did $\overline{2xy \cdots zt}$. By applying the same reasoning as before we find that, necessarily, $z = 3$. Continuing this line of reasoning, and moving from right to left, we find that the tens position of the number must have the digit 3. But the initial number must commence with the digit 2. This is obviously impossible.

23. *First solution.* Let us designate the digits of the integer sought by $x, y, \cdots, z, t$. Then proceeding as in problem 19 (a), we obtain

$$\overline{7xy \cdots zt} \cdot \frac{1}{3} = \overline{xy \cdots zt7} ,$$

or

$$\overline{xy \cdots zt7} \cdot 3 = \overline{7xy \cdots zt} .$$

It is clear that $t = 1$; we can determine the digit z (17·3 yields 51; this means that $z = 5$), and, moving from right to left we obtain, successively, the digits forming the integer sought. The process ends when we obtain the digit 7. It is convenient to display the result in the following form:

$$\underset{\cdots\cdots\cdots\cdots\cdots\cdots\cdots\cdots\cdots\cdots\cdots}{24137931034482758620689655 1} \cdot 7 \cdot 3 = \underset{\cdots\cdots\cdots\cdots\cdots\cdots\cdots\cdots\cdots\cdots\cdots}{724137931034482758620689655 1}$$

(the calculations are made from right to left). Thus, the least integer satisfying the conditions of the problem is 7,241,379,310,344,827,586, 206,896,551.

If in the course of these calculations we do not stop when we first obtain a 7, we can find additional numbers which satisfy the conditions of the problem. All such numbers will be of the form

$$\underbrace{724137931034482758620689655 1 \cdots 724137931034482758620689655 1}_{k \text{ times}}$$

Second solution. Let $\overline{7xyz \cdots t}$ be the integer sought. Division by 3 must yield the integer $\overline{xyz \cdots t7}$. We write this requirement in the form

$$\frac{\overline{7xyz \cdots t} \ \lfloor 3}{\overline{xyz \cdots t7}}.$$

It is clear that $x = 2$ is necessary. If we replace x by 2 in the dividend and quotient, we can find the next digit of the quotient, and this is also the third digit of the dividend; we can then determine the third digit of the quotient, which is the fourth digit of the dividend, and so on. The process is complete when the final digit of the quotient is 7 and simultaneously the dividend is exactly divisible by 3. This dividend is then the number sought and is the least integer having the required property.

The following arrangement is a convenient scheme for carrying out this computation (the numbers in the second row are written last):

7	24	1	3	79	31	0	3	4	48	2	7	5	8	6	2			
7	12	4	11	23	27	9	3	1	10	13	14	24	8	22	17	25	18	6	2
2	41	3	7	9	3	1	0	3	4	4	82	7	5	8	6	2	0	

							0	6	8	9	6	55	1
							20	26	28	19	16	15	5	21
							6	8	9	6	5	51	7

The least integer is found to be 7,241,379,310,344,827,586,206,896,551.

Third solution. As in the first solution of problem 19 (a), set up the formula

$$(7 \cdot 10^m + X) \cdot \frac{1}{3} = 10X + 7 ,$$

from which

$$X = \frac{7 \cdot 10^m - 21}{29} .$$

The problem then becomes that of finding an integer of form 70,000,··· which upon division by 29 yields the remainder 21. It is left to the reader to verify that the same solution is obtained as before.

Remark: A similar algorithm may be employed to solve the following generalized problem:

Find the least integer having a given first digit which is diminished to $\frac{1}{8}$ its original size when the first digit is transferred to the end. To make the solution possible for integers having 1 or 2 as a first digit, let us agree that a zero may be put as the first digit of the quotient (and this is then the second digit of the dividend). The only nonzero integers satisfying the requirements are (in addition to that produced above):

1,034,482,758,620,689,655,172,413,793;

2,068,965,517,241,379,310,344,827,586;

3,103,448,275,862,068,965,517,241,379;

4,137,931,034,482,758,620,689,655,172;

5,172,413,793,103,448,275,862,068,965;

6,206,896,551,724,137,931,034,482,758;

8,275,862,068,965,517,241,379,310,344;

9,310,344,827,586,206,896,551,724,137.

The same procedure will solve the following problem:

Find the least integer commencing with a given digit a which is decreased by a factor l when the first digit is transferred to the end.

24. (a) The conditions of the problem may be expressed by the equality

$$\overline{xy \cdots zt} \cdot a = \overline{tz \cdots yx} ,$$

where a is one of the numbers 2, 3, 5, 6, 7 or 8.

If $a = 5$, then x must be 1; otherwise the number on the right will contain one digit too many. (The case in which x is 0 may be excluded, since upon multiplying each side by 2 and deleting the final zero we obtain $\overline{y \cdots zt} = 2 \cdot \overline{tz \cdots y}$; that is, we arrive at the same problem as for $a = 2$.) But the integer $\overline{tz \cdots y1}$ is not divisible by 5. Analogous reasoning will prove that a cannot be 6 or 8.

If $a = 7$, then x again must be 1. In this case, t must be the digit 3, otherwise $\overline{1y \cdots zt} \cdot 7$ fails to end with the digit 1. But the equation $\overline{1y \cdots z3} \cdot 7 = \overline{3z \cdots y1}$ is clearly impossible, since the left number is greater than the right number.

If $a = 2$, then x cannot exceed 4. Since the integer $\overline{tz \cdots yx}$ must be even, x is either 2 or 4. If $x = 4$, the digit t (the first digit of $\overline{4y \cdots zt} \cdot 2$) can be only 8 or 9, and neither $\overline{4y \cdots z8} \cdot 2$ nor $\overline{4y \cdots z9} \cdot 2$ ends with the digit 4. If $x = 2$, then t (the first digit of $\overline{2y \cdots zt} \cdot 2$) can only be 4 or 5; but neither $\overline{2y \cdots z4} \cdot 2$ nor $\overline{2y \cdots z5} \cdot 2$ can end in 2.

Finally, if $z = 3$, then x cannot exceed 3. If $x = 1$, then t must be 7 ($t \cdot 3$ ends in 1); if $x = 2$, then t must be 4; if $x = 3$, then t must be 1. But in the first case, $\overline{tz \cdots yx}$ is certainly greater than $\overline{xy \cdots zt} \cdot 3$, and in the second and their cases it is clearly smaller.

(b) Let $xy \cdots zt$ be an integer which is $\frac{1}{4}$ its inversion. We then have

$$\overline{xy \cdots zt} \cdot 4 = \overline{tz \cdots yx}.$$

Since $\overline{xy \cdots zt} \cdot 4$ is to contain the same number of digits as does $\overline{xy \cdots zt}$, the digit x can be only 0, 1, or 2. Since $\overline{tz \cdots yx}$ must be divisible by 4, it follows that x must be even, and so the only possibilities are $x = 0$ (if we allow 0 to be counted as a digit at the beginning of the integer), or $x = 2$.

Suppose $x = 0$. Then $t = 0$ or else $t = 5$. Write $\overline{0y \cdots yt} \cdot 4 = \overline{tz \cdots y0}$; we note that t cannot be the digit 5 (regardless what digit y is), and so the only possibility is $t = 0$. We can now write $\overline{y \cdots z} \cdot 4 = \overline{z \cdots y}$; that is, if a number meeting the condition of the problem begins with 0, then it also ends in 0, and the integer obtained by deleting these two zeros will also meet the conditions of the problem.

It suffices therefore to investigate $x = 2$, for which $\overline{2y \cdots zt} \cdot 4 = \overline{tz \cdots y2}$. Since $2 \cdot 4 = 8$, t can be only 8 or 9, but $t = 9$ can immediately be eliminated as a possibility. Thus, $t = 8$, and we can write $\overline{2y \cdots z8} \cdot 4 = \overline{8z \cdots y2}$. Since $23 \cdot 4 > 90$, y can be only 0, 1, or 2; also, the tens digit of any product of form $\overline{z8} \cdot 4$ must be odd. Thus, $y = 1$. Since we know the final two digits (12) in the product $\overline{21 \cdots z8} \cdot 4$, we can easily find the only two possibilities for the next successive digit: z can be only 2 or 7. But $21 \cdot 4 > 82$, which means that $z = 7$.

Hence, the number must have the form $\overline{21\cdots78}$. We note that the integer 2178 satisfies the condition of the problem, and hence it is the only such four-digit number. We now consider solutions in terms of integers having more than four digits. We must have

$$\overline{21uv\cdots rs78}\cdot 4 = \overline{87\,sr\cdots vu12} ,$$

which may be written

$$84\cdot 10^{k+2} + 312 + \overline{uv\cdots rs\,00}\cdot 4$$
$$= 87\cdot 10^{k+2} + 12 + \overline{sr\cdots vu00} ,$$

or

$$\overline{uv\cdots rs}\cdot 4 + 3 = \overline{3sr\cdots vu}.$$

Since when the number $uv\cdots rs$ is multiplied by 4, and 3 is added, the digit 3 is produced as the leading digit of the resulting calculation, the digit u must be at least as great as 6, but since $\overline{3sr\cdots vu}$ must be odd, the only possibilities are 9 or 7. We shall investigate both possibilities.

When $u = 9$ we have

$$\overline{9v\cdots rs}\cdot 4 + 3 = \overline{3sr\cdots v9} ,$$

from which it follows that $s = 9$ ($s\cdot 4$ must terminate in 6; if $s = 4$, then $\overline{34r\cdots v9}$ is smaller than $\overline{9v\cdots r4}\cdot 4 + 3$), and so

$$\overline{9v\cdots r9}\cdot 4 + 3 = \overline{39r\cdots v9} .$$

That is (for $u = 9$), in order for $\overline{21uv\cdots rs78}$ to meet the conditions of the problem, the final digit of $\overline{uv\cdots rs}$ must also be 9. In particular, $\overline{uv\cdots rs}$ must be 9, 99, 999, and so on. We obtain the numbers

$$21,978;\ 219,978;\ 2199;\ 978;\ \cdots .$$

We can easily verify that all such numbers satisfy the conditions of the problem.

When $u = 7$ we have

$$\overline{7v\cdots rs}\cdot 4 + 3 = \overline{3sr\cdots v7} .$$

Reasoning as we did initially in this problem, we find that $s = 1$, $v = 8, r = 2$; that is, the digit sequence $\overline{uv\cdots rs}$ must have the form $\overline{78\cdots 21}$. We can easily verify that if the digit pairs 78 and 21 are inserted between 21 as the first digit pair, and 78 as the last digit pair, the resulting integer satisfies the conditions of the problem: For example,

$$21\ 78\ 21\ 78\ 21\ 78 \cdot 4 = 87\ 12\ 87\ 12\ 87\ 12\ .$$

But according to the treatment for $u = 9$ we will also obtain solutions if we insert the digit 9 in the appropriate positions. For example, the following numbers are solutions:

$$0;\ 2178;\ 21{,}978;\ 219{,}978;\ \cdots\ ,$$

$$2199 \underbrace{\cdots}_{k \text{ times}} 978,\ 2199 \underbrace{\cdots}_{(k+1) \text{ times}} 9978,\ \cdots\ . \qquad (1)$$

The insertion of the digit 9 may be made after any sequence $\overline{21}$, provided it is also made in the equivalent location counted from the other end of the number. Thus, any sequential array of digits of the form

$$P_1 P_2 \cdots P_{n-1} P_n P_{n-1} \cdots P_2 P_1\ ,$$

where each letter P_i is one of the numbers from the display (1). For example

$$2{,}197{,}821{,}978\ ,$$

$$2{,}199{,}782{,}178{,}219{,}978\ ,$$

$$21{,}978{,}021{,}997{,}800{,}219{,}978{,}021{,}978\ ,$$

$$02{,}199{,}999{,}780\ .$$

(The last number can be considered a solution if we allow zeros at the beginning of the number.)

It may be proven in an analogous way that all integers which are increased by a factor of 9 upon reversal of the digits can be obtained by sequentially placing together integers of the form

$$0;\ 1089;\ 10{,}989;\ 109{,}989;\ \cdots;\ 1099 \underbrace{\cdots}_{k \text{ times}} 989;\ 1099 \underbrace{\cdots}_{(k+1) \text{ times}} 9989;\ \cdots$$

in the same fashion as done for (1).

25. (a) Let N be the number sought. Designate the number formed by the first three digits of N by p, and the number formed by the final three digits of N by q. Then the conditions of the problem yield

$$6(1000p + q) = 1000q + p \qquad (= 6N)\ ,$$

or

$$(1000q + p) - (1000p + q) = 999(q - p) = 5N;$$

that is, N is divisible by 999.

Further, $p + q = (1000p + q) - 999p = N - 999p$, and it follows that $p + q$ also must be divisible by 999. But p and q are each three-digit numbers, and obviously neither is 999; consequently, $p + q = 999$.

We find, without difficulty, that

$$(1000q + p) + (1000p + q) = 1001(p + q) = 7N,$$

which yields $7N = 999,999$, or $N = 142,857$.

(b) Reasoning as we did in problem (a), and designating by p the integer formed by the first four digits of the derived number N, and by q the integer formed by the final four digits of N, we obtain

$$7N = 10,001(p + q) = 99,999,999,$$

which fails to yield an integer for N, since 99,999,999 is not exactly divisible by 7.

26. Let x be a number satisfying the condition of the problem. Since $6x$ must be, along with x, a six-digit integer, the first digit of x must be 1 (and the following digit cannot exceed 6). Hence:

(1) The leading digits of the numbers $x, 2x, 3x, 4x, 5x$, and $6x$ are all different and, as a result, must comprise all the digits contained in the integer x (each of these digits must appear in the original number x).

(2) All the digits of x are different.

None of these digits is 0 (otherwise one of the above products starts with 0), and so the final digit of x must be odd (otherwise $5x$ ends with an 0); also, the final digit must differ from 5 (otherwise $2x$ ends in 0). Consequently, the final digits of the numbers $x, 2x, 3x, 4x, 5x$, and $6x$ are all different, which implies that these digits comprise all the digits appearing in x. Therefore, one of these final digits is 1. Since only the number $3x$ can terminate with a 1 ($2x, 4x, 6x$ being even, $5x$ having 5 as a final digit, and x itself already having 1 as its first digit), it follows that x must end with the digit 7; $2x$ ends with the digit 4; $3x$ ends with the digit 1; $4x$ ends with the digit 8; $5x$ ends with the digit 5; and $6x$ ends with the digit 2.

Now, the first digits of these numbers are the individual digits of the digit set comprising x; we display them in increasing order, using (asterisks to represent unknown digits):

$$x \cdot 1 = 1 ****7,$$
$$x \cdot 2 = 2 ****4,$$
$$x \cdot 3 = 4 ****1,$$

$$x \cdot 4 = 5 \ast\ast\ast\ast 8 \, ,$$
$$x \cdot 5 = 7 \ast\ast\ast\ast 5 \, ,$$
$$x \cdot 6 = 8 \ast\ast\ast\ast 2 \, .$$

In this display, not only must each row contain (on the right) all six distinct digits $1, 2, 4, 5, 7$, and 8, but each column must contain these six distinct digits, in some order. Suppose that $x \cdot 2$ and $x \cdot 5$ have the same digit a in a certain position, say in the third position (a can have only one of the two values not assumed by either the first or the last digit of either of the two investigated numbers). Since the difference $x \cdot 5 - x \cdot 2 = x \cdot 3$ will be a six-digit number, either the digit 0 or the digit 9 will stand in its third position (since we can take, at most, a unit carry-over into this place for the subtraction). But this is an impossibility, since we already know that the number $x \cdot 3$ will contain neither an 0 nor a 9 as one of its digits.

Therefore, in the above display of $x \cdot 1, x \cdot 2, \cdots$ the sum of the digits in any column is $1 + 2 + 4 + 5 + 7 + 8 = 27$. We can therefore add the right member of this display and obtain

$$x \cdot 21 = 2{,}999{,}997 \, ,$$

whence $x = 142{,}857$, which is the integer sought. As a check we have:

$$x = 142{,}857 \, , \qquad 4x = 571{,}428 \, ,$$
$$2x = 285{,}714 \, , \qquad 5x = 714{,}285 \, ,$$
$$3x = 428{,}571 \, , \qquad 6x = 857{,}142 \, .$$

27. (a) By factorization, $n^3 - n = (n - 1)n(n + 1)$. The factors on the right represent three consecutive integers, whence one of them is divisible by 3.

(b) $n^5 - n = n(n-1)(n+1)(n^2+1)$. If the integer n terminates with one of the digits $0, 1, 4, 5, 6,$ or 9, then one of the first three factors on the right is divisible by 5. If n ends in one of the digits $2, 3, 7$ or 8, then n^2 ends in 4 or 9, and in this event $n^2 + 1$ is divisible by 5.

(c) $n^7 - n = n(n-1)(n+1)(n^2 - n + 1)(n^2 + n + 1)$. If n is divisible by 7, or yields a remainder of 1 or 6 upon division by 7, then one of the first three factors on the right is divisible by 7. If n yields a remainder of 2 when divided by 7 (that is, $n = 7k + 2$), then n^2 yields a remainder of 4 when divided by 7 (that is, $n^2 = 49h^2 + 28k + 4$), and so $n^2 + n + 1$ is divisible by 7. Similar reasoning shows that if

$n = 7k + 4$, then again $n^2 + n + 1$ is divisible by 7. For the remainders 3 or 5, $n^2 - n + 1$ is divisible by 7.

(d) $n^{11} - n = n(n-1)(n+1)(n^8 + n^6 + n^4 + n^2 + 1)$. If n is divisible by 11, or yields the remainder 1 or 10 upon division by 11, then one of the first three factors is divisible by 11. If this remainder is either 2 or $9(n = 11k \pm 2)$, then n^2 clearly yields a remainder of 4; n^4 yields a remainder of $5 = 16 - 11$; n^6 yields a remainder of $9 = 20 - 11[n^6 = n^4 \cdot n^2 = (11k+5)(11k_2+4) = (11k_1+5)(11k_2+4) = 121k_1k_2 + 11(4k_1 + 5k_2) + 20]$, and n^8 yields a remainder of $3 = 25 - 22$. It follows, in either case, that $n^8 + n^6 + n^4 + n^2 + 1$ is divisible by 11. In the same manner we can easily verify that if one of the remainders is $\pm 3, \pm 4,$ or ± 5 (the only remaining possibilities upon division of n by 11), then $n^8 + n^6 + n^4 + n^2 + 1$ is divisible by 11.

(e) $n^{13} - n = n(n-1)(n+1)(n^2+1)(n^4 - n^2 + 1)(n^4 + n^2 + 1)$. The procedure is analogous to that of problem (d). If n is divisible by 13, or yields upon division by 13 the remainder ± 1, then one of the first three factors is divisible by 13; if n yields the remainder ± 5, then $n^2 + 1$ is divisible by 13; if the remainder is ± 2 or ± 6, then $n^4 - n^2 + 1$ is divisible by 13; if the remainder is ± 3 or ± 4, then $n^4 + n^2 + 1$ is divisible by 13.

28. (a) The difference of like even powers of any two numbers is divisible by the sum of the bases (one of the factors is this sum). Hence, $3^{6n} - 2^{6n} = 27^{2n} - 8^{2n}$ is divisible by $27 + 8 = 35$.

(b) It is readily verified that

$$n^5 - 5n^3 + 4n = n(n^2 - 1)(n^2 - 4)$$
$$= (n - 2)(n - 1)n(n + 1)(n + 2) .$$

The factorization displays five consecutive integers. One of them must be divisible by 5; at least one of them is divisible by 3; at least two of them are divisible by 2; and one of these two is divisible by 4. Thus, the product of the five consecutive numbers is divisible by $5 \cdot 3 \cdot 2 \cdot 4 = 120$ [see the solution of problem 27 (a)].

(c) Prime factorization of the given number yields

$$56,786,730 = 2 \cdot 3 \cdot 5 \cdot 7 \cdot 11 \cdot 13 \cdot 31 \cdot 61 .$$

We must show that $mn(m^{60} - n^{60})$ is divisible by each of these relatively prime numbers. If m and n are both odd, then $m^{60} - n^{60}$ is even; consequently, $mn(m^{60} - n^{60})$ is also even, so it is divisible by 2. Further, it follows from problem 27 that if k is equal to $3, 5, 7,$ 11, or 13, and if n is not divisible by k, then the difference $n^{k-1} - 1$

must be divisible by k. In particular, if neither n nor m is divisible by 3, then $m^{60} - 1 = (m^{30})^2 - 1$ and $n^{60} - 1 = (n^{30})^2 - 1$ are divisible by 3; that is, m^{60} and n^{60} yield the same remainder, 1, upon division by 3. Hence, if mn is not divisible by 3, then $m^{60} - n^{60}$ is divisible by 3, which means that in all cases $mn(m^{60} - n^{60})$ is divisible by 3. It can be shown, in the same way, that the difference

$$m^{60} - n^{60} = (m^{15})^4 - (n^{15})^4 = (m^{10})^6 - (n^{10})^6$$
$$= (m^6)^{10} - (n^6)^{10} = (m^5)^{12} - (n^5)^{12}$$

is divisible by 5 in the event that neither m nor n is divisible by 5, and is divisible by 7 if 7 fails to divide either m or n, and the analogous conclusion holds for 11 and for 13. Thus, $mn(m^{60} - n^{60})$ is divisible by $2 \cdot 3 \cdot 5 \cdot 7 \cdot 11 \cdot 13$.

Divisibility of $mn(m^{60} - n^{60})$ by 31 and by 61 is demonstrated in similar fashion (since $n^{31} - n$ is, for all integral n, divisible by 31, and $n^{61} - n$ is, for all integral n, divisible by 61; problem 240).

29. We shall use the identity

$$n^2 + 3n + 5 = (n + 7)(n - 4) + 33 .$$

If this number is to be divisible by 11, then for the suitable n, $(n+7)(n - 4)$ must be divisible by 11. Since $(n + 7)-(n-4) = 11$, either both terms are divisible by 11 or neither is. Hence, if $(n + 7)(n - 4)$ is divisible by 11, then it is divisible also by 121, and $(n + 7)(n - 4) + 33$ fails to be divisible by 121.

30. The given expression factors to

$$(m - 2n)(m - n)(m + n)(m + 2n)(m+3n) .$$

If $n \neq 0$, no two of these factors are equal. However, the integer 33 can be factored only as a product of at most four factors:

$$33 = (-11) \cdot 3 \cdot 1 \cdot (-11) ,$$

or

$$33 = 11 \cdot (-3) \cdot 1 \cdot (-1) .$$

If $n = 0$, the given expression becomes m^5, which cannot equal 33 for any integral value of m.

31. Every integer is either divisible by 5 or else can be represented in one of the forms $5k + 1, 5k + 2, 5k - 2$, or $5k - 1$. If the number is divisible by 5, then its 100th power is divisible by $5^3 = 125$; hence, we need only investigate the case for integers not divisible by 5. According to the binomial theorem,

$$(5k \pm 1)^{100} = (5k)^{100} \pm \cdots + \frac{100 \cdot 99}{1 \cdot 2}(5k)^2 \pm 100 \cdot 5k + 1 \ ,$$

where every term except the final one contains 5^3 as a factor, and so numbers of this form leave a remainder of 1 upon division by 125. Also,

$$(5k \pm 2)^{100} = (5k)^{100} \pm \cdots$$

$$\pm \frac{100 \cdot 99}{1 \cdot 2}(5k)^2 \cdot 2^{98} \pm 100 \cdot 5k \cdot 2^{99} + 2^{100} \ .$$

Again, each term, except the final one, contains 125 as a factor. The number 2^{100} can be represented in the form

$$(5 - 1)^{50} = 5^{50} - \cdots + \frac{50 \cdot 49}{1 \cdot 2} \cdot 5 - 50 \cdot 5 + 1 \ ,$$

from which the remainder 1 is obtained upon division by 125.

Therefore, the only two remainders possible when the 100th power of an integer is divided by 5 are 0 (if the integer itself is divisible by 5) and 1.

32. The problem may be characterized as follows. If n is relatively prime to 10, then $n^{101} - n = n(n^{100} - 1)$ is divisible by 1000; that is, $n^{100} - 1$ is divisible by 1000. First, it is obvious that if n is an odd integer, then $n^{100} - 1 = (n^{50} + 1)(n^{25} + 1)(n^{25} - 1)$ is divisible by 8. Further, from the result of the preceding problem we know that if n is not divisible by 5, then $n^{100} - 1$ is divisible by 125. Thus, we see that $n^{100} - 1$ is divisible by $8 \cdot 125 = 1000$ if n is odd and not divisible by 5, and these conditions are satisfied if n is relatively prime to 10.

33. Let N be the integer sought. The condition of the problem requires that $N^2 - N$ end in three zeros, that is, that it be divisible by 1000. Since $N^2 - N = N(N - 1)$, and since N and $N - 1$ are relatively prime, divisibility by 1000 is possible only if one of these factors is divisible by 8 and the other by 125 (neither N nor $N - 1$ is divisible by 1000, since N is a three-digit integer).

If N is a three-digit integer divisible by 125, then $N - 1$ is divisible by 8 only if $N = 625$ (as we can easily verify), whence $N - 1 = 624$. It is also easily verified that if $N - 1$ is a three-digit integer divisible by 125, then N is divisible by 8 only if $N - 1 = 375$, or $N = 376$.

Now, since $N^{k-1} - 1$ is (for $k \geq 2$) divisible by $N - 1$, it follows that $N^k - N = N(N^{k-1} - 1)$ is for all integral k divisible by $N(N-1) = N^2 - N$. Therefore, if $N^2 - N$ ends in three zeros, then $N^k - N$ will, for all $k \geqq 2$, end with three zeros; that is, N^k will end in the same three digits as does N. It follows that the numbers 625 and 376 (and only these) satisfy the conditions of the problem.

34. The two final digits of N^{20} can be found in the following manner. The number N^{20} is divisible by 4 (since N is even); further, N is not divisible by 5 (since then it would be divisible by 10, which denies the hypothesis). Hence, N is representable in the form $5k \pm 1$ or $5k \pm 2$ (see the solution of problem 31). The number

$$(5k \pm 1)^{20} = (5k)^{20} \pm \frac{20 \cdot 19}{1 \cdot 2}(5k)^{19} + \cdots$$

$$+ \frac{20 \cdot 19}{1 \cdot 2}(5k)^2 \pm 20 \cdot 5k + 1$$

yields the remainder 1 when divided by 25, and the number

$$(5k \pm 2)^{20} = (5k)^{20} \pm \frac{20 \cdot 19}{1 \cdot 2}(5k)^{10} \cdot 2 + \cdots$$

$$+ \frac{20 \cdot 19}{1 \cdot 2}(5k)^2 \cdot 2^{18} \pm 20 \cdot 5k \cdot 2^{19} + 2^{20}$$

yields the same remainder upon division by 25 as does

$$2^{20} = (2^{10})^2 = (1024)^2 = (1025 - 1)^2 ,$$

that is, 1. Since N^{20} yields the remainder 1 upon division by 25, it follows that the final two digits of this number can be only 01, 26, 51, or 76. In asmuch as N^{20} is divisible by 4, the possibilities narrow down to the number 76 (since a number is divisible by 4 if, and only if, the number formed by its final two digits is so divisible). This yields 7 as the digit standing in the tens place of N^{20}.

We shall now find the final three digits of N^{200}. The number N^{200} is divisible by 8. Further, since N and 5 are relatively prime, it follows that N^{100} yields the remainder 1 upon division by 125 (see the solution of problem 31); that is, $N^{100} = 125k + 1$. But $N^{200} = (125k + 1)^2 = (125k)^2 + 250k + 1$ also yields the remainder 1 upon division by 125. Therefore, the only possibilities for the final three digits of N^{200} are 126, 251, 376, 501, 626, 751, and 876. Since N^{200} is divisible by 8, it is clear that N^{200} ends in 376. Thus, the digit in the hundreds place of N^{200} is 3.

Remark: It is easily reasoned that the number N^{100} must end with the digits 376.

35. The series $1 + 2 + 3 + \cdots + n$ is equal to $\frac{n(n+1)}{2}$. Hence, we must show that if k is odd, then $S_k = 1^k + 2^k + 3^k + \cdots + n^k$ is divisible by $\frac{n(n+1)}{2}$.

We first note that for odd k, $a^k + b^k$ is divisible by $a + b$. Two cases will now be examined.

n is an even integer. Here, the sum S_k is divisible by $n + 1$, since each of the sums

$$1^k + n^k, 2^k + (n-1)^k, 3^k + (n-2)^k, \cdots, \left(\frac{n}{2}\right)^k + \left(\frac{n}{2}+1\right)^k$$

is divisible by

$$1 + n\left[= 2 + (n-1) = 3 + (n-2) = \cdots = \frac{n}{2} + \left(\frac{n}{2}+1\right)\right].$$

The sum S_k is divisible also by $\frac{n}{2}$, since

$$1^k + (n-1)^k, 2^k + (n-2)^k, 3^k + (n-3)^k, \cdots ,$$
$$\left(\frac{n}{2}-1\right)^k + \left(\frac{n}{2}+1\right)^k , \left(\frac{n}{2}\right)^k, n^k$$

are all divisible by $\frac{n}{2}$.

n is an odd integer. Here the sum S_k is divisible by $\frac{n+1}{2}$, since

$$1^k + n^k, 2^k + (n-1)^k, 3^k + (n-2)^k, \cdots, \left(\frac{n-1}{2}\right)^k + \left(\frac{n+3}{2}\right)^k \left(\frac{n+1}{2}\right)^k$$

are all divisible by $\frac{n+1}{2}$. Also, S_k is divisible by n, since

$$1^k + (n-1)^k, 2^k + (n-2)^k, 3^k + (n-3)^k, \cdots,$$
$$\left(\frac{n-1}{2}\right)^k + \left(\frac{n+1}{2}\right)^k, n^k$$

are all divisible by n.

36. Let N be written in the form

$$N = a_n \cdot 10^n + a_{n-1}10^{n-1} + a_{n-2}10^{n-2} + \cdots + a_1 10 + a_0$$

(the a_k are, of course, the digits of N). Subtract from N the number

$$M = a_0 - a_1 + a_2 - a_3 + \cdots \pm a_n$$

(that is, the algebraic sum, taken with alternating signs, of the digits of N). A simple regrouping of terms yields

$$N - M = a_1(10 + 1) + a_2(10^2 - 1) + a_3(10^3 + 1)$$
$$+ a_4(10^4 - 1) + \cdots + a_n(10 \pm 1) \,,$$

which is divisible by 11 since each term on the right is divisible by 11. [In fact, upon division by 11, $10^k = (11 - 1)^k$ yields the remainder -1 if k is odd and the remainder 1 if k is even, as binomial expansion will show.] The number N is divisible by 11 if, and only if, the number M is divisible by 11 (zero, of course, is considered divisible by all nonzero integers). A criterion, then, is as follows: A number N is divisible by 11 if, and only if, the difference of the sum of digits in the odd-number (1st, 3rd, 5th, $\cdots$) positions and the sum of digits in the even-number (2nd, 4th, $\cdots$) positions is divisible by 11.

37. The number 15 yields the remainder 1 upon division by 7; it follows that

$$15^2 = (7\cdot 2 + 1)\cdot(7\cdot 2 + 1) = 7n_1 + 1$$

also yields the remainder 1 upon division by 7, as does

$$15^3 = 15^2\cdot 15 = (7n_1 + 1)\cdot(7\cdot 2 + 1) = 7n_2 + 1 \,.$$

It is now easily verified that *every* power of 15 yields the remainder 1 upon division by 7. Now if the sum $1 + 2 + 3 + 4 + \cdots + 14 = 105$ is subtracted from the given number, the difference can, after some simple regrouping and factoring, be displayed in the form

$$13(15 - 1) + 12(15^2 - 1) + 11(15^3 - 1) + \cdots$$
$$+ 2(15^{12} - 1) + 1(15^{13} - 1) \,,$$

that is, each term is divisible by 7. But since the difference between the given number and the (decimal) number $105(= 7\cdot 15)$ is divisible by 7, it follows that the given number is also divisible by 7.

38. Let K be an n-digit integer. In the set of natural numbers containing $n + 2$ digits and beginning with the digits 10 (that is, with numbers of aspect $\overline{10a_1a_2 \cdots a_n}$, this notation indicating the integer $1\cdot 10^{n+2} + a_1\cdot 10^n + a_2\cdot 10^{n-1} + \cdots + a_n$) there can always be found at least one digit which is divisible by the n-digit number K. Suppose this number is $\overline{10b_1b_2 \cdots b_n}$. Then, to satisfy the conditions of the problem, both the numbers $\overline{b_1b_2\cdots b_n10}$ and $\overline{b_1b_2 \cdots b_n01}$ would have to be divisible by K. Their difference is 9, which then also must be divisible by K. Since the only divisors of 9 are 1, 3, and

9, these are the only numbers which can satisfy the conditions of the problem.

39. We must show that the number

$$N = 27{,}195^8 - 10{,}887^8 + 10{,}152^8$$

is divisible by $26{,}460 = 2^2 \cdot 3^3 \cdot 5 \cdot 7^2$. The proof will be given in two steps:

(1) $N = 27{,}195^8 - (10{,}887^8 - 10{,}152^8)$. Now, $27{,}195 = 3 \cdot 5 \cdot 7^2 \cdot 37$, and so this number is divisible by $5 \cdot 7^2$ The difference shown in the parentheses is divisible by

$$10{,}887 - 10{,}152 = 735 = 3 \cdot 5 \cdot 7^2$$

(since $a^{2n} - b^{2n}$ is divisible by $a - b$). Hence, N is divisible by $5 \cdot 7^2$.

(2) $N = (27{,}195^8 - 10{,}887^8) + 10{,}152^8$. Now, $10{,}152 = 2^3 \cdot 3^3 \cdot 47$ is divisible by $2^2 \cdot 3^3$. The difference shown in the parentheses is divisible by

$$27{,}195 - 10{,}887 = 16{,}308 = 2^2 \cdot 3^3 \cdot 151 \ .$$

Thus, N is divisible by $2^2 \cdot 3^3$.

Since N is divisible by $5 \cdot 7^2$ and by $2^2 \cdot 3^3$, it follows that N is divisible by the product of these (relatively prime) numbers, and this product is $26{,}460$.

40. It is readily verified that

$$11^{10} - 1^{10} = (11 - 1)(11^9 + 11^8 + 11^7 + \cdots + 11^2 + 11 + 1) \ .$$

The second factor of the right number is divisible by 10, since it is the sum of ten integers each ending with the digit 1. Inasmuch as both factors on the right are divisible by 10, their product is divisible by 100. Therefore, $11^{10} - 1$ is divisible by 100. Therefore, $11^{10} - 1$ is divisible by 100.

41. We have

$$2222^{5555} + 5555^{2222} = (2222^{555} + 4^{555})$$
$$+ (5555^{2222} - 4^{2222}) - (4^{5555} - 4^{2222}) \ .$$

Consider the three terms enclosed by parentheses. The first is divisible by $2222 + 4 = 2226 = 7 \cdot 318$ (since $a^n + b^n$ is divisible by $a+b$ if n is odd), and so this term is divisible by 7. The second term is also divisible by 7, since it is divisible by $5555 - 4 = 5551 = 7 \cdot 793$ ($a^n - b^n$ is always divisible by $a - b$). The third term may be written

$$4^{2222}(4^{3333} - 1) = 4^{2222}(64^{1111} - 1) \ ;$$

clearly it is divisible by $64 - 1 = 63$, and hence by 7.

42. We use mathematical induction. The number $\overline{aaa}$, consisting of three identical digits (the overbar indicating, as before, that the integer is given by the succession of digits shown), is divisible by 3 (since the sum of these digits is $3a$, which is divisible by 3). Assume that the proposition has been proved for any integer consisting of 3^n identical digits. The expanded integer consisting of 3^{n+1} identical digits can be written in the following form:

$$\underbrace{aa \cdots a}_{3^n \text{ times}} \underbrace{aa \cdots a}_{3^n \text{ times}} \underbrace{aa \cdots a}_{3^n \text{ times}} = \underbrace{aa \cdots a}_{3^n \text{ times}} \cdot 1\underbrace{00 \cdots 0}_{3^n \text{ digits}}1\underbrace{00 \cdots 0}_{3^n \text{ digits}}1 .$$

There are two factors on the right. The first factor is divisible by 3^n, according to the induction hypothesis. The second factor is divisible by 3 (the sum of its digits being 3). Therefore, the product is divisible by 3^{n+1}.

43. First note that $10^6 - 1 = 999\ 999$ is divisible by 7 (in fact, $999\ 999 = 7 \cdot 142{,}857$). It follows that 10^N (for any integer N) yields upon division by 7 the same remainder as does 10^r, where r is the remainder obtained by division of N by 6, since if $N = 6k + r$, then

$$10^N - 10^r = 10^{6k+r} - 10^r = 10^r(10^{6k} - 1) ,$$

and since $10^{6k} - 1 = (10^6)^k - 1$ is divisible by $10^6 - 1$, which in turn is divisible by 7, then $10^N - 10^r$ is divisible by 7. This means that 10^N and 10^r yield the same remainder upon division by 7.[†]

Now, it is readily verified that every integral power of 10 yields a remainder of 4 upon division by 6 (that is, $10 \equiv 4 \bmod 6$). The exponents of each term of the sum given in the problem are all powers of 10, hence each exponent is congruent to 4 modulo 6. This means that we can replace each of the ten terms by 10^4 in order to find the remainder upon division by 7. We have

$$\underbrace{10^4 + 10^4 + \cdots + 10^4}_{10 \text{ terms}} = 10^5 \, 1{,}000{,}000 = 7 \cdot 14{,}285 + 5 .$$

Therefore, the remainder is 5.

44. (a) Any even power of 9 may be expressed in the form

$$9^{2n} = 81^n = \underbrace{81 \cdot 81 \cdot \cdots \cdot 81}_{n \text{ times}}$$

[†] In more familiar terminology, 10^N is *congruent to* 10^r, *modulo* 7, or $10^N = 10^r \pmod 7$ [*Editor*].

and therefore ends with the digit 1. Any odd power of 9 can be written as $9^{2n+1} = 9 \cdot 81^n$ and therefore ends with the digit 9. Since $9^{(9^9)}$ is an odd power of 9, it must end with the digit 9.

It is obvious that 16^n ends with the digit 6 for all $n \geqq 1$. Hence any power of 2 whose exponent is a multiple of 4 (that is, 2^{4n}) ends with 6, since $2^{4n} = 16^n$. Now, $3^4 - 1$ is divisible by $3 + 1 = 4$, and so $2^{(3^4)} = 2 \cdot 2^{(3^4-1)}$, which is the product of 2 and an integer ending with 6, must end with 2.

(b) If we find the remainder yielded upon division of 2^{999} by 100, it will be the number formed by the two final digits of 2^{999}. We first show, that the number 2^{1000} yields the remainder 1 upon division by 25. In fact, $2^{10} + 1 = 1024 + 1 = 1025$ is divisible by 25; and so $2^{20}-1 = (2^{10} + 1)(2^{10}-1)$ is divisible by 25. Thus, $2^{1000}-1=(2^{20})^{50}-1$, being divisible by $2^{20} - 1$, is also divisible by 25, and so upon division by 25 the number 2^{1000} yields the remainder 1.

It follows that the final two digits of 2^{1000} can be only 01, or $01 + 25 = 26$, or $01 + 50 = 51$, or $01 + 75 = 76$. Since 2^{1000} is divisible by 4, the only possibility among these four numbers is 76. Thus, 2^{999} is the quotient obtained by dividing an integer ending in 76 by 2. The only possibilities are 38 and 88. Since 2^{999} is divisible by 4, there remains the one possibility, 88, for the final two digits.

As above, we investigate the remainder obtained upon dividing 3^{999} by 100. We recall that every even power of 9 ends with the digit 1 and that every odd power ends with the digit 9 [see the solution to problem (a)]. Now consider the remainder obtained upon division of $9^5 + 1$ by 100. We have

$$9^5 + 1 = (9 + 1) \cdot (9^4 - 9^3 + 9^2 - 9 + 1)$$
$$= 10 \cdot (9^4 - 9^3 + 9^2 - 9 + 1) \,.$$

The numbers 9^4, 9^2, and 1 all end with the digit 1, and the numbers 9^3 and 9 end with 9. Thus, $9^4 + 9^2 + 1$ ends with 3, and $9^3 + 9$ ends with 8, which means that the number $9^4 - 9^3 + 9^2 - 9 + 1$ must end with 5. Accordingly, $9^5 + 1$ must yield upon division by 100 the remainder $10 \cdot 5 = 50$. It follows that $9^{10} - 1 = (9^5 + 1) \cdot (9^5 - 1)$ is divisible by 100, and since $3^{1000} - 1 = 9^{500} - 1 = (9^{10})^{50} - 1$ is divisible by $9^{10} - 1$, it follows that $3^{1000} - 1$ is divisible by 100. Thus, 3^{1000} ends with the digits 01. But this number is, of course, divisible by 3; consequently, the carry-over from the hundreds place of 3^{1000} to the tens place must be 2 (if it were 0 or 1, then 3^{1000} would not be divisible by 3). Therefore, the number $3^{999} = 3^{1000}/3$ must end with the same two digits as the number $201/3 = 67$.

(c) We must find the remainder, upon division by 100, of the number $14^{(14^{14})} = (7 \cdot 2)^{(14^{14})}$. First we find the remainders, upon division by 100, of $7^{(14^{14})}$ and $2^{(14^{14})}$.

The number $7^4 - 1 = 2401 - 1 = 2400$ is divisible by 100. It follows that if $n = 4k$ (k being an integer), then $7^n - 1$ is divisible by 100 (since $7^{4k} - 1 = (7^4)^k - 1^k$ is divisible by $7^4 - 1$). Now, $14^{14} = 2^{14} \cdot 7^{14}$ is divisible by 4; consequently, $7^{(14^{14})} - 1$ is divisible by 100, which means that $7^{(14^{14})}$ ends with the digits 01.

In the solution of part (b) it was shown that $2^{20} - 1$ is divisible by 25; hence, if $n = 20k$, then $2^n - 1$ is divisible by 25. We will now find the remainder obtained from division of 14^{14} by 20. Clearly, $14^{14} = 2^{14} \cdot 7^{14}$. But $2^{14} = 4 \cdot 2^{12}$. Since $2^{12} - 1 = (2^4)^3 - 1$ is divisible by $2^4 - 1 = 16 - 1 = 15$, it follows that $4(2^{12} - 1)$ is divisible by 20, and consequently $2^{14} = 4 \cdot 2^{12}$ yields upon division by 20 a remainder of 4. Further, $7^{14} = 49 \cdot 7^{12}$. Since 7^{12} yields a remainder of 1 upon division by 20 (12 is divisible by 4, whence $7^{12} - 1$ is divisible by 100), it follows that $49 \cdot 7^{12}$ yields upon division by 20 the same remainder as does 49, that is, 9. Similarly, $14^{14} = 2^{14} \cdot 7^{14}$ yields upon division by 20 the same remainder as does the product $4 \cdot 9 = 36$, that is, 16; or $14^{14} = 20k + 16$. It follows that $2^{(14^{14})} = 2^{16} \cdot 2^{20k}$ yields upon division by 25 the same remainder as does $2^{16} = 65,536$; that is, $2^{(14^{14})}$ can end only with one of the numbers 11, 36, 61, or 86. Since $2^{(14^{14})}$ is divisible by 4, the final two digits must be 36.

Therefore, since $7^{(14^{14})}$ ends with the digits 01, and $2^{(14^{14})}$ has as its two final digits 36, the product $7^{(14^{14})} \cdot 2^{(14^{14})} = 14^{(14^{14})}$ ends with 36.

45. (a) We make use of the fact that the product of two numbers ending respectively with the digits a and b will have the same final digit as does the product $a \cdot b$. This provides a simple solution for the problem. We consider successively greater powers of 7, keeping track of the final digit only: 7^2 ends with the digit 9; $7^3 = 7 \cdot 7^2$ with the digit 3; $7^4 = 7 \cdot 7^3$ with the digit 1; and $7^7 = 7^4 \cdot 7^3$ with the digit 3.

Moreover, we find that $(7^7)^7$ ends with the digit 7 [$(7^7)^2$ ends with 9; $(7^7)^3$ ends with 7; $(7^7)^4$ ends with 1, and, finally, $(7^7)^7$ ends with 7]. We find, at the next stage, that the number $((7^7)^7)^7$ ends with the same digit as does 7^7 (the digit 3), and the number $(((7^7)^7)^7)^7$ ends again with the digit 7, and so on. Continuation of this process must then yield the following rule. For an odd number of exponents 7 we obtain a final digit 3, and for an even number of exponents we obtain a final digit 7. Since 1000 is an even integer, the number sought has 7 as its final digit.

If an integer ends with a two-digit number A, and another integer ends with a two-digit number B, then the product of the two integers ends with the same two-dight number as does the product $A \cdot B$. This fact allows us to find the final two digits for the number mentioned by the problem. We easily verify, by the methods used above, that 7^7 ends with the two-digit number 43, and $(7^7)^7$ ends with the same two digits as does 43^7, namely 07. It follows that in taking successive 7th powers, $7, 7^7, (7^7)^7, \cdots$, we obtain for an odd number of such "raises" a number with final digits 43, and for an even number a number with final digits 07. Therefore, the number in which we are interested must end with 07.

    (b) In the solution of problem (a) we saw that 7^4 ends with the digit 1. Therefore, $7^{4k} = (7^4)^k$ also ends with the dight 1, and 7^{4k+r}, where k is one of the numbers $0, 1, 2$, or 3, ends with the same digit as does 7^r (since $7^{4k+r} = 7^{4k} \cdot 7^r$). Thus, the problem reduces to finding the remainder, modulo 4 (that is, after division by 4), of the "exponential part" of the given number.

The power to which 7 is raised is again a power of 7. We must determine the remainder obtained by dividing the latter power by 4. Now $7 = 8 - 1$, and it follows that: $7^2 = (8 - 1) \cdot (8 - 1)$ yields upon division by 4 the remainder 1; $7^3 = 7^2 \cdot (8 - 1)$ yields upon division by 4 the remainder -1 (equivalent, upon division by 4, to the remainder 3); and, in general, every even power of 7 yields upon division by 4 the remainder 1, and odd powers yield the remainder -1 (that is, $+3$). For the number in question, we are concerned with an odd power of 7, since the exponential part is itself a power of 7, and, consequently, owing to the conditions of the problem, it is of form 7^{4k+3}. Therefore, it ends with the same digit as does 7^3, that is, with the digit 3.

Since 7^4 ends with 01, 7^{4k+r} ends with the same two-digit number as does 7^r. Therefore, the given number ends with the same two digits as does 7^3, that is, 43.

46. Consider the following five numbers:

(1) $Z_1 = 9$.

(2) $Z_2 = 9^{Z_1} = (10 - 1)^{Z_1}$

                 $= 10^{Z_1} - C^1_{Z_1} \cdot 10^{Z_1 - 1} + \cdots + C^1_{Z_1} \cdot 10 - 1$

(where the integers not explicitly displayed are obviously divisible by 100). Now, $C^1_{Z_1} = 9$, and hence the two final digits of Z_2 are the same as the final two digits of $9 \cdot 10 - 1 = 89$.

(3) $Z_3 = 9^{Z_2} = (10-1)^{Z_2}$
$$= 10^{Z_2} - C_{Z_2}^1 \cdot 10^{Z_2-1} + \cdots - C_{Z_2}^2 \cdot 10^2 + C_{Z_2}^1 \cdot 10 - 1 \ .$$

Now, Z_2 ends with 89; consequently, $C_{Z_2}^1 = Z_2$ ends with 89, and

$$C_{Z_2} = \frac{Z_2(Z_2-1)}{1 \cdot 2} = \frac{\overline{\cdots 89} \cdot \overline{\cdots 88}}{1 \cdot 2}$$

(the dots designating unknown digits) ends with the digit 6. Accordingly, the final three digits of the number Z_3 will be the same as the final three digits of the number $-600 + 890 - 1 = 289$.

(4) $Z_4 = 9^{Z_3} = (10-1)^{Z_3}$
$$= 10^{Z_3} - C_{Z_3}^1 \cdot 10^{Z_3-1} + \cdots + C_{Z_3}^3 \cdot 10^3 - C_{Z_3}^2 \cdot 10^2 + C_{Z_3}^1 \cdot 10 - 1 \ .$$

Since Z_3 ends with 289, $C_{Z_3}^1 = Z_3$ ends with 289;

$$C_{Z_3}^2 = \frac{Z_3(Z_3-1)}{1 \cdot 2} = \frac{\cdots 289 \cdot \cdots 288}{1 \cdot 2}$$

ends with 16, and

$$C_{Z_3}^3 = \frac{Z_3(Z_3-1)(Z_3-2)}{1 \cdot 2 \cdot 3} = \frac{\cdots 289 \cdot \cdots 288 \cdot \cdots 287}{1 \cdot 2 \cdot 3}$$

ends with the digit 4. Hence, the final four digits of Z_4 will form the same number as do the final four digits of the number $4000 - 1600 + 2890 - 1 = 5289$.

(5) $Z_5 = 9^{Z_4} = (10-1)^{Z_4} = 10^{Z_4} - C_{Z_4}^1 \cdot 10^{Z_4-1} + \cdots$
$$- C_{Z_4}^4 \cdot 10^4 + C_{Z_4}^3 \cdot 10^3 - C_{Z_4}^2 \cdot 10^2 + C_{Z_4}^1 \cdot 10 - 1 \ .$$

Since Z_4 ends with 5289, $C_{Z_4}^1 = Z_4$ ends with 5289;

$$C_{Z_4}^2 = \frac{Z_4(Z_4-1)}{1 \cdot 2} = \frac{\cdots 5289 \cdot \cdots 5288}{1 \cdot 2}$$

ends with 116;

$$C_{Z_4}^3 = \frac{Z_4(Z_4-1)(Z_4-2)}{1 \cdot 2 \cdot 3} = \frac{\cdots 5289 \cdot \cdots 5288 \cdot \cdots 5287}{1 \cdot 2 \cdot 3}$$

ends with 64; and

$$C_{Z_4}^4 = \frac{Z_4(Z_4-1)(Z_4-2)(Z_4-3)}{1 \cdot 2 \cdot 3 \cdot 4} = \frac{\cdots 5289 \cdot \cdots 5288 \cdot 5287 \cdot \cdots 5286}{1 \cdot 2 \cdot 3 \cdot 4}$$

ends with the digit 6. Therefore, Z_5 ends with the same digit as does the number

$$-60{,}000 + 64{,}000 - 11{,}600 + 52{,}890 - 1 = 45{,}289 \ .$$

Further, since the final four digits of the number Z_5 coincide with the final four digits of Z_4, it follows that the final five digits of the number $Z_6 = 9^{Z_5} = (10 - 1)^{Z_5}$ coincide with the final five digits of the number Z_5 $(= 9^{Z_4})$. It can be shown, in exactly the same way, that all the numbers of the sequence

$$Z_5;\ Z_6 = 9^{Z_5};\ Z_7 = 9^{Z_6};\ \cdots;\ Z_{1000} = 9^{Z_{999}};\ Z_{1001} = 9^{Z_{1000}}$$

end with the same four digits, namely the digits forming the number 45,289. Thus, Z_{1001} is the number N called for by our problem.

47. Using the formula for the sum of a geometric progression, we find

$$N = \frac{50^{1000} - 1}{50 - 1} = \frac{50^{1000} - 1}{49}.$$

Now 1/49 forms a periodic decimal; it is found (by tedious but straightforward division) to have a period of 42 digits:

$$\frac{1}{49} = 0.(020408163265306122448979591836734693877551),$$

or, in abbreviated form

$$\frac{1}{49} = 0 \cdot P,$$

where P expresses the 42 digits written above.

The multiple of 42 nearest 1000 is $1008 = 24 \cdot 42$. Consequently,

$$\frac{10^{1008}}{49} = 10^{1008} \cdot \frac{1}{49} = \underbrace{PP \cdots P}_{24 \text{ times}}.$$

Similarly,

$$M = \frac{10^{1008} - 1}{49} = 10^{1008} \cdot \frac{1}{49} - \frac{1}{49} = \underbrace{PP \cdots P}_{24 \text{ times}}$$

is an integer consisting of 1008 digits, which can be arranged in 24 repeating groups of 42 digits each (the number M consists not of 1008 digits but of 1007, since the number P begins with a zero).

We construct the difference between the number N in which we are interested and the number M:

$$N - M = \frac{5^{1000} \cdot 10^{1000} - 1}{49} - \frac{10^{1008} - 1}{49} = \frac{5^{1000} - 10^8}{49} \cdot 10^{1000}.$$

Since the difference $N - M$ of two whole numbers is an integer, and 10^{1000} is relatively prime to 49, it follows that $5^{1000} - 10^8$ must be divisible by 49. Hence, the number $x = \dfrac{5^{1000} - 10^8}{49}$ is an integer, and the difference $N - M = 10^{1000} \cdot x$ terminates with 1000 zeros. Hence the final 1000 digits of N coincide with those of M, namely

$$q\underbrace{PP \cdots P}_{23 \text{ times}},$$

where q is a group of 34 digits consisting of the last 34 digits of the number P.

48. The number of zeros at the end of a number indicates how many times the number 10 enters as a factor. Now, $10 = 2 \cdot 5$. In the product of all the integers from 1 to 100 (that is, in 100!) the factor 2 enters to a higher power than does 5. Hence 100! is divisible by 10 as many times as the factor 5 appears (and will terminate in this many zeros). Up to and including 100, there are 20 integers which are multiples of 5. Four of these (25, 50, 75, and 100) are also multiples of 25, that is each contains 5 twice as a factor. Therefore, in the number 100! the factor 5 is encountered 24 times, and so there will be 24 zeros at the end of this integer.

49. *First solution of problems (a) and (b).*

(a). Let $t + 1, t + 2, \cdots, t + n$ be n consecutive integers for some arbitrary integer t. We first determine, for a prime number p, to what degree p is a factor in $n!$, and to what degree this prime enters as a factor in the product $(t + 1) \cdots (t + n)$. Designate by m_1 the number of integers in the sequence $1, 2, \cdots, n$ for which p is at least a simple factor, by m_2 the number of integers for which p is at least a twofold factor, and so on. Then the degree to which p enters as a factor in $n!$ is given by $m = m_1 + m_2 + \cdots$.

If s_1 is the number of integers of the sequence $t + 1, t + 2, \cdots, t + n$ which are divisible by p_1 and s_2 is the number of integers in this sequence which are divisible by p^2, and so on, then the degree s to which p enters as a factor of $(t + 1) \cdots (t + a)$ will be $s = s_1 + s_2 + \cdots$.

Now, the number of integers in the sequence $t + 1, \cdots, t + n$ which are divisible by p is not less than m_1, since among the integers $t + 1, \cdots, t + n$ are the numbers $t + p, t + 2p, \cdots, t + m_1 p$, and in each interval between $t + kp$ to $t + (k + 1)p$ $(k = 0, 1, 2, \cdots, m_1 - 1)$ there is at least one integer which is divisible by p. Thus, $s_1 \geqq m_1$, and, analogously, $s_2 \geqq m_2$, and so on, and so $s \geqq m$. This means that every prime factor of $n!$ enters as a factor of $(t + 1) \cdots (t + n)$ to a degree not less than it enters as a factor of $n!$. That is, the number $(t + 1) \cdots (t + n)$ is divisible by $n!$

(b)   The product of the first a factors of $n!$ is, of course, $a!$. The product of the following six consecutive integers of $n!$ is, according to problem (a), divisible by $b!$. The product of the next successive c consecutive integers of $n!$ is divisible by $c!$, and so on. Since $a + b + \cdots + k \leqq n$, it follows that $n!$ is divisible by $a! b! \cdots k!$.

Alternate solution of problems (a) and (b) using the result of problem 101. Problem (b) will be considered first. The power m to which the prime p is a factor of $a!$ is, as we have seen, equal to $m = m_1 + m_2 + \cdots$, where m_1 is the number of integers of the sequence $1, 2, \cdots, a$ of which p is at least a simple factor; m_2 is the number of integers of which p^2 is a factor; and so on. The number of integers in this sequence which are multiples of p is given by $\left[\dfrac{a}{p} \right]$; the the number of integers which are multiples of p^2 is given by $\left[\dfrac{a}{p^2} \right]$; and so on, where $\left[\dfrac{a}{p} \right], \left[\dfrac{a}{p^2} \right], \cdots$ are the greatest integers in $\dfrac{a}{p}$, $\dfrac{a}{p^2}, \cdots$ (see the remark just prior to the statement of problem 101). Thus, $m = \left[\dfrac{a}{p} \right] + \left[\dfrac{a}{p^2} \right] + \cdots$. Let p be a prime number; then the degree to which p enters as a factor of the numerator is equal to $\left[\dfrac{n}{p} \right] + \left[\dfrac{n}{p^2} \right] + \cdots$. The degree to which p is a factor of the denominator is

$$\left[\frac{a}{p} \right] + \left[\frac{a}{p^2} \right] + \cdots + \left[\frac{b}{p} \right] + \left[\frac{b}{p^2} \right] + \cdots + \left[\frac{k}{p} \right] + \left[\frac{k}{p^2} \right] + \cdots .$$

Since $n \geqq a + b + \cdots + k$, we have (using the result of problem 101, part 1)

$$\left[\frac{n}{p} \right] + \left[\frac{n}{p^2} \right] + \cdots$$
$$\geqq \left(\left[\frac{a}{p} \right] + \left[\frac{b}{p} \right] + \cdots \right) + \left(\left[\frac{a}{p^2} \right] + \left[\frac{b}{p^2} \right] + \cdots \right) + \cdots ,$$

that is, p enters the numerator as a factor to a higher degree than it enters the denominator and so the given fraction is an integer.

The proposition of problem (a) immediately follows. Consider the product $(t + a)!$. According to what has just been proven,

$$\frac{(t + a) \cdots (t + 1)t(t - 1) \cdots 1}{a!t(t - 1) \cdots 1} = \frac{(a + t)!}{a!t!} = \frac{(t + 1) \cdots (t + a)}{a!}$$

is an integer.

(c) $(n!)!$ is the product of the first $n!$ integers. These $n!$ integers can be written as the product of $(n - 1)!$ product sets each containing $n!$ successive integers. Each of these sets is, according to the solution of problem (a), divisible by $n!$.

(d) Designate the integers by $a, a + d, a + 2d, \cdots, a + (n - 1)d$. We first show that there exists an integer k such that the product kd yields a remainder 1 when divided by $n!$. Consider the $(n! - 1)$ numbers $d, 2d, 3d, \cdots, (n! - 1)d$. None of these numbers is divisible by $n!$ (since d and $n!$ are, by hypothesis, relatively prime). Further, no two products pd and qd, where p and q are distinct integers less than $n!$, can yield the same remainder upon division by $n!$ (otherwise $pd - qd = (p - q)d$ would be divisible by $n!$). Hence the $n! - 1$ integers all yield different remainders upon division by $n!$, and so, for some k, the remainder 1 appears upon division by $n!$, that is, $kd = r \cdot n! + 1$.

If now we designate ka by A, we have

$$ka = A$$
$$k(a + d) = A + kd = (A + 1) + r \cdot n! \, ,$$
$$k(a + 2d) = A + 2kd = (A + 2) + 2r \cdot n! \, ,$$
$$\cdots\cdots\cdots\cdots\cdots\cdots\cdots\cdots\cdots\cdots\cdots\cdots\cdots\cdots\cdots , $$
$$k[a + (n - 1)d] = A + (n - 1)kd = [A + (n - 1)] + (n - 1)r \cdot n! \, .$$

It follows that

$$k^n(a + d)(a + 2d) \cdots [a + (n - 1)d]$$

gives the same remainder upon division by $n!$ as does

$$A(A + 1)(A + 2) \cdots [A + (n - 1)] \, .$$

The latter product is divisible by $n!$, in view of the result of problem (a); also k^n is relatively prime to $n!$ (since if k is not relatively prime to $n!$, then neither is kd). Therefore, $n!$ divides $a(a + d)(a + 2d) \cdots [a + (n - 1)d]$.

50. The number of combinations of 1000 elements taken 500 at a time is given by

$$\frac{1000!}{(500!)^2} \cdot$$

Since 7 is a prime number, the highest power of 7 which is a factor of 1000! [see the second solution of problem 49 (b)] is equal to

$$\left[\frac{1000}{7}\right] + \left[\frac{1000}{49}\right] + \left[\frac{1000}{343}\right] = 142 + 20 + 2 = 164 .$$

The highest power of 7 in 500! is equal to

$$\left[\frac{500}{7}\right] + \left[\frac{500}{49}\right] + \left[\frac{500}{343}\right] = 71 + 10 + 1 = 82 ,$$

and so the degree to which 7 enters the denominator of the fraction is $82 \cdot 2 = 164$. Thus, both numerator and denominator contain the factor 7 exactly the same number of times. When 7^{164} is cancelled out of numerator and denominator, no multiple of 7 remains in the resulting number. Therefore, the integer represented by C_{1000}^{500} is not divisible by 7.

51. (a) It is readily seen that every prime number satisfies the given condition, since p does not appear as a factor in $(p - 1)!$. If n is a composite number which can be written as the product of two unequal factors, a and b, then both a and b are less than $n - 1$, and consequently, both appear in the composition of $(n - 1)!$. This means that $(n - 1)!$ is divisible by $ab = n$. If n is the square of a prime $p > 2$, then $n - 1 = p^2 - 1 > 2p$, which implies that both p and $2p$ enter into the product composition of $(n - 1)!$. Hence, $(n - 1)!$ is divisible by $p \cdot 2p = 2p^2 = 2n$. Thus, all the composite numbers except $2^2 = 4$ may be eliminated. However, 4 satisfies the condition of the problem, as well as do all the prime numbers less than 100:

$$2, 3, 4, 7, 11, 13, 17, 19, 23, 29, 31, 37, 41,$$
$$43, 47, 53, 59, 61, 67, 71, 73, 79, 83, 89, 97 .$$

(b) It will be shown that $(n - 1)!$ fails to be divisible by n^2 in the following cases only: n is prime, n is twice a prime, n is the square of a prime, $n = 8$, $n = 9$.

If n is neither a prime number, nor twice a prime, nor the square of a prime, nor the numbers 8 or 16, then n may be written as a product $a \cdot b$, where a and b are distinct numbers, neither of which is less than 3. Assume $b > a \geq 3$. Then the numbers $a, b, 2a, 2b, 3a$ are all less than $n - 1$; moreover, $a, b,$ and $2b$ are clearly distinct from each other, and at least one of the numbers $2a$ or $3a$ differs from $a, b,$ and $2b$. Hence, in $(n - 1)!$ there appear the distinct factors $a, b, 2b, 2a,$ or else $a, b, 2b, 3a$ (and possibly all of $a, b, 2a, 2b,$ and $3a$

appear separately as factors of that product). In every case, $(n-1)!$ is divisible by $a^2b^2 = n^2$.

Moreover, if $n = p^2$, where $p > 4$ is prime, then $n - 1 > 4p$, and $(n-1)!$ contains as factors all of the numbers $p, 2p, 3p, 4p$ and hence is divisible by $p^4 = n^2$. If $n = 2p$, then $(n-1)!$ is not divisible by p^2, and so it is not divisible by n^2; when $n = 8$ or $n = 9$, $(n-1)!$ is not divisible by n^2 (7! is not divisible by 8^2, nor is 8! divisible by 9^2); if $n = 16$, $(n-1)!$ is divisible by n^2 (since 15! contains as factors the numbers $2, 4 = 2^2, 6 = 3 \cdot 2, 8 = 2^3, 10 = 2 \cdot 5, 12 = 2^2 \cdot 3, 14 = 2 \cdot 7$, and so is divisible by $2^{1+2+1+3+1+2+1} = 2^{11} = 16^2 \cdot 2^3$).

Thus, the condition of problem (b) is satisfied by all the numbers which satisfy the condition of problem (a), and in addition by the integers $6, 8, 9, 10, 14, 22, 26, 34, 38, 46, 58, 62, 74, 82, 86, 94$, that is, all primes, doubles of primes, and the integers 8 and 9.

52. Assume that the integer n is divisible by all numbers $m \leq \sqrt{n}$, and consider the least common multiple K of all these numbers m. Of course, all prime numbers $p \leq \sqrt{n}$ are included, as well as powers up to $p^k \leq \sqrt{n}$, but $p^{k+1} > \sqrt{n}$. Assume that there are l primes which are less than $\sqrt{n}$, and designate them by $p_1, p_2, \cdots, p_l$. The least common multiple of all the integers less than $\sqrt{n}$ will be the product $K = p_1^{k_1} p_2^{k_2} \cdots p_l^{p_l}$, where k_i is the integer such that

$$p_i^{k_i} \leq \sqrt{n} < p_i^{k_i+1} \qquad (i = 1, 2, \cdots, l).$$

From the l inequalites

$$\sqrt{n} < p_1^{k_1+1},$$
$$\sqrt{n} < p_2^{k_2+2},$$
$$\cdots\cdots\cdots\cdots,$$
$$\sqrt{n} < p_l^{k_l+1}$$

we obtain

$$(\sqrt{n})^l < p_1^{k_1+1} \cdot p_2^{k_2+1} \cdot \cdots \cdot p_l^{k_l+1}.$$

However,

$$p_1^{k_1+1} p_2^{k_2+1} \cdots p_l^{k_l+1} = p_1^{k_1} p_2^{k_2} \cdots p_l^{k_l} \cdot p_1 p_2 \cdots p_l \leq K^2$$

(since $p_1^{k_1} p_2^{k_2} \cdots p_l^{k_l} = K$), and, consequently, $p_1 p_2 \cdots p_l \leq K$. Hence, we have

$$(\sqrt{n})^l < K^2.$$

Inasmuch as n is divisible by K, we must have $K \leqq n$, whence $(\sqrt{n})^l < n^2$. Therefore, $l < 4$. Since $p_1, p_2, \cdots p_l$, are primes less than $\sqrt{n}$, $p_4 = 7 > \sqrt{n}$ (the fourth prime number is 7), and so $n < 49$.

If we examine the integers less than 49, we readily ascertain that only the following integers satisfy the conditions of the problem:

$$24, 12, 8, 6, 4, 3, 2 .$$

53. (a) Designate five consecutive integers by

$$n - 2, n - 1, n, n + 1, n + 2 .$$

Then

$$(n - 2)^2 + (n - 1)^2 + n^2 + (n + 1)^2 + (n + 2)^2 = 5n^2 + 10 = 5(n^2 + 2) .$$

If $5(n^2 + 2)$ is a perfect square, then it mush be divisible by 25 (it has the prime factor 5, which must appear twice), hence $(n^2 + 2)$ must be divisible by 5. This is possible only if the final digit of n^2 is 8 or 3, and no square of an integer ends in either of these digits.

(b) Of three consecutive integers, one is divisible by 3, another yields a remainder of 1 upon division by 3, and the third yields remainder of 2 (or, equivalently, a remainder of -1). Upon multiplication of two such integers, the remainders obtained from division by any number are also multiplied; actually,

$$(pk + r)(qk + s) = pqk^2 + pks + qkr + rs = k(pqk + ps + qr) + rs .$$

Hence, if a number yields the remainder 1 upon division by 3, then all of its powers yield the remainder 1 upon division by 3; if the remainder is -1, then all of the odd powers of the number will yield the remainder -1, and the even powers will yield the remainder 1.

'Thus, given three even powers of consecutive integers, we have, upon division by 3, the remainder 0 for one power and the remainder 1 for the other two.

Therefore, the sum of even powers of three consective integers yields the remainder 2 when divided by 3 (or, equivalently, the remainder -1); but no even power of any integer, as we have just shown, can yield this remainder upon division by 3.

Remark: The even powers referred to in problem (b) need not be the same. In problem (c) note that the even powers referred to are all the same.

(c) As we saw in the solution of problem (a), the sum of even powers of three consecutive integers yields a remainder of 2 when divided by 3. It follows that the sum of even powers of nine consecutive numbers yield a "remainder" of 6 upon division by 3, that is, this sum is divisible by 3. We must show that such a sum (wherein the even powers are the same) is not divisible by $3^2 = 9$.

Of nine consecutive integers, one is divisible by 9, and the others yield remainders from 1 to 8. If $2k$ is the (even) power to which the nine consecutive integers are raised, then the sum yields the same remainder upon division by 9 as does

$$0 + 1^{2k} + 2^{2k} + 3^{2k} + 4^{2k} + 5^{2k} + 6^{2k} + 7^{2k} + 8^{2k}$$

or the sum

$$2(1^k + 4^k + 7^k)$$

(since 3^2 and 6^2 are divisible by 9; 1^2 and $8^2 = 64$ each yield a remainder 1; $2^2 = 4$ and $7^2 = 49$ yield remainders of 4; $4^2 = 16$ and $5^2 = 25$ yield remainders of 7).

Now note that $1^3 = 1$, $4^3 = 64$, and $7^3 = 343$ all yield the remainder 1 upon division by 9. It follows that if $k = 3l$, then $1^k + 4^k + 7^k = 1^l + 64^l + 343^l$ yields the same remainder upon division by 9 as does $1^l + 1^l + 1^l = 3$; it is not divisible by 9. If $k = 3l + 1$, then $1^k + 4^k + 7^k = 1^l \cdot 1 + 64^l \cdot 4 + 343^l \cdot 7$ yields the same remainder as does the sum $1 \cdot 1 + 1 \cdot 4 + 1 \cdot 7 = 12$; it is not divisible by 9. If $k = 3l + 2$, then $1^k + 4^k + 7^k = 1^l \cdot 1 + 64^l \cdot 4^2 + 343^l \cdot 7^2$ yields the same remainder upon division by 9 as does the sum $1 \cdot 1 + 1 \cdot 16 + 1 \cdot 49 = 66$; it is not divisible by 9.

54. (a) The sum of the digits of each number is $1 + 2 + 3 + 4 + 5 + 6 + 7 = 28$. It follows that both numbers yield a remainder of 1 upon division by 9 (an integer yields the same remainder upon division by 9 as does the sum of its digits). But if $A/B = n$, or $A = nB$, where n is an integer different from 1, then, since $B = 9N + 1$, it follows from $A = nB = 9M + n$ that n must yield a remainder of 1 upon division by 9. The least value which n can assume is 10. However, $A/B < 10$, inasmuch as A and B are both seven-digit numbers.

(b) Designate the integers sought by $N, 2N, 3N$. Since an integer yields the same remainder upon division by 9 as does the sum of its digits, the sum $N + 2N + 3N$ must yield the same remainder upon division by 9 as does $1 + 2 + 3 + \cdots + 9 = 45$ in order to meet the condition imposed by the problem. Hence, $6N$ (and consequently $3N$) is divisible by 9.

Since $3N$ is to be a three-digit number, the first digit of N cannot exceed 3. It follows that the last digit of N cannot be 1, since the integer $2N$ would end with 2 and $3N$ would end with 3, and then none of these three digits is available to begin N. The integer N cannot terminate with 5, since $2N$ would end with 0. Assume now that the final digit of N is 2; then the final digits of $2N$ and $3N$ are, respectively, 4 and 6. The remaining two digits for $3N$ can be chosen only from 1, 3, 5, 7, 8, and 9. Since the sum of all the digits of $3N$ must be a multiple of 9, the first two digits of $3N$ are either 3 and 9 or 5 and 7. By checking all the possibilities, we find that the following three-digit numbers satisfy the condition of the problem: 192, 384, 576. Analogously, we can investigate the cases for which N terminates with 3, 4, 6, 7, 8, or 9. This procedure will produce three additional solutions: 273, 546, 819; 327, 654, 981; and 219, 438, 657.

55. A perfect square can terminate in only one of the digits 0, 1, 4, 9, 6, or 5. Moreover, the square of an even integer is obviously divisible by 4, and the square of an odd integer yields the remainder 1 upon division by 4 [since $(2k + 1)^2 = 4(k^2 + k) + 1$]. Hence, no square can end with any of the pairs 11, 99, 66, or 55, since numbers ending in the digits 11, 99, 66, or 55 yield upon division by 4 the respective remainders 3, 3, 2, and 3).

We now investigate which remainders are possible when a perfect square is divided by 16. Every integer can be represented in one of the following forms:

$$8k , \qquad 8k \pm 3 ,$$
$$8k \pm 1 , \qquad 8k \pm 4 .$$
$$8k \pm 2 ,$$

The squares of these numbers have the following forms:

$$16(4k^2) , \qquad 16(4k^2 \pm 3k) + 9 ,$$
$$16(4k^2 \pm k) + 1 , \qquad 16(4k^2 \pm 4k + 1) .$$
$$16(4k^2 \pm 2k) + 4 ,$$

These forms show that the square of an integer is either divisible by 16 or will yield a remainder of 1, 4, or 9 when divided by 16. The possibility of ending with 1 or 9 has been excluded. A number ending with the succession of digits 4444 yields a remainder of 12 upon division by 16 and therefore must also be eliminated as a possibility for a perfect square.

Therefore, if a perfect square ends with four identical digits, then these digits must be zeros (for example, $100^2 = 10,000$).

56. *First solution.* We designate the sides of the rectangle by x and y, and the diagonal by z. According to the Pythagorean Theorem,

$$x^2 + y^2 = z^2$$

We are to prove that the product xy is divisible by 12. We shall first show that xy is divisible by 3, then that it is divisible by 4. Since

$$(3k + 1)^2 = 3(3k^2 + 2k) + 1 ,$$

and

$$(3k + 2)^2 = 3(3k^2 + 4k + 1) + 1 ,$$

the square of every integer which is not a multiple of 3 yields a remainder of 1 upon division by 3. Therefore, if neither x nor y is divisible by 3, then the sum $x^2 + y^2$ will yield a remainder of 2 when divided by 3 and thus cannot be the square of any integer. Hence, a necessary condition for $x^2 + y^2$ to be the square of an integer z is that at least one of x or y be a multiple of 3, which in turn means that xy is divisible by 3.

Further, not both x and y can be odd numbers, since if $x = 2m + 1$ and $y = 2n + 1$, then

$$x^2 + y^2 = 4m^2 + 4m + 1 + 4n^2 + 4n + 1$$
$$= 4(m^2 + n + n^2 + n) + 2 ,$$

which cannot be the square of an integer (the square of an odd number is odd, and the square of an even number is divisible by 4). If both x and y are even, then their product is certainly divisible by 4. Assume then that x is even and y is odd. We have $x = 2m$, $y = 2n + 1$. The number z^2 (and hence z) is then odd (the sum of even and odd). If we write $z = 2p + 1$, we have

$$(2m)^2 = (2p + 1)^2 - (2n + 1)^2$$
$$= 4p^2 + 4p + 1 - 4n^2 - 4n - 1 ,$$

or

$$m^2 = p(p + 1) - n(n + 1) .$$

It follows that m^2 is an even number (each term of the above difference is the product of two consecutive integers and so is even). Therefore, since m is even, $x = 2m$ is divisible by 4, and so the product xy is divisible by 4.

Second solution. It follows from the formulas of the solution of problem 128 (a) that the sides x and y of such a rectangle can have lengths expressible as $x = 2tab$, $y = t(a^2 - b^2)$, where t, a, b are any integers for which a and b are relatively prime. [The diagonal length is then the integer $t(a^2 + b^2)$.] If at least one of the integers a or b is even, then x is divisible by 4. If both a and b are odd, then x is divisible by 2 and y is divisible by 2, hence xy is divisible by 4. Further, if either a or b is divisible by 3, then x is divisible by 3; if neither a nor b is divisible by 3, then: one of them yields a remainder of 1 when divided by 3 and the other a remainder of 2, or else both yield the same remainder. In both cases $y = t(a + b)(a - b)$ is divisible by 3. Therefore, in every case the product xy is divisible by 12.

57. We see from the formula giving the roots of a quadratic equation,

$$x = \frac{-b \pm \sqrt{b^2 - 4ac}}{2a} \,,$$

that the roots of the given quadratic equation will be rational if, and only if, the discriminant $b^2 = 4ac$ is a perfect square. Let $b = 2n + 1$, $a = 2p + 1$, $c = 2q + 1$. Then we can write:

$$
\begin{aligned}
b^2 - 4ac &= (2n + 1)^2 - 4(2p + 1)(2q + 1) \\
&= 4n^2 + 4n - 16pq - 8p - 8q - 3 \\
&= 8\left(\frac{n(n + 1)}{2} - 2pq - p - q - 1 \right) + 5 \,.
\end{aligned}
$$

Since this number is odd $\left[\dfrac{n(n + 1)}{2} \right.$ is integral, since one of the factors of the numerator is necessarily even$\left. \right]$, it can be the square of an odd number only. Now, every odd number can be written as $4k \pm 1$, and so the square of an odd number has the form

$$(4k \pm 1)^2 = 16k^2 \pm 8k + 1 = 8(2k^2 \pm k) + 1 \,.$$

That is, the square of an odd number always yield the remainder 1 upon division by 8. Therefore, since $b^2 - 4ac$ is odd, but yields a remainder of 5 upon division by 8, it cannot be a perfect square.

58. We have

$$\frac{1}{n} + \frac{1}{n + 1} + \frac{1}{n + 2} = \frac{3n^2 + 6n + 2}{n(n + 1)(n + 2)} \,.$$

The numerator of this fraction is clearly not divisible by 3, but the denominator is divisible by 3 (being the product of three consecutive integers). Hence, since there is an uncancelled factor in the denominator of the reduced fraction which differs from 2 and 5, the decimal representation certainly is nonterminating. We shall show that the denominator is not relatively prime to 10, hence that the period of the decimal expansion is a deferred one.

Of the two integers n and $n + 1$, one must be even and the other odd. If n is odd, then $3n^2$ is odd, and so the numerator of the fraction is odd; hence it has no factor 2. If n is even, then $n + 2$ is divisible by 2, and the denominator is divisible by 2^2. But the numerator is divisible only by 2, since if $n = 2k$, then

$$3n^2 + 6n + 2 = 12k^2 + 12k + 2 = 2(6k^2 + 6k + 1) \; ;$$

and so the denominator has a factor of 2 not shared with the numerator. Therefore, the denominator of the reduced fraction is not relatively prime to 10, and so its representation as a decimal must have deferred periodicity.

59. (a) and (b). Of the fractions composing the sum

$$M = \frac{1}{2} + \cdots + \frac{1}{m} \left(\text{or } N = \frac{1}{n} + \frac{1}{n + 1} + \cdots + \frac{1}{n + m} \right)$$

we select that one whose denominator contains the highest power of 2 as a factor; there can be only one such term. Now, if we rewrite each term of the sum so as to have as denominator the least common multiple of all the denominators, then each of them, save the selected fraction, will acquire the factor 2 in its numerator, but the selected fraction will acquire only odd factors. Therefore, when the fractions are added in this form, the resulting numerator will be the sum of several even numbers and exactly one odd number, but the (common) denominator will be even. Hence the numerator will be odd and the denominator even, and so the sum cannot be an integer.

(c) Consider that term of the summation whose denominator contains as a factor the highest power (say n) of 3. Since all the denominators are odd, no fraction of form $\frac{1}{2 \cdot 3^n}$ can appear as a term of the sum K. If we obtain the least common multiple of all the denominators, and express all the fractions with this denominator, then each of them, except the selected fraction, will acquire a factor 3 in its numerator, but the numerator of the selected fraction will not have a factor 3. Consequently, we obtain for K a fraction whose denominator is divisible by 3 but whose numerator is not divisible by 3. This cannot be an integer.

60. (a) We consider the sum using denominator $(p-1)!$ For the numerator of the sum we obtain the sum of all possible products of the numbers $1, 2, \cdots, p-1$, taken $p-2$ at a time. Since the denominator $(p-1)!$ of the sum is not divisible by p, we need only show that the sum of all distinct products of $1, 2, \cdots, p-1$, taken $p-2$ at a time, is divisible by p^2.

We designate the sum of all possible products of the numbers $1, 2, \cdots, n$, taken k at a time, by Π_n^k:

$$\Pi_n^1 = 1 + 2 + 3 + \cdots + n,$$
$$\Pi_n^2 = 1\cdot 2 + 1\cdot 3 + \cdots + 1\cdot n + 2\cdot 3 + 2\cdot 4 + \cdots + 2\cdot n$$
$$\qquad + 3\cdot 4 + \cdots + 3\cdot n + 4\cdot 5 + \cdots + (n-1)\cdot n,$$
$$\dotfill,$$
$$\Pi_n^n = 1\cdot 2\cdot 3 \cdots \cdot n = n!$$

We shall show that if $n + 1 = p$ is a prime number, then all the sums

$$\Pi_n^1,\ \Pi_n^2,\ \cdots,\ \Pi_n^{n-1}$$

are divisible by p, and that Π_n^{n-1} itself is divisible by p^2. The assertion of the problem will follow directly from the latter statement.

Consider the polynomial

$$P(x) = (x-1)(x-2)(x-3) \cdots (x-n).$$

If this product is multiplied out, we obtain

$$P(x) = x^n - \Pi_n^1 x^{n-1} + \Pi_n^2 x^{n-2} - \cdots + \Pi_n^n$$

(by our hypothesis, n is even).

Consider, further, the expression $P(x)[x - (n+1)]$. This can be expanded in two ways:

$$P(x)[x-(n+1)] = (x^n - \Pi_n^1 x^{n-1} + \Pi_n^2 x^{n-2} - \cdots + \Pi_n^n)[x-(n+1)],$$

and

$$P(x)[x-(n+1)] = (x-1)(x-2)(x-3) \cdots (x-n)(x-n-1)$$
$$= (x-1)\{[(x-1)-1][(x-1)-2] \cdots [(x-1)-n]\}$$
$$= (x-1)\cdot P(x-1) = (x-1)[(x-1)^n - \Pi_n^1(x-1)^{n-1}$$
$$+ \Pi_n^2(x-1)^{n-2} - \cdots + \Pi_n^n].$$

We then have the equality

$$(x^n - \Pi_n^1 x^{n-1} + \pi_n^2 x^{n-2} - \cdots + \Pi_n^n)[x - (n+1)]$$
$$= (x-1)^{n+1} - \Pi_n^1 (x-1)^n + \Pi_n^2 (x-1)^{n-1} - \cdots + \Pi_n^n (x-1) . \quad (1)$$

If two polynomials are equal for all x, then they are identical—we can equate the coefficients of like powers of x from both sides (designating by $C_n^m = \binom{n}{m}$ the $(m+1)$st binomial coefficient), and obtain the following system of equations:

$$\Pi_n^1 + (n+1) = C_{n+1}^1 + \Pi_n^1 ,$$
$$\Pi_n^2 + (n+1)\Pi_n^1 = C_{n+1}^2 + C_n^1 \Pi_n^1 + \Pi_n^2 ,$$
$$\Pi_n^3 + (n+1)\Pi_n^2 = C_{n+1}^3 + C_n^2 \Pi_n^1 + C_{n-1}^1 \Pi_n^2 + \Pi_n^3 ,$$
$$\cdots\cdots\cdots\cdots\cdots\cdots\cdots\cdots\cdots ,$$
$$\Pi_n^n + (n+1)\Pi_n^{n-1} = C_{n+1}^n + C_n^{n-1} \Pi_n^1 + C_{n-1}^{n-2} \Pi_n^2$$
$$\qquad\qquad\qquad + C_{n-2}^{n-3} \Pi_n^3 + \cdots + C_2^1 \Pi_n^{n-1} + \Pi_n^n ,$$
$$(n+1)\cdot\Pi_n^n = 1 + \Pi_n^1 + \Pi_n^2 + \cdots + \Pi_n^{n-1} + \Pi_n^n .$$

The first of these equalities is obvious. From the second, third, $\cdots$, nth, we derive

$$(n+1) - C_n^1 = 1, (n+1) - C_{n-1}^1 = 2, \cdots, (n+1) - C_2^1 = n-1 ;$$
$$\Pi_n^1 = C_{n+1}^2 ,$$
$$2\Pi_n^2 = C_n^3 + 1 + C_n^2 \Pi_n^1 ,$$
$$3\Pi_n^3 = C_{n+1}^4 + C_n^3 \Pi_n^1 + C_{n-1}^2 \Pi_n^2 ,$$
$$\cdots\cdots\cdots\cdots\cdots\cdots\cdots ,$$
$$(n-1)\Pi_n^{n-1} = C_{n+1}^n + C_n^{n-1} \Pi_n^1 + C_{n-1}^{n-2} \Pi_n^2 + \cdots + C_3^2 \Pi_n^{n-2} . \tag{2}$$

Since, by assumption, $n + 1 = p$ is prime,

$$C_{n+1}^k = C_p^k = \frac{p(p-1)(p-2)\cdots(p-k+1)}{1\cdot 2\cdot 3\cdots k}$$

is, for $k < p$, divisible by p (since the numerator of this fraction is divisible by p and the denominator is not). Therefore we see from the first formula of (2) that Π_n^1 is divisible by p, from the second formula that Π_n^2 is divisible by p, and so on up to divisibility of Π_n^{n-1} by p.

Finally, we substitute $x = p$ into the basic equation

$$(x-1)(x-2)(x-3)\cdots(x-p+1)$$
$$= x^{p-1} - \Pi_{p-1}^1 x^{p-2} + \Pi_{p-1}^2 x^{p-3} - \Pi_{p-1}^3 x^{p-4} + \cdots + \Pi_{p-1}^{p-1} .$$

We obtain

$$(p-1)! = p^{p-1} - \Pi_{p-1}^1 p^{p-2} + \Pi_{p-1}^2 p^{p-3}$$
$$- \Pi_{p-1}^3 p^{p-4} + \cdots + \Pi_{p-1}^{p-3} p^2 - \Pi_{p-1}^{p-2} p + \Pi_{p-1}^{p-1}.$$

But $\Pi_{p-1}^{p-1} = (p-1)!$. If we cancel $(p-1)!$ from both sides, factor out p, transpose Π_{p-1}^{p-2} to the left side, and so on, we obtain

$$\Pi_{p-1}^{p-2} = p(p^{p-3} - \Pi_{p-1}^1 p^{p-4} + \Pi_{p-1}^2 p^{p-5} - \cdots + \Pi_{p-1}^{p-3}),$$

from which we find that Π_{p-1}^{p-2} is divisible by p^2 (the expression in parentheses is, as we have shown above, divisible by p if $p > 3$.

(b) Using the common denominator $[(p-1)!]^2$, we arrive at a sum

$$\frac{A}{[(p-1)!]^2}$$

where A is the sum of all possible products containing $p - 2$ distinct factors taken from the numbers $1^2, 2^2, 3^2, \cdots, (p-1)^2$ (or, as characterized before, "taken $p - 2$ at a time"). To show that A is divisible by p, we shall consider the square of the sum Π_{p-1}^{p-2} [the terminology has the same meaning as in problem (a)]. Since the square of a polynomial is equal to the sum of the squares of its individual terms plus twice the sum of all possible pairwise products of the terms, the sum $(\Pi_{p-1}^{p-2})^2$ consists of the terms of A plus a series of numbers (all the possible doubled products). We consider one of these doubled products:

$$2[1\cdot2 \cdots (i-1)(i+1) \cdots (p-1)] \times [1\cdot2 \cdots(j-1)(j+1) \cdots (p-1)].$$

This may be written in the form

$$2[1\cdot2\cdot3 \cdots (p-1)\cdot1\cdot2 \cdots (i-1)(i+1) \cdots (j-1)(j+1) \cdots (p-1)].$$

Summing all such terms shows that

$$(\Pi_{p-1}^{p-2})^2 = A + 2(p-1)!\Pi_{p-1}^{p-3},$$

whence

$$A = (\Pi_{p-1}^{p-2})^2 - 2(p-1)!\Pi_{p-1}^{p-3}.$$

In view of what has been proved in problem (a), A is divisible by p, as was to be shown.

61. The fraction is reducible if, and only if, its reciprocal is reducible. Hence we may consider $\dfrac{a^4 + 3a^2 + 1}{a^3 + 2a} = a + \dfrac{a^2 + 1}{a^3 + 2a}$. This simplifies the problem to proving irreducibility for $\dfrac{a^2+1}{a^3+2a}$, which is reducible if, and only if, its reciprocal is reducible. We have

$$\frac{a^3+2a}{a^2+1}=a+\frac{a}{a^2+1}\,.$$

Continuation of this procedure leads to examination of $1/a$, which clearly is irreducible for any integer a.

62. We first show that, for any integer b, the number of differences a_k-a_l which are divisible by b is not less than the number of differences $k-l$ which are divisible by b. We first determine how many differences a_k-a_l are divisible by b.

Assume that n_0 of the integers $a_1, a_2, \cdots, a_n$ are divisible by b, that n_1 of them yield the remainder 1 upon division by b, that n_2 of them yield the remainder 2, and so on up to n_{b-1} integers that yield a remainder of $b-1$ when divided by b. Since an integer yields precisely one of the remainders $0, 1, 2, 3, \cdots, b-1$ upon division by b, it is clear that

$$n_0+n_1+n_2+\cdots+n_{b-1}=n\,.$$

The difference a_k-a_l is divisible by b if, and only if, the two terms yield the same remainder upon division by b. The number of differences a_k-a_l divisible by b since both terms are divisible by b will be designated by $C_{n_0}^2=\dfrac{n_0(n_0-1)}{2}$ [clearly, there are $1+2+\cdots+(n_0-1)$ of them]; the number of differences a_k-a_l divisible by b since both terms yield a remainder of 1 upon division by b will be designated by $C_{n_1}^2=\dfrac{n_1(n_1-1)}{2}$, and so on up to $C_{n_{b-1}}^2=\dfrac{n_{b-1}(n_{b-1}-1)}{2}$ differences a_k-a_l divisible by b since both terms yield the remainder $b-1$ upon division by b. It follows that the total number of differences a_k-a_l divisible by b is exactly

$$N=\frac{n_0(n_0-1)}{2}+\frac{n_1(n_1-1)}{2}+\cdots+\frac{n_{b-1}(n_{b-1}-1)}{2}\,.$$

This expression may by rewritten as

$$N=\frac{n_0^2+n_1^2+n_2^2+\cdots+n_{b-1}^2}{2}-\frac{n_0+n_1+n_2+\cdots+n_{b-1}}{2}$$

$$=\frac{n_0^2+n_1^2+n_2^2+\cdots+n_{b-1}^2}{2}-\frac{n}{2}\,.$$

The following expansion can be made:

$$\frac{n_0^2 + n_1^2 + n_2^2 + \cdots + n_{b-1}^2}{2}$$

$$= \frac{(n_0 + n_1 + n_2 + \cdots + n_{b-1})^2 - 2n_0 n_1 - 2n_0 n_2 - \cdots - 2n_{b-2}n_{b-1}}{2}$$

$$= \frac{n^2}{2} + \frac{1}{2}\{[(n_0 - n_1)^2 - n_0^2 - n_1^2] + [(n_0 - n_2)^2 - n_0^2 - n_2^2]$$

$$+ \cdots + [(n_{b-2} - n_{b-1})^2 - n_{b-2}^2 - n_{b-1}^2]\}$$

$$= \frac{n^2}{2} + \frac{1}{2}[(n_0 - n_1)^2 + (n_0 - n_2)^2 + \cdots + (n_{b-2} - n_{b-1})^2$$

$$- (n_0^2 + n_1^2) - (n_0^2 + n_2^2) - \cdots - (n_{b-2}^2 + n_{b-1}^2)]$$

$$= \frac{n^2}{2} + \frac{1}{2}[(n_0 - n_1)^2 + (n_0 - n_2)^2 + \cdots + (n_{b-2} - n_{b-1})^2$$

$$- (b - 1)(n_0^2 + n_1^2 + \cdots + n_{b-1}^2)].$$

[Note that in the last term within parentheses, $n_0^2, n_1^2, \cdots, n_{b-1}^2$ each appear $(b - 1)$ times.] We transfer to the left member all the terms containing squares of n_k, and divide both members by b, obtaining

$$\frac{1}{2}(n_0^2 + n_1^2 + n_2^2 + \cdots + n_{b-1}^2)$$

$$= \frac{n^2}{2b} + \frac{(n_0 - n_1)^2(n_0 - n_2)^2 + \cdots + (n_{b-2} - n_{b-1})^2}{2b},$$

from which we obtain

$$N = \frac{(n_0 - n_1)^2 + (n_0 - n_2)^2 + \cdots + (n_{b-1} - n_b)^2}{2b} + \frac{n^2}{2b} - \frac{n}{2}.$$

We can show, in the same way, that the number N' of differences $k - l$ (where k and l are integers such that $n \geq k > l \geq 1$) which are divisible by b is exactly equal to

$$N' = \frac{(n_0' - n_1')^2 + (n_0' - n_2')^2 + \cdots + (n_{b-2}' - n_{b-1}')^2}{2b} + \frac{n^2}{2b} - \frac{n}{2},$$

where n_k' is the number of integers of the sequence $1, 2, 3, \cdots, n$ which give a remainder of k upon division by b.

It follows immediately from the formula just obtained that if $n = mb$ (that is, if n is a multiple of b), then the number N is not less than N'. The numbers $n_0', n_1', \cdots, n_{b-1}'$ are all equal to m, and consequently the sum of the squares of the pairwise differences of the numbers vanishes. It is less obvious that if n yields the remainder $r \neq 0$ upon division by b (that is, $n = mb + r$, $0 < r < b$), we obtain the inequality $N \geq N'$. Here, r of the numbers $n_0', n_1', \cdots, n_{b-1}'$ (in particular, the numbers $n_1', n_2', \cdots, n_r')$ are equal to $m + 1$, and the remaining numbers, $n_0', n_{r+1}', \cdots, n_{b-1}'$, are equal to m. In order to prove that N cannot be less than N', we employ the following formal method.

Since the sum of the b numbers $n_0, n_1, \cdots, n_{b-1}$ is equal to $n = mb + r$, at least one of these numbers, say n_t, does not exceed m [otherwise the sum of these numbers would be not less than $b(m + 1) > n$]. We now add one more number, a_{n+1}, to the numbers $a_1, a_2, \cdots, a_n$ — a number giving a remainder of t upon division by b. Then the number of differences $a_k - a_l$ is augmented by the n differences: $a_{n+1} - a_1, a_{n+1} - a_2, \cdots, a_{n+1} - a_n$. Of these new differences exactly n_t will be divisible by b. Now, the number of differences $k - l$ is augmented by the n differences $(n + 1) - 1, (n + 1) - 2, \cdots, (n + 1) - n$, and it is clear that $m \geqq n_t$ of the differences will be divisible by b. Therefore, if we prove that at least as many of the C_{n+1}^2 differences, $a_k - a_l$, for $k > l$ and $k, l = 1, 2, \cdots, n + 1$, are divisible by b as before (that is, as among the differences $a_k - a_l$ for $k, 1 = 1, 2, \cdots, n$), then it will follow that of the number $C_{n_1}^2$ of differences $a_k - a_l$ ($k > l$; $k, l = 1, 2, \cdots, n$) no fewer will be divisible by b than were divisible among the differences $k - l$, where $k, l = 1, 2, \cdots, n$. If $n + 1$ is divisible by b, then our quest is ended: the result sought follows from what we have done (it is analagous to the case $n = mb$). If $n + 1$ fails to be divisible by b, then we adjoin another integer to the sequence $a_1, a_2, \cdots, a_n, a_{n+1}$, —and we may continue adjoining additional integers until their number does form a multiple of b. This completes the proof of the initial assertion.

The proposition of the problem follows immediately. If $b = p$ is any prime number, then p divides at least as many factors $a_k - a_l$ as it does factors $k - l$; the same assertion will be true for $p^2, p^3, \cdots$. Thus, every prime p will enter the numerator as a factor to an order at least as great as it enters the denominator; therefore, the denominator of the fraction obtained by multiplying together all numbers of the form $\dfrac{a_k - a_l}{k - l}$ will divide the numerator, and so that product will be an integer.

63. The numbers of our sequence may be expressed in the form $1 + 10^4 + 10^8 + \cdots + 10^{4k}$. We shall investigate, along with these numbers, the integers of form $1 + 10^2 + 10^4 + 10^6 + \cdots + 10^{2k}$. It is readily shown that

$$10^{4k+4} - 1 = (10^4 - 1) \cdot (1 + 10^4 + 10^8 + \cdots + 10^{4k}),$$
$$10^{2k+2} - 1 = (10^2 - 1) \cdot (1 + 10^2 + 10^4 + \cdots + 10^{2k}).$$

Moreover, it is clear that

$$10^{4k+4} - 1 = (10^{2k+2} - 1)(10^{2k+2} + 1) .$$

Comparison of these equalities yields

$$10^{4k+4} - 1 = (10^4 - 1)(1 + 10^4 + 10^8 + \cdots + 10^{4k})$$
$$= (10^2 - 1)(1 + 10^2 + 10^4 + \cdots + 10^{2k})(10^{2k+2} + 1) ,$$

or, since $\dfrac{10^4 - 1}{10^2 - 1} = 10^2 + 1 = 101$,

$$(1 + 10^4 + 10^8 + \cdots + 10^{4k}) \cdot 101$$
$$= (1 + 10^2 + 10^4 + \cdots + 10^{2k})(10^{2k+2} + 1) .$$

Since 101 is a prime number, either $1 + 10^2 + 10^4 + \cdots + 10^{2k}$ or $10^{2k+2} + 1$ is divisible by 101. If $k > 1$, then whichever of these two numbers is divisible by 101, the quotient will exceed 1; hence $1 + 10^4 + 10^8 + \cdots + 10^{4k}$ is, for $k > 1$, expressible as the product of two (nontrivial) factors. If $k = 1$, we have the number $10^4 + 1 = 10,001$, which is a composite number $(10,001 = 73 \cdot 137)$.

Remark: It is possible to prove in a similar way that the following numbers are all composite:

$$\underbrace{100\cdots0}_{(2k+1)}\underbrace{100\cdots0}_{(2k+1)}1, \underbrace{100\cdots0}_{(2k+1)}\underbrace{100\cdots0}_{(2k+1)}\underbrace{100\cdots0}_{(2k+1)}1, \cdots .$$

64. (a) We have

$$a^{128} - b^{128} = (a^{64} + b^{64})(a^{64} - b^{64})$$
$$= (a^{64} + b^{64})(a^{32} + b^{32})(a^{32} - b^{32})$$
$$= (a^{64} + b^{64})(a^{32} + b^{32})(a^{16} + b^{16})(a^{16} - b^{16}) = \cdots$$
$$= (a^{64} + b^{64})(a^{32} + b^{32})(a^{16} + b^{16})(a^8 + b^8)$$
$$\times (a^4 + b^4)(a^2 + b^2)(a + b)(a - b) .$$

Consequently, the required quotient is equal to $a - b$.

(b) As in part (a),

$$\frac{a^{2^{k+1}} - b^{2^{k+1}}}{(a + b)(a^2 + b^2)(a^4 + b^4)(a^8 + b^8) \cdots (a^{2^{k-1}} + b^{2^{k-1}})(a^{2^k} + b^{2^k})} = a - b .$$

65. We note that

$$2^{2^n} - 1 = (2^{2^{n-1}} + 1)(2^{2^{n-1}} - 1)$$
$$= (2^{2^{n-1}} + 1)(2^{2^{n-2}} + 1)(2^{2^{n-2}} - 1) = \cdots$$
$$= (2^{2^{n-1}} + 1)(2^{2^{n-2}} + 1)(2^{2^{n-3}} + 1) \cdots (2^2 + 1)(2 + 1)(2 - 1).$$

(See problem 64. The last factor, $2 - 1$, can be disregarded.) Thus, the integer $2^{2^n} - 1 = (2^{2^n} + 1) - 2$ is divisible by all the numbers of the given sequence which it exceeds. It follows that if $2^{2^n} + 1$ and $2^{2^k} + 1$, where $k < n$, have a common nontrivial divisor, then this common divisor must also divide 2, and hence must be 2. Since all the integers of the sequence are odd, it follows that there exists no common divisor for any two of them.

66. The number 2^n is not divisible by 3. If 2^n yields the remainder 1 upon division by 3, then $2^n - 1$ is divisible by 3; if 2^n yields the remainder 2 upon division by 3, then $2^n + 1$ is divisible by 3. Therefore, in all cases one of the two numbers, $2^n - 1$ or $2^n + 1$, is divisible by 3; hence, if both integers exceed 3, they cannot both be primes.

67. (a) If a prime number $p > 3$ yields the remainder 2 when divided by 3, then $8p - 1$ is divisible by 3. Hence, to meet the condition of the problem, p must yield the remainder 1 when divided by 3. But $8p + 1$ is divisible by 3. If $p = 3$, then $8p + 1 = 25$, which is composite.

   (b) If p is not divisible by 3, then p^2 yields the remainder 1 when divided by 3 [see the solution of problem 53 (b)], and then $8p^2 + 1$ is divisible by 3. Hence, the condition of the problem can be met only if $p = 3$. But then $8p^2 - 1 = 71$ is a prime number.

68. If $p > 3$ is a prime, it can yield only 1 or 5 as a remainder upon division by 6 (if $p = 6k + 2$, or $p = 6k + 4$, then it is an even number; if $p = 6k + 3$, then it is divisible by 3). Hence, the square of the prime p must have one of the two forms $36k^2 + 12n + 1$ or $36n^2 + 60n + 25$. Each of these integers yields the remainder 1 when divided by 12.

69. As explained in the solution of problem 68, a prime number $p > 3$ must have one of the two forms $6n + 1$ or $6n + 5$. Given three distinct primes, all exceeding 3, at least two of them must yield the same remainder upon division by 6. The difference between these two numbers, which is d or $2d$, where d is the common difference of the arithmetic progression, is then divisible by 6, whence, necessarily, the common difference d is divisible by 3. But since d is the difference of two odd numbers, it is also divisible by 2. Therefore, in every case, d must be divisible by 6. [See also the solution of problem 70 (a).]

70. (a) Since all primes exceeding 2 are odd, the common difference of the arithmetic progression sought must be an even number;
hence, we may eliminate 2 as a possible term of the progression.
Also, since there are certainly three successive terms of the progression, all of which exceed 3 and which by themselves must form
an arithmetic progression, the common difference d must be (according to problem 69) divisible by 6, that is, divisible by both 2
and by 3.

We now show that d must be divisible by 5. Assume d is not
divisible by 5. Then the numbers

$$a, a + d, a + 2d, a + 3d, a + 4d$$

all yield different remainders upon division by 5 (if two of the remainders are equal, then it is easily shown that d is divisible by 5,
a contradiction of the assumption just made). Thus, one of the
numbers of the progression is then divisible by 5. Since all the
term of the progression are prime, this is a contradiction. Hence
d must be divisible by 5. We can show, in the same manner, that
d must be divisible by 7. (This conclusion cannot be reached for
11, since there are to be only ten terms in the progression, and
$a, a + d, \cdots, a + 9d$ would not necessarily provide a number divisible
by 11.) Therefore, the common difference d of the progression must
be a multiple of $2 \cdot 3 \cdot 5 \cdot 7 = 210$; that is, $d = 210k$.

According to the conditions of the problem,

$$a_{10} = a_1 + 9d = a_1 + 1890k < 3000 .$$

This inequality is impossible for $k \geqq 2$, hence, necessarily, $k = 1$. It
follows that

$$a_1 < 3000 - 9d = 3000 - 1890 = 1110 .$$

Now, $210 = 11 \cdot 19 + 1$; consequently, the $(m + 1)$st term of the progression may be represented in the form

$$a_{m+1} = a_1 + (11 \cdot 19 + 1) \cdot m = 11 \cdot 19m + (a_1 + m) .$$

It follows that if a_1 yields a remainder of 2 upon division by 11,
then a_{10} is divisible by 11. If a_1 yields a remainder of 3 when divided
by 11, then a_9 is divisible by 11, and so on. Therefore, a_1 cannot
yield upon division by 11 any of the remainders $2, 3, 4, \cdots$, or 10.
If $a_1 \neq 11$, then since a_1 is prime it cannot be divisible by 11; this
means that either $a_1 = 11$ or a_1 yields a remainder of 1 upon division
by 11. Further, since $210 = 13 \cdot 16 + 16 + 2$, and so

$$a_{m+1} = a_1 + (13 \cdot 16 + 2)m = 13 \cdot 16m + (a_1 + 2m) \, ,$$

it may be shown that if a_1 is divisible by 13, it can yield as a remainder only one of the numbers 2, 4, 6, 8, 10, or 12. Since a_1 is odd (as are all the terms of the progression), either $a_1 = 11$ or it can be written in one of the following forms:

$$2 \cdot 11 \cdot 13l + 23 = 286l + 23 \, , \qquad 286l + 155 \, ,$$
$$286l + 45 \, , \qquad 286l + 177 \, ,$$
$$286l + 67 \, , \qquad 286l + 199 \, ,$$

Since $a_1 < 1110$, the possible values for a_1 are limited to the integers

$$11; 23, 309, 595, 881; 45, 331, 615, 903; 67, 353, 637, 925;$$
$$155, 441, 727, 1013; 177, 463, 749, 1035; 199, 485, 771, 1057 \, ,$$

of which the following are prime:

$$11, 23, 881, 331, 67, 353, 727, 1013, 463, 199 \, .$$

We have found the necessary conditions for the existence of the progression sought; namely, $d = 210$, and a_1 equal to one of the prime numbers listed. We must test each of the possibilities (for example, $a_1 = 11$ is quickly found untenable, since then $a_2 = 221 = 13 \cdot 17$, which is not prime). Exactly one of the above primes, $a_1 = 199$, will produce an acceptable progression:

$$199, 409, 619, 829, 1039, 1249, 1459, 1669, 1879, 2089 \, .$$

(b) This problem is solved in a manner analogous to the solution of problem (a); however, the progression found there cannot be extended since the following term fails to be prime.

If $a_1 \neq 11$, then, proceeding exactly as in problem (a), we find that the common difference d of the progression sought must be a multiple of $2 \cdot 3 \cdot 5 \cdot 7 \cdot 11 = 2310$ ($d = 2310k$). It follows that

$$a_{11} = a_1 + 23\,100k > 20{,}000 \, .$$

Hence we need investigate only the case for which $a_1 = 11$. Here, we can have only $d = 210k$. Since $210 = 13 \cdot 16 + 2$, we can write for the general term of the desired progression

$$a_{n+1} = 11 + (13 \cdot 16 + 2)kn = 13(16kn + 1) + 2(kn - 1) \, .$$

However, for any $k = 1, 2, 3, 4, 5, 7, 8, 9,$ or 10, we can always find an $n \leq 10$ such that $kn - 1$ is divisible by 13, whence a_{n+1} fails to be a prime number. (These values of n are, respectively, 1, 7, 9, 10, 8, 2, 5, 3, 4.) If $k = 6$ and $d = 210 \cdot 6 = 1260$, then

$$a_4 = 11 + 3 \cdot 1260 = 3791 \, ,$$

which is divisible by 17. Therefore, if $a_1 = 11$, then necessarily $k > 10$, and so $d \geqq 2100$, but then $a_{10} > 20,000$.

71. (a) If the difference of two odd numbers does not exceed 4, then they cannot have a common divisor which exceeds 4. Thus, two of the five consecutive numbers can either have at most a common divisor of 2, 3, or 4 or be relatively prime. At least two of the five consecutive numbers must be odd, and of two consecutive odd numbers at least one will fail to be divisible by 3. Hence there is at least one odd number among the five consecutive integers which fails to be divisible by 3. This integer will necessarily be relatively prime to the remaining four integers.

(b) The reasoning employed here closely resembles that used in problem (a), but it is much more involved. If the difference of two odd numbers does not exceed k, then they cannot have a common divisor which exceeds k. To determine whether two integers are relatively prime, it suffices to consider only prime factors; hence it suffices to show that, given sixteen consecutive integers, it is always possible to find one of them which fails to have in common with any one of the other integers a divisor of 2, 3, 5, 7, 11, or 13. That integer will be relatively prime to all the others.

First we discard the even numbers of the sixteen successive integers. There remain eight consecutive odd numbers. Divisibility by 3 clearly holds for either

(1) the first, fourth, and seventh of these eight numbers,
(2) for the second, fifth, and eighth numbers, or
(3) for the third and sixth numbers.

Divisibility by 5 holds for either the first and sixth, or for the second and seventh, or for the third and eighth, or for one number only (the fourth or the fifth). Divisibility by 7 holds either for the first and eighth or for only one of the other integers. One, and only one, of the odd numbers can be divisible by 11, and only one by 13.

If not more than five of the eight consecutive odd numbers are divisible by one of the primes 3, 5, or 7, then there must exist among the remaining three (or more) odd numbers at least one which is divisible neither by 11 nor by 13. Since that number will fail to have 2, 3, 5, 7, 11, or 13 as a factor, it will be relatively prime to all the other integers of the original sequence of sixteen numbers.

We now consider the case in which the number of odd integers divisible by 3, 5, or 7 does not exceed six (which is the maximal number of odd integers of the sequence which can be so divisible). We first assume that three of the eight odd numbers are divisible by 3. Then, depending upon which three numbers these are (first and fourth and seventh, or second and fifth and eighth), two remaining numbers can be divisible by 5 (third and eighth, or first and sixth), and one of the remaining numbers might be divisible by 7.

If we strike out the (at most) five numbers divisible by 3 or by 5, there will remain either the second and fifth and sixth or the third and fourth and seventh of the eight odd numbers. We consider the first case. The second, fifth and sixth odd numbers stand either in the fourth, tenth, and twelfth positions of the original sequence of sixteen numbers or in the third ninth and eleventh positions of that sequence. In the first-named positioning two of these odd numbers must fail to have 7 as a divisor; and, of these two, neither can have a common divisor of 13 with any other number of the original sequence, since both differ from all the other numbers by less than 13. Since at most one of these two numbers is divisible by 11, at least one remains which cannot be divisible by 2, 3, 5, 7, 11, or 13 and so must be relatively prime to all the other numbers of the original sequence. In the second-named positioning (third, ninth, and eleventh), if one of these odd numbers has the factor 13 in common with another number of the original sequence of sixteen consecutive numbers, it can be only that number standing in the third position. If we throw out that number, we are left with numbers in the ninth and eleventh positions. Only one of these two numbers can be divisible by 7; whichever it is, the remaining one cannot have a factor of 11 in common with any other number of the original sequence, since it differs from all of them by a number less than 11. Hence at least one number will be relatively prime to all the others of the original sequence. The argument for the case in which the third, fourth, and seventh numbers of the sequence of odd integers remain after those divisible by 3 or by 5 are thrown out is quite analogous, and is left for the reader.

If only two numbers of the sequence of eight odd numbers are divisible by 3 (the third and sixth), then it is possible for two of the remaining numbers (the first and eighth) to be divisible by 7, and two more can be divisible by 5 (the second and seventh). If these six numbers are struck out, only the fourth and fifth of the eight odd numbers are retained, and these two are not divisible by 3, 5, or 7. Each of these remaining two numbers will be relatively prime to the other fifteen numbers of the original sequence, since each of them will differ from the remaining fifteen numbers by less than 11, and hence could not share with any of them a common divisor of 11 or 13. This completes the proof.

Remark: The proposition may be proved for any sequence of successive integers fewer than sixteen (say, ten or twelve) by techniques similar to that used above. The proposition does not hold for a sequence of seventeen numbers.[†] Whether such a proposition is true for $k > 17$ numbers, or for special numbers k, is not known.

72. Since $6 = 3 \cdot 2$, the product in question will be the same as that obtained by multiplying the integer A_1 consisting of 666 digits 9 by the integer B_1 composed of 666 digits 2. But A_1 is 1 less than 10^{666} (the digit 1 with 666 zeros following); and so if B_1 is multiplied by A_1, the result is the same as multiplying B_1 by 10^{666} (which yields an integer composed of 666 digits 2 followed by 666 zeros) and subtracting the integer B_1. It clearly follows that the result will be a number of form

$$\underbrace{22 \cdots 21}_{665} \underbrace{77 \cdots 78}_{665} .$$

73. The number 777,777 is exactly divisible by 1001, yielding the quotient 777. Hence the number

$$\underbrace{777 \cdots 700000}_{996} ,$$

yields, upon division by 1001, a quotient of

$$\underbrace{777\,000\,777\,000 \cdots 777\,000\,00}_{\text{the grouping 777,000 repeated 166 times}} .$$

Moreover, the number 77,777 yields a quotient of 77 and a remainder of 700 upon division by 1001; and so the quotient obtained by dividing A by 1001 has the form

$$\underbrace{777\,000\,777\,000 \cdots 777\,000\,77}_{\text{the grouping 777,000 repeated 166 times}} ,$$

and there is a remainder of 700.

74. Since the integer 222,222 is not a perfect square, the integer

---

[†] The "counter-example" displayed in the original Russion text is incorrect and so not given here [*Editor*].

sought has the form 222,222 $a_7a_8 \cdots a_n$, where $a_7, a_8, \cdots, a_n$ are to be determined.

First assume that the integer n is even: $n = 2k$. We shall employ the usual process for finding the square root:

$$\sqrt{22\,22\,22\,a_7a_8 \cdots a_{2k-1}a_{2k}} = 471\,405$$

```
        16
    87 | 6 22
    7  | 6 09
   941 | 13 22
    1  |  9 41
  9424 | 3 81 a₇a₈
    4  | 3 76 9 6
942805 | x₁ x₂x₃ a₉a₁₀a₁₁a₁₂
    5  | 4 7 1 4 0 2 5
```

(the fifth digit of the result is 0, since x_1 can be only 4 or 5, and analogous reasoning shows that the sixth digit, if it is to terminate the square root and produce a least number, will be 5).

The remainder now vanishes if $a_9 = 4$, $a_{10} = 0$, $a_{11} = 2$, $a_{12} = 5$ and $x_1 = 4$, $x_2 = 7$, $x_3 = 1$; it is easily deduced that $a_8 = 6 + 1 = 7$ and $a_7 = (7 + 9) - 10 = 6$. Hence the smallest integer with an even number of digits and satisfying the conditions of the problem is

$$222,222,674,025 = 471,405^2 .$$

Now we assume that n is odd, $2k + 1$, and we obtain

$$\sqrt{2\,22\,22\,2a_7a_8a_9 \cdots a_{2k}a_{2k+1}} = 149071 \cdots$$

```
         1
    24 | 1 22
    4  |   96
   289 | 26 22
    9  | 26 01
 29807 | 21 2a₇ a₈a₉
    7  | 20 86  4 9
298141 | x₁x₂x₃x₄a₁₀a₁₁
    1  | 2 9 8 1 4 2
298142   x₅x₆x₇x₈x₉x₁₀a₁₂a₁₃
```

Since the number formed by the digits x_1, x_2 is not less than 33 $(=119 - 86)$ and does not exceed 43 $(=129 - 86)$, it follows that the sixth digit of the root is 1. Hence the extraction of the root does not end here, but continues. Consequently, the smallest number having an odd number of digits, and satisfying the conditions of the problem, is not less than the twelve-digit number 222,222,674,025. Therefore, this is the number sought.

75. If the positive number α is less than 1, then $\sqrt{\alpha}$ is less than 1. Assume that the decimal representation of $\sqrt{\alpha}$ mentioned in the problem has fewer than 100 digits 9 at its beginning. This implies that $\sqrt{\alpha} < 1 - \left(\dfrac{1}{10}\right)^{100}$. If we square both sides of this inequality, we have

$$\alpha < 1 - 2\left(\frac{1}{10}\right)^{100} + \left(\frac{1}{10}\right)^{200}.$$

But $1 - 2\left(\dfrac{1}{10}\right)^{100} + \left(\dfrac{1}{10}\right)^{200} < 1 - \left(\dfrac{1}{10}\right)^{100}$; therefore, $\alpha < 1 - \left(\dfrac{1}{10}\right)^{100}$, which means that the decimal representing α cannot begin with 100 digits 9. This contradiction proves the proposition of the problem.

76. The desired number commencing with the digits 523 and divisible by $7 \cdot 8 \cdot 9 = 504$ can be written in the form $523{,}000 + X$, where X is a three-digit number. Ordinary division yields $523{,}000 = 504 \cdot 1037 + 352$; that is, $523{,}000$ yields a remainder of 352 when divided by 504. Since the sum of $523{,}000$ and the three-digit number X must be divisible by 504, it follows that X can be equal to

$$504 - 352 = 152,$$

or to

$$2 \cdot 504 - 352 = 656$$

(the number $3 \cdot 504 - 352$ has four digits). Hence the two numbers $523{,}152$ and $523{,}656$ satisfy the conditions of the problem.

77. Let N be the desired four-digit number. The conditions of the problem gives

$$N = 131k + 112 = 132l + 98,$$

where k and l are positive integers. Since N is a four-digit number, it is clear that

$$l = \frac{N - 98}{132} < \frac{10{,}000 - 98}{132} < 75 \cdot 02$$

that is, $l \leq 75$. Further,

$$131k + 112 = 132l + 98;$$
$$131(k - l) = l - 14.$$

It is evident that if $k - l$ is not zero, then $l - 14$ exceeds 130 in absolute value, which is impossible if $l \leq 75$. Therefore, necessarily, $k - l = 0$, or $k = l$. This yields

$$l - 14 = 0 \,,$$
$$k = l = 14 \,;$$
$$N = 131 \cdot 14 + 112 \ (=132 \cdot 14 + 98) = 1946 \,.$$

78. (a) The number given in this problem has $2n$ digits and may be written as follows:

$$4 \cdot 10^{2n-1} + 9 \cdot 10^{2n-2} + 4 \cdot 10^{2n-3} + 9(10^{2n-4} + 10^{2n-5} + \cdots + 10^n) + 5 \cdot 10^{n-1}$$

$$+ 5 \cdot 10^{n-2} = 4 \cdot 10^{2n-1} + 9 \cdot 10^{2n-2} + 4 \cdot 10^{2n-3} + 9 \cdot 10^n \frac{10^{n-3} - 1}{9}$$

$$+ 5 \cdot 10^{n-1} + 5 \cdot 10^{n-2} = 4 \cdot 10^{2n-1} + 9 \cdot 10^{2n-2} + 5 \cdot 10^{2n-3} - 10^n$$

$$+ 5 \cdot 10^{n-1} + 5 \cdot 10^{n-2} = \frac{1}{2} (8 \cdot 10^{2n-1} + 18 \cdot 10^{2n-2} + 10 \cdot 10^{2n-3}$$

$$- 2 \cdot 10^n + 10^n + 10^{n-1}) = \frac{1}{2} (9 \cdot 10^{2n-1} + 9 \cdot 10^{2n-2} - 9 \cdot 10^{n-1})$$

$$= \frac{[(10^n - 1) + 10^{n-1}] \cdot 9 \cdot 10^{n-1}}{2} \,.$$

This number is equal to the sum of the arithmetic progression having common difference 1 and having 10^{n-1} as its first term and $10^n - 1$ as its last term (the number of terms of this progression is equal to $10^n - 10^{n-1} = 9 \cdot 10^{n-1}$), which is the sum of all the n-digit integers.

(b) The number of integers (in the sum) which have a given digit a as first digit (a may be $1, 2, 3, 4,$ or 5) is $6 \cdot 6 \cdot 3 = 108$ (since any of the six digits $0, 1, 2, 3, 4, 5$ may stand in the second or in the third position, and any one of the three digits $0, 2, 4$ may terminate the even integer). Consider the contribution made to the sum by the thousands column: when the digit 1 stands here the sum is $1 \cdot 108 \cdot 1000$; when the digit 2 stands here, the contribution of this column to the sum is $2 \cdot 108 \cdot 1000$; and so on. Thus, the contribution made by the thousands column to the sum is the total

$$(1 + 2 + 3 + 4 + 5) \cdot 108 \cdot 1000 = 1,620,000 \,.$$

Now we consider, in analogous fashion, the contribution to the sum made by the hundreds place of the numbers. For any digit b in this (second) position there are $5 \cdot 6 \cdot 3 = 90$ possible numbers in the other positions. This column contributes to the sum the total

$$(1 + 2 + 3 + 4 + 5) \cdot 90 \cdot 100 = 135,000 \,.$$

(We need not consider here the numbers in which the digit 0 stands in the hundreds position; they contribute $0 \cdot 90 \cdot 100 = 0$ to the total for this column.)

In a similar manner we find that the sum contributed by the tens column is $(1 + 2 + 3 + 4 + 5) \cdot 90 \cdot 10 = 13{,}500$; and the sum contributed by the units column is $(2 + 4) \cdot 5 \cdot 6 \cdot 6 = 1080$. Therefore, the sum sought is equal to

$$1{,}620{,}000 + 135{,}000 + 13{,}500 + 1080 = 1{,}769{,}580 \ .$$

79. We first investigate the integers from 0 to 99,999,999. In so doing we "fill in" with zeros at the beginning, all integers having fewer than eight digits so that these integers have the format of eight-digit numbers. We then have 100,000,000 eight-digit "numbers." To write all of these, we shall need a stockpile containing 800,000,000 digits. Now, if we write 00,000,000 at the top of a sheet and 99,999,999 at the bottom and, moving from the top down and from the bottom up, fill in the successive numbers (always filling in with zeros where necessary), we readily see that each digit (including zero) will be used exactly the same number of times. That is, if the stockpile of 800,000,000 digits is partitioned into ten bins, each labeled for a different digit, then each bin will contain 80,000,000 digits all of the same kind.

Now we calculate how many zeros we have used to fill out integers containing fewer than eight meaningful digits. There are nine one-digit numbers (omitting zero), $99 - 9 = 90$ two-digit numbers, $999 - 99 = 900$ three digit-numbers, and so on. To fill out the one-digit numbers, we used seven zeros; to fill out the two-digit numbers, we needed six zeros; and so on. Hence the total number of extraneous zeros required for filling out the numbers (disregarding the zeros used for the first "number," 00,000,000) is given by the series

$$7 \cdot 9 + 6 \cdot 90 + 5 \cdot 900 + 4 \cdot 9000 + 3 \cdot 90{,}000$$
$$+ \ 2 \cdot 900 \cdot 000 + 1 \cdot 9{,}000{,}000 = 11{,}111{,}103 \ .$$

We now append the digit 1 to the first number 00,000,000, thus obtaining all the integers from 1 to 100,000,000. In order to write all these numbers we need 80,000,000 twos and the same number of threes, fours, and so on up to and including nines; we also need 80,000,001 digits 1 and $80{,}000{,}000 - 11{,}111{,}103 = 68{,}888{,}897$ zeros.

80. In all, there are nine one-digit numbers, $99 - 9 = 90$ two-digit numbers, $999 - 99 = 900$ three-digit numbers; and, in general, there are $9 \cdot 10^n$ n-digit numbers.

One-digit numbers occupy nine positions in the sequential array written in the problem. Two-digit numbers occupy the next $90 \cdot 2 = 180$ positions; three-digit numbers occupy the next $900 \cdot 3 = 2700$ positions; four-digit numbers occupy the next $9000 \cdot 4 = 36,000$ positions; and five-digit numbers occupy the next $90,000 \cdot 5 = 450,000$ positions. It is clear that the digit of interest to us appears in one of the five-digit numbers, since we have filled only 38,889 positions before starting to write five-digit numbers, and when we have appended the five-dight numbers we will have filled 488,889 places.

To find out how many five-digit numbers are in the interval from the 38,889th to the 206,788th position, we divide the difference $206,788 - 38,889 = 167,889$ by 5:

$$206,788 - 38,889 = 5 \cdot 33,579 + 4 .$$

Thus, the digit sought must belong to the 33,580th five-digit number, that is, to the number 33,579 (since the five-digit numbers began with 10,000). The digit sought is the fourth digit of this number: 7.

81. Assume that the decimal $0 \cdot 1234 \cdots$ is periodic, that n is the periodicity (number of digits in a period), and that k is the number of digits encountered before the periodic position starts. Consider the integer 10^m (the digit 1 followed by m zeros), where m is not less than $n + k$. In composing the decimal we wrote in succession all the integers; hence any chosen number N will appear somewhere (it will surely appear when we append the nth integer). Since in the sequence of numbers written in to make up the infinite decimal $m \geqq n + k$ zeros must be encountered, it follows that the only possible period consists of one zero—a situation which does not hold for this decimal. Hence the decimal is not periodic.

82. Let us take nine weights weighing $n^2, (n + 1)^2, (n + 2)^2, \cdots,$ $(n + 8)^2$ units (for a suitable n) and group them in three sets as follows:

Set I: $n^2, (n + 5)^2, (n + 7)^2$.

Then, $n^2 + (n + 5)^2 + (n + 7)^2 = 3n^2 + 24n + 74$.

Set II: $(n + 1)^2, (n + 3)^2, (n + 8)^2$.

Then, $(n + 1)^2 + (n + 3)^2 + (n + 8)^2 = 3n^2 + 24n + 74$.

Set III: $(n + 2)^2, (n + 4)^2, (n + 6)^2$.

Then, $(n + 2)^2 + (n + 4)^2 + (n + 6)^2 = 3n^2 + 24n + 56$.

For any allowable n the total weights of the first and second sets are equal, and the total weight of the third set is lighter by 18 units.

If we do this for $n = 1$, we will have taken the first nine weights and grouped them as follows: Set I: 1, 6, 8; Set II: 2, 4, 9; Set III: 3, 5, 7. The first two sets are equal in weight, and the third is 18 units lighter. We now do this for $n = 10$ (thus choosing the next nine weights), but after the groupings we interchange the second and third sets, which means that the second set is 18 units lighter than the other two sets. We do this for the third set of nine weights ($n = 19$), and after grouping as for the first trial we interchange the first and third sets, whence the first set is 18 units lighter than the other two sets. It is now clear how the three final groupings may be made.

83. Draw a ray from the pin through one of the vertices of the polygon. If this ray turns through $25\frac{1}{2}°$, then it must contain another vertex of the polygon. Now, $25\frac{1}{2}°$ is 17/240 of the circumference. Since 17 and 240 are relatively prime, if the ray is turned through an angle of $25\frac{1}{2}°$ any number of times up to 239 times, it will not duplicate a prior position. If the ray is turned through this angle m times, it turns through $17m/240$ circumferences. In order for the kth and lth turns of the ray to fall on a prior position, $17k/240$ circumferences must differ from $17l/240$ circumferences by an integral number; that is, $\dfrac{17(k - l)}{240}$ must be an integer. Therefore, $k - l$ must be divisible by 240. It follows that either $k = l$ or $k \geq 240$. Counting in the initial position we obtain 240 different rays. One vertex lies on each of these rays, and hence there can be no fewer than 240 vertices. On the other hand, the 240-sided polygon, upon rotation through 17/240 of a circumference, coincides with its initial position. Hence the least number of sides which the polygon can have is 240.

84. (a) The first digit of each of the three-digit numbers must be the smallest possible one; hence we may assume that the three numbers have the form

$$\overline{1Aa}, \ \overline{2Bb}, \ \overline{3Cc} ,$$

where the overbar indicates a succession of digits; for example, $\overline{2Bb} \equiv 2\cdot10^2 + B\cdot10 + b$.

We shall show that necessarily: $A < B < C$; $a < b < c$; each of the digits $a, b,$ and c is greater than any of the digits $A, B,$ and C.

(1) Assume $A > B$. Then $\overline{Aa} > \overline{Bb}$, and so

$$\overline{1Aa}\cdot\overline{2Bb} - \overline{2Aa}\cdot\overline{1Bb}$$
$$= (100 + \overline{Aa})(200 + \overline{Bb}) - (200 + \overline{Aa})(100 + \overline{Bb})$$
$$= 100\,(\overline{Aa} - \overline{Bb}) > 0\,,$$

which would mean that

$$\overline{1Bb}\cdot\overline{2Aa}\cdot\overline{3Cc} < \overline{1Aa}\cdot\overline{2Bb}\cdot\overline{3Cc}\,,$$

whence the product on the right would not be the least, as required. The assumption $B > C$ will produce a similar contradiction.

(2) Assume $a > b$. Then

$$\overline{1Aa}\cdot\overline{2Bb} - \overline{1Ab}\cdot\overline{2Ba}$$
$$= (10\cdot\overline{1A} + a)(10\cdot\overline{2B} + b) - (10\cdot\overline{1A} + b)(10\cdot\overline{2B} + a)$$
$$= (10\cdot\overline{2B} - 10\cdot\overline{1A})(a - b) > 0\,,$$

which would mean that

$$\overline{1Ab}\cdot\overline{2Ba}\cdot\overline{3Cc} < \overline{1Aa}\cdot\overline{2Bb}\cdot\overline{3Cc}\,,$$

which again is a contradiction. The result $b < c$ is similarly shown.

(3) Assume $C > a$, or $C = a + x$, where $x > 0$. According to the first demonstration C is the largest of the digits A, B, C; and according to the second demonstration a is the smallest of the digits a, b, c. In this case we would have

$$\overline{1Aa}\cdot\overline{3Cc} - \overline{1AC}\cdot\overline{3ac} = \overline{1Aa}\cdot(\overline{3ac} + 10x) - (\overline{1Aa} + x)(\overline{3ac})$$
$$= x(10\cdot\overline{1Aa} - \overline{3ac}) > 0\,,$$

which yields the contradiction

$$\overline{1AC}\cdot\overline{2Bb}\cdot\overline{3ac} < \overline{1Aa}\cdot\overline{2Bb}\cdot\overline{3Cc}\,.$$

It follows from (1), (2), and (3) that

$$A < B < C < a < b < c\,,$$

whence the product sought is composed as follows:

$$147\cdot258\cdot369\,.$$

(b) It is clear that the largest digits must be the initial digits of each number; hence we may write the product in the form

$$\overline{9Aa}\cdot\overline{8Bb}\cdot\overline{7Cc}\,.$$

Employing techniques analogous to those used in problem (a) we readily prove that: $A < B < C$; $a < b < c$; each of the digits a, b, c is smaller than any of the digits A, B, C. That is,

$$a < b < c < A < B < C ,$$

and so the product sought is

$$941 \cdot 852 \cdot 763 .$$

85. Write $m + (m + 1) + \cdots + (m + k) = 1000$. Using the formula for the sum of an arithmetic progression, we have

$$\frac{2m + k}{2} \cdot (k + 1) = 1000 ,$$

or

$$(2m + k)(k + 1) = 2000 .$$

Since $(2m + k) - (k + 1) = 2m - 1$ is odd, one of the two terms on the left is even and the other is odd. Moreover, it is clear that $2m + k > k + 1$. Since $2000 = 2^4 \cdot 5^3$, its odd factors are only $1, 5, 25$, and 125. For odd $(k + 1)$ we need consider only $1, 5$, and 25; for odd $(2m + k)$ we can have only 125. Hence the problem has the following readily found solutions:

$$2m + k = 2000, \quad k + 1 = 1, \quad m = 1000, \quad k = 0 ;$$
$$2m + k = 400, \quad k + 1 = 5, \quad m = 198, \quad k = 4 ;$$
$$2m + k = 80, \quad k + 1 = 25, \quad m = 28, \quad k = 24 ;$$
$$2m + k = 125, \quad k + 1 = 16, \quad m = 55, \quad k = 15 .$$

86. (a) Let N be an integer which is not a power of 2. Then the following equation can be written:

$$N = 2^k(2l + 1) ,$$

where 2^k is the greatest power of 2 appearing as a factor of N, $k \geqq 0$, $l \geqq 1$, and $2l + 1$ is the greatest odd divisor of N. Consider the arithmetic progression

$$(2^k - l) + (2^k - l + 1) + \cdots + (2^k - l + 2l - 1) + (2^k - l + 2l)$$
$$= \frac{(2l + 1)(2^k - l + 2^k - l + 2l)}{2} = 2^k(2l + 1) = N .$$

If some of the $2l + 1$ consecutive integers which form the progression are negative (that is, $l > 2^k$), then it is possible to cancel them with the first-appearing positive integers. It is readily shown that at least the final two terms of the progression must remain uncancelled. (If only the final term of the progression were to remain, we could set up the equation $2^k + l = N = 2^k(2l + 1)$, which would imply that $k = -1$.)

Assume now that some number of form 2^k can be written as the sum of m consecutive positive integers $n, n + 1, \cdots, n + m - 2, n + m - 1$. Then

$$2^{k+1} = 2[n + (n + 1) + \cdots + (n + m - 2) + (n + m - 1)]$$
$$= m(n + n + m - 1) = m(2n + m - 1).$$

But the difference $(2n + m - 1) - m = 2n - 1$ is an odd number, and, consequently, one of the numbers, m or $2n + m - 1$, must be odd (and both differ from 1 since, by hypothesis, $m > 1$ and $n > 0$). This means that the equality derived above, $2^{k+1} = m(2n + m - 1)$, is impossible since 2^{k+1} cannot have an odd divisor other than 1.

(b) We have, for any $m > n + 1$,

$$(2n + 1) + (2n + 3) + (2n + 5) + \cdots + (2m - 1)$$
$$= \frac{(2n + 1) + (2m - 1)}{2} \cdot (m - n) = (m + n) \cdot (m - n).$$

[If $m = n + 1$, then there is only one term; there are $(m - n)$ terms.] Hence if a number N can be written as the sum of consecutive odd numbers, then it is a composite number (the product of numbers $m + n$ and $m - n$). Now, any composite odd number N can be written as the product of two odd factors a and b, $(a \geq b > 1)$, and so we can write

$$N = a \cdot b = (m + n)(m - n),$$

where we set $m = \frac{a + b}{2}$ and $n = \frac{a - b}{2}$. (Note that for $b > 1$, $m > n + 1$). Hence, $N = (m + n)(m - n)$ is the sum of a sequence of consecutive odd numbers, the odd numbers from $a - b + 1$ to $a + b - 1$. Clearly, no prime number can be represented in this form, since then the prime would be the product $(m + n)(m - n)$, whence $m - n = 1$, and so the series reduces to one term, the prime number itself. This proves the first assertion.

Now, in the formula $N = (m + n)(m - n)$, the factors $m + n$ and $m - n$ are either both even or else both are odd (their difference is even). Hence if N is an even integer, both of these factors must be even. In this case N is divisible by 4; therefore, an even number N which fails to be divisible by 4 cannot be written as a sum of consecutive odd numbers. On the other hand, if $N = 4n$, then N can be written as the sum of the two consecutive odd numbers $2n - 1$ and $2n + 1$.

(c) It is readily seen that

$$(n^{k-1} - n + 1) + (n^{k-1} - n + 3) + \cdots + (n^{k-1} - 1)$$
$$+ (n^{k-1} + 1) + \cdots + (n^{k-1} + n - 3) + (n^{k-1} + n - 1)$$
$$= \frac{(n^{k-1} - n + 1) + (n^{k-1} + n - 1)}{2} \cdot n = n^k$$

(all the terms of the sum are odd, since n^{k-1} and n are simultaneously even or odd).

87. Designate the four consecutive integers by $n, n + 1, n + 2, n + 3$. If 1 is added to their product, we have

$$n(n + 1)(n + 2)(n + 3) + 1$$
$$= [n(n + 3)][(n + 1)(n + 2)] + 1 = (n^2 + 3n)(n^2 + 3n + 2) + 1$$
$$= (n^2 + 3n)^2 + 2(n^2 + 3n) + 1 = (n^2 + 3n + 1)^2 .$$

Therefore, the product of the four numbers is one less than the square of the integer $n^2 + 3n + 1$.

88. We shall show that the set of integers can contain only four different values. Assume the contrary—that among the $4n$ integers there are five of them, a_1, a_2, a_3, a_4, a_5, all distinct. Let us agree that $a_1 < a_2 < a_3 < a_4 < a_5$.

Consider the integers a_1, a_2, a_3, a_4. Under the conditions of the problem it is possible to form a proportion out of these integers. Hence the product of the extremes will be equal to the product of the means. This is feasible only if

$$a_1 a_4 = a_2 a_3$$

(the equation $a_1 a_3 = a_2 a_4$ is impossible, since $a_1 < a_2$ and $a_3 < a_4$; it is clear also that $a_1 a_2 = a_3 a_4$ is impossible).

Now consider the integers a_1, a_2, a_3, a_5. Again, if a proportion is to be formed, the only possibility is $a_1 a_5 = a_2 a_3$. Consequently, we must have $a_1 a_4 = a_1 a_5$, which is a contradiction, since $a_4 \neq a_5$.

Therefore, the set of $4n$ numbers cannot contain more than four distinct integers, and so at least one of the integers must appear n times.

89. We note, first, that the difference of two positive integers is odd if, and only if, one of the integers is odd and the other even; we obtain an even difference only from two even or two odd integers.

Let us designate an even number by the letter e, and an odd number by the letter o. For four arbitrary integers $A, B, C,$ and D, we have the following "essentially different" possible even-odd combinations and arrangements:

(1)	e, e, e, e ;	(4)	e, o, e, o ;
(2)	e, e, e, o ;	(5)	e, o, o, o ;
(3)	e, e, o, o .	(6)	o, o, o, o ;

All other arrangements can be obtained by "cyclic permutation" of these six arrangements (that is, by moving an arrangement to the right, and putting the fourth integer in the first place, and so on, not changing the order of appearance). We shall show that, in every case, after not more than four steps (as described in the problem) we must arrive at four even numbers. First, combination (1) already consists of four even numbers; combination (6) achieves the desired effect in the first step; combination (4) becomes combination (6) in one step, so it arrives in the desired form in two steps; combination (3) becomes combination (4) in one step, so it becomes combination (1) in three steps; finally, combinations (2) and (5) become, in one step, combination (3) [for combination (5) we first employ cyclic permutation], so in four steps we achieve the sought combination (1). Thus, in every case, we arrive at a quadruple of four even numbers in at most four steps.

We continue the process of forming new quadruples of numbers, and now we are working entirely with even integers. It is readily reasoned that after at most four more steps, we obtain numbers divisible by 4 (if some of the even numbers are not already divisible by 4, we divide all numbers by 2, and in four steps we have numbers divisible again by 2); in at most four more steps we obtain integers divisible by 8, and so on. If the process is continued long enough, we must be able to arrive at a quadruple of numbers divisible by any desired power of 2. Since the numbers we obtain are decreasing in absolute value, we must arrive at a point where we have at least one zero, and finally at a point where we have four zeros (this will occur in at most 4^n steps, albeit we will usually arrive at this point much earlier).

Remark: The analogous theorem may be proved for any number 2^k of positive integers. For n numbers, where n is not a power of 2, the proof does not carry through. For example, if the members are $1, 1, 0$, we never do arrive at the triple $0, 0, 0$. We have:

1, 1, 0,

0, 1, 1,

1, 0, 1,

1, 1, 0,

....

We note, further, that the numbers A, B, C, and D can be rational numbers—case not essentially different from the case in which $A, B, C,$ and D are positive integers. Here the fractions may be written with a common denominator and the procedure carried out as before. The proposition fails to hold for irrational numbers.

90. (a) It is readily seen that if the first 100 integers are displayed in the following order, the condition of the problem is satisfied.

```
10  9  8  7  6  5  4  3  2  1    20 19 18 17 16 15 14 13 12 11
30 29 28 27 26 25 24 23 22 21    40 39 38 37 36 35 34 33 32 31
50 49 48 47 46 45 44 43 42 41    60 59 58 57 56 55 54 53 52 51
70 69 68 67 66 65 64 63 62 61    80 79 78 77 76 75 74 73 72 71
90 89 88 87 86 85 84 83 82 81   100 99 98 97 96 95 94 93 92 91
```

(b) From whatever arrangement of the 101 integers has been presented we select the first integer, labeling it $a_1^{(1)}$; then (always moving left to right) we select the next integer $a_2^{(1)}$ which follows and exceeds $a_1^{(1)}$, and so on. This produces an increasing sequence $a_1^{(1)}, a_2^{(1)}, \cdots, a_i^{(1)}$ (which may conceivably end at the first integer). If more than ten numbers appear in this sequence ($i > 10$), the problem is solved. If, however, $i < 10$, we cross out the integers already used, and from the remaining $101 - i$ integers we begin the construction of a new sequence, following the same procedure. We then obtain a new increasing sequence $a_1^{(2)}, a_2^{(2)}, \cdots, a_{i_2}^{(2)}$. Continuation of this process creates from the ordered set of 101 integers a number of increasing sequences. If any one of them contains more than ten integers, the problem has heen solved. Hence we need consider only the case in which none of the sequence we have made up contains more than ten integers.

Since there are 101 integers in all, the number k of increasing sequences must in this case be at least equal to 11. But then it is possible to select from the 101 integers a *decreasing* sequence of eleven integers. A procedure for doing this follows.

We select as the final element of the sequence to be constructed the final element of the final sequence $a_{i_k}^{(k)}$. Then if we select from the previous sequence that number among those of the original sequence not yet crossed out, which just precedes $a_{i_k}^{(k)}$, that number will exceed $a_{i_k}^{(k)}$; otherwise $a_{i_k}^{(k)}$ would have appeared in the preceding sequence. This selected number $a_j^{(k-1)}$ is placed to the left of $a_{i_k}^{(k)}$ in the new sequence. Now we work with $a_j^{(k-1)}$. In the sequence just preceding the $(k-1)$st, the element which had appeared just before $a_j^{(k-1)}$ (among those not already crossed out in the original sequence) will exceed $a_j^{(k-1)}$. This number is placed to the left $a_j^{(k-1)}$ in the new sequence being formed. We can continue in this manner, selecting larger and larger numbers, each appearing earlier in the initial array of 101 numbers, and it is clear that we can make at least eleven such selections.

Remark: It can be proved in analogous way that $(n-1)^2$ positive integers can be arranged in such an order that no n of them will follow sequentially in either an increasing or a decreasing order, but that among $k > (n-1)^2$ integers a sequence of n increasing, or decreasing, numbers can be selected sequentially from the initial array.

91. (a) Reduce each of the 101 selected numbers by the greatest power of 2 which divides it, thus obtaining 101 odd numbers. Since there are precisely 100 distinct odd numbers from 1 to 200, at least two of the 101 odd numbers must be identical. This means that among the 101 numbers originally selected there exist two whose factorizations differ only by a power of 2. The smaller of these two numbers must divide the other.

Remark: A proof by mathematical induction is also possible.

(b) The desired numbers can be selected in the following manner: The fifty odd numbers from 101 to 199; the odd numbers from 51 to 99, each multiplied by 2 (twenty-five numbers); the odd numbers from 27 to 49, each multiplied by 4 (twelve numbers); the odd numbers from 13 to 25, each multiplied by 8 (seven numbers); the odd numbers from 7 to 11, each multiplied by 16 (three numbers); and finally, the three numbers $3 \cdot 32, 5 \cdot 32$, and $1 \cdot 64$. These total 100 numbers, all less than 200, and none of which is divisible by any other.

(c) Assume that it is possible to select 100 integers, none exceeding 200, such that no one of them is divisible by any other. We shall show that none of the numbers 1–15 inclusive can appear in the selected set.

As in the solution of problem (a), we divided out of each number the largest power of 2 which appears as a factor. We obtain a second set consisting of 100 odd numbers, none exceeding 200. No two of these odd numbers belonging to the second set can be equal (otherwise the corresponding original numbers would differ from each other only by a factor which is a power of 2, and so one would be divisible by the other). Accordingly, the 100 odd numbers of the second set are $1, 3, 5, \cdots, 199$.

Since 15 divides 45, the integer $2^{\alpha} \cdot 15$ of the initial set must have $\alpha > 0$, since otherwise 15 would divide $45 \cdot 2^{\beta}$, which also appears in the initial set. This means that 15 cannot have appeared in the initial set. The same line of reasoning eliminates the possibility that 13, 11, or 9 appeared in the initial set. We now consider the integer 7. It is clear that $7 \cdot 2^{\alpha}$, which is a number of the initial set, must have $\alpha \geqq 1$. But suppose that $\alpha = 1$. Since 7 divides 49, which in turn divides 147, and since both $49 \cdot 2^{\beta}$ and $147 \cdot 2^{\gamma}$ belong to the initial set, it is clear that $\beta = \gamma$ is impossible, hence not both β and γ can be zero. But γ is certainly zero, since the initial set centains no integer exceeding 200. Therefore, $7 \cdot 2$ must divide $49 \cdot 2^{\beta}$; this eliminates both 7 and 14 from appearance in the initial set. The same line of reasoning shows that the integers 5, 10, 3, 6, and 12 cannot belong to the initial set. The integer 1 is automatically rejected, by hypothesis, and the integer 2 is rejected, since its inclusion would give precisely the set of 100 odd numbers between 1 and 200, a set which clearly fails to meet the conditions of the problem.

There remain for consideration only the integers 4 and 8. Were 4 to belong to the initial set, then 2^{α} could not be a factor of any other member of the initial set unless $\alpha \leqq 1$. But this is impossible, since it would eliminate the appearance of the integer 3 from the second set (neither 3 nor 6, as has been shown, can belong to the initial set). By similar reasoning the integer 8 is shown to be impossible of inclusion in the initial set. Therefore, none of the integers from 1 to 15, inclusive, can appear in the initially selected set.

Remark: It can proved, in general, that out of the first $2n$ (or fewer) integers it is impossible to select $(n + 1)$ integers having the property that no one of them is divisible by another, but that it is possible to select n (or fewer) such integers. If $3^k < 2n < 3^{k+1}$, then out of the first $2n$ integers it is impossible to select n integers, one of them less than 2^k, such that no number is divisible by another. But n such numbers can be selected, provided the least is equal to 2^k. For example, from 200 numbers it is possible to select 100, the least equal to 16, not one of which is divisible by another.

92. (a) Consider the absolute values of the "least" remainders obtained upon divisible by 100. That is, if a number a yields a positive remainder exceeding 50, then increase the quotient by 1 to obtain a negative remainder; $a = 100q - r$, where now $0 < r < 50$. Since (counting zero) there exist 51 possible distinct remainders not exceeding 50 in absolute value, and since there will be 52 such remainders obtained upon dividing 52 numbers by 100, at least two of these remainders are equal in absolute value. If these two remainders have the same sign ($+$ or $-$), then the difference of the corresponding dividends is divisible by 100; if they are opposite in sign, their sum is divisible by 100.

     (b) Let $a_1, a_2, a_3, \cdots, a_{100}$ be given integers (in any order). Consider the sums

$$s_1 = a_1 ;$$
$$s_2 = a_1 + a_2 ;$$
$$s_3 = a_1 + a_2 + a_3 ;$$
$$\cdots\cdots\cdots\cdots\cdots ;$$
$$s_{100} = a_1 + a_2 + a_3 + \cdots + a_{100} .$$

There are 100 such sums; therefore, unless one or more are divisible by 100, at least two of the sums must yield the same remainder upon division by 100 (there being only 99 different positive remainders possible). If we subtract the smaller of the two sums yielding equal remainders from the larger, we obtain a sum of the form $a_{k+1} + a_{k+2} + \cdots + a_m$, which is divisible by 100.

93. Starting with some initial day, say Monday, assume that the chess player plays a_1 games; on Monday and Tuesday he plays a_2 games; in the three-day period Monday through Wednesday he plays a_3 games, and so on, by the end of the 77th day he has played a_{77} games. Consider the following sequence of integers:

$$a_1, a_2, a_3, \cdots, a_{77} ;$$
$$a_1 + 20, a_2 + 20, \cdots, a_{77} + 20 .$$

We have, in all, $2 \cdot 77 = 154$ integers, none of which exceeds $132 + 20 = 152$, inasmuch as a_{77} is not, according to the imposed conditions, to exceed $11 \cdot 12 = 132$ games played in eleven weeks. Hence at least two of these 154 integers must be equal. However, no two integers of the sequence $a_1, a_2, \cdots, a_{77}$ can be equal, since the chess player has played at least one game every day; similarly, no two of the integers $a_1 + 20, a_2 + 20, \cdots, a_{77} + 20$ can be equal. Therefore, we must have, for some k and some l, the equality

$$a_k = a_l + 20 \ .$$

This equation states that $a_k - a_l = 20$: on a succession of $k - l$ days—from the $(l + 1)$st to the kth, inclusive—the chess master played exactly 20 games.

94. *First solution.* Consider the remainders obtained upon dividing the following numbers by N:

$$1, 11, 111, \cdots, \underbrace{1111 \cdots 1}_{N \text{ times}} \ .$$

Since it is possible to obtain at most $N - 1$ different nonzero remainders from these N numbers, then one of these numbers is divisible by N (in which case the proof is completed), or else two of them, say

$$K = \underbrace{11 \cdots 1}_{k \text{ times}} \text{ and } L = \underbrace{1111 \cdots 1}_{l \text{ times}} \qquad (l > k) \ ,$$

must yield the same remainder upon division by N. But in the latter case the difference

$$L - K = \underbrace{11 \cdots \underset{\substack{\\ \text{times}}}{100 \cdots 0}}_{\substack{(l - k) \qquad k \text{ times}}}$$

is divisible by N.

If N is relatively prime to 10, then, since $L - K = \underbrace{11 \cdots 1}_{(l-k) \text{ times}} \cdot 10^k$ is

divisible by N, it follows that $\underbrace{11 \cdots 1}_{\substack{(l - k) \\ \text{times}}}$ is divisible by N.

Second solution. Consider the decimal expansion of $1/N$:

$$\frac{1}{N} = 0 \cdot b_1 b_2 \cdots b_k a_1 a_2 \cdots a_l \cdots \ ,$$

where $a_i b_j$ represents the succession of digits, and where $a_1 a_2 \cdots a_l$ is the "periodic" part of the decimal expansion. This can be rewritten in the following form:

$$\frac{1}{N} = \frac{b_1 b_2 \cdots b_k a_1 a_2 \cdots a_l - b_1 b_2 \cdots b_k}{\underbrace{99 \cdots 9}_{l \text{ times}}\underbrace{00 \cdots 0}_{k \text{ times}}} \ .$$

It follows that the number $A = \underbrace{99 \cdots 9}_{l \text{ times}}\underbrace{00 \cdots 0}_{k \text{ times}}$ is divisible by N.

But $A = 9A_1$, where $A_1 = \underbrace{11 \cdots 1}_{l \text{ times}}\underbrace{00 \cdots 0}_{k \text{ times}}$. We now consider the number

$$B = \underbrace{11 \cdots 1}_{l \text{ times}}\underbrace{00 \cdots 0}_{k \text{ times}}\underbrace{11 \cdots 1}_{l \text{ times}}\underbrace{00 \cdots 0}_{k \text{ times}} \cdots \underbrace{11 \cdots 1}_{l \text{ times}}\underbrace{00 \cdots 0}_{k \text{ times}},$$

which is obtained if the number A_1 is written nine times successively. It is clear that the number B, equal to the product of A_1 by

$$\underbrace{\underbrace{100 \cdots 01}_{\substack{(l+k) \\ \text{digits}}}\underbrace{00 \cdots 0}_{\substack{(l+k) \\ \text{digits}}} \cdots \underbrace{100 \cdots 01}_{\substack{(l+k) \\ \text{digits}}}}_{8 \text{ times}},$$

is divisible by 9 (the sum of its digits being divisible by 9). Therefore, the number B, consisting entirely of ones and zeros, is divisible by $9A_1 = A$, and hence by N.

If N is relatively prime to 10, then $1/N$ yields a decimal which is periodic, and so B will consist entirely of ones.

95. We shall consider only the final four digits of the integers of the Fibonacci sequence wherever these integers contain five or more digits; that is, we shall deal with a sequence of integers all less than 10^4. Let a_k designate the (four-digit) number appearing in the kth place of this sequence (that is, the final four digits of the kth term of the Fibonacci sequence). If we know the integers a_{k+1} and a_k, we can easily find a_{k-1}, since the four digits composing a_{k-1} will depend only on the corresponding four digits of a_{k+1} and a_k. Now, if for two natural numbers k and n we find that $a_k = a_{n+k}$ and $a_{k+1} = a_{n+k+1}$, then it will follow that

$$a_{k-1} = a_{n+k-1},$$
$$a_{k-2} = a_{n+k-2},$$
$$\cdots\cdots\cdots\cdots,$$
$$a_1 = a_{n+1}$$

(since any number of a Fibonacci sequence is the positive difference between the two succeeding numbers). However, since $a_1 = 0$, it must then follow that $a_{n+1} = 0$; an integer terminating in four zeros will stand in the $(n + 1)$st place of the Fibonacci sequence.

It will suffice to show, then, that there exists an identical pair among the $10^8 + 1$ pairs:

$$a_1, \qquad a_2,$$
$$a_2, \qquad a_3,$$
$$\cdots\cdots\cdots\cdots,$$
$$a_{10^8}, \qquad a_{10^8+1},$$
$$a_{10^8+1}, \qquad a_{10^8+2}$$

But this certainly must occur in the set of numbers $a_1, a_2, a_3, \cdots,$ a_{10^8+2}, since none of them exceeds 10^4, and from the 10^4 integers $0, 1, 2, 3, 4, \cdots, 999$ we can obtain only $10^4 \cdot 10^4 = 10^8$ distinct pairs (the first number can assume at most 10^4 distinct values, and the second number can take on only 10^4 distinct values).

Remark: It can be shown that the first integer of this Fibonacci sequence which will end in four zeros stands in the 7501st place (see the book by B. B. Dynkin and V. A. Uspensky, *Mathematical Conversations*, Issue 6, Library of the USSR Mathematical Society, especially problem 174 and the discussion of it).

96. Consider the decimal parts of the 1001 numbers:

$$0 \cdot \alpha = 0, \alpha, 2\alpha, 3\alpha, \cdots, 1000\alpha$$

(the difference between the given number and the largest number not exceeding the given number). This yields 1001 positive numbers, is decimal form, all less than 1. Partition the interval between 0 and 1 on the real-number axis into 1000 intervals (we shall assume that each interval contains its left end point but not its right end point). We shall investigate how the points of the above sequence are distributed among these intervals of length 1/1000. Since there are 1000 intervals and 1001 points, at least one of the intervals must contain two (or more) points. This shows that there exist two distinct natural numbers, say p and q (neither exceeding 1000) such that the difference between the numbers $p\alpha$ and $q\alpha$ is less than 1/1000.

Say that $p > q$. Consider the number $(p - q)\alpha = p\alpha - q\alpha$. Since $p\alpha = P + d_1$, $q\alpha = Q + d_2$, where P and Q are integers, and d_1 and d_2 are decimals $p\alpha$ and $q\alpha$, it follows that $(p - q)\alpha = (P - Q) + d_2 - d_1$ differs from the integer $P - Q$ by less than 1/1000. This implies that the fraction $\dfrac{P - Q}{p - q}$ differs from α by less than $0.001 \left(\dfrac{1}{p - q}\right)$.

97. First, it is readily seen that none of the fractions under consideration is an integer. Further, no two of these fractions can be equal, for if, say,

$$\frac{k(m + n)}{m} = \frac{l(m + n)}{n}$$

(where k is one of the numbers $1, 2, \cdots, m - 1$, and l is one of the integers $1, 2, \cdots, n - 1$), we would have

$$\frac{k}{m} = \frac{l}{n},$$

$$m = \frac{k}{l}n,$$

which contradicts the fact that m and n are relatively prime (inasmuch as $l < n$ and so cannot be divisible by n).

Consider now a natural number A less than $m + n$. The fractions

$$\frac{m + n}{m}, \frac{2(m + n)}{m}, \cdots, \frac{k(m + n)}{m}$$

will be less than A, or $k(m + n)$ is less than Am, if k is less than $\frac{Am}{m + n}$. Clearly, the number of such fractions is equal to the integral part $\left[\dfrac{Am}{m + n}\right]$ of the fractions $\dfrac{Am}{m + n}$. Similarly, the fractions

$$\frac{m + n}{n}, \frac{2(m + n)}{n}, \cdots, \frac{l(m + n)}{n}$$

will be less than A for $l < \dfrac{An}{m + n}$. The number of such fractions is equal to the greatest integer $\left[\dfrac{An}{m + n}\right]$ in the fractions $\dfrac{An}{m + n}$. The numbers $\dfrac{Am}{m + n}$ and $\dfrac{An}{m + n}$ are nonintegral, since m, n, and $m + n$ are pairwise relatively prime, and the sum of these two numbers is A:

$$\frac{Am}{m + n} + \frac{An}{m + n} = A.$$

But if the sum of two numbers α and β, neither an integer, is equal to an integer A, then $[\alpha] + [\beta] = A - 1$. (The proof of this follows immediately from Figure 7.) Thus,

$$\left[\frac{Am}{m + n}\right] + \left[\frac{An}{m + n}\right] = A - 1,$$

which implies that in the interval $(0, A)$ on the real-number axis there exist precisely $A - 1$ of the fractions.

Figure 7

The proposition of the problem follows immediately. Let us first assume that $A = 1$; we see that none of the fractions is in the interval $(0, 1)$. Let $A = 2$; since there is one fraction in the interval $(0, 2)$, it follows that this fraction must be in the interval $(1, 2)$. Let $A = 3$; since the interval $(0, 3)$ contains two of the fractions, it follows that the interval $(2, 3)$ contains just one of them. Continuation of this reasoning completes the proof of the proposition.

98. *First Solution.* If a number a is in the interval

$$\frac{1000}{m} \geqq a > \frac{1000}{m + 1} \,,$$

then there are obviously, m integral multiples of a which do not exceed 1000: $a, 2a, 3a, \cdots, ma$. Now, if we designate by k_1 the number of given integers which lie between 1000 and $\frac{1000}{2}$, by k_2 the number which lie in the interval $\left(\frac{1000}{2}, \frac{1000}{3}\right)$, by k_3 the number which lie in the interval $\left(\frac{1000}{3}, \frac{1000}{4}\right)$, and so on, then we have, in all, $k_1 + 2k_2 + 3k_3 + \cdots$ numbers, not exceeding 1000, which are multiples of at least one of the given numbers. But, according to the conditions of the problem, all these multiples are different, and so

$$k_1 + 2k_2 + 3k_3 + \cdots < 1000 \,.$$

It remains to be shown that the sum of the reciprocals of all the given integers is less than

$$k_1 \frac{1}{\frac{1000}{2}} + k_2 \frac{1}{\frac{1000}{3}} + k_3 \frac{1}{\frac{1000}{4}} + \cdots = \frac{2k_1 + 3k_2 + 4k_3 + \cdots}{1000}$$

$\Bigg($ here we have replaced k_1 of the largest of the numbers by $\dfrac{1000}{2}$, the following k_2 numbers by $\dfrac{1000}{3}$, the following k_3 numbers by $\dfrac{1000}{4}$, and so on $\Bigg)$. Now we have

$$2k_1 + 3k_2 + 4k_3 + \cdots$$
$$= (k_1 + 2k_2 + 3k_3 + \cdots) + (k_1 + k_2 + k_3 + \cdots)$$
$$= (k_1 + 2k_2 + 3k_3 + \cdots) + n < 1000 + n < 2000 ;$$

consequently, the sum of the reciprocals of the numbers is less than 2.

Second Solution. We introduce an important variation of the foregoing type of reasoning. The number of terms of the sequence $1, 2, \cdots, 1000$ which are divisible by an integer a_k is obviously the greatest integer $\left[\dfrac{1000}{a_k}\right]$ in $\dfrac{1000}{a_k}$. Since the least common multiple of any two of the integers $a_1, a_2, \cdots, a_n$ exceeds 1000, not one of the numbers $1, 2, 3, \cdots, 1000$ can be divisible by two of the integers $a_1, a_2, \cdots, a_n$. It follows that the number of terms of the sequence $1, 2, 3, \cdots, 1000$ which are divisible by at least one of the integers $a_1, a_2, \cdots, a_n$ is equal to the sum

$$\left[\frac{1000}{a_1}\right] + \left[\frac{1000}{a_2}\right] + \left[\frac{1000}{a_3}\right] + \cdots + \left[\frac{1000}{a_n}\right].$$

Since there are 1000 numbers in the sequence $1, 2, \cdots, 1000$, it is clear that

$$\left[\frac{1000}{a_1}\right] + \left[\frac{1000}{a_2}\right] + \cdots + \left[\frac{1000}{a_n}\right] \leqq 1000 .$$

But the greatest integer in a fraction differs from the fraction itself by less than 1; that is,

$$\left[\frac{1000}{a_1}\right] > \frac{1000}{a_1} - 1, \left[\frac{1000}{a_2}\right] > \frac{1000}{a_2} - 1, \cdots, \left[\frac{1000}{a_n}\right] > \frac{1000}{a_n} - 1 .$$

Consequently,

$$\left(\frac{1000}{a_1} - 1\right) + \left(\frac{1000}{a_2} - 1\right) + \cdots + \left(\frac{1000}{a_n} - 1\right) < 1000 ;$$

that is,

$$\frac{1000}{a_1} + \frac{1000}{a_2} + \frac{1000}{a_3} + \cdots + \frac{1000}{a_n} < 1000 + n < 2000 \,,$$

and so

$$\frac{1}{a_1} + \frac{1}{a_2} + \cdots + \frac{1}{a_n} < 2 \,.$$

Remarks: The estimation of this bound can be made very precise. Consider all the multiples of those given integers which do not exceed 500. Now, k_1 of the given integers will exceed 500; $k_2 + k_3$ of the integers will not exceed 500, but will exceed $\dfrac{500}{2}$; $k_4 + k_5$ of them will not exceed $\dfrac{500}{2}$, but will exceed $\dfrac{500}{3}$; and so on. It follows, as in the first solution, that the total number of integers not exceeding 500, and which are multiples of at least one of the given n numbers, is equal to

$$(k_2 + k_3) + 2(k_4 + k_5) + 3(k_6 + k_7) + \cdots \,.$$

Therefore,

$$(k_2 + k_3) + 2(k_4 + k_5) + 3(k_6 + k_7) + \cdots < 500 \,.$$

We note now that the difference

$$500 - [(k_2 + k_3) + 2(k_4 + k_5) + \cdots]$$

expresses the number of integers, not exceeding 500, which are not multiples of any one of the given integers; and the difference

$$1000 - (k_1 + 2k_2 + 3k_3 + \cdots)$$

is the number of integers not exceeding 1000 which fail to be multiples of any one of the given integers. Consequently,

$$500 - [(k_2 + k_3) + 2(k_4 + k_5) + 3(k_6 + k_7) + \cdots] < 1000 - (k_1 + 2k_2 + 3k_3 + \cdots),$$

from which we obtain

$$(k_1 + k_2) + 2(k_3 + k_4) + 3(k_5 + k_6) + \cdots < 500 \,.$$

It remains to note that

$$2k_1 + 3k_2 + 4k_3 + 5k_4 + 6k_5 + 7k_6 + \cdots$$
$$< (k_1 + 2k_2 + 3k_3 + 4k_4 + 5k_5 + 6k_6 + \cdot \,\cdot)$$
$$+ [(k_1 + k_2) + 2(k_3 + k_4) + 3(k_5 + k_6) + \cdots]$$
$$< 1000 + 500 = 1500 \,.$$

Thus, the sum of the reciprocals of all of the numbers is less than

$$\frac{2k_1 + 3k_2 + 4k_3 + \cdot\cdot}{1000} \,,$$

or less than $1\frac{1}{2}$.

Analogously, if initially we consider the multiples of those given integers not exceeding 333, we can prove that the sum of the reciprocals of the given numbers is less than $1\frac{1}{4}$.

Note that the number 1000 in this problem can be replaced by any other integer.

99. Consider the conversion of $\frac{q}{p}$ to a (periodic) decimal (see the footnote accompanying the solution of problem 38).

$$\frac{q}{p} = \overline{A \cdot a_1 a_2 \cdots a_k a_1 a_2 \cdots a_k a_1 a_2 \cdots} \ ,$$

Here, A is the whole part of the quotient obtained upon division of q by p. We can write

$$q = Ap + q_1 \ ,$$

where the remainder q_1 is less than p. Moreover, $\overline{Aa_1}$ will be the whole part of the quotient obtained upon division of $10q$ by p ($\overline{Aa_1}$ is the integer composed of the digits of A followed by the digit a_1):

$$10q = \overline{Aa_1} \cdot p + q_2 \ ,$$

where $q_2 < p$. Similarly,

$$10^2 \cdot q = \overline{Aa_1 a_2} \cdot p + q_3, \ \cdots, \ 10^k \cdot q = \overline{Aa_1 a_2 \cdots a_k} \cdot p + q_k, \ \cdots \ .$$

The periodic part of the decimal begins again at the point where division of $10^k q$ by p yields the remainder first obtained, $q_k = q_1$, upon division of q by p. Thus, the number k of digits composing a period of the decimal is determined as the least power of 10 such that $10^k \cdot q$ yields on division by p the same remainder as did q. For this k it is clear that $10^k q - q = (10^k - 1)q$ is divisible by p whence $10^k - 1$ is divisible by p, inasmuch as p and q are relatively prime (p itself is prime).

Assume now that k is even: $k = 2l$. Since $10^{2l} - 1 = (10^l - 1) \times (10^l + 1)$ is divisible by p, either $10^l - 1$ or $10^l + 1$ is divisible by the prime p. But $10^l - 1$ cannot be divisible by p, since if it were, it would yield the same remainder as does q upon division by p, and the period of the fraction $\frac{q}{p}$ would be l instead of $k = 2l$. Hence, we must conclude that $10^l + 1$ is divisible by p.

It follows from this conclusion that the sum $\frac{10^l q}{p} + \frac{q}{p}$ is an integer. But

$$\frac{10^l q}{p} + \frac{q}{p} = \overline{Aa_1 a_2 \cdots a_l. \ a_{l+1} a_{l+2} \cdots a_{2l} a_1 a_2 \cdots a_l \cdots}$$
$$+ \ \overline{A. \ a_1 a_2 \cdots a_l a_{l+1} \cdots a_{2l} \cdots} \ .$$

Hence, the sum of the decimal fractions

$$\overline{0, a_{l+1}a_{l+2}\cdots a_{2l}a_1a_2\cdots a_l\cdots} + \overline{0, a_1a_2\cdots a_la_{l+1}\cdots a_{2l}\cdots}$$

is an integer. Since each of these positive decimals has a value less than 1, this sum must be $1 = 0.999\cdots$, which is possible only if

$$a_1 + a_{l+1} = 9,$$
$$a_2 + a_{l+2} = 9,$$
$$\cdots\cdots\cdots\cdots,$$
$$a_l + a_{2l} = 9.$$

It immediately follows that

$$\frac{a_1 + a_2 + \cdots + a_{2l}}{2l} = \frac{9}{2}.$$

If k is odd, the equation

$$\frac{a_1 + a_2 + \cdots + a_k}{k} = \frac{9}{2}$$

will not be possible, since k is not divisible by 2.

100. The numbers of digits in the period of the fractions $\dfrac{a_n}{p^n}$ and $\dfrac{a_{n+1}}{p^{n+1}}$ are equal to the least positive integers k and l, respectively, such that $10^k - 1$ is divisible by p^n and $10^l - 1$ is divisible by p^{n+1} (see the solution of the preceding problem). Consider the difference

$$(10^l - 1) - (10^k - 1) = 10^k(10^{l-k} - 1).$$

Since this difference is divisible by p^n, $10^{l-k} - 1$ is divisible by p^n. We shall show that $10^d - 1$, where d is the greatest common divisor of $l - k$ and k, is also divisible by p^n.

Write $l - k = qk + r$; we can then write

$$10^{l-k} - 1 = 10^{qk+r} - 1 = 10^r(10^{qk} - 1) + (10^r - 1).$$

But $10^{qk} - 1 = (10^k)^q - 1^q$ is divisible by $10^k - 1$, whence by p^n, and therefore $10^r - 1$ is divisible by p^n. Similarly, it may be shown that: $10^{r_1} - 1$ (where r_1 is the remainder obtained upon division of k by r) is also divisible by p^n; that $10^{r_2} - 1$ (where r_2 is the remainder obtained upon division of r by r_1) is divisible by p^n; that $10^{r_3} - 1$ (where r_3 is the remainder obtained upon division of r_1 by r_2) is divisible by p^n; and so on. (This process, by which a diminishing sequence of remainders is obtained, is called the Euclidean Algorithm.)

It is readily shown that the sequence of positive integers $r, r_1, r_2, \cdots$ must include the number d; since both $l - k$ and k are divisible by d so is $r = (l - k) - qk$; since both k and r are divisible by d, so is r_1; since both r and r_1 are divisible by d, so is r_2, and so on. Consequently, all the (remainders) numbers of the sequence we have been forming are divisible by d; further, this sequence of positive integers is decreasing and so must terminate with 0. If the final nonzero remainder is r_k, then r_{k-1} is divisible by r_k (since the following remainder is zero); r_{k-2} is divisible by r_k (since now both r_{k-1} and r_k are divisible by r_k); r_{k-3} is divisible by r_k (since both r_{k-2} and r_{k-1} are divisible by r_k), and so on. Finally, k and $l - k$ are divisible by r_k. The inequality $r_k > d$ would contradict the fact that d is the greatest common divisor of $l - k$ and k; hence, $r_k = d$. (Clearly, $r_k < d$ is impossible, since d divides $r_k \neq 0$.)

Now, k has been defined as the least integer such that $10^k - 1$ is divisible by p^n. Therefore, since $10^d - 1$ is also divisible by p^n, it follows that $d = k$, and $l - k$ is a multiple of k, which means that l is a multiple of k; that is, $l = kr$.

We can expand $10^l - 1$ as follows:

$$10^l - 1 = 10^{kr} - 1$$
$$= (10^k - 1)[10^{(r-1)k} + 10^{(r-2)k} + \cdots + 10^k + 1] .$$

Since $10^k - 1$ is divisible by p^n, 10^k yields a remainder of 1 when divided by p^n. It follows that $10^{2k} = 10^k \cdot 10^k$ yields a remainder of 1 when divided by p^n; $10^{3k} = 10^{2k} \cdot 10^k$ yields a remainder of 1 when divided by p^n; and so on. Therefore, each term of the parenthetical sum on the right, above, yields a remainder of 1 when divided by p^n, and hence the remainder of this sum is r. It follows that if $10^k{}^{1}$ fails to be divisible by p^{n+1}, then the least value of l such that $10^l - 1$ is divisible by p^{n+1} is pk, inasmuch as $10^{pk} - 1$ is divisible by p^{n+1}, but is not divisible by p^{n+2} (since the expression in parentheses is not divisible by p^2).

The assertion of the problem follows.

101. (1) If x is any real number, it can be written in the form $x = [x] + \alpha$, where α is a nonnegative number less than 1.

We write the number y as $y = [y] + \beta$ $(0 \leq \beta < 1)$. Then $x + y = [x] + [y] + \alpha + \beta$. Since $\alpha + \beta \geq 0$, it is clear from this equation that $[x] + [y]$ is an integer which does not exceed $x + y$. Since $[x + y]$ is the largest integer which fails to exceed $x + y$, it follows that $[x + y] \geq [x] + [y]$.

(2) *First Solution.* Write $x = [x] + \alpha$, where $0 \leq \alpha < 1$. Assume that the integer $[x]$, when divided by n, yields the quotient q and

the remainder r:

$$[x] = qn + r \quad (0 \leqq r \leqq n - 1) .$$

We then have

$$\frac{[x]}{n} = q + \frac{r}{n} \,;$$

$$\left[\frac{[x]}{n} \right] = q \,;$$

$$x = qn + r + \alpha = qn + r_1 \,;$$

where $r_1 = r + \alpha < n$; $\dfrac{x}{n} = q + \dfrac{r_1}{n}$; $\left[\dfrac{x}{n} \right] = q = \left[\dfrac{[x]}{n} \right]$; which proves the assertion.

Second Solution. Consider all integers divisible by n but not exceeding x. Clearly, there are $\left[\dfrac{x}{n} \right]$ of them. Consider, also, all integers divisible by n but not exceeding $[x]$. There are $\left[\dfrac{[x]}{n} \right]$ of these. Since these two integers are the same, we have

$$\left[\frac{[x]}{n} \right] = \left[\frac{x}{n} \right] .$$

(3) *First Solution.* Let $x = [x] + \alpha$. Since $0 \leqq \alpha < 1$, then α is equal to one of, or lies between two successive fractions of, the sequence

$$\frac{0}{n}, \frac{1}{n}, \frac{2}{n}, \ldots, \frac{n-1}{n}, \frac{n}{n} .$$

Assume that the fractions in question are $\dfrac{k}{n}$ and $\dfrac{k+1}{n}$; that is, that

$$\frac{k}{n} \leqq \alpha < \frac{k+1}{n} .$$

Now, we have

$$x + \frac{n-k-1}{n} = [x] + \alpha + \frac{n-k-1}{n} < [x] + \frac{k+1}{n} + \frac{n-k-1}{n}$$
$$= [x] + 1 ,$$
$$x + \frac{n-k}{n} = [x] + \alpha + \frac{n-k}{n} \geqq [x] + \frac{k}{n} + \frac{n-k}{n} = [x] + 1 ,$$

and

$$x + \frac{n-1}{n} = [x] + \alpha + \frac{n-1}{n} < [x] + \frac{k+1}{n} + \frac{n-1}{n}$$

$$= [x] + \frac{n+k}{n} \leq x + 2 .$$

It follows that

$$[x] \leq \left[x + \frac{1}{n} \right] \leq \cdots \leq \left[x + \frac{n-k-1}{n} \right] < [x] + 1 ,$$

$$[x] + 1 \leq \left[x + \frac{n-k}{n} \right] \leq \cdots \leq \left[x + \frac{n-1}{n} \right] < [x] + 2 ;$$

that is,

$$[x] = \left[x + \frac{1}{n} \right] = \cdots = \left[x + \frac{n-k-1}{n} \right] ,$$

$$\left[x + \frac{n-k}{n} \right] = \cdots = \left[x + \frac{n-1}{n} \right] = [x] + 1 .$$

Since after the first $n - k$ numbers, there remain k, we have

$$[x] + \cdots + \left[x + \frac{n-1}{n} \right] = (n - k)[x] + k([x] + 1) = n[x] + k .$$

But this is can be shown to be equal to the integral part of nx. Since $k \leq n\alpha < k + 1$, then $n\alpha = k + \beta$, where $0 \leq \beta < 1$. As a result,

$$[nx] = [n[x] + n\alpha] = [n[x] + k + \beta] = n[x] + k .$$

This proves that

$$[x] + \left[x + \frac{1}{n} \right] + \cdots + \left[x + \frac{n-1}{n} \right] = [nx] .$$

Second Solution. Consider the left member of the given equation. If $0 \leq x < \frac{1}{n}$, then all the numbers $x, x + \frac{1}{n}, \cdots, x + \frac{n-1}{n}$ are less than 1, and so the integral part of each is 0; also $[nx] = 0$. Thus the equation holds for such values of x.

Now let x be arbitrary. If we multiply x by $\frac{1}{n}$, all the terms on the left are "shifted" once to the right, and the final term,

$$\left[x + \frac{n-1}{n} \right] ,$$

Page 184

becomes $[x + 1]$, which exceeds $[x]$ by 1. This means that multiplication of x by $\frac{1}{n}$ increases the left member of the equation by 1. The right member is also increased by 1 when multiplied by $\frac{1}{n}$. For x it is possible to find a number α, where

$$0 \leqq \alpha < \frac{1}{n},$$

such that x differs from α by $\frac{m}{n}$, where m is an integer. It follows that the equality is preserved for all x.

102. Consider all those points of the Cartesian plane having integer coordinates ("integer points") x and y, where $1 \leqq x \leqq q - 1$ and $1 \leqq y \leqq p - 1$. These points lie within a rectangle $OABC$ the lengths of whose sides are $OA = q$ and $OC = p$ (Figure 8). There are $(q - 1)(p - 1)$ points. Draw the diagonal OB of the rectangle. It is

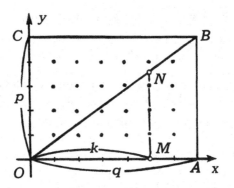

Figure 8

clear that no integer point will lie on this diagonal, since the coordinates (x, y) of all the points of the diagonal OB are related by the proportion

$$\frac{x}{y} = \frac{OA}{AB} = \frac{q}{p},$$

and since q and p are relatively prime, there do not exist integers $x < p$ and $y < q$ such that $\frac{x}{y} = \frac{q}{p}$.

We note now that the number of integer points whose abscissa is equal to k, where k is some positive integer less than q and which appear under the diagonal OB, is equal to the integral part of the length of the segment MN shown in Figure 8. Since

$$MN = \frac{OM}{OA} \cdot AB = \frac{kp}{q},$$

this number is $\left[\frac{kp}{q}\right]$. Hence, the sum

$$\left[\frac{p}{q}\right] + \left[\frac{2p}{q}\right] + \left[\frac{3p}{q}\right] + \cdots + \left[\frac{(q-1)p}{q}\right]$$

is equal to the total number of all integer points appearing under the diagonal OB. There are, in all, $(q-1)(p-1)$ integer points in the rectangle; because of the symmetrical pattern of these points and the fact that OB bisects the rectangle, exactly half of these points lie under the diagonal (none lies on the diagonal). Therefore,

$$\left[\frac{p}{q}\right] + \left[\frac{2p}{q}\right] + \left[\frac{3p}{q}\right] + \cdots + \left[\frac{(q-1)p}{q}\right] = \frac{(q-1)(p-1)}{2}.$$

It is shown in the same way that

$$\left[\frac{p}{q}\right] + \left[\frac{2p}{q}\right] + \left[\frac{3p}{q}\right] + \cdots + \left[\frac{(p-1)q}{p}\right] = \frac{(p-1)(q-1)}{2}.$$

103. *First Solution.* The equation is valid, trivially, if $n = 1$. Proceeding by mathematical induction, we shall show that if the equality is assumed to hold for an integer n, then it must hold for $n + 1$.

If $n + 1$ is not divisible by k; then

$$n + 1 = qk + r \qquad (1 \leqq r \leqq k - 1),$$

and $n = qk + r'$, where $r' = r - 1$; that is, $0 \leqq r' \leqq k - 2$. It follows that the integers $\left[\frac{n+1}{k}\right]$ and $\left[\frac{n}{k}\right]$ coincide (both are equal to q). If, then, $n + 1$ is divisible by k, or $n + 1 = qk$, then

$$\left[\frac{n+1}{k}\right] = q, \quad \text{and} \quad \left[\frac{n}{k}\right] = q - 1;$$

that is,

$$\left[\frac{n+1}{k}\right] = \left[\frac{n}{k}\right] + 1.$$

Hence,

$$\left[\frac{n+1}{k}\right] = \left[\frac{n}{k}\right], \text{ if } k \text{ is not a divisor of } n+1;$$

$$\left[\frac{n+1}{k}\right] = \left[\frac{n}{k}\right] + 1, \text{ if } k \text{ is a divisor of } n+1.$$

It follows that:

(a) $\left[\dfrac{n+1}{1}\right] + \left[\dfrac{n+1}{2}\right] + \cdots + \left[\dfrac{n+1}{n+1}\right]$

$$= \left[\frac{n}{1}\right] + \left[\frac{n}{2}\right] + \cdots + \left[\frac{n}{n+1}\right] + t_{n+1}.$$

That is, if

$$\left[\frac{n}{1}\right] + \left[\frac{n}{2}\right] + \cdots + \left[\frac{n}{n}\right] = t_1 + t_2 + \cdots + t_n,$$

then

$$\left[\frac{n+1}{1}\right] + \left[\frac{n+1}{2}\right] + \cdots + \left[\frac{n+1}{n+1}\right] = t_1 + t_2 + \cdots + t_n + t_{n+1}.$$

(b) $\left[\dfrac{n+1}{1}\right] + 2\left[\dfrac{n+1}{2}\right] + \cdots + (n+1)\left[\dfrac{n+1}{n+1}\right]$

$$= \left[\frac{n}{1}\right] + 2\left[\frac{n}{2}\right] + \cdots + (n+1)\left[\frac{n}{n+1}\right] + s_{n+1}.$$

That is, if

$$\left[\frac{n}{1} + 2\left[\frac{n}{2}\right] + \cdots + n\left[\frac{n}{n}\right]\right] = s_1 + s_2 + \cdots + s_n,$$

then

$$\left[\frac{n+1}{1}\right] + 2\left[\frac{n+1}{2}\right] + \cdots + (n+1)\left[\frac{n+1}{n+1}\right]$$

$$= s_1 + s_2 + \cdots s_n + s_{n+1}.$$

Second Solution. The number of integers of the sequence $1, 2, 3,$ $\cdots, n$ which are divisible by any chosen integer k is equal to $\left[\dfrac{n}{k}\right]$ $\left(\text{these will be the integers } k, 2k, 3k, \cdots, \left[\dfrac{n}{k}\right]k\right)$. The sum obtained by totaling this divisor each time it appears is $k\left[\dfrac{n}{k}\right]$ $\left(\text{that is,}\right.$

$k + k + \cdots + k;$ $\left[\dfrac{n}{k}\right]$ times$\Big)$. We have the following results.

(a) The sum $\left[\dfrac{n}{1}\right] + \left[\dfrac{n}{2}\right] + \cdots + \left[\dfrac{n}{k}\right] + \cdots + \left[\dfrac{n}{n}\right]$ is equal to the number of integers of the sequence $1, 2, 3, \cdots, n$ which are divisible by 1, plus the number of integers of this sequence which are divisible by 2, $\cdots$, plus the number which are divisible by n. But this is precisely the sum $t_1 + t_2 + t_3 + \cdots + t_n$ given in the problem.

(b) The sum $1\left[\dfrac{n}{1}\right] + 2\left[\dfrac{n}{2}\right] + \cdots + k\left[\dfrac{n}{k}\right] + \cdots + n\left[\dfrac{n}{n}\right]$ is equal to the sum obtained by adding the integer 1 each time it appears as a divisor of a number of the sequence $1, 2, 3, \cdots, n$, plus the integer 2 taken each time it appears as a divisor of one of the number of this sequence, plus the integer 3 taken each time it appears as a divisor, and so on. But this is precisely the series $s_1 + s_2 + s_3 + \cdots + s_n$.

Third Solution. An elegant geometric solution suggested by the second solution will be given. Consider the equilateral hyperbolas $y = \dfrac{k}{x}$, or, equivalently, $xy = k$ (of which we shall take only the branches in the first quadrant, see Figure 9).

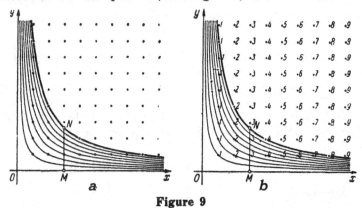

Figure 9

We note all the points in the first quadrant which have integral coordinates (integer point). Now, if an integer x is a divisor of the integer k, then the point (x, y) is a point on the graph of the hyperbola $xy = k$. Conversely, if the hyperbola $xy = k$ contains an integer point, then the x-coordinate is a divisor of k. Hence, the number of integers t_k which are divisors of the integer k is equal to the number of integer points lying on the hyperbola $xy = k$. The number t_k of divisors of k is thus equal to the number of abscissas of integer points lying on the hyperbola $xy = k$. Now, we make use of the fact that the hyperbola $xy = n$ lies "farther out" in the quadrant than do $xy = 1$, $xy = 2$, $xy = 3$, $\cdots$, $xy = n - 1$. The following implications hold.

(a) The sum $t_1 + t_2 + \cdots + t_n$ is equal to the number of integer points lying under (or on) the hyperbola $xy = n$. Each such point will lie on a hyperbola $xy = k$, where $k \leq n$. The number of integer points with abscissa k located under the hyperbola is equal to the integral part of the length of the segment MN [Figure 9 (a)]. That is, $\left[\dfrac{n}{k}\right]$, since $MN = \dfrac{n}{k}$ (compare with the solution of problem 102). Thus, we obtain

$$t_1 + t_2 + t_3 + \cdots + t_n = \left[\frac{n}{1}\right] + \left[\frac{n}{2}\right] + \left[\frac{n}{3}\right] + \cdots + \left[\frac{n}{n}\right].$$

(b) Write alongside each integer point [Figure 9 (b)] the number equal to its abscissa; then, $s_1 + s_2 + \cdots + s_n$ is equal to the sum of the integers $\left[\dfrac{n}{k}\right]$ for the integer points located under (or on) the hyperbola $xy = n$. But the sum of the integers for all such points for a specific abscissa k is equal to $k\left[\dfrac{n}{k}\right]$. Hence, we have

$$s_1 + s_2 + s_3 + \cdots + s_n = \left[\frac{n}{1}\right] + 2\left[\frac{n}{2}\right] + 3\left[\frac{n}{3}\right] + \cdots + n\left[\frac{n}{n}\right].$$

104. It is readily shown that $(2 + \sqrt{2})^n + (2 - \sqrt{2})^n$ is an integer: if $(2 + \sqrt{2})^n = a_n + b_n\sqrt{2}$, where a_n and b_n are integers, then $(2 - \sqrt{2})^n = a_n - b_n\sqrt{2}$ (this follows from the binomial formula, but may also be shown by mathematical induction). Since $(2 - \sqrt{2})^n < 1$, it follows that

$$[(2 + \sqrt{2})^n] = (2 + \sqrt{2})^n + (2 - \sqrt{2})^n - 1,$$

and, consequently,

$$(2 + \sqrt{2})^n - [(2 + \sqrt{2})^n] = 1 - (2 - \sqrt{2})^n.$$

But since $(2 - \sqrt{2}) < 1$, we can, by taking the power n sufficiently high, make $(2 - \sqrt{2})^n$ as small as we wish. If $(2 - \sqrt{2})^n < 0 \cdot 000001$, then

$$(2 + \sqrt{2})^n - [(2 + \sqrt{2})^n] = 1 - (2 - \sqrt{2})^n > 0.999999.$$

105. (a) *First Solution.* Since $(2 + \sqrt{3})^n + (2 - \sqrt{3})^n$ is an integer (see the solution to problem 104), and since $(2 - \sqrt{3})^n < 1$, it follows that

$$[(2 + \sqrt{3})^n] = (2 + \sqrt{3})^n + (2 - \sqrt{3})^n - 1 .$$

If we expand $(2 \pm \sqrt{3})^n$ by the binomial formula, we obtain

$$(2 + \sqrt{3})^n + (2 - \sqrt{3})^n = 2(2^n + C_n^2 2^{n-2} \cdot 3 + C_n^4 2^{n-4} \cdot 3^2 + \cdots) ,$$

which is divisible by 2. Therefore, $[(2 + \sqrt{3})]^n = (2 + \sqrt{3})^n + (2 - \sqrt{3})^n - 1$ is odd.

Second Solution. The number $(2 + \sqrt{3})^n$ can be written in the form $a_n + b_n\sqrt{3}$, where a_n and b_n are integers. We shall show that

$$a_n^2 - 3b_n^2 = 1 .$$

The proof will be carried out by mathematical induction. First, if $n = 1$, then $a_1 = 2$, $b_1 = 1$, and $2^2 - 3 \cdot 1 = 1$.

Now, assume that for some n

$$(2 + \sqrt{3})^n = a_n + b_n\sqrt{3} ,$$

where $a_n^2 - 3b_n^2 = 1$. Then for $(2 + \sqrt{3})^{n+1}$ we have

$$(2 + \sqrt{3})^{n+1} = (a_n + b_n\sqrt{3})(2 + \sqrt{3}) = (2a_n + 3b_n) + (a_n + 2b_n)\sqrt{3} .$$

We see that $a_{n+1} = 2a_n + 3b_n$ and $b_{n+1} = a_n + 2b_n$. Consequently,

$$a_{n+1}^2 - 3b_{n+1}^2 = (2a_n + 3b_n)^2 - 3(a_n + 2b_n)^2 = a_n^2 - 3b_n^2 = 1 .$$

Therefore, $a_n^2 - 3b_n^2$ for any n.

Thus, we have

$$\begin{aligned}
[a_n + b_n\sqrt{3}] &= a_n + [b_n\sqrt{3}] = a_n + [\sqrt{3b_n^2}] \\
&= a_n + [\sqrt{a_n^2 - 1}] = a_n + (a_n - 1) = 2a_n - 1 .
\end{aligned}$$

That is, $[(2 + \sqrt{3})^n] = [a_n + b_n\sqrt{3})]$ is odd.

(b) We note that

$$[(1 + \sqrt{3})^n] = \begin{cases} (1 + \sqrt{3})^n + (1 - \sqrt{3})^n - 1 , & \text{if } n \text{ is even}, \\ (1 + \sqrt{3})^n + (1 - \sqrt{3})^n , & \text{if } n \text{ is odd}. \end{cases}$$

Consequently, the expressions on the right are integers [see the solution to problem (a)]. For n even, $0 < (1 - \sqrt{3})^n < 1$; for n odd, $-1 < (1 - \sqrt{3})^n < 0$.

We shall investigate separately the cases for even n and odd n.

n is even: $n = 2m$. Then

$$[1 + \sqrt{3}\,)^{2m}] = (1 + \sqrt{3}\,)^{2m} + (1 - \sqrt{3}\,)^{2m} - 1$$
$$= \{(1 + \sqrt{3}\,)^2\}^m + \{(1 - \sqrt{3}\,)^2\}^m - 1$$
$$= (4 + 2\sqrt{3}\,)^m + (4 - 2\sqrt{3}\,)^m - 1$$
$$= 2^m\{(2 + \sqrt{3}\,)^m + (2 - \sqrt{3}\,)^m\} - 1 .$$

But the number in the braces is clearly an integer, and so the number $[(1 + \sqrt{3}\,)^{2m}] = 2^m N - 1$ is always odd. Hence, if n is even, the highest power of 2 by which $[(1 + \sqrt{3}\,)^n]$ is divisible is zero.

n is odd. $n = 2m + 1$. Then

$$[(1 + \sqrt{3}\,)^{2m+1}] = (1 + \sqrt{3}\,)^{2m+1} + (1 - \sqrt{3}\,)^{2m+1}$$
$$= (4 + 2\sqrt{3}\,)^m(1 + \sqrt{3}\,) + (4 - 2\sqrt{3}\,)^m(1 - \sqrt{3}\,)$$
$$= 2^m\{(2 + \sqrt{3}\,)^m(1 + \sqrt{3}\,) + (2 - \sqrt{3}\,)^m(1 - \sqrt{3}\,\}$$
$$= 2^m\{[(2 + \sqrt{3}\,)^m + (2 - \sqrt{3}\,)^m]$$
$$+ \sqrt{3}\,[(2 + \sqrt{3}\,)^m - (2 - \sqrt{3}\,)^m]\} .$$

Let $(2 + \sqrt{3}\,)^m = a_m + b_m\sqrt{3}$, where a_m and b_m are integers; then $(2 - \sqrt{3}\,)^m = a_m + b_m\sqrt{3}$. Substitution yields

$$[(1 + \sqrt{3}\,)^{2m+1}] = 2^m\{a_m + b_m\sqrt{3} + a_m - b_m\sqrt{3}$$
$$+ \sqrt{3}\,(a_m + b_m\sqrt{3} - a_m + b_m\sqrt{3}\,)\}$$
$$= 2^m(2a_m + 6b_m) = 2^{m+1}(a_m + 3b_m) .$$

But $a_m + 3b_m$ is odd. In fact,

$$(a_m + 3b_m)(a_m - 3b_m) = a_m^2 - 9b_m^2 = (a_m^2 - 3b_m^2) - 6b_m^2 = 1 - 6b_m^2$$

[see the second solution to problem (a)]. Since $1 - 6b_m^2$ is odd, both factors, $(a_m + 3b_m)$ and $(a_m - 3b_m)$, are odd. Hence the greatest power of 2 by which $[(1 + \sqrt{3}\,)^n]$ is divisible, for $n = 2m + 1$, is equal to

$$m + 1 = \frac{n+1}{2} = \left[\frac{n}{2}\right] + 1 .$$

106. Since $(2 + \sqrt{5}\,)^p - (2 - \sqrt{5}\,)^p$ is an integer, and since $-1 < (2 - \sqrt{5}\,)^p < 0$ (for p odd; see the solutions to problems 104 and 105), it follows that

$$[(2 + \sqrt{5}\,)^p] = (2 + \sqrt{5}\,)^p + (2 - \sqrt{5}\,)^p .$$

Use of the binomial formula yields

$$(2 + \sqrt{5}\,)^p + (2 - \sqrt{5}\,)^p$$
$$= 2(2^p + C_p^2 2^{p-2} 5 + C_p^4 2^{p-4} 5^2 + \cdots + C_p^{p-1} 2 \cdot 5^{(p-1)/2}) ,$$

so that

$$[(2 + \sqrt{5})^p] - 2^{p+1}$$
$$= 2(C_p^2 2^{p-2} 5 + C_p^4 2^{p-4} 5^2 + \cdots + C_p^{p-1} 2 \cdot 5^{(p-1)/2}).$$

All the binomial coefficients

$$C_p^2 = \frac{p(p-1)}{1 \cdot 2},$$

$$C_p^4 = \frac{p(p-1)(p-2)(p-3)}{1 \cdot 2 \cdot 3 \cdot 4},$$

$$\cdots\cdots\cdots\cdots\cdots\cdots,$$

$$C_p^{p-1} = p$$

are divisible by the prime p (they are integers, and the numerators are divisible by the prime p, but the denominators are not). This means that the difference $[(2 + \sqrt{5})^p] - 2^{p+1}$ is divisible by p.

107. $C_n^p = \dfrac{n(n-1)(n-2)\cdots(n-p+1)}{p!}$. Exactly one of the p consecutive integers $n, n-1, n-2, \cdots, n-p+1$ is divisible by p; let us designate that integer by N. Then $\left[\dfrac{n}{p}\right] = \dfrac{N}{p}$, and the difference given in the problem takes on the form

$$\frac{n(n-1)\cdots(N+1)N(N-1)\cdots(n-p+1)}{p!} - \frac{N}{p}.$$

We observe now that the integers $n, n-1, \cdots, N+1, N-1, \cdots, n-p+1$ (the integer N being omitted) yield all the possible positive remainders $1, 2, \cdots, p-1$ upon division by p, inasmuch as the p successive integers from n to $n-p+1$ must yield all these remainders including 0, and we have left out the sole integer divisible by p. Now we shall show that

$$n(n-1)\cdots(N+1)(N-1)\cdots(n-p+1) - (p-1)!$$

is divisible by p. We write the equations

$$n = k_1 p + a_1,$$
$$n - 1 = k_2 p + a_2,$$
$$\cdots\cdots\cdots\cdots\cdots,$$
$$N + 1 = k_i p + a_i,$$
$$N - 1 = k_{i+1} p + a_{i+1},$$
$$\cdots\cdots\cdots\cdots\cdots,$$
$$n - p + 1 = k_{p-1} p + a_{p-1},$$

where $a_1, a_2, \cdots, a_{p-1}$ are the integers $1, 2, \cdots, p-1$ in some order. It is clear that if all the integers on the right are multiplied together, every term of the expansion will contain p as a factor, except the term $a_1 \cdot a_2 \cdots a_{p-1}$, and this term is equal to $(p-1)!$. Thus, the difference shown is divisible by p.

If we multiply this difference by the integer $\dfrac{N}{p}$, the product still remains divisible by p; that is,

$$\frac{n(n-1)\cdots(n-p+1)}{p} - \frac{N(p-1)!}{p}$$

is divisible by p. If we divide this difference by $(p-1)!$ [which does divide it; see problem 49 (a)], we obtain

$$\frac{n(n-1)\cdots(n-p+1)}{p!} - \frac{N}{p} = C_n^p - \left[\frac{n}{p}\right].$$

This still is divisible by p, inasmuch as p and $(p-1)!$ are relatively prime.

108. Let α and β be two numbers satisfying the conditions of the problem. For some selected positive integer N we consider all the integers of the sequence $[\alpha], [2\alpha], [3\alpha], \cdots$, which do not exceed N; we shall write these as $[\alpha], [2\alpha], \cdots, [n\alpha]$. Since $[n\alpha] \leq N$, where n is maximal, we have $[(n+1)\alpha] > N$, and hence $n\alpha < N+1$, $(n+1)\alpha \geq N+1$, $N+1 > n\alpha \geq N+1-\alpha$, or $n\alpha = N+l$, where $1 > l \geq 1-\alpha$.

We may show, in a similar manner, that if m is the maximal integer such that the integers $[\beta], [2\beta], [3\beta], \cdots, [m\beta]$ all fail to exceed N, then $m\beta = N+l'$, where $1 > l' \geq 1-\beta$. Since, according to the conditions of the problem, we encounter in the sequences $[\alpha], [2\alpha], \cdots, [n\alpha]$ and $[\beta], [2\beta], \cdots, [m\beta]$ each positive integer $1, 2, 3, \cdots, N$ exactly once, it follows that $n+m=N$. This implies that

$$\frac{N+l}{\alpha} + \frac{N+l'}{\beta} = N$$

or

$$N\left[1 - \left(\frac{1}{\alpha} + \frac{1}{\beta}\right)\right] = \frac{l}{\alpha} + \frac{l'}{\beta};$$

and

$$\frac{1}{N}\left(\frac{l}{\alpha} + \frac{l'}{\beta}\right) = 1 - \left(\frac{1}{\alpha} + \frac{1}{\beta}\right).$$

From the inequalities involving l and l' it follows that

$$\frac{1}{\alpha} + \frac{1}{\beta} > \frac{l}{\alpha} + \frac{l'}{\beta} \geqq \frac{1}{\alpha} + \frac{1}{\beta} - 2 \, .$$

Designating the sum $\frac{1}{\alpha} + \frac{1}{\beta}$ by t, and dividing all the numbers of the preceding inequality by N, we arrive at the inequality

$$\frac{t}{N} > 1 - t \geqq \frac{t-2}{N} \, ,$$

which must be satisfied for any selected positive integer N, no matter how large. But since the first and third members of this inequality can be made as small as we wish, it must follow that $1 - t = 0$, from which it follows that $t = \frac{1}{\alpha} + \frac{1}{\beta} = 1$. Further, it is readily seen that the numbers α and β must be irrational, since if $\alpha = \frac{p}{q}$, then we would have $\beta = \frac{p}{p-q}$, and this would imply that $[q\alpha] = [(p-q)\beta]$, which contradicts the conditions of the problem.

Assume now that α and β are irrational numbers, and that $\frac{1}{\alpha} + \frac{1}{\beta} = 1$. Since necessarily $\alpha > 1$ and $\beta > 1$, it will not be possible to find two equal numbers among the integers $[\alpha]$, $[2\alpha]$, $[3\alpha]$, $\cdots$, n or among the integers $[\beta]$, $[2\beta]$, $[3\beta]$, $\cdots$.

We shall now show that, given any positive integer N, either an integer n can be found such that $[n\alpha] = N$ or an integer m can be found such that $[m\beta] = N$, and both possibilities cannot exist simultaneously. Let

$$[(n-1)\alpha] < N \leqq [n\alpha] \, ,$$
$$[(m-1)\beta] < N \leqq [m\beta] \, .$$

These inequalities imply that

$$n\alpha - \alpha < N \, , \qquad n\alpha > N \, ;$$
$$m\beta - \beta < N \, , \qquad m\beta > N$$

(neither $n\alpha$ nor $m\beta$ can be exactly equal to N, since α and β are irrational); that is,

$$n\alpha = N + d \, ,$$
$$m\beta = N + d' \, ,$$

where $0 < d < \alpha$, $0 < d' < \beta$. It follows that

$$n + m = \frac{N}{\alpha} + \frac{d}{\alpha} + \frac{N}{\beta} + \frac{d'}{\beta}$$

$$= N\left(\frac{1}{\alpha} + \frac{1}{\beta}\right) + \left(\frac{d}{\alpha} + \frac{d'}{\beta}\right) = N + \left(\frac{d}{\alpha} + \frac{d'}{\beta}\right),$$

and, consequently,

$$\frac{d}{\alpha} + \frac{d'}{\beta} = n + m - N$$

is an integer. But since $0 < \frac{d}{\alpha} < 1$, $0 < \frac{d'}{\beta} < 1$, we have

$$0 < \frac{d}{\alpha} + \frac{d'}{\beta} < 2.$$

Hence we arrive at the equality

$$\frac{d}{\alpha} + \frac{d'}{\beta} = 1,$$

which, since

$$\frac{1}{\alpha} + \frac{1}{\beta} = 1,$$

is possible only in the event that one of the numbers d or d' is less than 1 and the other is greater than 1. But this means that of the integers $[n\alpha] = [N + d]$ and $[m\beta] = [N + d']$, neither of which is less than N, one must be equal to N and the other must exceed N.

109. *First Solution.* It is readily seen that $(a) = \left[a + \frac{1}{2}\right]$; Hence we can put the equation we wish to derive into the following form.

$$N = \left[\frac{N}{2} + \frac{1}{2}\right] + \left[\frac{N}{4} + \frac{1}{2}\right] + \left[\frac{N}{8} + \frac{1}{2}\right] + \cdots.$$

Now let

$$N = a_n \cdot 2^n + a_{n-1} \cdot 2^{n-1} + \cdots + a_1 \cdot 2 + a_0$$

$(a_n, a_{n-1}, \cdots, a_1, a_0$ are either 0 or 1) be the expansion of N in powers of 2 (as in the binary number system). We then have

$$\left[\frac{N}{2} + \frac{1}{2}\right] = \left[a_n \cdot 2^{n-1} + a_{n-1} \cdot 2^{n-2} + \cdots + a_1 + \frac{a_0 + 1}{2}\right]$$

$$= a_n \cdot 2^{n-1} + a_{n-1} \cdot 2^{n-2} + \cdots + a_1 + a_0,$$

$$\left[\frac{N}{4} + \frac{1}{2}\right] = \left[a_n \cdot 2^{n-2} + a_{n-1} \cdot 2^{n-3} + \cdots + \frac{a_1 + 1}{2} + \frac{a_0}{4}\right]$$

$$= a_n \cdot 2^{n-2} + a_{n-1} \cdot 2^{n-3} + \cdots + a_1,$$

. . . . ,

$$\left[\frac{N}{2^n} + \frac{1}{2}\right] = \left[a_n + \frac{a_{n-1}+1}{2} + \frac{a_{n-2}}{4} + \cdots + \frac{a_0}{2^n}\right] = a_n + a_{n-1},$$

$$\left[\frac{N}{2^{n+1}} + \frac{1}{2}\right] = \left[\frac{a_n+1}{2} + \frac{a_{n-1}}{4} + \cdots + \frac{a_0}{2^{n+1}}\right] = a_n,$$

and

$$\left[\frac{N}{2^{n+2}} + \frac{1}{2}\right] = \left[\frac{N}{2^{n+3}} + \frac{1}{2}\right] = \cdots = 0$$

(recalling that the a_i are either 0 or 1). Hence we obtain

$$\left[\frac{N}{2} + \frac{1}{2}\right] + \left[\frac{N}{4} + \frac{1}{2}\right] + \cdots + \left[\frac{N}{2^{n+1}} + \frac{1}{2}\right] + \cdots$$

$$= a_n(2^{n-1} + 2^{n-2} + \cdots + 1 + 1)$$
$$+ a_{n-1}(2^{n-2} + 2^{n-3} + \cdots + 1 + 1) + \cdots + a_1(1 + 1) + a_0$$
$$= a_n \cdot 2^n + a_{n-1} \cdot 2^{n-1} + \cdots + a_1 \cdot 2 + a_0 = N,$$

which is what we wished to show.

Second Solution. It is obvious that the number of odd numbers exceeding N is equal to $\frac{N}{2}$ if N is even, and is $\left[\frac{N+1}{2}\right] = \left[\frac{N}{2}\right] + 1$ if N is odd—or, equivalently, $\left(\frac{N}{2}\right)$. Thus, the number of even numbers not divisible by 4, and not exceeding N, is equal to $\left[\frac{N}{4}\right]$ if N is either divisible by 4 or yields the remainder 1, and is equal to $\left[\frac{N}{4}\right] + 1$ if N yields a remainder of 2 or 3 upon division by 4. In either event this number is $\left(\frac{N}{4}\right)$. The number of integers not exceeding N which are divisible by 4 but not by 8 is: equal to $\left[\frac{N}{8}\right]$ if N is either divisible by 8 or yields one of the remainders 1, 2, or 3; or else is equal to $\left[\frac{N}{8}\right] + 1$. In either event the number is equal to $\left(\frac{N}{8}\right)$.

In a similar manner it is shown that: $\left(\frac{N}{16}\right)$ is the number of integers not exceeding N which are divisible by 8 but not by 16; $\left(\frac{N}{32}\right)$ is the number of integers not exceeding N which are divisible by 16 but not by 32; and so on. If we use these results to examine all the numbers in which we are interested, we find that

$$\left(\frac{N}{2}\right) + \left(\frac{N}{4}\right) + \left(\frac{N}{8}\right) + \cdots = N .$$

which is what we wished to show.

110. (a) Let a be the first digit, and b the last digit, of the desired integer N. Then the integer can be written as

$$N = 1000a + 100a + 10b + b ,$$

or as

$$N = 1100a + 11b = 11(100a + b) .$$

Since this integer is to be a perfect square, and since it clearly must be divisible by 11, it must also be divisible by 121; that is, $\frac{N}{11} =$ $100a + b$ must be divisible by 11. But

$$100a + b = 99a + (a + b) = 11 \cdot 9a + (a + b) .$$

Hence $a + b$ must be divisible by 11. Since neither a nor b exceeds 9, and since a is not 0, it follows that $1 \leqq a + b \leqq 18$, whence $a + b = 11$.

This implies that

$$100a + b = 11 \cdot 9a + 11 = 11(9a + 1) ,$$
$$\frac{N}{121} = \frac{100a + b}{11} = 9a + 1 .$$

Since N is a perfect square, $\frac{N}{121}$ is also a square. But among the integers of form $9a + 1$, where a ranges through the integer values 1 to 9, only $9 \cdot 7 + 1 = 64$ is a perfect square. This means that $N = 121 \cdot 64 = 7744 = 88^2$.

(b) Let a be the digit in the tens place of the desired integer, and let b be the digit in the units place; that is, the number is $10a + b$. If the order of digits is reversed, the integer becomes $10b + a$. The conditions of the problem yield

$$10a + b + 10b + a = 11(a + b) = k^2 ,$$

where k is an integer.

It follows that k^2 (hence k) must be divisible by 11; also, $a + b$ must be divisible by 11. Since $a + b \leqq 18$, this implies that $a + b = 11$, or $k^2 = 121$. Therefore, the only possibilities for the integers sought are

29, 38, 47, 56, 65, 74, 83, 92,

all of which satisfy the conditions of the problem.

111. Designate by a the two-digit integer formed by the first two digits of the number N sought, and by b the two-digit integer formed by the last two digits of N. Then, $N = 100a + b$, and the conditions of the problem yield

$$100a + b = (a + b)^2 ,$$

which may be written

$$99a = (a + b)^2 - (a + b) = (a + b)(a + b - 1) . \qquad (1)$$

Thus, the product $(a + b)(a + b - 1)$ must be divisible by 99. We shall investigate the possible values for a and b.

(1) $a + b = 99k$, $a + b - 1 = \dfrac{a}{k}$. Since a and b are two-digit numbers, $k \leqq 2$. The case $k = 2$, when used with equation (1), leads immediately to $a = 99$ and $b = 99$. Similarly,

$$k = 1 ,$$
$$a + b = 99 ,$$
$$a = a + b - 1 = 98 ,$$
$$N = 9801 = (98 + 1)^2 .$$

(2) $a + b = 11m$, $a + b - 1 = 9n$, $mn = a$. Here we obtain $9n = 11m - 1$. Since $11m - 1$ is thus shown to be divisible by 9, it follows that m yields the remainder 5 upon division by 9 (it is easily verified that if there were any other remainder, then $11m - 1$ would not be divisible by 9). Hence, $m = 9t + 5$, and so $9n = 99t + 54$, $n = 11t + 6$. We now have

$$a = mn = (9t + 5)(11t + 6) = 99t^2 + 109t + 30 .$$

Since a is to be a two-digit integer, we must have $t = 0$; and, consequently, $a = 30$, $a + b = 11m = 55$, $b = 25$, $N = 3025 = (30 + 25)^2$.

(3) $a + b = 9m$, $a + b - 1 = 11n$, $mn = a$. Reasoning as we did above in part (2), we have the single possibility $N = 2025 = (20 + 25)^2$.

(4) $a + b = 33m$, $a + b - 1 = 3n$, or $a + b = 3m$, $a + b - 1 = 33n$, which is untenable, inasmuch as $a + b$ and $a + b - 1$ are relatively prime numbers.

(5) $a + b - 1 = 99k$, $a + b = \dfrac{a}{k}$. Here we will have

$$a + b - 1 = 99 \, ,$$
$$a + b = 100 \, ,$$
$$a = \frac{(a + b)(a + b - 1)}{99} = 100 \, ,$$

which is an impossibility.

Therefore, the conditions of the problem can be satisfied only by the numbers 9801, 3025, and 2025, and these numbers do satisfy the conditions.

112. (a) Since the numbers are to contain only even digits, they can begin only with 2, 4, 6, or 8; hence we need examine only those integers between 1999 and 3000, 3999 and 5000, 5999 and 7000, and 7999 and 9000. Accordingly, the square roots of such possible four-digit integers must be between 45 and 55, or 63 and 71, or 77 and 84, or 89 and 95. Further, since $(10x + y)^2 = 100x^2 + 20xy + y^2$, it follows, for $0 \leq y \leq 9$, that the tens digit of the number $(10x + y)^2$ is odd or even simultaneously with the tens digit of y^2 (the $20xy$ term contributes an even digit to that place, and $100x^2$ contributes a zero digit). Hence the square root of the numbers sought cannot end with the digit 4 or the digit 6.

Since the square roots of the numbers sought must be even, we are left with only the following four possibilities:

$$68^2 = 4624 \, ; \qquad 80^2 = 6400 \, ;$$
$$78^2 = 6084 \, ; \qquad 92^2 = 8464 \, .$$

These four integers satisfy the conditions of the problem.

(b) Reasoning analogous to that used in problem (a) shows that there does not exist any number composed of four odd digits and which is a perfect square.

113. (a) We shall designate the hundreds, tens, and units digits of the integers N sought by x, y, and z, respectively, so that $N = 100x + 10y + z$. The condition of the problem, then, is

$$100x + 10y + z = x! + y! + z! \, .$$

Since $7! = 5040$ is already a four-digit number, none of the digits can exceed 6. Consequently, N cannot exceed 700, which implies that no one of the digits can exceed 5 (since $6! = 720$). At least one of the digits must be 5, otherwise the greatest value obtainable for $x! + y! + z!$ would be $3 \cdot 4! = 72 < 100$, whereas N is a three-digit number. The possibility $x = 5$ can be eliminated, since in that event

$N \geqq 500$, but the greatest value obtainable for $x! + y! + z!$ is $3 \cdot 5! = 360$. This, in turn, implies that x cannot exceed 3. Further, it can easily be reasoned that x cannot exceed 2; in fact, $3! + 2 \cdot 5! = 242 < 300$. Now, the number 255 does not fulfill the conditions of the problem; if only one digit of the number sought is 5, then x cannot exceed 1, since $2! + 5! + 4! = 146 < 200$. Moreover, since $1! + 5! + 4! = 145 < 150$, we must conclude that y cannot exceed 4. Consequently, $z = 5$, since we have shown that at least one of the digits has to be 5. Therefore, we must have $x = 1$, $4 \geqq y \geqq 0$, and $z = 5$. This allows us to find the one possibility for the solution of the problem—a number satisfying the problem's condition is $N = 145$.

(b) The desired numbers N cannot have more than three digits, since in the extreme case $(9^2 + 9^2 + 9^2 + 9^2 = 4 \cdot 9^2 = 324)$ we obtain only a three-digit number. Hence we can write, for the integers, $N = 100x + 10y + z$, where x, y, and z are the respective digits of the number (from left to right, and allowing the possibility of $x = 0$, or even both $x = 0$ and $y = 0$).

The problem then imposes the condition

$$100x + 10y + z = x^2 + y^2 + z^2 \,,$$

or, equivalently,

$$(100 - x)x + (10 - y)y = z(z - 1). \tag{1}$$

The last equation implies that, necessarily, $x = 0$; if this is not the case, we have on the left an integer not less than 90 $[x \geqq 1, 100 - x \geqq 90, (10 - y)y \geqq 0]$, and the integer on the right is not more than $9 \cdot 8 = 72$. Consequently, equation (1) can be replaced by

$$(10 - y)y = z(z - 1) \,.$$

It is readily verified that this equation cannot be satisfied by any nonzero y (recall that y and z are digits). If $y = 0$, then there remains only the trivial possibilities $z = 0$ or $z = 1$.

Hence, except for $N = 0$, there is only the solution $N = 1$.

114. (a) Clearly, the numbers N which are sought cannot have more than four digits, since the sum of the digits of a five-digit number cannot exceed $5 \cdot 9 = 45$, and the square of this number is 2025, a four-digit number. Further, since $4 \cdot 9 = 36$, and $36^2 = 1296$, if N is a four-digit number satisfying the condition of the problem, its first digit cannot exceed 1. But $1 + 3 \cdot 9 = 28$, and $28^2 = 784$, whence even four-digit numbers are excluded from consideration. Thus N can have at most three digits. Assume, now, that $N =$

$100x + 10y + z$, where x, y, and z are digits (the possibility $x = 0$ is allowed, as is $x = y = 0$).

Now the condition of the problem can be written in the form

$$100x + 10y + z = (x + y + z)^2 ,$$

or

$$99x + 9y = (x + y + z)^2 - (x + y + z)$$
$$= (x + y + z)(x + y + z - 1) .$$

It follows from the above equality that one of $(x + y + z)$ or $(x + y + z - 1)$ is divisible by 9 (not both can be divisible by 3 since they are relatively prime). Also, $1 \leqq x + y + z \leqq 27$. We shall investigate all possible cases.

(1) $x + y + z - 1 = 0$; $99x + 9y = 0$, $x = y = 0$, $z = 1$; $N = 1$.

(2) $x + y + z = 9$; $99x + 9y = 9 \cdot 8 = 72$, $x = 0$, $9y = 72$, $y = 8$, $z = 1$; $N = 81 \ [= (8 + 1)^2]$.

(3) $x + y + z - 1 = 9$; $99x + 9y = 9 \cdot 10 = 90$, $x = 0$, $9y = 90$, which is impossible.

(4) $x + y + z = 18$; $99x + 9y = 18 \cdot 17 = 306$, $x = 3$, $y = 1$, $z = 18 - (3 + 1) = 14$, which is impossible.

(5) $x + y + z - 1 = 18$; $99x + 9y = 19 \cdot 18 = 342$, $x = 3$, $y = 5$, $z = 19 - (3 + 5) = 11$, which is impossible.

(6) $x + y + z = 27$; $99x + 9y = 27 \cdot 26 = 702$, $x = 7$, $y = 1$, $z = 27 - (7 + 1) = 19$, which is impossible.

Therefore, the conditions of the problem are satisfied only by the numbers 1 and 81.

(b) The cube of a three-digit integer can contain no more than nine digits; hence, the sum of the digits of the cube of a three-digit number cannot exceed $9 \cdot 9 = 81 < 100$. This implies that the numbers sought cannot be three-digit numbers; and the same kind of reasoning will show that such a number cannot have more than three digits. The integer, or integers, sought can have at most two digits.

The cube of a two-digit integer cannot have more than six digits, and the sum of the digits of such a cube cannot exceed $6 \cdot 9 = 54$, whence the numbers sought cannot exceed 54. However, if a number not exceeding 54 is cubed, the first digit of the cube cannot be greater than 1; but then the sum of the digits of the cube cannot be greater than $1 + 5 \cdot 9 = 46$. Thus, the numbers sought cannot exceed 46.

Proceeding as before, we find that if an integer does not exceed

46, its cube contains at most five digits, and since this cube is less than 99,999, the sum of the digits is at most equal to $4 \cdot 9 + 8 = 44$. The cube of 44 is a five-digit number ending with the digit 4. Hence the number 44 must be thrown out as a possibility. The numbers sought cannot exceed 43.

Now we make use of the fact that the sum of the digits of every positive integer yields the same remainder upon division by 9 as does the number itself upon division by 9. It follows that any of the integers sought must yield the same remainder upon division by 9 as does its cube. This is possible only if this remainder is -1, 0, or 1 (solution of $r^3 = r$).

Thus, the numbers sought do not exceed 43, and they can yield upon division by 9 only the remainders -1, 0, or 1. Only the following thirteen integers can satisfy these conditions:

$$1;$$
$$8, \quad 9, \quad 10;$$
$$17, \quad 18, \quad 19;$$
$$26, \quad 27, \quad 28;$$
$$35, \quad 36, \quad 37.$$

Of these possibilities, the following integers satisfy the condition of the problem:

$$1 \ (1^3 = 1) \ ;$$
$$8 \ (8^3 = 512) \ ;$$
$$17 \ (17^3 = 4193) \ ;$$
$$18 \ (18^3 = 5832) \ ;$$
$$26 \ (26^3 = 17{,}576) \ ;$$
$$27 \ (27^3 = 19{,}683) \ .$$

115. (a) Direct verification assures us that for $x < 5$ the only positive integers solving the given equation are $x = 1$, $y = 1$ and $x = 3$, $y = 3$. We shall show that no solution exists for $x \geqq 5$. We note, first, that

$$1! + 2! + 3! + 4! = 33$$

ends with the digit 3; and $5!, 6!, 7!, \cdots$ all end with the digit 0. Therefore, if $x \geqq 5$, the sum $1! + 2! + \cdots + x!$ terminates with the digit 3 and therefore cannot be the square of any integer y.

(b) Two cases will be considered.

z = 2n is an even integer. This case is readily reduced to problem (a), inasmuch as $y^{2n} = (y^n)^2$. For even integers z the solutions are
$$x = 1; \quad y = 1; \quad z \text{ is any even number;}$$
$$x = 3; \quad y = \pm 3; \quad z = 2.$$

z is an odd number. For $z = 1$ any value of x, and the corresponding obvious value of y, will suffice. Let $z \geqq 3$. We note that
$$1! + 2! + 3! + \cdots + 8! = 46,233 .$$

This number is divisible by 9, but not by 27, whereas for all $n \geqq 9$ the number 9! is divisible by 27. The sum $9! + 10! + \cdots + x!$ is therefore divisible by 27; and so, for $x \geqq 8$, the sum $1! + 2! + \cdots + x!$ is divisible by 9, but not by 27. In order for y^z to be divisible by 9 it is necessary that y be divisible by 3, but then y^z must be divisible by 27 (for $z \geqq 3$). Consequently, for $x \geqq 8$ and $z \geqq 3$ the given equation has no solution in integers.

It remains to consider the case for $x < 8$. We have $1! = 1 = 1^z$, where z can be any integer $1! + 2! = 3$ (which does not provide an integer solution for $z \geqq 3$), $1! + 2! + 3! = 3^2$ (a solution arrived at earlier, but not for the case of odd z) and

$$1! + 2! + 3! + 4! = 33 ,$$
$$1! + 2! + \cdots + 5! = 153 ,$$
$$1! + 2! + \cdots + 6! = 873 ,$$
$$1! + 2! + \cdots + 7! = 5913 .$$

As can be easily verified, none of these integers is an integral power ($\geqq 3$) of any natural number. Thus, for odd z we have only the following additional solutions:
$$x = 1; \quad y = 1; \quad z = \text{any odd number;}$$
$$x = \text{any natural number;} \quad y = 1! + 2! + \cdots + x!; \quad z = 1.$$

116. Let
$$a^2 + b^2 + c^2 + d^2 = 2^n .$$

We shall designate by p the greatest power of 2 which divides all four integers $a, b, c,$ and d. Upon dividing both sides of the above equation by $(2^p)^2 = 2^{2p}$ (an even power of 2), we obtain

$$a_1^2 + b_1^2 + c_1^2 + d_1^2 = 2^{n-2p} ,$$

where at least one of the four integers a_1, b_1, c_1, d_1 is odd. If exactly one, or if exactly three, of the integers a_1, b_1, c_1, d_1 are odd, then $a_1^2 + b_1^2 + c_1^2 + d_1^2$ is odd, and in this event the equality is impossible.

If two of the integers are odd, say $a_1 = 2k + 1$ and $b_1 = 2l + 1$, and two are even, say $c_1 = 2m$ and $d_1 = 2n$, then we may write

$$a_1^2 + b_1^2 + c_1^2 + d_1^2 = 4k^2 + 4k + 1 + 4l^2 + 4l + 1 + 4m^2 + 4n^2$$

$$= 2[2(k^2 + k + l^2 + l + m^2 + n^2) + 1],$$

which is a contradiction that $a_1^2 + b_1^2 + c_1^2 + d_1^2 = 2^{n-2p}$ cannot have an odd divisor (the expression in brackets cannot be 1 except for $k = l = m = n = 0$; $c_1 = d_1 = 0$ and $c = d = 0$). If all four integers are odd, that is, $a_1 = 2k + 1$, $b_1 = 2l + 1$, $c_1 = 2m + 1$, $d_1 = 2n + 1$, we have

$$a_1^2 + b_1^2 + c_1^2 + d_1^2 = 4k^2 + 4k + 1 + 4l^2 + 4l + 1$$

$$+ 4m^2 + 4m + 1 + 4n^2 + 4n + 1$$

$$= 4[k(k + 1) + l(l + 1) + m(m + 1) + n(n + 1) + 1].$$

Now, the product of two consecutive integers must be an even integer, and so the expression in the brackets immediately above is odd. Under the circumstances, its value can be only $2^0 = 1$. This implies that $n - 2p = 2$, $n = 2p + 2$, and $k = l = m = n = 0$, $a_1 = b_1 = c_1 = d_1 = 1$, $a = b = c = d = 2^p$.

Therefore, if n is an odd number, then 2^n cannot be written as the sum of four squares; if $n = 2p$ is even, then 2^n can be expressed as the sum of four squares only in the following way:

$$2^{2p} = (2^{p-1})^2 + (2^{p-1})^2 + (2^{p-1})^2 + (2^{p-1})^2.$$

117. (a) *First Solution.* The equation

$$x^2 + y^2 + z^2 = 2xyz$$

is satisfied by $x = y = z = 0$. It is obvious that no other solution is available which involves zero value for any one of $x, y,$ or z; hence we may assume that none of them can vanish. We can write

$$x = 2^\alpha x_1,$$

$$y = 2^\beta y_1,$$

$$z = 2^\gamma z_1,$$

where x_1, y_1, z_1 are odd (if any one of x, y, z is initially odd, then the exponent for 2 can be taken to be 0).

Since $x, y,$ and z enter the equation symmetrically, we may assume, with generality, that $\alpha \leqq \beta \leqq \gamma$. We shall now determine by what power of 2 the left member of the equation is divisible.

(1) If $\alpha < \beta \leqq \gamma$, or else if $\alpha = \beta = \gamma$, then after factoring out

$(2\alpha)^2 = 2^{2\alpha}$, there will remain the sum of an odd and two even numbers, or else the sum of three odd numbers—that is, an odd number.

(2) If $\alpha = \beta < \gamma$, then it is possible to write

$$x = 2^{\alpha}(2k + 1),$$
$$y = 2^{\alpha}(2l + 1),$$
$$z = 2^{\alpha} \cdot 2m.$$

In this case,

$$
\begin{aligned}
x^2 + y^2 + z^2 &= 2^{2\alpha}[(2k + 1)^2 + (2l + 1)^2 + (2m)^2] \\
&= 2^{2\alpha}(4k^2 + 4k + 1 + 4l^2 + 4l + 1 + 4m^2) \\
&= 2^{2\alpha+1}[2(k^2 + k + l^2 + l + m^2) + 1] ;
\end{aligned}
$$

after factoring out $2^{2\alpha+1}$, there remains in the brackets an odd number.

On the other hand, the right member of the equation is divisible by $2^{\alpha+\beta+\gamma+1}$. Also, the right member must be divisible by the same power of 2 as is the left member.

For case (1) we must have $2\alpha = \alpha + \beta + \gamma + 1$. Since $\alpha \leqq \beta \leqq \gamma$, we arrive at the untenable inequality $2\alpha \geqq 3\alpha + 1$.

For case (2) it follows that $2\alpha + 1 = \alpha + \beta + \gamma + 1$. Since $\alpha = \beta < \gamma$, the following inequality is implied, which again is impossible: $2\alpha + 1 > 3\alpha + 1$.

Therefore, there fails to exist any solution in integers for $x^2 + y^2 + z^2 = 2xyz$, except the trivial one $x = y = z = 0$.

Second Solution. Since the sum of the squares is to be an even number, it may be reasoned that either all three of the numbers x^2, y^2, z^2 (hence also x, y, and z) are even, or one of them is even and two are odd. But in the last event, the sum would be divisible only by 2 and the product $2xyz$ would be divisible by 4. Hence we must conclude that x, y, and z must all be even: $x = 2x_1, y = 2y_1, z = 2z_1$. If we substitute these into the given equation and divide through by 4, we obtain

$$x_1^2 + y_1^2 + z_1^2 = 4x_1 y_1 z_1.$$

As above, this equation implies that x_1, y_1, and z_1 are all even numbers, and so we can write $x_1 = 2x_2,\ y_1 = 2y_2,\ z_1 = 2z_2$, which yields the equation

$$x_2^2 + y_2^2 + z_2^2 = 9x_2 y_2 z_2,$$

which, in turn, implies that also x_2, y_2, and z_2 are all even numbers.

Continuation of this process leads to the conclusion that the following set of numbers are all even:

$$x, y, z;$$

$$x_1 = \frac{x}{2}, \quad y_1 = \frac{y}{2}, \quad z_1 = \frac{z}{2};$$

$$x_2 = \frac{x}{4}, \quad y_2 = \frac{y}{4}, \quad z_2 = \frac{z}{4};$$

$$x_3 = \frac{x}{8}, \quad y_3 = \frac{y}{8}, \quad z_3 = \frac{z}{8}, \quad \cdots;$$

$$x_k = \frac{x}{2^k}, \quad y_k = \frac{y}{2^k}, \quad z_k = \frac{z}{2^k}; \quad \cdots$$

(the numbers x_k, y_k, z_k satisfy the equation $x_k^2 + y_k^2 + z_k^2 = 2^{k+1} x_k y_k z_k$). But this is possible only if $x = y = z = 0$.

(b) As above, it may be proved that the only integer solution of the equation $x^2 + y^2 + z^2 + v^2 = 2xyzv$ is $x = y = z = v = 0$. Here, it suffices to make the special investigation for the case where the highest power of 2 which divides x, y, z, and v is the same; that is, when

$$x = 2^\alpha(2k + 1), \qquad y = 2^\alpha(2l + 1),$$
$$z = 2^\alpha(2m + 1), \qquad v = 2^\alpha(2n + 1),$$

where α is a nonnegative integer, and k, l, m, and n are integers. Then

$$x^2 + y^2 + z^2 + v^2 = 2^{2\alpha}[(4k^2 + 4k + 1) + (4l^2 + 4l + 1)$$
$$+ (4m^2 + 4m + 1) + (4n^2 + 4n + 1)]$$
$$= 2^{2\alpha+2}(k^2 + k + l^2 + l + m^2 + m + n^2 + n + 1)$$
$$= 2^{2\alpha+2}[k(k + 1) + l(l + 1) + m(m + 1) + n(n + 1) + 1].$$

Now, the last expression in brackets must be odd (the terms which are products of two successive numbers are even). Therefore, the greatest power of 2 which divides the left member of the equation has exponent $2\alpha + 2$. But, in the original equation, the highest power of 2 which divides the right member is $2^{4\alpha+1}$. Accordingly, we must conclude that $2\alpha + 2 = 4\alpha + 1$, which is untenable for any integer α.

Remark: A second solution of problem (b), which is analogous to that of problem (a), is left to the reader.

118. (a) Let x, y, z be any three nonnegative integers satisfying the equation

$$x^2 + y^2 + z^2 = kxyz .\qquad(1)$$

We shall show, first, that it is always possible to assume

$$x \leq \frac{kyz}{2}, \quad y \leq \frac{kxz}{2}, \quad z \leq \frac{kxy}{2}\qquad(2)$$

(none of the integers on the left of (1) can exceed half the value of the right member). To show this, assume, for example, that $z > \frac{kxy}{2}$. Then consider not x, y, z, but the lesser integers x, y, and $z_1 = kxy - z$, which clearly also satisfy equation (1):

$$x^2 + y^2 + (kxy - z)^2 = kxy(kxy - z) .$$

If any integer of this new triple is greater than the product of the other two multiplied by $\frac{k}{2}$, we again make a similar substitution until the conditions of (2) are satisfied (after which the process will fail to yield decreasing numbers for x, y, z).

Assume now that $x \leq y \leq z$. First, since $y \leq z \leq \frac{kxy}{2}$, it follows that

$$1 \leq \frac{kx}{2} ;$$

$$kx \geq 2 .$$

Equation (1) can be restated in the form

$$x^2 + y^2 + \left(\frac{kxy}{2} - z\right)^2 = \left(\frac{kxy}{2}\right)^2 .$$

Since $z \leq \frac{kxy}{2}$, if in the left member z is replaced by $y \leq z$, the numerical value is increased (or remains the same if $y = z$); consequently,

$$x^2 + y^2 + \left(\frac{kxy}{2} - y\right)^2 \geq \frac{k^2 x^2 y^2}{4} .$$

This yields the inequality

$$x^2 + 2y^2 \geq kxy^2 .$$

Further, it follows from $x \leq y$ that

$$y^2 + 2y^2 \geq kxy^2 ;$$

that is, $kx \leq 3$.

Hence, we have $2 \leq kx \leq 3$; that is, kx is equal to 2 or 3. But if $kx = 2$, then equation (1) assumes the form

$$x^2 + y^2 + z^2 = 2yz ,$$

or

$$x^2 + (y - z)^2 = 0 ;$$

thus $x = 0$ and $kx = 0$ instead of 2. Therefore, $kx = 3$, which means that k can have only value 1 or 3. It is readily verified by examples that these values for k are possible [see the solution of problem (b)].

(b) Reasoning similar to that employed for problem (a) is used here. There we had $x^2 + 2y^2 \geq kxy^2$; since $kx = 3$, this inequality can be rewritten in the form

$$x^2 + 2y^2 \geq 3y^2 ,$$

or

$$x^2 \geq y^2 .$$

However, we assumed $x \leq y$; consequently, $x = y$. Assume now that in the basic equation, (1), $x = y$, $kx = 3$. We obtain

$$2x^2 + z^2 = 3xz ,$$

or

$$(z - x)(z - 2x) = 0 .$$

Thus, $z = x$ or $z = 2x$. Since

$$z \leq \frac{kxy}{2} = \frac{3y}{2} = \frac{3x}{2} ,$$

it is impossible for z to be equal to $2x$. Therefore, $z = x$.

Accordingly, for condition (2) to be fulfilled, we must have $x = y = z$. But, inasmuch as $kx = 3$, x can be only 1 or 3, and we obtain two solutions for equation (1):

$$x = y = z = 1 \quad (k = 3) ;$$
$$x = y = z = 3 \quad (k = 1) .$$

In the solution of problem (a) we saw that any three integers x, y, z which satisfy condition (1) can produce other solutions, by successive replacements of the form $z_1 = kxy - z$ of (2). But if $z_1 = kxy - z$, then $z = kxy - z_1$; hence every solution of equation (1) can be obtained from the least solution by successive substitutions of form $z_1 = kxy - z$. In particular, we obtain in this way the following solutions for equation (1) (up to 1000):

(1) For $k = 3$:

$\dfrac{x}{1}$	$\dfrac{y}{1}$	$\dfrac{z}{1}$
1	1	3
1	1	2
1	2	5
1	5	13
1	13	34
1	34	89
1	89	233
1	233	610
2	5	29
2	29	169
2	169	985
5	13	194
5	29	433

(2) For $k = 1$:

$\dfrac{x}{3}$	$\dfrac{y}{3}$	$\dfrac{z}{3}$
3	3	6
3	6	15
3	15	39
3	39	102
3	102	267
3	267	699
6	15	87
6	87	507
15	39	582

The fact that solutions corresponding to $k = 1$ can be obtained from those corresponding to $k = 3$ by simple multiplication by 3 follows from the fact that the least triple $(3, 3, 3)$ satisfying the equation for $k = 1$ is related to the least triple for $k = 3$ by the factor 3.

119. If the relatively prime pair x, y is to meet the conditions of the problem, the following equations must be solvable in integers:

$$x^2 + 125 = uy ,$$
$$y^2 + 125 = vx .$$

We shall show that a solution of these equations produces not merely the relatively prime pair x and y satisfying the conditions, but two other pairings x, u and y, v which satisfy the same conditions. That is, we shall show that $x^2 + 125$ is divisible by u and $u^2 + 125$ is divisible by x, and also that $y^2 + 125$ is divisible by v and $v^2 + 125$ is divisible by y.

First, $x^2 + 125 = uy$ means that $x^2 + 125$ is divisible by u. The equalities yield

$$u^2 y^2 = x^4 + 250 x^2 + 125^2 ;$$
$$u^2 (vx - 125) = x^4 + 250 x^2 + 125^2 ,$$

and, finally,

$$x(u^2 v - x^3 - 250 x) = 125(u^2 + 125) ,$$

from which it follows that $u^2 + 125$ is divisible by x (which is relatively prime to 125, otherwise x and y would not be relatively prime). Also, the integers x and u are relatively prime, for if x and u had a common divisor of d, then $125 = uy - x^2$ would also be divisible by d (that is, $d = 5$), and, consequently, y would be divisible by d (since $y^2 = vx - 125$). It may be shown, in exactly the same way, that the integers y and v are relatively prime, and the square of either when increased by 125 is divisible by the other.

We redesignate our integers x, y, u, v as follows: $x = x_0$, $y = x_1$, $u = x_{-1}$, $v = x_2$. If $u^2 + 125 = x_{-2} x$, then the pair of integers u, x_{-2}, that is, x_{-1}, x_{-2}, also satisfy the conditions of the problem. Thus, if $v^2 + 125 = x_3 y$, then the pair of integers v, x_3, that is, x_2, x_3, also satisfy this condition. Hence, beginning with an integer pair satisfying the conditions of the problem, it is possible to construct an infinite double-end array of integers,

$$\cdots, x_{-2}, x_{-1}, x_0, x_1, x_2, \cdots ,$$

where neighboring pairs satisfy the equation

$$\frac{x_\alpha^2 + 125}{x_{\alpha+1}} = x_{\alpha-1} ;$$
$$\frac{x_{\alpha+1}^2 + 125}{x_\alpha} = x_{\alpha+2} .$$

Further, in the given array the ratio of the sum of a pair neighboring an integer to that integer is constant:

$$\frac{x_{a-1} + x_{a+1}}{x_a} = \frac{x_{a-1}x_{a+1} + x_{a+1}^2}{x_a x_{a+1}} = \frac{(x_a^2 + 125) + x_{a+1}^2}{x_a x_{a+1}}$$

$$= \frac{x_a^2 + (x_{a+1}^2 + 125)}{x_a x_{a+1}} = \frac{x_a + x_a x_{a+2}}{x_a x_{a+1}} = \frac{x_a + x_{a+2}}{x_{a+1}}.$$

This observation is helpful in extending the double-end array of integers above: it we designate $\frac{x_{a-1} + x_{a+1}}{x_a}$ by t, we will have

$$x_{a+1} = t x_a - x_{a-1},$$

$$x_{a-1} = t x_a - x_{a+1}.$$

Now we must develop a procedure which will help us determine all such arrays which yield solutions of the problem. We note, first, that if $x_{a+1} \geqq x_a$, then

$$x_{a+2} = \frac{x_{a+1}^2 + 125}{x_a} > \frac{x_{a+1}^2}{x_a} \geqq x_{a+1},$$

and if $x_{a-1} \geqq x_a$, then $x_{a-2} > x_{a-1}$; hence every chain of integers augments in both directions from some least integer (or equal pair of least integers, and two neighboring numbers of the chain can be equal only if both are 1).

Let x_0 be the least integer of the solution-chain:

$$x_1 \geqq x_0, \qquad x_{-1} \geqq x_0.$$

We shall show that $x_0 < \sqrt{125} < 12$.

We first observe that the number t, referred to above, is an integer exceeding 2. In fact, it follows from the equation

$$t = \frac{x_{a-1} + x_{a+1}}{x_a} = \frac{x_a + x_{a+2}}{x_{a+1}}$$

that t is an integer. Hence x_a and x_{a+1} are relatively prime, and consequently

$$t = \frac{x_{a-1} + x_{a+1}}{x_a}$$

cannot have in its denominator a divisor of x_{a+1}, and

$$t = \frac{x_a + x_{a+2}}{x_{a+1}}$$

cannot have in its denominator any divisor of x_a. Further,

$$t = \frac{x_a^2 + x_{a+1}^2 + 125}{x_a x_{a+1}} > \frac{x_a^2 + x_{a+1}^2}{x_a x_{a+1}} \geq 2 \,,$$

or $x_a^2 + x_{a+1}^2 \geq 2x_a x_{a+1}$ (since $x_a^2 + x_{a+1}^2 - 2x_a x_{a+1} = (x_a - x_{a+1})^2 \geq 0$).
Thus, $t \geq 3$. Now, since

$$\begin{aligned}
(x_1 - x_0)(x_{-1} - x_0) &= x_1 x_{-1} + x_0^2 - x_0 x_1 - x_0 x_{-1} \\
&= x_1 x_{-1} - x_0^2 - x_0(x_1 - x_0) - x_0(x_{-1} - x_0) \\
&\leq x_1 x_{-1} - x_0^2 = 125 \,,
\end{aligned}$$

it follows that at least one of the (nonnegative) integers $x_1 - x_0$ and $x_{-1} - x_0$ fails to exceed 11. Suppose that $x_1 - x_0 \leq 11$. We have

$$t = \frac{x_0^2 + x_1^2 + 125}{x_0 x_1} \,;$$

$$t - 2 = \frac{x_0^2 + x_1^2 + 125 - 2x_0 x_1}{x_0 x_1} = \frac{(x_1 - x_0)^2 + 125}{x_0[x_0 + (x_1 - x_0)]} \,.$$

But $t \geq 3$ and so $t - 2 \geq 1$; consequently,

$$(x_1 - x_0)^2 + 125 \geq x_0^2 + x_0(x_1 - x_0) \,,$$

which is impossible if $x_0 \geq 12 > (x_1 - x_0)$ [since here $x_0^2 > 125$, $x_0(x_1 - x_0) > (x_1 - x_0)^2$].

It remains now to test for x_0 all integers that are less than 12 and which are relatively prime to 125; for such an x_0, x_1 is an integer such that $x_0 + 125$ is divisible by x_1 (that is, x_1 is a divisor of $x_0^2 + 125$), and $x_1^2 + 125$ is divisible by x_0. Since $x_0^2 + 125 = x_1 x_{-1}$, we may use the fact that x_1 is the smaller of the neighbors of x_0 to obtain

$$x_0 \leq x_1 \leq \sqrt{x_0^2 + 125} \,.$$

It is not difficult to verify that all pairs x_0, x_1 satisfying the conditions of the problem can be expressed as follows:

$$1, 1 \left(t = \frac{1^2 + 1^2 + 125}{1 \cdot 1} = 127 \right) ; \qquad 1, 2 \left(t = \frac{1^2 + 2^2 + 125}{1 \cdot 2} = 65 \right) ;$$

$$1, 3 \left(t = \frac{1^2 + 3^2 + 125}{1 \cdot 3} = 45 \right) ; \qquad 1, 6 \left(t = \frac{1^2 + 6^2 + 125}{1 \cdot 6} = 27 \right) ;$$

$$1, 7 \left(t = \frac{1^2 + 7^2 + 125}{1 \cdot 7} = 25 \right) ; \qquad 1, 9 \left(t = \frac{1^2 + 9^2 + 125}{1 \cdot 9} = 23 \right) ;$$

$$2, 3 \left(t = \frac{2^2 + 3^2 + 125}{2 \cdot 3} = 23 \right) ; \qquad 6, 7 \left(t = \frac{6^2 + 7^2 + 125}{6 \cdot 7} = 5 \right) .$$

This yields the following solution arrays:

$\cdots$, 15,001, 126, 1, 1, 126, 15,001, $\cdots$;
$\cdots$, 4094, 63, 2, 129, 8383, $\cdots$;
$\cdots$, 1889, 42, 1, 3, 134, 6027, $\cdots$;
$\cdots$, 15,261, 566, 21, 1, 6, 161, 4341, $\cdots$;
$\cdots$, 10,826, 449, 18, 1, 7, 174, 4343, $\cdots$;
$\cdots$, 7369, 321, 14, 1, 9, 206, 4729, $\cdots$;
$\cdots$, 22,658, 987, 43, 2, 3, 67, 1538, $\cdots$;
$\cdots$, 2501, 522, 109, 23, 6, 7, 29, 138, 661, 3167, $\cdots$.

All possible pairs of relatively prime positive integers x, y, less than 1000, such that $x^2 + 125$ is divisible by y and $y^2 + 125$ is divisible by x, are given by the following tables.

x	y	x	y	x	y
126	1	449	18	522	109
1	1	18	1	109	23
63	1	1	7	23	6
1	2	7	174	6	7
2	129	321	14	7	29
42	1	14	1	29	138
1	3	1	9	138	661
3	134	9	206		
566	21	987	43		
21	1	43	2		
1	6	2	3		
6	161	3	67		

120. The problem gives rise to the following system of equations to be solved in integers (where x, y, z, u are the integers sought):

$$x^2 + y + z + u = (x + v)^2 ,$$
$$y^2 + x + z + u = (y + w)^2 ,$$
$$z^2 + x + y + u = (z + t)^2 ,$$
$$u^2 + x + y + z = (u + s)^2$$

or,

$$y + z + u = 2vx + v^2 ,$$
$$x + z + u = 2wy + w^2 ,$$
$$x + y + u = 2tz + t^2 ,$$
$$z + y + z = 2su + s^2 .$$

(*1*)

If we add the equations of (*1*), we obtain

$$(2v - 3)x + (2w - 3)y + (2t - 3)z + (2s - 3)u$$
$$+ v^2 + w^2 + t^2 + s^2 = 0 . \qquad (2)$$

Note that in equation (*2*) at least one of the numbers $2v - 3$, $2w - 3$, $2t - 3$, $2s - 3$ is negative; otherwise there would be on the left of the equation the sum of positive integers. Let us assume that $2v - 3 < 0$; This is possible only if $v = 0$ or $v = 1$. In the first event, the first equation of system (*1*) yields $y + z + u = 0$, which is untenable for y, z, u all positive. Hence it must be assumed that all the integers v, w, t, s are positive, and that $v = 1$. In this event equation (*2*) can be rewritten in the form

$$x = (2w - 3)y + (2t - 3)z + (2s - 3)u + w^2 + t^2 + s^2 + 1 . \qquad (3)$$

We now consider the several possibilities.

The numbers x, y, z, u are all distinct. Here, the integers v, w, t, s are also all different; if, for example, $v = w$, the difference of the first two of the equations (*1*) yields $y - x = 2v(x - y)$, which is impossible for positive v and $x \neq y$. Further, if we assume $v = 1$, the first of equations (*1*) yields $2x = y + z + u - 1$,

$$x = \frac{1}{2}y + \frac{1}{2}z + \frac{1}{2}u - \frac{1}{2} ,$$

which is inconsistent with equation (*3*), where the coefficients of y, z, u in the right member are positive integers (since $w, t,$ or s cannot be equal to 1 because they are distinct from v which is equal to 1). Therefore, this case is not possible.

Precisely two of the integers x, y, z, u are equal. Here we must separately investigate two cases.

If $z = u$, then $t = s$. Equation (*3*) and the first of equations (*1*) now yield:

$$x = (2w - 3)y + 2(2t - 3)z + w^2 + 2t^2 + 1 ;$$
$$2x = y + 2z - 1 .$$

As before, these equations are inconsistent.

If $x = y$, then $w = v = 1$. Equation (*2*) and the first of equations (*1*) yield, respectively,

$$2x = (2t - 3)z + (2s - 3)u + t^2 + s^2 + 2 ;$$
$$x = z + u - 1 .$$

Substituting the second equality in the first, we have

$$(2t - 5)z + (2s - 5)u + t^2 + s^2 + 4 = 0 , \qquad (4)$$

from which it follows that at least one of the two members $2t - 5$ or $2s - 5$ must be negative. Assume $2t - 5 < 0$; since $t > 0$ and $t \neq 1$ (for $v = 1$, $t \neq v$, since $z \neq x$), it follows that $t = 2$. Now, if twice the first of equations (*1*) is added to the third equation, we obtain

$$4z + 4x + 6 = 4x + 2z + 3u ,$$

that is, $z = \dfrac{3}{2}u - 3$. Substituting this into equation (*4*), along with $t = 2$, we have

$$(4s - 13)u + 2s^2 + 22 = 0 .$$

Clearly, $4s - 13 < 0$. Since $s > 0$, $s \neq 1$, $s \neq 2$, we must have $s = 3$. If these values are now substituted into equations (*1*), there results a system of three linear equations in three unknowns:

$$x + z + u = 2s + 1 ,$$
$$2x + u = 4z + 4 ,$$
$$2x + z = 6u + 9 .$$

We easily find that $x(= y) = 96$, $z = 57$, $u = 40$.

The integers x, y, z, u are two pairs of equal numbers. Assume that $x = y$ and $z = u$. In this event, the first of equations (*1*) yields $x = 2z - 1$; if this is substituted in (*2*), we obtain

$$x = (2t - 3)z + t^2 + 1 ,$$

and so

$$(2t - 5)z + t^2 + 2 = 0 .$$

It follows that $2t - 5 < 0$, and since $t > 0$, $t \neq 1$, we have $t = 2$. Equations (*1*) may now be written

$$x + 2z = 2x + 1 ,$$
$$2x + 5 = 4z + 4 ,$$

whence $x(= y) = 11$, $z(= u) = 6$.

Three of the integers x, y, z, u are equal. It is necessary to consider two cases.

If $y = z = u$, then equations (*3*) and the first of equations (*1*) take on the form

$$x = 3(2w - 3)y + 3w^2 + 1 \; ;$$
$$2x = 3y - 1 \; ,$$

and these, clearly, are inconsistent.

If $x = y = z$, then the first of equations (*1*) is

$$2x + u = 2x + 1 \; ,$$

from which we find $u = 1$. The last of equations (*1*) becomes

$$3x = 2su + s^2 = 2s + s^2 \; ;$$
$$x = \frac{s(s + 2)}{3} \; .$$

But x must be an integer; hence either s or $s + 2$ must be divisible by 3. That is, $s = 3k$, $x = k(3k + 2)$, or $s = 3k - 2$, $x = (3k - 2)k$. Here k is an arbitrary integer.

All the numbers x, y, z, u are the same. In this case, the first of equations (*1*) yields $3x = 2x + 1$, $x = 1$. Hence we have the following solutions for the problem:

$$x = y = 96 \; , \quad z = 57 \; , \quad u = 40 \; ; \quad x = y = 11 \; , \quad z = u = 6 \; ;$$
$$x = y = z = k(3k \pm 2) \; , \quad u = 1 \; ; \quad x = y = z = u = 1 \; .$$

121. Let x and y be the numbers sought:

$$x + y = xy \; ,$$

or,

$$xy - x - y + 1 = 1 \; ,$$
$$(x - 1)(y - 1) = 1 \; .$$

Since 1 may be expressed as the product of two whole numbers in only two ways, we must have

$$x - 1 = 1 \; , \quad y - 1 = 1 \; ;$$

that is, $x = 2$, $y = 2$, or else,

$$x - 1 = -1 \; , \quad y - 1 = -1 \; ;$$

that is, $x = 0$, $y = 0$.

122. We shall show, first, that at least one of the positive integers x, y, z the sum of whose reciprocals is equal to 1, must be less than 4. In the equation

$$\frac{1}{x} + \frac{1}{y} + \frac{1}{z} = 1,$$

if all of the integers x, y, and z exceed 4, then the sum $\frac{1}{x} + \frac{1}{y} + \frac{1}{z}$ must be $\leqq \frac{1}{4} + \frac{1}{4} + \frac{1}{4} = \frac{3}{4}$. Assuming now that $x \leqq y \leqq z$, we have at most two possible values for x: $x = 2$ or $x = 3$ (clearly, $x > 1$). These two possibilities will be considered separately.

If $x = 2$, then $\frac{1}{y} + \frac{1}{z} = 1 - \frac{1}{x} = \frac{1}{2}$. If we use a common denominator for $\frac{1}{y} + \frac{1}{z} - \frac{1}{2} = 0$, we find the following necessary condition:

$$yz - 2y - 2z = 0,$$
$$yz - 2y - 2z + 4 = 4,$$

or,

$$(y - 2)(z - 2) = 4.$$

Since y and z must exceed 1, neither $y - 2$ nor $z - 2$ can be negative, and only the following cases are possible:

$$y - 2 = 2, z - 2 = 2; \quad y = 4, z = 4.$$
$$y - 2 = 1, z - 2 = 4; \quad y = 3, y = 6.$$

If $x = 3$, then $\frac{1}{y} + \frac{1}{z} = 1 - \frac{1}{x} = \frac{2}{3}$, or

$$2yz - 3y - 3z = 0,$$
$$4yz - 6y - 6z + 9 = 9,$$
$$(2y - 3)(2z - 3) = 9.$$

Since $y \geq x = 3$, $2y - 3 \geq 3$, and $2z - 3 \geq 3$, there is only one possibility:

$$2y - 3 = 3, \quad 2z - 3 = 3; \quad y = 3, \quad z = 3.$$

Therefore, all solutions of the problem are given by the equations

$$\frac{1}{2} + \frac{1}{4} + \frac{1}{4} = 1;$$
$$\frac{1}{2} + \frac{1}{3} + \frac{1}{6} = 1;$$
$$\frac{1}{3} + \frac{1}{3} + \frac{1}{3} = 1.$$

123. (a) If $\dfrac{1}{x} + \dfrac{1}{y} = \dfrac{1}{a}$, then we must have as a necessary condition

$$ax + ay = xy ,$$

or,

$$xy - ax - ay + a^2 = a^2 ,$$
$$(x - a)(y - a) = a^2 .$$

The last equation has $2\mu - 1$ solutions, where μ is the number of divisors of the integer a^2 (including 1 and a^2 as divisors). To obtain all of these solutions, we may write the 2μ possible systems of form $x - a = d$, $y - a = \dfrac{a^2}{d}$ and $x - a = -d$, $y - a = -\dfrac{a^2}{d}$ (where d is a divisor of a^2), and discard the system $x - a = -a$, $y - a = -a$, which leads to the unsatisfactory result $x = y = 0$.

If $a = 14$, then $a^2 = 196$, and the divisors of $a^2 = 196$ are

$$1, 2, 4, 7, 14, 28, 49, 98, 196.$$

We obtain the following seventeen solutions to the problem, which correspond to the above.

x	y	x	y
15	210	13	−182
16	112	12	−84
18	63	10	−35
21	42	7	−14
28	28	−14	7
42	21	−35	10
63	18	−84	12
112	16	−182	13
210	15		

(b) The given equation may be converted, as in the solution to problem (a) to the form

$$(x - z)(y - z) = z^2 . \tag{1}$$

Let t represent the greatest common divisor of the integers x, y, and z; that is, $x = x_1 t$, $y = y_1 t$, $z = z_1 t$, whence x_1, y_1, and z_1 are a relatively prime set (that is, there is no nontrivial divisor for all three). Further, let us designate by m the greatest common divisor of the integers x_1 and z_1, and by n the greatest common divisor of y_1 and

z_1. That is, we write $x_1 = mx_2$, $z_1 = mz_2$; $y_1 = ny_2$, $z_1 = nz_2$, where x_2 and z_2, y_2 and z_2 are relatively prime. The integers m and n are relatively prime, since x_1, y_1, and z_1 have no common divisor. Since z_1 is divisible both by m and by n, we may write $z_1 = mnp$.

If we now substitute into the basic equation (1) $x = mx_2t$, $y = ny_2t$, $z = mnpt$, and divide by mnt^2, we obtain

$$(x_2 - np)(y_2 - mp) = mnp^2 . \tag{2}$$

But x_2 is relatively prime to p, for m is the greatest common divisor of the numbers $x_1 = mx_2$ and $z_1 = mnp$; analogously, y_2 and p are relatively prime. Upon expanding the left member of (2), we see that $x_2y_2 = x_2mp + y_2np$ is divisible by p. It follows that $p = 1$, and the equation takes on the form

$$(x_2 - n)(y_2 - m) = mn .$$

Now x_2 is relatively prime to n, for the three integers $x_1 = mx_2$, $y_1 = ny_2$, and $z_1 = mn$ are relatively prime. Consequently, $x_2 - n$ is relatively prime to n, whence $y_2 - m$ is divisible by n. Analogously, $x_2 - n$ is divisible by m. Thus, $x_2 - n = \pm m$, $y_2 - m = \pm n$; $x_2 = \pm y_2 = \pm m + n$. Therefore,

$$x = m(m + n)t ,$$
$$y = \pm n(m + n)t ,$$
$$z = mnt ,$$

where m, n, t are arbitrary integers.

124. (a) It is readily reasoned from the equation $x^y = y^x$ that the prime divisors of x and y must be the same:

$$x = p_1^{\alpha_1} p_2^{\alpha_2} \cdots p_k^{\alpha_k} ;$$
$$y = p_1^{\beta_1} p_2^{\beta_2} \cdots p_k^{\beta_k} ;$$

where $p_1, p_2, \cdots, p_k$ are prime numbers. In view of this, and the equation, we have

$$\alpha_1 y = \beta_1 x, \ \alpha_2 y = \beta_2 x, \ \cdots, \ \alpha_k y = \beta_k x .$$

We shall assume $y > x$. It follows that

$$\alpha_1 < \beta_1, \ \alpha_2 < \beta_2, \ \cdots, \ \alpha_k < \beta_k .$$

Consequently, y is divisible by x, or $y = kx$, for some integer k. We may rewrite our equation as

$$x^{kx} = (kx)^x .$$

If we take the xth root on both sides, we obtain $x^k = kx$, or $x^{k-1} = k$. Since $y > x$, then $k > 1$, which implies $x > 1$. But $2^{2-1} = 2$, and for $k > 2$ we always have $x^{k-1} > k$. In fact, if $k > 2$ and $x \geq 2$, then

$$x^{k-1} \geq 2^{k-1} > k ,$$

since already $2^{3-1} > 3$; and for $k = 2$, $x > 2$,

$$x^{k-1} = x > 2 = k .$$

Therefore, the only solution of the problem, in positive integers, is $x = 2$, $k = 2$, $y = kx = 4$.

(b) We shall designate the ratio $\dfrac{y}{x}$ by k, whence $y = kx$. If we substitute this for y in the equation, we obtain

$$x^{kx} = (kx)^x ,$$

or, taking the xth root on each side and then dividing by x, we obtain

$$x^{k-1} = k ,$$

from which we obtain

$$x = k^{1/(k-1)} ,$$
$$y = kk^{1/(k-1)} = k^{k/(k-1)} .$$

Write the rational number $\dfrac{1}{k-1}$ as a fraction $\dfrac{p}{q}$ in lowest terms. Substituting this expression for $\dfrac{1}{k-1}$ into the formula, we have

$$k - 1 = \frac{q}{p} , \quad k = 1 + \frac{q}{p} = \frac{p+q}{p} , \quad \frac{k}{k-1} = \frac{p+q}{q} ;$$

$$x = \left(\frac{p+q}{p}\right)^{p/q} , \quad y = \left(\frac{p+q}{p}\right)^{(p+q)/q} .$$

Since p and q are relatively prime, in order for x and y to be rational numbers we must be able to extract the qth roots of p and of $p + q$. But since, for $q \geq 2$ and $p = n^q$, we have the following inequality

$$n^q < p + q < (n+1)^q = n^q + pn^{q-1} + \frac{q(q-1)}{2}n^{q-2} + \cdots ,$$

we must conclude that $q = 1$ is necessary.

Therefore, all positive rational numbers satisfying the given equation are given by the formula

$$x = \left(\frac{p+1}{p}\right)^p , \qquad y = \left(\frac{p+1}{p}\right)^{p+1} ,$$

where p is an arbitrary integer other than 0 or 1.

125. Let n be the number of eighth-grade students who participated, and let m be the number of points assumed by each of them. The total number of points won in the tournament is then $nm + 8$. This number clearly is also equal to the number of games played. Since there were, altogether, $n + 2$ players in the tournament, and each played $(n + 1)$ games (one with each remaining player), there was a total of $\dfrac{(n + 2)(n + 1)}{2}$ games played (two players to each game, hence we divide by 2). Therefore, we have

$$mn + 8 = \frac{(n + 2)(n + 1)}{2} ,$$

or, after simplification,

$$n(n + 3 - 2m) = 14 .$$

Now, n is an integer, as is the number in parentheses (since m is either an integer or a fraction having denominator 2).

Since n must divide 14, n can be only one of the numbers 1, 2, 7, or 14. We must discard the possibilities $n = 1$ and $n = 2$, since in either of these cases the total number of participants could not have exceeded 4 and the two seventh graders could not have amassed as many as 8 points.

There remain the possibilities $n = 7$ and $n = 14$. If $n = 7$, then $7(7 + 3 - 2m) = 14$; $m = 4$. If $n = 14$, then $14(14 + 3 - 2m) = 14$; $m = 8$. Hence, there are two answers: $n = 7$ and $n = 14$.

126. Let n be the number of ninth graders who participated, and let m be the number of points won by them. Then there were $10n$ tenth-grade students, and they won a total of $\dfrac{9}{2}m$ points. In all, then, there were $11n$ participants in the tournament, and $\dfrac{11}{2}m$ points were won.

The total number of points won is, clearly, equal to the number of games played. But since each of the $11n$ players played each other once, there was a total of $\dfrac{11n(11n - 1)}{2}$ games played, and so we have the equation:

$$\frac{11}{2}m = \frac{11n(11-1)}{2} \ ,$$

and thus,

$$m = n(11n - 1) \ .$$

But each ninth-grade student played $11n - 1$ games (since there were $11n$ participants in all), and the n ninth-grade players could have amassed $n(11n - 1)$ points only in the event that each of them won every game he played. But this is possible only if there were just one ninth-grader playing; that is, necessarily, $n = 1$. Therefore, since this is a possible solution, one ninth-grade student participated and he won ten points.

127. The conditions of the problem are (using Heron's formula for the area of a triangle in terms of the lengths of the sides)

$$\sqrt{p(p-a)(p-b)(p-c)} = 2p \ ,$$

where a, b, and c are positive integers, and $p = \dfrac{a+b+c}{2}$. Let $p - a = x$, $p - b = y$, and $p - c = z$. Then the above equation becomes

$$\sqrt{(x+y+z)xyz} = 2(x+y+z) \ ,$$

or, upon squaring both sides and simplifying,

$$xyz = 4(x + y + z) \ .$$

Here, x, y, and z are all integers, or else all of them are fractions with denominator 2 (depending upon whether p is integral or not), that is, all are half of odd integers. In the latter case, however, the left member of the equation is actually a fraction and the right member is an integer. Hence, x, y, and z are necessarily integers.

Let us assume that $x \geq y \geq z$. The equation yields

$$x = \frac{4y + 4z}{yz - 4} \ ;$$

consequently,

$$\frac{4y + 4z}{yz - 4} \geq y \ .$$

We multiply this inequality by $yz - 4$ (it is clear that $yz - 4 > 0$, since otherwise x would be negative) and investigate the resulting inequality in y:

$$y^2z - 8y - 4z \leqq 0,   \tag{1}$$

that is,

$$(y - y_1)(y - y_2) \leqq 0,$$

where y_1 and y_2 are the roots of the quadratic equation in y, $zy^2 - 8y - 4z = 0$ (z being assumed fixed):

$$y_1 = \frac{4 + \sqrt{16 + 4z^2}}{z},$$

$$y_2 = \frac{4 - \sqrt{16 + 4z^2}}{z}.$$

However, y_2 is clearly negative here, which means that $y - y_2 > 0$ (y itself being positive). Consequently, a necessary condition for validity of inequality (1) is that

$$y - y_1 \leqq 0,$$

$$y \leqq \frac{4 + \sqrt{16 + 4z^2}}{z}.$$

Hence, we must have $yz \leqq 4 + \sqrt{16 + 4z^2}$, which implies $z^2 - 4 \leqq \sqrt{16 + 4z^2}$ (for $z \leqq y$). Squaring both members of this last inequality, we obtain

$$z^4 - 8z^2 + 16 \leqq 16 + 4z^2;$$

$$z^4 \leqq 12z^2,$$

which can hold only for $z \leqq 3$.

We shall consider the three possibilities.

(1) $z = 1$, $y \leqq \dfrac{4 + \sqrt{16 + 4}}{1} < 9$;

$$x = \frac{4y + 4z}{yz - 4} = \frac{4y + 4}{y - 4}.$$

We will obtain a positive integer for x only in the following cases (for $y < 9$):

$$y = 5 \text{ (here, } x = 24),$$
$$y = 6 \text{ (here, } x = 14),$$
$$y = 8 \text{ (here, } x = 9).$$

(2) $z = 2$, $y \leq \dfrac{4 + \sqrt{16 + 4 \cdot 4}}{2} < 5$;

$$x = \frac{4y + 4z}{yz - 4} = \frac{4y + 8}{2y - 4} = \frac{2y + 4}{y - 2}.$$

This produces the integral x only if $y = 3$ (yielding $x = 10$) or $y = 4$ (yielding $x = 6$).

(3) $z = 3$, $y = \dfrac{4 + \sqrt{16 + 4 \cdot 9}}{3} < 4$. For $z = y = 3$, $x = \dfrac{4y + 4z}{yz - 4}$ will fail to be integer.

These possibilities yield the following five solutions of the problem.

x	y	z	$x + y + z = p$	a	b	c
24	5	1	30	6	25	29
14	6	1	21	7	15	20
9	8	1	18	9	10	17
10	3	2	15	5	12	13
6	4	2	12	6	8	10

128. (a) The problem clearly involves the solution, in integers, of the equation

$$x^2 + y^2 = z^2 .$$

If t is the greatest common divisor of x, y, and z, we may factor out t^2 from both sides of the equation to obtain an equivalent equation (having the same solution triples x, y, z); hence, we shall assume from the beginning that there is no common divisor for x, y, z. This will imply that each pair of integers is relatively prime, since if two of the integers have a common divisor, that number must also divide the remaining integer.

Since now x, y, z are assumed to be relatively prime, it follows that at most one of these unknowns can be even; if two them, for example, are even, then the third must be even, and hence they would not be relatively prime. Further, if x and y are both odd, say $x = 2k + 1$ and $y = 2l + 1$, we must have

$$x^2 + y^2 = (2k + 1)^2 + (2l + 1)^2 = 2[2(k^2 + l^2 + k + l) + 1] .$$

Now, the square of the integer z is odd if z is odd, or divisible by 4 if z is even. Since the expression on the right for $x^2 + y^2$ is not divisible by 4, we must conclude that $x^2 + y^2$ must be an odd number, and so one of the integers x, y is even and the other odd; also z is odd. We may, with generality, assume that $x = 2x_1$ is even.

The equation may now be written in the form

$$(2x_1)^2 = z^2 - y^2 ,$$

or,

$$x_1^2 = \frac{z+y}{2} \cdot \frac{z-y}{2} .$$

Let $\dfrac{z+y}{2} = u$ and $\dfrac{z-y}{2} = v$; then $z = u + v$ and $y = u - v$. The integers u and v (recall that z and y are both odd) must be relatively prime (otherwise z and y would not be relatively prime). Hence since their product is a perfect square, each of them is a perfect square; that is, $u = a^2$ and $v = b^2$ for some integers a and b. Finally, we have

$$z = u + v = a^2 + b^2 ,$$
$$y = u - v = a^2 - b^2 ,$$
$$x_1 = \sqrt{uv} = ab ,$$

or, if now we relieve the condition that x, y, z be relatively prime

$$x = 2tab ,$$
$$y = t(a^2 - b^2) ,$$
$$z = t(a^2 + b^2) ,$$

where a and b arbitrary relatively prime integers, $a > b$, and t is an arbitrary integer.

These formulas yield all solutions of the problem.

(b) We shall designate the sides of the triangle by x, y, and z (z being the side opposite the 60° angle). Using the law of cosines from trigonometry, we have $z^2 = x^2 + y^2 - xy$.

We must solve this equation in integers. It is convenient to use a rather indirect method; we can put the equation into the format

$$[4z + (x + y)]^2 = [2z + 2(x + y)]^2 + [3(x - y)]^2 ,$$

or,

$$w^2 = u^2 + v^2 ,$$

where $u = 2z + 2(x + y)$, $v = 3(x - y)$, and $w = 4z + (x + y)$. Now the result of problem (a) can be used to obtain

$$u = 2tab ,$$
$$v = t(a^2 - b^2) ,$$
$$w = t(a^2 + b^2) ,$$

where a and b are some relatively prime numbers and t is an arbitrary integer.

Therefore, we obtain

$$4z + (x + y) = t(a^2 + b^2) ,$$
$$2z + 2(x + y) = 2tab ,$$
$$3(x - y) = t(a^2 - b^2) .$$

Solution of this system of equations in three unknowns yields

$$6z = 2t(a^2 + b^2) - 2tab ,$$
$$3(x + y) = 4tab - t(a^2 + b^2) ,$$
$$3(x - y) = t(a^2 - b^2) ,$$

and finally,

$$x = \frac{1}{3}tb(2a - b) ,$$

$$y = \frac{1}{3}ta(2b - a) ,$$

$$z = \frac{1}{3}t(a^2 + b^2 - ab) .$$

In order for the values of x, y, and z in these equations to be integers, it is necessary that at least one of the numbers, t or $a + b$, be divisible by 3. [If $t = 3t_1$, then the equations may be expressed in the form

$$x = t_1 b(2a - b) ,$$
$$y = t_1 a(2b - a) ,$$
$$z = t_1(a^2 + b^2 - ab) .$$

If $a + b$ is divisible by 3, then either $a = 3a_1 + 1$, $b = 3b_1 + 2$, and

$$x = t(3b_1 + 2)(2a_1 - b_1) ,$$
$$y = t(3a_1 + 1)(2b_1 - a_1 + 1) ,$$
$$z = t(3a_1^2 + 3b_1^2 - 3a_1 b_1 + 3b_1 + 1) ,$$

or else $a = 3a_1 + 2$, $b = 3b_1 + 1$, and

$$x = t(3b_1 + 1)(2a_1 - b_1 + 1) ,$$
$$y = t(3a_1 + 2)(2b_1 - a_1) ,$$
$$z = t(3a_1^2 + 3b_1^2 - 3a_1 b_1 + 3a_1 + 1) .]$$

If the equations we obtain are to be meaningful, it is necessary that $2a > b$ and $2b > a$; that is, $\dfrac{a}{2} < b < 2a$. In order for these conditions to hold, the largest of the three numbers x, y, z (if $a > b$, the largest will be the number x) will be less than the sum of the other two. That is, it will be possible to construct the triangle using segments of lengths x, y, and z.

(c) Using the law of cosines, we have

$$z^2 = x^2 + y^2 + xy ,$$

where z is the side of the triangle opposite the 120° angle and x and y are the remaining sides.

This relationship can be rewritten in the form

$$[4z + (x - y)]^2 = [2z + 2(x - y)]^2 + [3(x + y)]^2 .$$

As in problem (b), we find

$$x = \frac{1}{3}ta(a - 2b) ,$$

$$y = \frac{1}{3}tb(2a - b) ,$$

$$z = \frac{1}{3}t(a^2 + b^2 - ab) ,$$

where a and b are relatively prime numbers such that $a > 2b$ and at least one of the integers t or $a + b$ is divisible by 3.

129. In the triangle ABC (whose sides are a, b, c and whose opposite angles are, respectively, A, B, C) suppose that $B = nA$. Then $C = 180° - (n + 1)A$, and, consequently, by the law of sines,

$$\frac{b}{a} = \frac{\sin nA}{\sin A} ,$$

$$\frac{c}{a} = \frac{\sin (n + 1)A}{\sin A} .$$

(a) $n = 2$. Since

$$\sin 2A = 2 \sin A \cos A ,$$
$$\sin 3A = 4 \cos^2 A \sin A - \sin A ,$$

we have

$$\frac{b}{a} = 2 \cos A ,$$

$$\frac{c}{a} = (2 \cos A)^2 - 1 . \qquad (1)$$

But $2 \cos A = \dfrac{b^2 + c^2 - a^2}{bc}$, and so in the integral triangle $2 \cos A$ will always be rational. Let $2 \cos A = \dfrac{p}{q}$, where p and q are integers. Then, by (1), we have

$$a : b : c = q^2 : pq : (p^2 - q^2) .$$

If p and q are relatively prime, then the three integers q^2, pq, and $p^2 - q^2$ do not have any common divisor other than 1. It follows that in all triangles satisfying the given condition $B = 2A$ and having least integral sides (not having a common divisor) the lengths of the sides are expressible by the formulas

$$a = q^2 ,$$
$$b = pq ,$$
$$c = p^2 - q^2 ,$$

where p and q are relatively prime integers.

In order actually to determine the triangle, where $B = 2A$, the numbers p and q must satisfy the following condition: the angle $A = \arccos \dfrac{p}{2q}$ must be such that $0 < A < 60°$ (A must be less than $60°$, since $A + B + C = 3A + C = 180°$). Since $\cos 0 = 1$ and $\cos 60° = \dfrac{1}{2}$, this condition can be rewritten as $2 > \dfrac{p}{q} > 1$. The least integers p and q satisfying this condition are $p = 3$, $q = 2$. It follows that the smallest triangle with integral sides satisfying the condition $B = 2A$ will be the one having sides $a = 4$, $b = 6$, $c = 5$.

We proceed now to problems (b) and (c). Here it will be necessary to use trigonometric functions of A to express $\sin 5A$, $\sin 6A$, and $\sin 7A$. Successive applications of the identity involving the sine of the sum of angles [or using the general formula of problem 222 (b) given in the "Problems" section] yields the following identities:

$$\sin 5A = (2 \cos A)^4 \sin A - 3(2 \cos A)^2 \sin A + \sin A ,$$
$$\sin 6A = [(2 \cos A)^2 - 1][(2 \cos A)^2 - 3]2 \cos A \sin A ,$$
$$\sin 7A = [(2 \cos A)^2 - 2][(2 \cos A)^2 - 3]4 \cos^2 A \sin A - \sin A .$$

The calculations are then carried out exactly as in problem (a).

Assume that $2 \cos A = (p/q)$, where p and q are relatively prime integers; It follows from the identities for $\sin 5A$, $\sin 6A$, and $\sin 7A$ that triangles with integral (nonreducible) sides whose angles satisfy the condition $B = nA$, where $n = 5$ or 6, have sides satisfying the following formulas:

   (b) For $n = 5$,

$$a = q^5 ,$$
$$b = q(p^4 - 3p^2q^2 + q^4) ,$$
$$c = p(p^2 - q^2)(p^2 - 3q^2);$$

   (c) For $n = 6$,

$$a = q^6 ,$$
$$b = pq(p^2 - q^2)(p^2 - 3q^2) ,$$
$$c = p^2(p^2 - 2q^2)(p^2 - 3q^2) - q^6 .$$

(Here, p and q are relatively prime integers.)

In order that a triangle be actually determined by these numbers, and where $B = nA$, the integers p and q must be such that:

   (b') For $n = 5$,

$$0 < \arccos \frac{p}{2q} < 30° ,$$

   (c') For $n = 6$,

$$0 < \arccos \frac{p}{2q} < \frac{180°}{7} \approx 25°43' .$$

Since $\cos 30° = \dfrac{\sqrt{3}}{2}$, the integers p and q, for $n = 5$, must satisfy the condition $2 > \dfrac{p}{q} > \sqrt{3} = 1.732 \cdots$. The least integers p and q which satisfy this condition are $p = 7$ and $q = 4$ (q cannot be less than 4 because $\dfrac{p}{q}$ differs from a whole number by less than $\dfrac{1}{3}$).

Therefore, the smallest triangle with integral sides, in which $B = 5A$, will be the triangle with sides

$$a = 1024 ,$$
$$b = 1220 ,$$
$$c = 231 .$$

For $n = 6$, the integers p and q must satisfy the condition

$$2 > \frac{p}{q} > 2 \cos \frac{180°}{7}.$$

We find from tables that

$$2 \cos \frac{180°}{7} \approx 2 \cos 25°43' \approx 1.802 \ .$$

Therefore, we find that, necessarily, $2 > \dfrac{p}{q} > 1.802$. The least integers p and q satisfying this condition are $q = 6$, $p = 11$. Substituting these values for p and q in the formulas, we find that the smallest triangle with integral sides, in which $B = 6A$, will be the triangle with sides

$$a = 46,656 \ ,$$
$$b = 72,930 \ ,$$
$$c = 30,421 \ .$$

130. Given a right triangle with integral sides x^2 and y^2 and with hypotenuse z, where z also is an integer. It is readily reasoned that x^2, y^2, and z are relatively prime, and so

$$x^2 = 2ab \ ,$$
$$y^2 = a^2 - b^2 \ ,$$
$$z = a^2 + b^2 \ ,$$

where a and b are relatively prime numbers, and $a > b$ [see the solution of problem 128 (a)]. The second of these equations can be rewritten as

$$a^2 = b^2 + y^2 \ ,$$

whence a, b, and y can be expressed by means of the formulas

$$b = 2tu \ ,$$
$$y = t^2 - u^2 \ ,$$
$$a = t^2 + u^2 \ ,$$

where t and u are relatively prime integers [again using the results of problem 128 (a)]. We obtain

$$x^2 = 2(t^2 + u^2) \, 2tu;$$
$$\left(\frac{x}{2}\right)^2 = tu(t^2 + u^2) \ .$$

But t and u are relatively prime, which means that they are also relatively prime to $t^2 + u^2$; consequently, since the product $tu(t^2 + u^2)$

is to be a perfect square, each of the factors must separately be perfect squares:

$$t = x_1^2 \,,$$
$$u = y_1^2 \,,$$
$$t^2 + u^2 = z_1^2 \,.$$

The last equation says that the under the initial assumption there exists a right-angle triangle with sides $t = x_1^2$ and $u = y_1^2$ and hypotenuse z_1, where x_1, y_1, and z_1 are again positive integers and, of particular importance, $z_1 < z$ [since, further, $z_1^4 = (t^2+u^2)^2 = a^2 < a^2+b^2 = z$]. Hence, if there exists any right-angle triangle each of whose legs are squares of integers, and whose hypotenuse is an integer, then there exists another such triangle which has a smaller hypotenuse. Employing the same reasoning we can construct a succession of these triangles with decreasing hypotenuse. Since all such hypotenuse lengths are integral, we must arrive at a triangle whose hypotenuse is of length 1. But this is a contradiction since 1 cannot be a sum of squares of two positive integers.

131. We shall designate the left member of the equation by A and the right member by B. Then $A = B$ follows from:

$$n! \cdot A = 1 \cdot 2 \cdot 3 \cdots n \cdot A = (2n)! \,,$$
$$n! \cdot B = (2^n \cdot n!) \cdot [1 \cdot 3 \cdot 5 \cdots (2n - 1)]$$
$$= (2 \cdot 4 \cdot 6 \cdots 2n) \cdot [1 \cdot 3 \cdots (2n - 1)] = (2n)! \,.$$

132. (a) If we employ the identity

$$\frac{1}{k(k + 1)} = \frac{1}{k} - \frac{1}{k + 1} \,,$$

we obtain

$$\frac{1}{1 \cdot 2} = 1 - \frac{1}{2} \,,$$
$$\frac{1}{2 \cdot 3} = \frac{1}{2} - \frac{1}{3} \,,$$
$$\frac{1}{3 \cdot 4} = \frac{1}{3} - \frac{1}{4} \,,$$
$$\frac{1}{(n - 1)n} = \frac{1}{n - 1} - \frac{1}{n} \,.$$

Adding together all these equations, we obtain

$$\frac{1}{1\cdot 2} + \frac{1}{2\cdot 3} + \cdots + \frac{1}{(n-1)n} = 1 - \frac{1}{n} \, .$$

(b) Using the identity

$$\frac{1}{k(k+1)(k+2)} = \frac{1}{2}\left[\frac{1}{k(k+1)} - \frac{1}{(k+1)(k+2)}\right]$$

(which can be readily verified), we obtain

$$\frac{1}{1\cdot 2\cdot 3} = \frac{1}{2}\left(\frac{1}{1\cdot 2} - \frac{1}{2\cdot 3}\right),$$

$$\frac{1}{2\cdot 3\cdot 4} = \frac{1}{2}\left(\frac{1}{2\cdot 3} - \frac{1}{3\cdot 4}\right),$$

$$\frac{1}{3\cdot 4\cdot 5} = \frac{1}{2}\left(\frac{1}{3\cdot 4} - \frac{1}{4\cdot 5}\right),$$

$$\cdots\cdots\cdots\cdots\cdots\cdots\cdots ,$$

$$\frac{1}{(n-2)(n-1)n} = \frac{1}{2}\left[\frac{1}{(n-2)(n-1)} - \frac{1}{(n-1)n}\right].$$

Adding together all these equations, we obtain

$$\frac{1}{1\cdot 2\cdot 3} + \frac{1}{2\cdot 3\cdot 4} + \frac{1}{3\cdot 4\cdot 5} + \cdots + \frac{1}{(n-2)(n-1)n}$$

$$= \frac{1}{2}\left[\frac{1}{2} - \frac{1}{(n-1)n}\right].$$

(c) We use the identity (whose validity is readily established):

$$\frac{1}{k(k+1)(k+2)(k+3)} = \frac{1}{3}\left[\frac{1}{k(k+1)(k+2)} - \frac{1}{(k+1)(k+2)(k+3)}\right].$$

This yields

$$\frac{1}{1\cdot 2\cdot 3\cdot 4} = \frac{1}{3}\left(\frac{1}{1\cdot 2\cdot 3} - \frac{1}{2\cdot 3\cdot 4}\right),$$

$$\frac{1}{2\cdot 3\cdot 4\cdot 5} = \frac{1}{3}\left(\frac{1}{2\cdot 3\cdot 4} - \frac{1}{3\cdot 4\cdot 5}\right),$$

$$\frac{1}{3\cdot 4\cdot 5\cdot 6} = \frac{1}{3}\left(\frac{1}{3\cdot 4\cdot 5} - \frac{1}{4\cdot 5\cdot 6}\right),$$

$$\cdots\cdots\cdots\cdots\cdots\cdots\cdots ,$$

$$\frac{1}{(n-3)(n-2)(n-1)n}$$

$$= \frac{1}{3}\left[\frac{1}{(n-3)(n-2)(n-1)} - \frac{1}{(n-2)(n-1)n}\right].$$

Adding together these equations, we obtain

$$\frac{1}{1\cdot2\cdot3\cdot4} + \frac{1}{2\cdot3\cdot4\cdot5} + \frac{1}{3\cdot4\cdot5\cdot6} + \cdots + \frac{1}{(n-3)(n-2)(n-1)n}$$

$$= \frac{1}{3}\left[\frac{1}{6} - \frac{1}{(n-2)(n-1)n}\right].$$

Remarks: If we "guess at" the results we expect to obtain (which is often quite feasible, if we try a few small values for n), we can often prove the general validity of a formula by using mathematical induction (see, for example, the solution of problem 133).

It is possible to prove, also, that in general

$$\frac{1}{1\cdot2\cdot3\cdots p} + \frac{1}{2\cdot3\cdot4\cdots(p+1)}$$

$$+ \cdots + \frac{1}{(n-p+1)(n-p+2)(n-p+3)\cdots n}$$

$$= \frac{1}{p-1}\left[\frac{1}{1\cdot2\cdot3\cdots(p-1)} - \frac{1}{(n-p+2))(n-p+3)(n-p+4)\cdots n}\right]$$

[the proof being analogous to that of problems (a – dash)].

133. These identities are most conveniently proved by mathematical induction. We leave the verification of (a) and (b) to the reader, but we give here the proof of the more general equation (c) of which (a) and (b) are merely special cases.

The equation is valid for $n = 1$, since

$$1\cdot2\cdot3\cdots p = \frac{1\cdot2\cdot3\cdots p(p+1)}{p+1}.$$

Assume now that the equation is valid for some n:

$$1\cdot2\cdot3\cdots p + \cdots + n(n+1)\cdots(n+p-1) = \frac{n(n+1)\cdots(n+p)}{p+1}.$$

Then we have

$$1\cdot2\cdot3\cdots p + 2\cdot3\cdots p(p+1)$$

$$+ \cdots + n(n+1)\cdots(n+p-1) + (n+1)\cdots(n+p-1)(n+p)$$

$$= \frac{n(n+1)\cdots(n+p)}{p+1} + (n+1)\cdots(n+p)$$

$$= \frac{n(n+1)\cdots(n+p) + (p+1)(n+1)\cdots(n+p)}{p+1}$$

$$= \frac{(n+1)\cdots(n+p)(n+p+1)}{p+1}.$$

By mathematical induction we conclude that the equation holds for all n.

134. *First Solution.* (a) We write the sequence of equations

$$1^3 = 1^3\,,$$
$$2^3 = (1+1)^3 = 1^3 + 3\cdot1^2 + 3\cdot1 + 1\,,$$
$$3^3 = (2+1)^3 = 2^3 + 3\cdot2^2 + 3\cdot2 + 1\,,$$
$$4^3 = (3+1)^3 = 3^3 + 3\cdot3^2 + 3\cdot3 + 1\,,$$
$$\cdots\cdots\cdots\cdots\cdots\cdots\cdots\cdots\cdots\cdots\cdots\cdots\cdots\,,$$
$$(n+1)^3 = n^3 + 3n^2 + 3n + 1\,.$$

Adding all these equalities, and cancelling equal terms on both sides, we arrive at

$$(n+1)^3 = 1^3 + 3(1^2 + 2^2 + 3^2 + \cdots + n^2)$$
$$+ 3(1 + 2 + 3 + \cdots + n) + n\,.$$

It follows that

$$1^2 + 2^2 + 3^2 + \cdots + n^2$$
$$= \frac{(n+1)^3 - 1 - 3(1+2+3+\cdots+n) - n}{3}$$
$$= \frac{2n^3 + 6n^2 + 6n - 3n^2 - 3n - 2n}{6} = \frac{n(2n^2 + 3n + 1)}{6}$$
$$= \frac{n(n+1)(2n+1)}{6}\,.$$

(b) We write the sequence of equations:

$$1^4 = 1^4\,,$$
$$2^4 = (1+1)^4 = 1^4 + 4\cdot1^3 + 6\cdot1^2 + 4\cdot1 + 1\,,$$
$$3^4 = (2+1)^4 = 2^4 + 4\cdot2^3 + 6\cdot2^2 + 4\cdot2 + 1\,,$$
$$4^4 = (3+1)^4 = 3^4 + 4\cdot3^3 + 6\cdot3^2 + 4\cdot3 + 1\,,$$
$$\cdots\cdots\cdots\cdots\cdots\cdots\cdots\cdots\cdots\cdots\cdots\cdots\cdots\cdots\,,$$
$$(n+1)^4 = n^4 + 4\cdot n^3 + 6\cdot n^2 + 4\cdot n + 1\,.$$

Adding all of these equalities, and cancelling equal terms on both sides, we obtain

$$(n + 1)^4 = 1^4 + 4(1^3 + 2^3 + \cdots + n^3)$$
$$+ 6(1^2 + 2^2 + \cdots + n^2) + 4(1 + 2 + \cdots + n) + n .$$

We now use the result of problem (a) to obtain

$1^3 + 2^3 + \cdots + n^3$

$$= \frac{1}{4}\left[(n + 1)^4 - 1 - 6\frac{n(n + 1)(2n + 1)}{6} - 4\frac{n(n + 1)}{2} - n\right]$$

$$= \frac{1}{4}\{[(n + 1)^2 - 1]\cdot[(n + 1)^2 + 1] - n(n + 1)(2n + 1)$$

$$- 2n(n + 1) - n\}$$

$$= \frac{n}{4}[(n + 2)(n^2 + 2n + 2) - (n + 1)(2n + 1) - 2(n + 1) - 1]$$

$$= \frac{n}{4}[(n + 1)(n^2 + 2n + 2 - 2n - 1 - 2) + (n^2 + 2n + 2 - 1)]$$

$$= \frac{n(n + 1)}{4}[(n^2 - 1) + (n + 1)] = \frac{n^2(n + 1)^2}{4}$$

     (c) Proceeding as in problems (a) and (b), and using the expansion $(k + 1)^5 = k^5 + 5k^4 + 10k^3 + 10k^2 + 5k + 1$, we obtain

$$(n + 1)^5 = 1^5 + 5(1^4 + 2^4 + \cdots + n^4)$$
$$+ 10(1^3 + 2^3 + \cdots + n^3) + 10(1^2 + 2^2 + \cdots + n^2)$$
$$+ 5(1 + 2 + \cdots + n) + n ,$$

from which we find, using the formulas from problems (a) and (b),

$$1^4 + 2^4 + \cdots + n^4 = \frac{n(n + 1)(2n + 1)(3n^2 + 3n - 1)}{30} .$$

     (d) We write

$$S_n = 1^3 + 2^3 + 3^3 + \cdots + n^3 = \frac{n^2(n + 1)^2}{4}$$

[see problem (b)]. We then obtain

$$1^3 + 3^3 + 5^3 + \cdots + (2n - 3)^3 + (2n - 1)^3$$
$$= [1^3 + 2^3 + \cdots + (2n)^3] - 2^3 [1^3 + 2^3 + \cdots + n^3]$$
$$= S_{2n} - 8S_n = \frac{(2n)^2(2n + 1)^2}{4} - 8 \frac{n^2(n + 1)^2}{4}$$
$$= n^2(2n + 1)^2 - 2n^2 (n + 1)^2 = n^2[(2n + 1)^2 - 2(n + 1)^2$$
$$= n^2(4n^2 + 4n + 1 - 2n^2 - 4n - 2) = n^2(2n^2 - 1) .$$

Second Solution. (a) Consider the following table.

1st row	1	2	3		k		n
2nd row	1	2	3		k		n
3rd row	1	2	3		k		n
kth row	1	2	3		k		n
nth row	1	2	3		k		n

The sum of all the integers of any row is equal to $1+2+3\cdots+n$, that is, $\dfrac{n(n+1)}{2}$, and so the sum of all the integers of the table is equal to $n\cdot\dfrac{n(n+1)}{2}$. Now let us sum up the numbers within any region bounded by lines. For the region bounded by the kth row and kth column we have the sum

$$1 + 2 + \cdots + (k-1) + k\cdot k$$
$$= \frac{(k-1)k}{2} + k^2 = \frac{3}{2}k^2 - \frac{1}{2}k \ .$$

Summing up all the regions in this way, we obtain

$$\frac{3}{2}(1^2 + 2^2 + \cdots + n^2) - \frac{1}{2}(1 + 2 + \cdots + n) = \frac{n^2(n+1)}{2} \ ,$$

from which we obtain

$$1^2 + 2^2 + \cdots n^2 = \frac{2}{3}\left[\frac{n^2(n+1)}{2} + \frac{n(n+1)}{4}\right] = \frac{n(n+1)(2n+1)}{6} \ .$$

(b) Consider the following table

1st row	1^2	2^2	3^2		k^2		n^2
2nd row	1^2	2^2	3^3		k^2		n^2
3rd row	1^2	2^2	3^2		k^2		n^2
kth row	1^2	2^2	3^2		k^2		n^2
nth row	1^2	2^2	3^2		k^2		n^2

The sum of all the integers of any one row of the table is
$$1^2 + 2^2 + 3^2 + \cdots + n^2 \, ,$$

that is, $\dfrac{n(n+1)(2n+1)}{6}$ [see the solution of problem (a)]. Hence the sum of all the integers of the table is equal to $\dfrac{n^2(n+1)(2n+1)}{6}$. However, the sum of the integers in the region bounded by the kth row and kth column is equal to

$$1^2 + 2^2 + \cdots + (k-1)^2 + k \cdot k^2 = \frac{(k-1)k(2k-1)}{6} + k^3$$

$$= \frac{4}{3}k^3 - \frac{1}{2}k^2 + \frac{1}{6}\,k \, .$$

This yields

$$\frac{4}{3}(1^3 + 2^3 + \cdots + n^3) - \frac{1}{2}(1^2 + 2^2 + \cdots + n^2)$$

$$+ \frac{1}{6}(1 + 2 + \cdots + n) = \frac{n^2(n+1)(2n+1)}{6} ,$$

from which, after some manipulation, and using the result of problem (a), we obtain

$$1^3 + 2^3 + \cdots + n^3 = \frac{n^2(n+1)^2}{4} \, .$$

(c) This problem can be solved in a method quite analogous to that used in problems (a) and (b) by employing the integers $1^3, 2^3, \cdots n^3$. It is left to the reader to carry out the details of the proof.

Remark: If we could "guess at" the results of problems 134 (a - d) by considering small values of n, mathematical induction would serve to establish the validity of the formulas.

135. If we add 1 to the left member of the given equation, we can write

$$[(1+a) + b(1+a)] + c(1+a)(1+b)$$
$$+ \cdots + l(1+a)(1+b)\cdots(1+k)$$
$$= [(1+a)(1+b)$$
$$+ c(1+a)(1+b)] + d(1+a)(1+b)(1+c)$$
$$+ \cdots + l(1+a)(1+b)\cdots(1+k)$$
$$= [(1+a)(1+b)(1+c)$$

$$+ d(1 + a)(1 + b)(1 + c)] + \cdots + l(1 + a)(1 + b) \cdots (1 + k)$$
$$= (1 + a)(1 + b)(1 + c)(1 + d)$$
$$+ \cdots + l(1 + a)(1 + b) \cdots (1 + k)$$
$$= (1 + a)(1 + b)(1 + c) \cdots (1 + l),$$

which proves the proposition.

If $a = b = c \cdots = l$, then we have

$$a + a(1 + a) + a(1 + a)^2 + a(1 + a)^3$$
$$+ \cdots + a(1 + a)^{n-1} = (1 + a)^n - 1 ,$$

where n is the number of integers $a, b, c, \cdots, l$; writing $1 + a = x$, whence $a = x - 1$, we have

$$(x - 1)(1 + x + x^2 + \cdots + x^{n-1}) = x^n - 1 ,$$

which is the formula for the sum of a geometric progression.

136. (a) We add 1 to the sum we wish to determine, and we obtain

$$(1! + 1 \cdot 1!) + 2 \cdot 2! + 3 \cdot 3! + \cdots + n \cdot n!$$
$$= (2! + 2 \cdot 2!) + 3 \cdot 3! + \cdots + n \cdot n!$$
$$= (3! + 3 \cdot 3!) + \cdots + n \cdot n! = 4! + \cdots + n \cdot n!$$
$$= (n! + n \cdot n!) = (n + 1)!;$$

therefore,

$$1 \cdot 1! + 2 \cdot 2! + 3 \cdot 3! + \cdots + n \cdot n! = (n + 1)! - 1 .$$

Remark: This result can be obtained from the formula of problem 135 by substituting $a = 1, b = 2, c = 3, \cdots, l = n$.

(b) We add to the summation under consideration the term $c_{n+1}^0 = 1$. If we employ the fact that

$$C_m^l + C_m^{l+1} = C_{m+1}^{l+1} ,$$

we obtain

$$(C_{n+1}^0 + C_{n+1}^1) + C_{n+2}^2 + C_{n+3}^3 + \cdots + C_{n+k}^k$$
$$= (C_{n+2}^1 + C_{n+2}^2) + C_{n+3}^3 + \cdots + C_{n+k}^k$$
$$= (C_{n+3}^2 + C_{n+3}^3) + \cdots + C_{n+k}^k$$
$$= C_{n+4}^3 + \cdots + C_{n+k}^k = \cdots = C_{n+k+1}^k;$$

therefore,

$$C_{n+1}^1 + C_{n+2}^2 + C_{n+3}^3 + \cdots + C_{n+k}^k = C_{n+k+1}^k - 1 .$$

Remark: This result can be obtained from the equation of problem 135 by letting $a = \dfrac{n+1}{1}$, $b = \dfrac{n+1}{2}$, $c = \dfrac{n+1}{3}$, $\cdots$, $l = \dfrac{n+1}{k}$.

137. From the definition of a logarithm we obtain

$$\frac{1}{\log_b a} = \log_a b \ .$$

$\left(\text{In fact, if } \log_b a = y, \text{ then } b^y = a, \text{ or } a^{1/y} = b, \text{ whence } \dfrac{1}{y} = \log_a b.\right)$

The equation can therefore be written in the form

$$\log_N 2 + \log_N 3 + \cdots + \log_N 100 = \log_N (2 \cdot 3 \cdots 100) \ ,$$

from which the desired conclusion immediately follows.

138. It will be shown by mathematical induction that the sum we seek to determine is equal to $\dfrac{1}{a_1 a_2 \cdots a_n}$. First, the proposition is obviously valid for $n = 1$. Now, assume that the assertion is valid for $n - 1$ positive integers, and consider the sum S of the given fractions for n integers. From each of the $n!$ terms of the sum S, we can factor out the fraction $\dfrac{1}{a_1 + a_2 + \cdots + a_n}$, since this final factor has as its denominator the sum of all the n positive integers (the order of the addition being unimportant, since no integer is omitted). Further, grouping in the parentheses separately the $(n-1)!$ terms corresponding to those permutations of indices where the index 1 is missing, the $(n-1)!$ terms corresponding to the permutations where the index 2 is missing, and so on, we arrive at n separate sums, to each of which we can apply the induction hypothesis, since each comprises $(n-1)$ integers; for example, $a_2, a_3, \cdots, a_n$, or $a_1, a_3, \cdots, a_n$, and so on.

Summing these, and using the induction hypothesis, we obtain

$$S = \frac{1}{a_1 + \cdots + a_n} \left(\frac{1}{a_2 a_3 \cdots a_n} + \frac{1}{a_1 a_3 \cdots a_n} + \cdots + \frac{1}{a_1 a_2 \cdots a_{n-1}} \right)$$

$$= \frac{1}{a_1 + a_2 + \cdots + a_n} \left(\frac{a_1 + a_2 + \cdots + a_n}{a_1 a_2 \cdots a_n} \right) = \frac{1}{a_1 a_2 \cdots a_n}$$

which is what we wished to show.

139. (a) Multiply the expression by $1 - \dfrac{1}{3}$ to obtain

$$\left[\left(1-\tfrac{1}{3}\right)\left(1+\tfrac{1}{3}\right)\right]\left(1+\tfrac{1}{9}\right)\left(1+\tfrac{1}{81}\right)\left(1+\tfrac{1}{3^8}\right)\cdots\left(1+\tfrac{1}{3^{2^n}}\right)$$

$$=\left[\left(1-\tfrac{1}{9}\right)\left(1+\tfrac{1}{9}\right)\right]\left(1+\tfrac{1}{81}\right)\left(1+\tfrac{1}{3^8}\right)\cdots\left(1+\tfrac{1}{3^{2^n}}\right)$$

$$=\left[\left(1-\tfrac{1}{81}\right)\left(1+\tfrac{1}{81}\right)\right]\left(1+\tfrac{1}{3^8}\right)\cdots\left(1+\tfrac{1}{3^{2^n}}\right)$$

$$=\cdots=\left(1-\tfrac{1}{3^{2^n}}\right)\left(1+\tfrac{1}{3^{2^n}}\right)=1-\tfrac{1}{3^{2^{n+1}}}.$$

Therefore

$$\left(1+\tfrac{1}{3}\right)\left(1+\tfrac{1}{9}\right)\left(1+\tfrac{1}{81}\right)\cdots\left(1+\tfrac{1}{3^{2^n}}\right)$$

$$=\frac{1-[1/(3^{2^{n+1}})]}{1-(1/3)}=\frac{3}{2}\left(1-\frac{1}{3^{2^{n+1}}}\right).$$

(b) If we multiply by sin α, we obtain

$$(\sin\alpha\cos\alpha)\cos 2\alpha\cos 4\alpha\cdots\cos 2^n\alpha$$

$$=\frac{1}{2}(\sin 2\alpha\cos 2\alpha)\cos 4\alpha\cdots\cos 2^n\alpha$$

$$=\frac{1}{4}(\sin 4\alpha\cos 4\alpha)\cdots\cos 2^n\alpha$$

$$=\cdots=\frac{1}{2^n}\sin 2^n\alpha\cos 2^n\alpha=\frac{\sin 2^{n+1}\alpha}{2^{n+1}}.$$

Therefore

$$\cos\alpha\cos 2\alpha\cos 4\alpha\cdots\cos 2^n\alpha=\frac{\sin 2^{n+1}\alpha}{2^{n+1}\sin\alpha}.$$

140. Since $2^{10}=1024$, we can write $2^{100}=1024^{10}$. Since $1000^{10}=10^{30}$, which is the number with 1 as a first digit followed by 30 zeros, and since $1024^{10}>1000^{10}$, it follows that the number $2^{100}=1024^{10}$ cannot have fewer than 31 digits. However,

$$\frac{1024^{10}}{1000^{10}}<\left(\frac{1025}{1000}\right)^{10}=\left(\frac{41}{40}\right)^{10}$$

$$=\frac{41}{40}\cdot\frac{41}{40}\cdot\frac{41}{40}\cdot\frac{41}{40}\cdot\frac{41}{40}\cdot\frac{41}{40}\cdot\frac{41}{40}\cdot\frac{41}{40}\cdot\frac{41}{40}\cdot\frac{41}{40}$$

$$<\frac{41}{40}\cdot\frac{40}{39}\cdot\frac{39}{38}\cdot\frac{38}{37}\cdot\frac{37}{36}\cdot\frac{36}{35}\cdot\frac{35}{34}\cdot\frac{34}{33}\cdot\frac{33}{32}\cdot\frac{32}{31}$$

$$=\frac{41}{31}<10,$$

since

$$\frac{41}{40} < \frac{40}{39} < \frac{39}{38} < \cdots$$

$\left(\frac{41}{40} = 1 + \frac{1}{40}; \frac{40}{39} = 1 + \frac{1}{39}; \text{ and so on}\right)$. Hence,

$$2^{100} = 1024^{10} < 10 \cdot 1000^{10} ,$$

which implies that 2^{100} contains fewer than 32 digits. Therefore, the integer 2^{100} contains exactly 31 digits.

Remark: This result is even more easily obtained by using logarithms. Since log 2 = 0.30103, log 2^{100} = 100 log 2 = 30.103, and so 2^{100} must have 31 digits. But the technique used in solving this problem without using logarithms has independent interest.

141. (a) *First solution*. We designate the product $\frac{1}{2} \cdot \frac{3}{4} \cdot \frac{5}{6} \cdot \cdots \cdot \frac{99}{100}$ by A, and consider also the product $B = \frac{2}{3} \cdot \frac{4}{5} \cdot \frac{6}{7} \cdot \cdots \cdot \frac{98}{99}$.
Since

$$\frac{2}{3} > \frac{1}{2}, \frac{4}{5} > \frac{3}{4}, \frac{6}{7} > \frac{5}{6}, \cdots, \frac{98}{99} > \frac{97}{98}, 1 > \frac{99}{100} ,$$

we have $B > A$. Now, clearly,

$$A \cdot B = \frac{1}{2} \cdot \frac{2}{3} \cdot \frac{3}{4} \cdot \frac{4}{5} \cdot \frac{5}{6} \cdot \frac{6}{7} \cdots \frac{98}{99} \cdot \frac{99}{100} = \frac{1}{100} .$$

It follows that

$$A^2 < AB = \frac{1}{100} ,$$

whence $A < \frac{1}{10}$. Further,

$$B < 2A = \frac{3}{4} \cdot \frac{5}{6} \cdot \frac{7}{8} \cdots \frac{99}{100} ,$$

since

$$\frac{2}{3} < \frac{3}{4}, \frac{4}{5} < \frac{5}{6}, \frac{6}{7} < \frac{7}{8}, \cdots, \frac{98}{99} < \frac{99}{100} .$$

Consequently,

$$A \cdot 2A > AB = \frac{1}{100}, \text{ which implies } A > \frac{1}{10\sqrt{2}} .$$

Second Solution. We write, as above,

$$\frac{1}{2}\cdot\frac{3}{4}\cdot\frac{5}{6}\cdots\frac{99}{100}=A\,.$$

Then,

$$A^2=\frac{1^2}{2^2}\cdot\frac{3^2}{4^2}\cdot\frac{5^2}{6^2}\cdots\frac{99^2}{100^2}\,,$$

from which we obtain

$$\frac{1^2}{2^2}\cdot\frac{3^2-1}{4^2}\cdot\frac{5^2-1}{6^2}\cdots\frac{99^2-1}{100^2}$$

$$<A^2<\frac{1^2}{2^2-1}\cdot\frac{3^2}{4^2-1}\cdot\frac{5^2}{6^2-1}\cdots\frac{99^2}{100^2-1}\,.$$

If we factor the factors of the numerator on the left and the denominator on the right as differences of two squares, we obtain

$$\frac{1}{2\cdot2}\cdot\frac{2\cdot4}{4\cdot4}\cdot\frac{4\cdot6}{6\cdot6}\cdots\frac{98\cdot100}{100\cdot100}<A^2<\frac{1}{1\cdot3}\cdot\frac{3\cdot3}{3\cdot5}\cdot\frac{5\cdot5}{5\cdot7}\cdots\frac{99\cdot99}{99\cdot101}\,,$$

or, after simplification,

$$\frac{1}{200}<A^2<\frac{1}{101}\,;$$

$$\frac{1}{10\sqrt{2}}<A<\frac{1}{\sqrt{101}}<\frac{1}{10}\,,$$

which is what we set out to prove.

Remark: A more general relationship may be proved in exactly the same way:

$$\frac{1}{2\sqrt{n}}<\frac{1}{2}\cdot\frac{3}{4}\cdot\frac{5}{6}\cdots\frac{2n-1}{2n}<\frac{1}{\sqrt{2n}}\,.$$

(b) We first show that if $n>1$, then

$$\frac{1}{2}\cdot\frac{3}{4}\cdot\frac{5}{6}\cdots\frac{2n-1}{2n}<\frac{1}{\sqrt{3n+1}}\,.$$

This may be done conveniently by mathematical induction. If $n=1$, we have

$$\frac{1}{2}=\frac{1}{\sqrt{3\cdot1+1}}\,.$$

Assume now that for some n

$$\frac{1}{2}\cdot\frac{3}{4}\cdot\frac{5}{6}\cdots\frac{2n-1}{2n}\leqq\frac{1}{\sqrt{3n+1}}\,.$$

If both sides of this inequality are multiplied by $\dfrac{2n+1}{2n+2}$, it becomes

$$\frac{1}{2} \cdot \frac{3}{4} \cdot \frac{5}{6} \cdots \frac{2n-1}{2n} \cdot \frac{2n+1}{2n+2} \leqq \frac{2n+1}{(2n+2)\sqrt{3n+1}} \cdot$$

Now,

$$\left[\frac{2n+1}{(2n+2)\sqrt{3n+1}}\right]^2 = \frac{(2n+1)^2}{12n^3 + 28n^2 + 20n + 4}$$

$$= \frac{(2n+1)^2}{(12n^3 + 28n^2 + 19n + 4) + n}$$

$$= \frac{(2n+1)^2}{(2n+1)^2(3n+4) + n} < \frac{1}{3n+4},$$

and it follows that

$$\frac{2n+1}{(2n+2)\sqrt{3n+1}} < \frac{1}{\sqrt{3n+4}} \cdot$$

Thus, we obtain

$$\frac{1}{2} \cdot \frac{3}{4} \cdot \frac{5}{6} \cdots \frac{2n-1}{2n} \cdot \frac{2n+1}{2n+2} < \frac{1}{\sqrt{3(n+1)+1}} \cdot$$

We conclude, by the principle of mathematical induction, that for every n

$$\frac{1}{2} \cdot \frac{3}{4} \cdot \frac{5}{6} \cdots \frac{2n-1}{2n} \leqq \frac{1}{\sqrt{3n+1}} \cdot$$

(We note that equality holds only for $n = 1$.)

If now we let $n = 50$ in this last inequality, we find

$$\frac{1}{2} \cdot \frac{3}{4} \cdot \frac{5}{6} \cdots \frac{99}{100} < \frac{1}{\sqrt{3 \cdot 50 + 1}} = \frac{1}{\sqrt{151}} = \frac{1}{12.288\ldots},$$

which proves the assertion of the problem.

142. We start with

$$\frac{1}{2^{100}} C_{100}^{50} = \frac{1 \cdot 2 \cdot 3 \ldots 100}{2^{50}(1 \cdot 2 \cdot 3 \ldots 50) \cdot 2^{50}(1 \cdot 2 \cdot 3 \ldots 50)}$$

$$= \frac{1 \cdot 2 \cdot 3 \ldots 100}{(2 \cdot 4 \cdot 6 \ldots 100) \cdot (2 \cdot 4 \cdot 6 \ldots 100)} = \frac{1 \cdot 3 \cdot 5 \ldots 99}{2 \cdot 4 \cdot 6 \ldots 100},$$

and apply the results of problem 141 (a).

143. It suffices to determine which is the larger: $101^n - 99^n$ or 100^n. Consider the relationship

$$\frac{101^n - 99^n}{100^n} = \frac{(100 + 1)^n - (100 - 1)^n}{100^n}$$

$$= \frac{2(C_n^1 \cdot 100^{n-1} + C_n^3 \cdot 100^{n-3} + \cdots)}{100^n}$$

$$= 2\left(\frac{n}{100} + \frac{n(n-1)(n-2)}{3! \cdot 100^3} + \cdots\right).$$

It is clear that the fraction on the left exceeds 1 if $n \geqq 50$. We show that this ratio exceeds 1 also for $n = 49$. We have

$$2\left(\frac{49}{100} + \frac{49 \cdot 48 \cdot 47}{3! \cdot 100^3} + \cdots\right) > 2\left(\frac{49}{100} + \frac{18,424}{100^3}\right)$$

$$> 2\left(\frac{49}{100} + \frac{100^2}{100^3}\right) = 1.$$

Now we show that if $n = 48$, the ratio under consideration is smaller than 1:

$$2\left(\frac{48}{100} + \frac{48 \cdot 47 \cdot 46}{3! \cdot 100^3} + \frac{48 \cdot 47 \cdot 46 \cdot 45 \cdot 44}{5! \cdot 100^5} + \cdots\right)$$

$$< 2\left[\frac{48}{100} + \frac{48^3}{(1 \cdot 2 \cdot 3) \cdot 100^3} + \frac{48^5}{(1 \cdot 2 \cdot 3)(2 \cdot 3)100^5}\right.$$

$$\left. + \frac{48^7}{(1 \cdot 2 \cdot 3)(2 \cdot 3)(2 \cdot 3) \, 100^7} + \cdots\right]$$

$$= 2\left[\frac{48}{100} + \frac{1}{6}\left(\frac{48}{100}\right)^3 + \frac{1}{6^2}\left(\frac{48}{100}\right)^5 + \cdots\right]$$

$$< 2\frac{\dfrac{48}{100}}{1 - \dfrac{1}{6}\left(\dfrac{48}{100}\right)^2} = \frac{9600}{9616} < 1.$$

Clearly, now, the ratio is also less than 1 for all positive integers less than 48.

Therefore, we finally obtain: $99^n + 100^n$ is greater than 101^n if $n \leqq 48$ and is less than 101^n if $n > 48$.

144. We first show that the product of n consecutive natural numbers is greater than the nth power of the square root of the product of the first and last of these numbers. Let the n integers be $a, a + 1, \cdots, a + n - 1$. Then the kth number from the beginning will be $a + k - 1$, and the kth number from the end will be $a + n - k$. Their product is

$$(a + k - 1)(a + n - k) = a^2 + an - a + (k - 1)(n - k)$$
$$\geqq a^2 + an - a = a(a + n - 1),$$

where the equality is obtained only for $k = 1$ or $k = n$. That is, the product of two positive integers equidistant respectively from each end of the sequence (for odd n, these two integers are taken to be the common middle one) always exceeds the product of the two extreme (first and last) integers. But then we have, for the product of all the numbers,

$$a(a + 1) \cdots (a + n - 1)$$
$$\geqq [a(a + n - 1)]^{n/2} = [\sqrt{a(a + n - 1)}]^n,$$

where the equality holds only if $n = 1$ or $n = 2$.

We shall show now that $300! > 100^{300}$. We have

$$1 \cdot 2 \cdots 25 > \sqrt{25^{25}} = 5^{25};$$
$$26 \cdots 50 > (\sqrt{26 \cdot 50})^{25} > 35^{25};$$
$$51 \cdots 100 > (\sqrt{51 \cdot 100})^{50} > 70^{50};$$
$$101 \cdots 200 > \sqrt{100^{100}} \cdot \sqrt{200^{100}} = 10^{200} \cdot 2^{50};$$
$$201 \cdots 300 > \sqrt{200^{100}} \cdot \sqrt{300^{100}} = 10^{200} \cdot 2^{50} \cdot 3^{50}.$$

If we multiply together all the left members of these inequalities and compare the result with the product of all the right members, we obtain

$$300! > 5^{25} \cdot 35^{25} \cdot 70^{50} \cdot 10^{400} \cdot 2^{100} \cdot 3^{50}$$
$$= 5^{50} \cdot 7^{25} \cdot 5^{50} \cdot 14^{50} \cdot 10^{400} \cdot 2^{100} \cdot 3^{50}$$
$$= 10^{500} \cdot 7^{25} \cdot 14^{50} \cdot 3^{50}$$
$$= 10^{500} \cdot 21^{25} \cdot 42^{25} \cdot 14^{25} > 10^{500} \cdot 20^{25} \cdot 40^{25} \cdot 14^{25}$$
$$= 10^{550} \cdot 2^{25} \cdot 4^{25} \cdot 14^{25} = 10^{550} \cdot 112^{25}$$
$$= 10^{600} \cdot 1{,}12^{25} > 10^{600} = 100^{300}.$$

Remark. A more general result is shown in problem 148.

145. We first show that, for any natural number $k \leqq n$,

$$1 + \frac{k}{n} \leqq \left(1 + \frac{1}{n}\right)^k < 1 + \frac{k}{n} + \frac{k^2}{n^2}.$$

We use mathematical induction. The proposition is obviously true for $k = 1$. Assume now that the proposition holds for a particular value of k. We shall show that it then holds for $k + 1$. We have

$$\left(1+\frac{1}{n}\right)^{k+1}=\left(1+\frac{1}{n}\right)^{k}\left(1+\frac{1}{n}\right)\geqq\left(1+\frac{k}{n}\right)\left(1+\frac{1}{n}\right)$$

$$=1+\frac{k+1}{n}+\frac{k}{n^{2}}>1+\frac{k+1}{n}.$$

We do not need here the fact that $k \leqq n$, hence the inequality is valid for any integral value of k. Assume now that $k \leqq n$. Then:

$$\left(1+\frac{1}{n}\right)^{k+1}=\left(1+\frac{1}{n}\right)^{k}\left(1+\frac{1}{n}\right)<\left(1+\frac{k}{n}+\frac{k^{2}}{n^{2}}\right)\left(1+\frac{1}{n}\right)$$

$$=1+\frac{k+1}{n}+\frac{k^{2}+2k+1}{n^{2}}-\frac{k+1}{n^{2}}+\frac{k^{2}}{n^{2}}$$

$$=1+\frac{k+1}{n}+\frac{(k+1)^{2}}{n^{2}}-\frac{n(k+1)-k^{2}}{n^{3}}$$

$$<1+\frac{k+1}{n}+\frac{(k+1)^{2}}{n^{2}},$$

since $n(k+1)>k^{2}$ if $n\geqq k$.

If we now substitute $k=n$ in the inequality, we obtain

$$2=1+\frac{n}{n}<\left(1+\frac{1}{n}\right)^{n}<1+\frac{n}{n}+\frac{n^{2}}{n^{2}}=3.$$

146. In view of the result of the preceding problem, we have

$$(1.000001)^{1,000,000}=\left(1+\frac{1}{1,000,000}\right)^{1,000,000}>2.$$

147. It is clear that

$$\frac{(1001)^{999}}{(1000)^{1000}}=\left(\frac{1001}{1000}\right)^{1000}\cdot\frac{1}{1001}=\left(1+\frac{1}{1000}\right)^{1000}\cdot\frac{1}{1001}<3\cdot\frac{1}{1001}<1$$

(see problem 145), and, consequently,

$$1000^{1000}>1001^{999}.$$

148. Assume that the given inequality is valid for some natural number n. To prove it valid for $n+1$, it suffices to establish the validity of the following inequality:

$$\frac{\left(\dfrac{n+1}{2}\right)^{n+1}}{\left(\dfrac{n}{2}\right)^{n}}\geqq n+1\geqq\frac{\left(\dfrac{n+1}{3}\right)^{n+1}}{\left(\dfrac{n}{3}\right)^{n}}.$$

Upon division by $n + 1$ these inequalities become

$$\frac{1}{2}\left(1 + \frac{1}{n}\right)^n \geq 1 \geq \frac{1}{3}\left(1 + \frac{1}{n}\right)^n,$$

which follow from the inequalities $2 \leq \left(1 + \frac{1}{n}\right)^n < 3$.

It remains only to note that for $n = 6$ the validity of the assertion of the problem follows, since

$$\left(\frac{6}{2}\right)^6 = 3^6 = 729,$$

$$6! = 720,$$

$$\left(\frac{6}{3}\right)^6 = 2^6 = 64.$$

149. (a) By the binomial theorem we have

$$\left(1 + \frac{1}{n}\right)^n = 1 + C_n^1 \frac{1}{n} + C_n^2 \frac{1}{n^2} + \cdots + C_n^{n-1} \frac{1}{n^{n-1}} + \frac{1}{n^n}$$

$$= 1 + n \cdot \frac{1}{n} + \frac{n(n-1)}{2!} \frac{1}{n^2} + \frac{n(n-1)(n-2)}{3!} \frac{1}{n^3}$$

$$+ \cdots + \frac{n(n-1)\cdots 2}{(n-1)!} \frac{1}{n^{n-1}} + \frac{n(n-1)\cdots 1}{n!} \frac{1}{n^n}$$

$$= 1 + 1 + \frac{1}{2!}\left(1 - \frac{1}{n}\right) + \frac{1}{3!}\left(1 - \frac{1}{n}\right)\left(1 - \frac{2}{n}\right)$$

$$+ \cdots + \frac{1}{n!}\left(1 - \frac{1}{n}\right)\left(1 - \frac{2}{n}\right)\cdots\left(1 - \frac{n-1}{n}\right),$$

and, analogously,

$$\left(1 + \frac{1}{n+1}\right)^{n+1} = 1 + 1 + \frac{1}{2!}\left(1 - \frac{1}{n+1}\right)$$

$$+ \frac{1}{3!}\left(1 - \frac{1}{n+1}\right)\left(1 - \frac{2}{n+1}\right)$$

$$+ \cdots + \frac{1}{n!}\left(1 - \frac{1}{n+1}\right)\left(1 - \frac{2}{n+1}\right)\cdots\left(1 - \frac{n-1}{n+1}\right)$$

$$+ \frac{1}{(n+1)!}\left(1 - \frac{1}{n+1}\right)\left(1 - \frac{2}{n+1}\right)\cdots\left(1 - \frac{n-1}{n+1}\right)\left(1 - \frac{n}{n+1}\right).$$

Comparison of these expressions shows that

$$\left(1 + \frac{1}{n+1}\right)^{n+1} > \left(1 + \frac{1}{n}\right)^{n},$$

and the assertion of the problem follows immediately.

(b) We write

$$\frac{\left(1 + \dfrac{1}{n}\right)^{n+1}}{\left(1 + \dfrac{1}{n-1}\right)^{n}} = \frac{\left(\dfrac{n+1}{n}\right)^{n+1}}{\left(\dfrac{n}{n-1}\right)^{n}}$$

$$= \frac{(n+1)^{n+1}(n-1)^{n}}{n^{2n+1}} = \left(\frac{n^2-1}{n^2}\right)^{n} \cdot \frac{n+1}{n} = \left(1 - \frac{1}{n^2}\right)^{n}\left(1 + \frac{1}{n}\right).$$

However, for $n \geqq 2$

$$\left(1 - \frac{1}{n^2}\right)^{n} = 1 - n \cdot \frac{1}{n^2} + \frac{n(n-1)}{2!}\frac{1}{n^4} - \frac{n(n-1)(n-2)}{3!}\frac{1}{n^6}$$

$$+ \frac{n(n-1)(n-2)(n-3)}{4!}\frac{1}{n^8} - \cdots$$

$$= 1 - \frac{1}{n} + \frac{1}{2}\frac{n-1}{n^3} - \left[\frac{1}{3!}\left(1 - \frac{1}{n}\right)\left(1 - \frac{2}{n}\right)\frac{1}{n^3}\right.$$

$$\left. - \frac{1}{4!}\left(1 - \frac{1}{n}\right)\left(1 - \frac{2}{n}\right)\left(1 - \frac{3}{n}\right)\frac{1}{n^4}\right] - \cdots$$

$$\leqq 1 - \frac{1}{n} + \frac{1}{2}\frac{1}{n^2} - \frac{1}{2}\frac{1}{n^3}.$$

On the other hand,

$$\left(1 - \frac{1}{n} + \frac{1}{2}\frac{1}{n^2} - \frac{1}{2}\frac{1}{n^3}\right)\left(1 + \frac{1}{n}\right) = 1 - \frac{1}{2}\frac{1}{n^2} - \frac{1}{2}\frac{1}{n^4} < 1.$$

Consequently, $\left(1 - \dfrac{1}{n^2}\right)^{n}\left(1 + \dfrac{1}{n}\right) < 1$, which means

$$\frac{\left(1 + \dfrac{1}{n}\right)^{n+1}}{\left(1 + \dfrac{1}{n-1}\right)^{n}} < 1,$$

$$\left(1 + \frac{1}{n}\right)^{n+1} < \left(1 + \frac{1}{n-1}\right)^{n},$$

from which the statement of the problem follows.

150. A proof by mathematical induction is given.

We show first that, for any natural number n,

$$n! > \left(\frac{n}{e}\right)^n . \tag{1}$$

This inequality clearly holds for $n = 1$: $1! = 1 > \frac{1}{e}$. Assume that inequality (1) holds for some positive integer n; we must show that

$$(n + 1)! > \left(\frac{n + 1}{e}\right)^{n+1} .$$

In view of problem 149 (a), we have

$$e > \left(1 + \frac{1}{n}\right)^n ,$$

$$\frac{e}{\left(1 + \frac{1}{n}\right)^n} > 1 .$$

From (1) we obtain

$$(n + 1)! = (n + 1)n! > \left(\frac{n}{e}\right)^n (n + 1) = \left(\frac{n + 1}{e}\right)^{n+1} \frac{n^n e}{(n + 1)^n}$$

$$= \left(\frac{n + 1}{e}\right)^{n+1} \frac{e}{\left(1 + \frac{1}{n}\right)^n} > \left(\frac{n + 1}{e}\right)^{n+1} .$$

It follows, by the principle of mathematical induction, that (1) is valid for all natural numbers n.

We now deal with the inequality

$$n! < n\left(\frac{n}{e}\right)^n . \tag{2}$$

We show that this inequality holds for all integers $n > 6$. With the aid of logarithm tables (natural logarithms are used here) it is readily verified that inequality (2) is valid for $n = 7$:

$$7! < 7\left(\frac{7}{e}\right)^7 ,$$

that is, $6! < \left(\frac{7}{e}\right)^7$ for $\ln 6! = \ln 720 \approx 6.58$, and $\ln\left(\frac{7}{e}\right)^7 = 7(\ln 7 - 1)$ ≈ 6.62.

Assume now the validity of (2). By the results of problem 149 (b),

$$\left(1 + \frac{1}{n}\right)^{n+1} > e;$$

that is,

$$\frac{e}{\left(1 + \dfrac{1}{n}\right)^{n+1}} < 1 .$$

But now, by (2), we have

$$(n + 1)! = (n + 1)n! < (n + 1)n \left(\frac{n}{e}\right)^n$$

$$= (n + 1)\left(\frac{n + 1}{e}\right)^{n+1} \frac{n^{n+1}e}{(n + 1)^{n+1}}$$

$$= (n + 1)\left(\frac{n + 1}{e}\right)^{n+1} \frac{e}{\left(1 + \dfrac{1}{n}\right)^{n+1}} < (n + 1)\left(\frac{n + 1}{e}\right)^{n+1} ;$$

that is, the analogous inequality, in which n is replaced by $n + 1$, will hold. Since inequality (2) holds for $n = 7$, it follows by mathematical induction that it will hold for all n exceeding 6. This completes the proof.

151. We note that in the sum

$$S = x^k + x^{k-1} + x^{k-2} + \cdots + x + 1 ,$$

if $x > 1$, then the first term is numerically the greatest, but if $x < 1$, then the last term is greatest. It follows that

$$(k + 1)x^k > S > k + 1, \text{ if } x > 1;$$
$$(k + 1)x^k < S < k + 1, \text{ if } x < 1 .$$

If both sides of these inequalities are multiplied by $x - 1$, it is found that for $x \neq 1$

$$(k + 1)x^k(x - 1) > x^{k+1} - 1 > (k + 1)(x - 1) .$$

Assume now that $x = \dfrac{p}{p - 1}$; then we find

$$\frac{(k + 1)p^k}{(p - 1)^{k+1}} > \frac{p^{k+1} - (p - 1)^{k+1}}{(p - 1)^{k+1}} > \frac{(k + 1)(p - 1)^k}{(p - 1)^{k+1}} .$$

Analogously, if we assume that $x = \dfrac{p + 1}{p}$, we obtain

$$\frac{(k+1)(p+1)^k}{p^{k+1}} > \frac{(p+1)^{k+1} - p^{k+1}}{p^{k+1}} > \frac{(k+1)p^k}{p^{k+1}} \, .$$

It follows that

$$(p+1)^{k+1} - p^{k+1} > (k+1)p^k > p^{k+1} - (p-1)^{k+1} \, ,$$

or, letting p successively have the values $1, 2, 3, \cdots, n$:

$$2^{k+1} - 1^{k+1} > (k+1)1^k > 1^{k+1} - 0 \, ,$$
$$3^{k+1} - 2^{k+1} > (k+1)2^k > 2^{k+1} - 1^{k+1} \, ,$$
$$4^{k+1} - 3^{k+1} > (k+1)3^k > 3^{k+1} - 2^{k+1} \, ,$$
$$\cdots\cdots\cdots\cdots\cdots\cdots\cdots\cdots\cdots\cdots\cdots\cdots \, ,$$
$$(n+1)^{k+1} - n^{k+1} > (k+1)n^k > n^{k+1} - (n-1)^{k+1} \, .$$

If these inequalities are added together, the following inequalities result:

$$(n+1)^{k+1} - 1 > (k+1)(1^k + 2^k + 3^k + \cdots + n^k) > n^{k+1} \, ,$$

or, dividing through these inequalities by $k + 1$,

$$\left[\left(1 + \frac{1}{n}\right)^{k+1} - \frac{1}{n^{k+1}}\right]\frac{1}{k+1}n^{k+1}$$

$$> 1^k + 2^k + 3^k + \cdots + n^k > \frac{1}{k+1}n^{k+1} \, .$$

This is essentially the set of inequalities sought.

152. (a) First, it is readily seen that

$$\frac{1}{n+1} + \frac{1}{n+2} + \cdots + \frac{1}{2n} > \underbrace{\frac{1}{2n} + \frac{1}{2n} + \cdots + \frac{1}{2n}}_{n \text{ times}} = \frac{1}{2} \, .$$

But also

$$\frac{1}{n} + \frac{1}{n+1} + \frac{1}{n+2} + \cdots + \frac{1}{2n} = \frac{1}{2}\left[\left(\frac{1}{n} + \frac{1}{2n}\right)\right.$$

$$+ \left(\frac{1}{n+1} + \frac{1}{2n-1}\right) + \left(\frac{1}{n+2} + \frac{1}{2n-2}\right) + \cdots + \left.\left(\frac{1}{2n} + \frac{1}{n}\right)\right]$$

$$= \frac{1}{2}\left[\frac{3n}{2n^2} + \frac{3n}{2n^2 + (n-1)} + \frac{3n}{2n^2 + 2(n-2)} + \cdots + \frac{3n}{2n^2}\right]$$

$$< \frac{1}{2}\left[\frac{3n}{2n^2}+\frac{3n}{2n^2}+\cdots+\frac{3n}{2n^2}\right]$$

$$\underbrace{\qquad\qquad\qquad\qquad}_{(n+1)\text{ times}}$$

$$= \frac{1}{2}\,(n+1)\,\frac{3}{2n}=\frac{3}{4}+\frac{1}{4n}<\frac{3}{4}+\frac{1}{n}\,,$$

which proves the assertions of the problem.

(b) It is first noted that

$$\frac{1}{3n}+\frac{1}{3n+1}<\frac{1}{2n}+\frac{1}{2n}=\frac{1}{n}\,.$$

It follows that

$$\frac{1}{n+1}+\frac{1}{n+2}+\cdots+\frac{1}{3n-1}+\left(\frac{1}{3n}+\frac{1}{3n+1}\right)$$

$$<\underbrace{\frac{1}{n}+\frac{1}{n}+\cdots+\frac{1}{n}}_{(2n-1)\text{ times}}+\frac{1}{n}=\frac{2n}{n}=2\,.$$

On the other hand, we have

$$\frac{1}{n+1}+\frac{1}{n+2}+\cdots+\frac{1}{3n+1}=\frac{1}{2}\Bigg[\left(\frac{1}{n+1}+\frac{1}{3n+1}\right)$$

$$+\left(\frac{1}{n+2}+\frac{1}{3n}\right)+\left(\frac{1}{n+3}+\frac{1}{3n-1}\right)+\cdots+\left(\frac{1}{3n+1}+\frac{1}{n+1}\right)\Bigg]$$

$$=\frac{1}{2}\Bigg[\frac{4n+2}{(2n+1)^2-n^2}+\frac{4n+2}{(2n+1)^2-(n-1)^2}$$

$$+\frac{4n+2}{(2n+1)^2-(n-2)^2}+\cdots+\frac{4n+2}{(2n+1)^2-n^2}\Bigg]$$

$$>\frac{1}{2}\underbrace{\left[\frac{4n+2}{(2n+1)^2}+\frac{4n+2}{(2n+1)^2}+\cdots+\frac{4n+2}{(2n+1)^2}\right]}_{(2n+1)\text{ times}}$$

$$=\frac{1}{2}\,(2n+1)\frac{4n+2}{(2n+1)^2}=1\,.$$

153. (a) We first prove that

$$2\sqrt{n+1}-2\sqrt{n}<\frac{1}{\sqrt{n}}<2\sqrt{n}-2\sqrt{n-1}\,.$$

We write

$$2\sqrt{n+1} - 2\sqrt{n} = \frac{2(\sqrt{n+1} - \sqrt{n})(\sqrt{n+1} + \sqrt{n})}{\sqrt{n+1} + \sqrt{n}}$$

$$= \frac{2}{\sqrt{n+1} + \sqrt{n}} < \frac{2}{\sqrt{n} + \sqrt{n}} = \frac{1}{\sqrt{n}} .$$

The second part of the inequality is shown in an analogous manner.
Now we have

$$1 + \frac{1}{\sqrt{2}} + \frac{1}{\sqrt{3}} + \cdots + \frac{1}{\sqrt{1,000,000}} > 1 + 2[(\sqrt{3} - \sqrt{2})$$

$$+ (\sqrt{4} - \sqrt{3}) + \cdots + (\sqrt{1,000,001} - \sqrt{1,000,000})]$$

$$= 1 + 2(\sqrt{1,000,001} - \sqrt{2}) > 2 \cdot 1000 - \sqrt{8} + 1$$

$$> 2000 - 3 + 1 = 1998 .$$

Analogously,

$$1 + \frac{1}{\sqrt{2}} + \frac{1}{\sqrt{3}} + \cdots + \frac{1}{\sqrt{1,000,000}} < 1 + 2[(\sqrt{2} - 1)$$

$$+ (\sqrt{3} - \sqrt{2}) + \cdots + \sqrt{1,000,000} - \sqrt{999,999})]$$

$$= 1 + 2(\sqrt{1,000,000} - 1) = 1 + 2 \cdot 999 = 1999 .$$

Consequently, the integral part of the sum

$$1 + \frac{1}{\sqrt{2}} + \frac{1}{\sqrt{3}} + \cdots + \frac{1}{\sqrt{1,000,000}}$$

is equal to 1998.

(b) A technique similar to that used in problem (a) yields

$$\frac{1}{\sqrt{10,000}} + \frac{1}{\sqrt{10,001}} + \cdots + \frac{1}{\sqrt{1,000,000}}$$

$$> 2[(\sqrt{10,001} - \sqrt{10,000}) + (\sqrt{10,002} - \sqrt{10,001})$$

$$+ \cdots + (\sqrt{1,000,001} - \sqrt{1,000,000})]$$

$$= 2(\sqrt{1,000,001} - \sqrt{10,000}) > 2(1000 - 100) = 1800$$

and

$$\frac{1}{\sqrt{10,000}} + \frac{1}{\sqrt{10,001}} + \cdots + \frac{1}{\sqrt{1,000,000}}$$

$$< 2[(\sqrt{10,000} - \sqrt{9999}) + (\sqrt{10,001} - \sqrt{10,000}$$

$$+ \cdots + \sqrt{1,000,000} - \sqrt{999,999})]$$

$$= 2(\sqrt{1,000,000} - \sqrt{9999})$$

$$= 2000 - \sqrt{39{,}996} < 2000 - 199.98 = 1800.02 \ .$$

Therefore, the sum

$$\frac{1}{\sqrt{10{,}000}} + \frac{1}{\sqrt{10{,}001}} + \cdots + \frac{1}{\sqrt{1{,}000{,}000}}$$

is equal to, with precision to within 0.02, the number 1800.

154. We note, by comparing the two equations

$$\left(1 + \frac{1}{n}\right)^2 = 1 + 2\frac{1}{n} + \frac{1}{n^2}$$

and

$$\left(1 + \frac{2}{3}\frac{1}{n}\right)^3 = 1 + 2\frac{1}{n} + \frac{4}{3}\frac{1}{n^2} + \frac{8}{27}\frac{1}{n^3}$$

that for every natural number n

$$\left(1 + \frac{2}{3} \cdot \frac{1}{n}\right)^3 > \left(1 + \frac{1}{n}\right)^2 \ .$$

From this we obtain $1 + \frac{2}{3} \cdot \frac{1}{n} > \left(1 + \frac{1}{n}\right)^{2/3}$; multiplication by $n^{2/3}$ yields

$$n^{2/3} + \frac{2}{3}n^{-(1/3)} > (n+1)^{2/3} \ ,$$

and finally,

$$\frac{1}{\sqrt[3]{n}} > \frac{3}{2}\left[\sqrt[3]{(n+1)^2} - \sqrt[3]{n^2}\right] \ .$$

Analogously,

$$\left(1 - \frac{2}{3}\frac{1}{n}\right)^3 = 1 - 2\frac{1}{n} + \frac{4}{3}\frac{1}{n^2} - \frac{8}{27}\frac{1}{n^3} > 1 - 2\frac{1}{n} + \frac{1}{n^2}$$

$$= \left(1 - \frac{1}{n}\right)^2$$

$\left(\text{since } \frac{1}{3} \cdot \frac{1}{n^2} - \frac{8}{27} \cdot \frac{1}{n^3} > \frac{1}{3}\frac{1}{n^2} - \frac{1}{3}\frac{1}{n^3} \geqq 0\right)$, from which it follows that

$$1 - \frac{2}{3}\frac{1}{n} > \left(1 - \frac{1}{n}\right)^{2/3} \ ,$$

$$n^{2/3} - \frac{2}{3}n^{-(1/3)} > (n-1)^{2/3} \ ,$$

$$\frac{1}{\sqrt[3]{n}} < \frac{3}{2} [\sqrt[3]{n^2} - \sqrt[3]{(n-1)^2}] .$$

Now we can write

$$\frac{1}{\sqrt[3]{4}} + \frac{1}{\sqrt[3]{5}} + \cdots + \frac{1}{\sqrt[3]{1,000,000}}$$

$$> \frac{3}{2} [(\sqrt[3]{5^2} - \sqrt[3]{4^2}) + (\sqrt[3]{6^2} - \sqrt[3]{5^2})$$

$$+ \cdots + (\sqrt[3]{1,000,001^2} - \sqrt[3]{1,000,000^2})]$$

$$= \frac{3}{2} (\sqrt[3]{1,000,002,000,001} - \sqrt[3]{16}) > \frac{3}{2} \cdot 10,000 - \sqrt[3]{54}$$

$$> 15,000 - 4 = 14,996 .$$

However,

$$\frac{1}{\sqrt[3]{4}} + \frac{1}{\sqrt[3]{5}} + \cdots + \frac{1}{\sqrt{1,000,000}}$$

$$< \frac{3}{2} [(\sqrt[3]{4^2} - \sqrt[3]{3^2}) + (\sqrt[3]{5^2} - \sqrt[3]{4^2})$$

$$+ \cdots + (\sqrt[3]{1,000,000^2} - \sqrt[3]{999,999^2})]$$

$$= \frac{3}{2} (\sqrt[3]{1,000,000,000,000} - \sqrt[3]{9}) < \frac{3}{2}(10,000 - 2) = 14,997 .$$

Thus, the integral part of the sum

$$\frac{1}{\sqrt[3]{4}} + \frac{1}{\sqrt[3]{5}} + \cdots + \frac{1}{\sqrt[3]{1,000,000}}$$

is equal to 14,996.

155. (a) It is readily seen that

$$\frac{1}{10^2} + \frac{1}{11^2} + \cdots + \frac{1}{1000^2} > \frac{1}{10 \cdot 11} + \frac{1}{11 \cdot 12} + \cdots + \frac{1}{1000 \cdot 1001}$$

$$= \left(\frac{1}{10} - \frac{1}{11}\right) + \left(\frac{1}{11} - \frac{1}{12}\right) + \cdots + \left(\frac{1}{1000} - \frac{1}{1001}\right)$$

$$= \frac{1}{10} - \frac{1}{1001} > 0.1 - 0.001 = 0.099 ,$$

and, analogously,

$$\frac{1}{10^2} + \frac{1}{11^2} + \cdots + \frac{1}{1000^2} < \frac{1}{9 \cdot 10} + \frac{1}{10 \cdot 11} + \cdots + \frac{1}{999 \cdot 1000}$$

$$= \left(\frac{1}{9} - \frac{1}{10}\right) + \left(\frac{1}{10} - \frac{1}{11}\right) + \cdots + \left(\frac{1}{999} - \frac{1}{1000}\right) = \frac{1}{9} - \frac{1}{1000}$$

$$< 0.112 - 0.001 = 0.111 .$$

Consequently, the sum $\dfrac{1}{10^2} + \dfrac{1}{11^2} + \cdots + \dfrac{1}{1000^2}$, with precision to 0.006, is equal to 0.105.

(b) We note, first, that

$$\frac{1}{10!} + \frac{1}{11!} + \frac{1}{12!} + \cdots + \frac{1}{1000!} > \frac{1}{10!} = \frac{1}{3,628,800} \approx 0.000000275 \ .$$

But, also

$$\frac{1}{10!} + \frac{1}{11!} + \frac{1}{12!} + \cdots + \frac{1}{1000!}$$

$$< \frac{1}{9}\left\{\frac{9}{10!} + \frac{10}{11!} + \frac{11}{12!} + \cdots + \frac{999}{1000!}\right\}$$

$$= \frac{1}{9}\left\{\frac{10-1}{10!} + \frac{11-1}{11!} + \frac{12-1}{12!} + \cdots + \frac{1000-1}{1000!}\right\}$$

$$= \frac{1}{9}\left\{\frac{1}{9!} - \frac{1}{10!} + \frac{1}{10!} - \frac{1}{11!} + \frac{1}{11!} - \frac{1}{12!} + \cdots + \frac{1}{999!} - \frac{1}{1000!}\right\}$$

$$= \left(\frac{1}{9!} - \frac{1}{1000!}\right) < \frac{1}{9} \cdot \frac{1}{9!} = \frac{1}{3,265,920} \approx 0.000000305 \ .$$

Therefore, the sum

$$\frac{1}{10!} + \frac{1}{11!} + \cdots + \frac{1}{1000!} \ ,$$

with precision to 0.00000015, is equal to 0.00000029.

156. We shall show that the sum

$$1 + \frac{1}{2} + \frac{1}{3} + \cdots + \frac{1}{n-1} + \frac{1}{n}$$

can be made greater than any given number N. Let N be some chosen integer, and take $n = 2^{2N}$. Then

$$1 + \frac{1}{2} + \frac{1}{3} + \frac{1}{4} + \cdots + \frac{1}{n-1} + \frac{1}{n} = 1 + \frac{1}{2} + \left(\frac{1}{3} + \frac{1}{4}\right)$$

$$+ \left(\frac{1}{5} + \frac{1}{6} + \frac{1}{7} + \frac{1}{8}\right) + \cdots + \left(\frac{1}{2^{2N-1}+1} + \frac{1}{2^{2N-1}+2}\right.$$

$$+ \cdots + \left.\frac{1}{2^{2N}-1} + \frac{1}{2^{2N}}\right) > 1 + \underbrace{\frac{1}{2} + \frac{1}{2} + \frac{1}{2} + \cdots + \frac{1}{2}}_{2N \text{ times}} > N + 1$$

[every sum in parentheses is greater than $\dfrac{1}{2}$; see problem 152 (a)].

Remark: This can also be proved as a consequence of problem 152 (b).

157. Designate by n_k the number of undeleted fractions between $\dfrac{1}{10^k}$ and $\dfrac{1}{10^{k+1}}$, including $\dfrac{1}{10^k}$ but not $\dfrac{1}{10^{k+1}}$. If the fraction $\dfrac{1}{q}$, lying between these two fractions, is one of the undeleted numbers, then of the numbers $\dfrac{1}{10q}, \dfrac{1}{10q+1}, \dfrac{1}{10q+2}, \cdots, \dfrac{1}{10q+8}, \dfrac{1}{10q+9}$ $\left(\text{all of which lie between } \dfrac{1}{10^k} \text{ and } \dfrac{1}{10^{k+1}}\right)$, only the final fraction will be deleted when those containing a digit 9 in the denominator are crossed out. If $\dfrac{1}{q}$ is one of the deleted numbers, then all of the additional fractions $\dfrac{1}{10q}, \dfrac{1}{10q+1}, \cdots, \dfrac{1}{10q+9}$ will also be deleted. It follows that

$$n_k = 9n_{k-1}.$$

Since $n_0 = 8$(of the fractions $1, \dfrac{1}{2}, \dfrac{1}{3}, \cdots, \dfrac{1}{8}, \dfrac{1}{9}$, only $\dfrac{1}{9}$ is deleted),

$$n_1 = 8 \cdot 9 = 72;$$
$$n_2 = 8 \cdot 9^2;$$
$$\cdots\cdots\cdots;$$
$$n_k = 8 \cdot 9^k.$$

Now consider, for $n < 10^{m+1}$, the sum

$$1 + \frac{1}{2} + \frac{1}{3} + \cdots + \frac{1}{n}.$$

Add this to the sum

$$1 + \frac{1}{2} + \frac{1}{3} + \cdots + \frac{1}{10^{m+1} - 1},$$

after throwing out all those fractions having a digit 9 in the denominator;

$$\left(1 + \frac{1}{2} + \frac{1}{3} + \cdots + \frac{1}{8}\right)$$
$$+ \left(\frac{1}{10} + \frac{1}{11} + \frac{1}{12} + \cdots + \frac{1}{18} + \frac{1}{20} + \cdots + \frac{1}{88}\right)$$
$$+ \left(\frac{1}{100} + \frac{1}{101} + \cdots + \frac{1}{888}\right) + \cdots + \left(\frac{1}{10^m} + \cdots + \frac{1}{\underbrace{88...8}}\right)$$
$$\underbrace{}_{(m+1) \text{ times}}$$

$$< 1 \cdot n_0 + \frac{1}{10} \cdot n_1 + \frac{1}{100} \cdot n_2 + \cdots + \frac{1}{10^{m-1}} \cdot n_{m-1} + \frac{1}{10^m} \cdot n_m \; .$$

If we replace each summation in parentheses by the product of the largest term contained therein and the number of terms in those parentheses, we obtain

$$1 \cdot n_0 + \frac{1}{10} \cdot n_1 + \frac{1}{100} \cdot n_2 + \cdots + \frac{1}{10^{m-1}} \cdot n_{m-1} + \frac{1}{10^m} \cdot n_m$$

$$= 8 \left(1 + \frac{9}{10} + \frac{9^2}{10^2} + \cdots + \frac{9^{m-1}}{10^{m-1}} + \frac{9^m}{10^m} \right)$$

$$= 8 \cdot \frac{1 - (9^{m+1}/10^{m+1})}{1 - \frac{9}{10}} < 8 \cdot \frac{1}{1 - \frac{9}{10}} = 8 \cdot 10 = 80 \; .$$

This verifies the assertion of the problem.

158. (a) Assume that in the summation $1 + \frac{1}{4} + \frac{1}{9} + \cdots + \frac{1}{n^2}$ the integer n is less than 2^{k+1}. Consider the summation

$$1 + \frac{1}{2^2} + \frac{1}{3^2} + \cdots + \frac{1}{(2^{k+1} - 1)^2} \; ,$$

and, as in the solution of problem 156, group the terms in the following manner:

$$1 + \left(\frac{1}{2^2} + \frac{1}{3^2} \right) + \left(\frac{1}{4^2} + \frac{1}{5^2} + \frac{1}{6^2} + \frac{1}{7^2} \right)$$

$$+ \cdots + \left[\frac{1}{(2^k)^2} + \frac{1}{(2^k + 1)^2} + \cdots + \frac{1}{(2^{k+1} - 1)^2} \right] < 1 + \left(\frac{1}{2^2} + \frac{1}{2^2} \right)$$

$$+ \left(\frac{1}{4^2} + \frac{1}{4^2} + \frac{1}{4^2} + \frac{1}{4^2} \right) + \cdots + \left[\frac{1}{(2^k)^2} + \frac{1}{(2^k)^2} + \cdots + \frac{1}{(2^k)^2} \right]$$

$$= 1 + \frac{1}{2} + \frac{1}{4} + \cdots + \frac{1}{2^k} = \frac{1 - (1/2^{k+1})}{1 - \frac{1}{2}} = 2 - \frac{1}{2^k} < 2 \; ,$$

This verifies the assertion of the problem.

Remark: It is possible to show, by similar techniques, that if α is any number exceeding 1, then for any natural number n

$$1 + \frac{1}{2^\alpha} + \frac{1}{3^\alpha} + \cdots + \frac{1}{n^\alpha} < \frac{2^\alpha - 1}{2^{\alpha-1} - 1} \; .$$

This sum is bounded, and its bound is independent of n; that is, n can be arbitrarily large. Problem 156 showed, on the other hand, that if $\alpha \leqq 1$, the sum $1 + \dfrac{1}{2^{\alpha}} + \dfrac{1}{3^{\alpha}} + \cdots + \dfrac{1}{n^{\alpha}}$ can be made as large as we wish by taking n large enough.

(b) It is readily seen that

$$\frac{1}{2^2} + \frac{1}{3^2} + \frac{1}{4^2} + \frac{1}{5^2} + \cdots + \frac{1}{n^2}$$

$$< \left(\frac{1}{1\cdot 2} - \frac{1}{4}\right) + \frac{1}{2\cdot 3} + \frac{1}{3\cdot 4} + \frac{1}{4\cdot 5} + \cdots + \frac{1}{(n-1)n}$$

$$= \left(\frac{1}{1\cdot 2} + \frac{1}{2\cdot 3} + \frac{1}{3\cdot 4} + \cdots + \frac{1}{(n-1)n}\right) - \frac{1}{4}\ .$$

However, by problem 132 (a), we have

$$\frac{1}{1\cdot 2} + \frac{1}{2\cdot 3} + \frac{1}{3\cdot 4} + \cdots + \frac{1}{(n-1)n} = 1 - \frac{1}{n} < 1\ ,$$

and, consequently,

$$1 + \frac{1}{2^2} + \frac{1}{3^2} + \frac{1}{4^2} + \cdots + \frac{1}{n^2} < 1 + \left(1 - \frac{1}{4}\right) = 1\frac{3}{4}\ ,$$

which proves the assertion of the problem.

159. we shall show, first, that

$$1 + \frac{1}{2} + \frac{1}{3} + \frac{1}{4} + \cdots + \frac{1}{n-1} + \frac{1}{n}$$

$$< \left(1 + \frac{1}{2} + \frac{1}{4} + \cdots + \frac{1}{2^k}\right)\left(1 + \frac{1}{3} + \frac{1}{9} + \cdots + \frac{1}{3^k}\right)$$

$$\cdots \left(1 + \frac{1}{p_l} + \frac{1}{p_l^2} + \cdots + \frac{1}{p_l^k}\right)\ ,$$

where k is an integer such that $2^k \leqq n < 2^{k+1}$, and p_l is the greatest prime not exceeding n. For this investigation we consider the various factors in parentheses of the right member of the inequality. Since every positive integer m from 1 to n can be written as the product of powers of primes $1, 3, 5, \cdots, p_l$, we may write

$$m = 2^{\alpha_1} \cdot 3^{\alpha_2} \cdot 5^{\alpha_3} \cdots p_l^{\alpha_l}\ ,$$

where all the exponents $\alpha_1, \alpha_2, \cdots, \alpha_l$ are nonnegative integers not exceeding k (zero exponents being, of course, allowable). We encounter as terms every fraction $1, \dfrac{1}{2}, \dfrac{1}{3}, \dfrac{1}{4}, \cdots, \dfrac{1}{n-1}, \dfrac{1}{n}$ as well as some additional positive numbers. This means that the right member of the inequality exceeds the left member.

If we take logarithms of both sides of the inequality, we find

$$\log \left(1 + \frac{1}{2} + \frac{1}{3} + \frac{1}{4} + \cdots + \frac{1}{n-1} + \frac{1}{n}\right)$$

$$< \log\left[\left(1 + \frac{1}{2} + \frac{1}{4} + \cdots + \frac{1}{2^k}\right)\left(1 + \frac{1}{3} + \frac{1}{9} + \cdots + \frac{1}{3^k}\right)\right.$$

$$\left. \times \cdots \times \left(1 + \frac{1}{p_l} + \frac{1}{p_l^2} + \cdots + \frac{1}{p_l^k}\right)\right] = \log\left(1 + \frac{1}{2} + \frac{1}{4}\right.$$

$$\left. + \cdots + \frac{1}{2^k}\right) + \log\left(1 + \frac{1}{3} + \frac{1}{9} + \cdots + \frac{1}{3^k}\right)$$

$$+ \cdots + \log\left(1 + \frac{1}{p_l} + \frac{1}{p_l^2} + \cdots + \frac{1}{p_l^k}\right).$$

But for any integers k and $p \geqq 2$,

$$\log\left(1 + \frac{1}{p} + \frac{1}{p^2} + \frac{1}{p^3} + \cdots + \frac{1}{p^k}\right) < \frac{2\log 3}{p}.$$

Consequently, we have

$$1 + \frac{1}{p} + \frac{1}{p^2} + \cdots + \frac{1}{p^k} = \frac{1 - (1/(p^{k+1}))}{1 - \frac{1}{p}} < \frac{1}{1 - \frac{1}{p}}$$

$$= \frac{p}{p-1} = 1 + \frac{1}{p-1}.$$

It follows from the results of problem 145 that

$$\left(1 + \frac{1}{p-1}\right)^{p-1} < 3,$$

$$1 + \frac{1}{p-1} < \sqrt[p-1]{3},$$

$$\log\left(1 + \frac{1}{p-1}\right) < \frac{\log 3}{p-1},$$

and, clearly,

$$\frac{2\log 3}{p} > \frac{\log 3}{p-1}.$$

Hence we conclude that

$$\log\left(1 + \frac{1}{2} + \cdots + \frac{1}{n}\right) < \frac{2\log 3}{2} + \frac{2\log 3}{3} + \frac{2\log 3}{5} + \cdots + \frac{2\log 3}{p_l}$$

$$= 2 \log 3 \left(\frac{1}{2} + \frac{1}{3} + \frac{1}{5} + \cdots + \frac{1}{p_l} \right) .$$

If there existed a natural number N such that for every positive integer l the sum $1 + \frac{1}{2} + \frac{1}{3} + \frac{1}{5} + \cdots + \frac{1}{p}$ would be less than N, then for all positive integers n the following inequality would have to hold,

$$\log \left(1 + \frac{1}{2} + \frac{1}{3} + \frac{1}{4} + \cdots + \frac{1}{n-1} + \frac{1}{n} \right)$$

$$< 2 \log 3 \left(\frac{1}{2} + \frac{1}{3} + \frac{1}{5} + \cdots + \frac{1}{p_l} \right) < 2(N - 1) \log 3 ,$$

from which it would follow that

$$1 + \frac{1}{2} + \frac{1}{3} + \frac{1}{4} + \cdots + \frac{1}{n-1} + \frac{1}{n} < 3^{2(N-1)} = N_1 ,$$

where N_1 is independent of n. But it was shown in problem 156 that such an N_1 does not exist; consequently, no number N can exist such that, for all l, $1 + \frac{1}{2} + \frac{1}{3} + \frac{1}{5} + \cdots + \frac{1}{p_l} < N$.

160. We have

$$\frac{b - c}{a} + \frac{c - a}{b} + \frac{a - b}{c} = \frac{b^2 c - bc^2 + ac^2 - a^2 c + a^2 b - ab^2}{abc}$$

$$= \frac{c^2(a - b) + ab(a - b) - (ac + bc)(a - b)}{abc}$$

$$= \frac{(a - b)(c^2 + ab - ac - bc)}{abc} = \frac{(a - b)[c(c - a) - b(c - a)]}{abc}$$

$$= \frac{(a - b)(c - b)(c - a)}{abc} = -\frac{(a - b)(b - c)(c - a)}{abc} .$$

We shall now investigate $\frac{a}{b - c} + \frac{b}{c - a} + \frac{c}{a - b}$. Let $a' = b - c$, $b' = c - a$, and $c' = a - b$. Then

$$b' - c' = c - a - (a - b) = b + c - 2a .$$

From the condition $a + b + c = 0$, we have $b + c = -a$, from which

$$b' - c' = -3a ,$$

$$a = -\frac{b' - c'}{3} .$$

In an analogous manner we also obtain

$$b = -\frac{c' - a'}{3},$$

$$c' = -\frac{a' - b'}{3}.$$

It follows that

$$\frac{a}{b-c} + \frac{b}{c-a} + \frac{c}{a-b} = -\frac{1}{3}\left(\frac{b'-c'}{a'} + \frac{c'-a'}{b'} + \frac{a'-b'}{c'}\right).$$

Using the above formula, we obtain

$$\frac{a}{b-c} + \frac{b}{c-a} + \frac{c}{a-b} = -\frac{1}{3}\left[-\frac{(a'-b')(b'-c')(c'-a')}{a'b'c'}\right]$$

$$= \frac{1}{3}\frac{(-3c)(-3a)(-3b)}{(b-c)(c-a)(a-b)} = -9\frac{abc}{(a-b)(b-c)(c-a)}.$$

Consequently, if $a + b + c = 0$, then

$$\left(\frac{b-c}{a} + \frac{c-a}{b} + \frac{a-b}{c}\right)\left(\frac{a}{b-c} + \frac{b}{c-a} + \frac{c}{a-b}\right)$$

$$= \left[-\frac{(a-b)(b-c)(c-a)}{abc}\right]\left[-9\frac{abc}{(a-b)(b-c)(c-a)}\right] = 9.$$

161. We have

$$0 = (a+b+c)^3 = a^3 + b^3 + c^3 + 3a^2b + 3a^2c$$
$$\qquad + 3b^2a + 3b^2c + 3c^2a + 3c^2b + 6abc$$
$$= a^3 + b^3 + c^3 + 3ab(a+b) + 3ac(a+c) + 3bc(b+c) + 6abc$$
$$= a^3 + b^3 + c^3 - 3abc - 3abc - 3abc + 6abc$$
$$= a^3 + b^3 + c^3 - 3abc.$$

It follows that $a^3 + b^3 + c^3 = 3abc$, which is what we set out to prove.

162. (a) *First Solution.* We have

$$a^3 + b^3 + 3^3 - 3abc$$
$$= a^3 + 3ab(a+b) + b^3 + c^3 - 3abc - 3ab(a+b)$$
$$= a^3 + 3a^2b + 3ab^2 + b^3 + c^3 - 3ab(c+a+b)$$
$$= (a+b)^3 + c^2 - 3ab(a+b+c)$$
$$= [(a+b)+c][(a+b)^2 - (a+b)c + c^2] - 3ab(a+b+c)$$

$$= (a + b + c)[(a + b)^2 - (a + b)c + c^2 - 3ab]$$
$$= (a + b + c)(a^2 + 2ab + b^2 - ac - bc + c^2 - 3ab)$$
$$= (a + b + c)(a^2 + b^2 + c^2 - ab - ac - bc) .$$

Second Solution. If in problem 161 we substitute x for a, we have $x^3 + b^3 + c^3 - 3xbc = 0$, if $x + b + c = 0$. Consequently, the equation $x^3 - 3bcx + b^3 + c^3 = 0$ has a root $x = -b - c$, from which it follows that the polynomial $x^3 - 3bcx + b^3 + c^3$ is divisible by $x - (-b - c) = x + b + c$. If in this result we resubstitute a for x, we find that $a^3 + b^3 + c^3 - 3abc$ is divisible by $a + b + c$. Ordinary division produces the other factor:

$$a^3 + b^3 + c^3 - 3abc = (a + b + c)(a^2 + b^2 + c^2 - ab - ac - bc) .$$

(b) *First solution.* We have

$$[(a + b + c)^3 - a^3] - (b^3 + c^3)$$
$$= [(a + b + c) - a][(a + b + c)^2 + a(a + b + c) + a^2]$$
$$\quad - (b + c)(b^2 - bc + c^2)$$
$$= (b + c)\{[(a + b + c)^2 - b^2] + a(a + c)$$
$$\quad + (ab + bc) + (a^2 - c^2)\}$$
$$= (b + c)\{[(a + b + c) - b][(a + b + c) + b] + a(a + c)$$
$$\quad + b(a + c) + (a + c)(a - c)\}$$
$$= (b + c)(a + c)(a + b + c + b + a + b + a - c)$$
$$= 3(b + c)(a + c)(a + b) .$$

Second Solution. Substitute, in the given expression, x for a:

$$(x + b + c)^3 - x^3 - b^3 - c^3.$$

If $x = -b$, the expression vanishes; consequently, the equation $(x + b + c)^3 - x^3 - b^3 - c^3 = 0$ has as a root $x = -b$, and so $(x+b+c)^3 - x^3 - b^3 - c^3$ is divisible by $x + b$. Resubstituting a for x, we can conclude that $(a + b + c)^3 - a^3 - b^3 - c^3$ is divisible by $a + b$.

It is similarly shown that $(a + b + c)^3 - a^3 - b^3 - c^3$ is also divisible by $a + c$ and by $b + c$. We can write (since the three factors are clearly relatively prime)

$$(a + b + c)^3 - a^3 - b^3 - c^3 = k(a + b)(a + c)(b + c) .$$

In order to determine the factor k, it suffices to equate, in this equality, the coefficients of any like term from each side: for example, the coefficient of a^2b. If we set $a = 0, b = c = 1$, then we find $k=3$.

163. In problem 162 (a) we found that $(\alpha^2+\beta^2+\gamma^2-\alpha\beta-\alpha\gamma-\beta\gamma) \times (\alpha+\beta+\gamma) = \alpha^3+\beta^3+\gamma^3-3\alpha\beta\gamma$. For α, β, and γ we substitute $\sqrt[3]{a}$, $\sqrt[3]{b}$, and $\sqrt[3]{c}$, respectively; we then have

$$(\sqrt[3]{a}+\sqrt[3]{b}+\sqrt[3]{c})(\sqrt[3]{a^2}+\sqrt[3]{b^2}+\sqrt[3]{c^2}-\sqrt[3]{ab}$$
$$-\sqrt[3]{ac}-\sqrt[3]{bc}) = a+b+c-3\sqrt[3]{abc}\ .$$

It follows that

$$\frac{1}{\sqrt[3]{a}+\sqrt[3]{b}+\sqrt[3]{c}} = \frac{\sqrt[3]{a^2}+\sqrt[3]{b^2}+\sqrt[3]{c^2}-\sqrt[3]{ab}-\sqrt[3]{ac}-\sqrt[3]{bc}}{a+b+c-3\sqrt[3]{abc}}\ .$$

Now it is not difficult to eliminate the radical from the denominator of the fraction on the right:

$$\frac{1}{\sqrt[3]{a}+\sqrt[3]{b}+\sqrt[3]{c}} = \frac{\sqrt[3]{a^2}+\sqrt[3]{b^2}+\sqrt[3]{c^2}-\sqrt[3]{ab}-\sqrt[3]{ac}-\sqrt[3]{bc}}{(a+b+c)^3-27abc}$$
$$\times [(a+b+c)^2+3(a+b+c)\sqrt[3]{abc}+9\sqrt[3]{a^2b^2c^2}\,]\ .$$

164. We saw in problem 162 (b) that $(a+b+c)^3-a^3-b^3-c^3$ differs from the product $(a+b)(a+c)(b+c)$ only by a constant factor; hence it suffices to show that

$$(a+b+c)^{3333}-a^{3333}-b^{3333}-c^{3333}$$

is divisible by $a+b$, $a+c$, and $b+c$. But this can be shown by exactly the same proof used in problem 162 (b).

165. We have

$$a^{10}+a^5+1 = \frac{(a^5)^3-1}{a^5-1} = \frac{a^{15}-1}{a^5-1}$$

$$= \frac{(a^3)^5-1}{(a-1)(a^4+a^3+a^2+a+1)} = \frac{(a^3-1)(a^{12}+a^9+a^6+a^3+1)}{(a-1)(a^4+a^3+a^2+a+1)}$$

$$= \frac{(a^2+a+1)(a^{12}+a^9+a^6+a^3+1)}{a^4+a^3+a^2+a+1}\ .$$

But division yields

$$\frac{a^{12}+a^9+a^6+a^3+1}{a^4+a^3+a^2+a+1} = a^8-a^7+a^5-a^4+a^3-a+1\ .$$

Consequently,

$$a^{10}+a^5+1 = (a^2+a+1)(a^8-a^7+a^5-a^4-a^3+a^3-a+1)\ .$$

166. *First Solution.* Designate the dividend polynomial by B and the divisor by A. Then

$$B - A = (x^{9999} - x^9)(x^{8888} - x^8) + (x^{7777} - x^7) + (x^{6666} - x^6)$$
$$+ (x^{5556} - x^5) + (x^{4444} - x^4) + (x^{3333} - x^3)$$
$$+ (x^{2222} - x^2) + (x^{1111} - x)$$
$$= x^9[(x^{10})^{999} - 1] + x^8[(x^{10})^{888} - 1] + x^7[(x^{10})^{777} - 1]$$
$$+ x^6[(x^{10})^{666} - 1] + x^5[(x^{10})^{555} - 1] + x^4[(x^{10})^{444} - 1]$$
$$+ x^3[(^{10})^{333} - 1] + x^2[(x^{10})^{222} - 1] + x[(x^{10})^{111} - 1] .$$

Each difference in parentheses is divisible by $x^{10} - 1$, and so by $A = \dfrac{x^{10} - 1}{x - 1}$. Therefore, $B - A$ is divisible by A, which means that B must be divisible by A.

Second Solution. We have

$$x^9 + x^8 + x^7 + x^6 + x^5 + x^4 + x^3 + x^2 + x + 1$$
$$= \frac{x^{10} - 1}{x - 1} = \frac{(x - 1)(x - a_1)(x - a_2)(x - a_3) \cdots (x - a_9)}{x - 1}$$
$$= (x - a_1)(x - a_2) \cdots (x - a_9) ,$$

where $a_k = \cos\dfrac{2k\Pi}{10} + i\sin\dfrac{2k\Pi}{10}$ $(k = 1, 2, \cdots, 9)$, since the roots of the equation $x^{10} - 1 = 0$, (that is, the ten tenth-roots of unity are of this form (see the discussion of Section 9, Complex Numbers, preceding the statement of problem 222). Consequently, in order to prove the assertion of the problem, it suffices to verify that

$$x^{9999} + x^{8888} + x^{7777} + x^{6666} + x^{5555} + x^{4444}$$
$$+ x^{3333} + x^{2222} + x^{1111} + 1$$

is divisible by each of the factors $(x - a_1), (x - a_2), \cdots, (x - a_9)$. This, however, is equivalent to the assertion that

$$x^{9999} + x^{8888} + x^{7777} + x^{6666} + x^{5555} + x^{4444}$$
$$+ x^{3333} + x^{2222} + x^{1111} + 1 = 0 \qquad (1)$$

has as roots $a_1, a_2, a_3, \cdots, a_9$. We shall verify that these are roots of equation (1). Since $a_k^{10} = 1 (k = 1, 2, 3, \cdots, 9)$, it follows that

$$a_k^{9999} = a_k^{9990+9} = (a_k^{10})^{999}a_k^9 = a_k^9;$$
$$a_k^{8888} = a_k^{8880+8} = (a_k^{10})^{888}a_k^8 = a_k^8; \text{ etc. } \cdots$$
$$a_k^{9999} + a_k^{8888} + a_k^{7777} + a_k^{6666} + a_k^{5555} + a_k^{4444} + a_k^{3333} + a_k^{2222} + a_k^{1111} + 1$$
$$= a_k^9 + a_k^8 + a_k^7 + a_k^6 + a_k^5 + a_k^4 + a_k^3 + a_k^2 + a_k + 1 = 0$$
$$(k = 1, 2, \cdots, 9) .$$

167. We shall find two numbers a and b to satisfy the equation

$$x^3 + px + q = x^3 + a^3 + b^3 - 3abx .$$

To do this, we must solve for a and b the following two equations in two unknowns:

$$a^3 + b^3 = q, \qquad ab = -\frac{p}{3} ,$$

or, equivalently,

$$a^3 + b^3 = q, \qquad a^3b^3 = -\frac{p^3}{27} .$$

Now, it is easily verified that a^3 and b^3 are roots of the quadratic equation $z^2 - qz - \frac{p^3}{27} = 0$, and, consequently, we will have[†]

$$a = \sqrt[3]{\frac{q}{2} + \sqrt{\frac{q^2}{4} + \frac{p^3}{27}}} , \qquad b = \sqrt[3]{\frac{q}{2} - \sqrt{\frac{q^2}{4} + \frac{p^3}{27}}} \qquad (1)$$

Now, in view of the result of problem 162 (a), we have

$$x^3 + px + q = x^3 + a^3 + b^3 - 3abx$$
$$= (a + b + x)(a^2 + b^2 + x^2 - ab - ax - bx) .$$

Therefore, the solution of the cubic equation reduces to the solution of the first-degree equation

$$a + b + x = 0 ,$$

from which we obtain

$$x_1 = -a - b ,$$

or,

$$x_1 = -\sqrt[3]{\frac{q}{2} + \sqrt{\frac{q^2}{4} + \frac{p^3}{27}}} - \sqrt[3]{\frac{q}{2} - \sqrt{\frac{q^2}{4} + \frac{p^3}{27}}} ,$$

[†] Formulas (1) are obtained from the formula displaying the roots of a quadratic equation. The roots are real if $\frac{q^2}{4} + \frac{p^3}{27} \geq 0$; if $\frac{q^2}{4} + \frac{p^3}{27} < 0$, then we will be involved with the cube roots of imaginary numbers, and a and b will be imaginary numbers. They may be found using the formula developed at the beginning of Section 9, which enables us to find the nth root of a complex number. For each of the three cube roots a, obtained from the number $\frac{q}{2} + \sqrt{\frac{q^2}{4} + \frac{p^3}{27}}$, the corresponding b can be obtained from the relation $ab = -\frac{p}{3}$.

and the quadratic equation

$$x^2 - (a + b)x + a^2 + b^2 - ab = 0 ,$$

from which it follows that

$$x_2 = \frac{a + b}{2} + \frac{(a - b)\sqrt{3}}{2} i ,$$

$$x_3 = \frac{a + b}{2} - \frac{(a - b)\sqrt{3}}{2} i ,$$

where a and b are determined by formula (1).

168. *First Solution.* We designate $\sqrt{a + x}$ by y, thereby obtaining a system of two equations:

$$\sqrt{a + x} = y , \qquad \sqrt{a - y} = x .$$

We square these equations to obtain

$$a + x = y^2 , \qquad a - y = x^2 .$$

If the second equation is subtracted from the first, the result is

$$x + y = y^2 - x^2 ,$$

or,

$$x^2 - y^2 + x + y = (x + y)(x - y + 1) = 0 .$$

Two possibilities arise. First,

$$x + y = 0;$$

then $y = - x$ and $x^2 - x - a = 0$, which yields

$$x_{1,2} = \frac{1}{2} \pm \sqrt{a + \frac{1}{4}} .$$

Or,

$$x - y + 1 = 0;$$

then $y = x + 1$ and $x^2 + x + 1 - a = 0$, from which we obtain

$$x_{3,4} = -\frac{1}{2} \pm \sqrt{a - \frac{3}{4}} .$$

These possibilities for the roots of the given equation must be tested in order to eliminate any extraneous roots.

Remark: If only positive roots are considered, it may be readily ascertained that the equation will have the single root $x_3 = -\frac{1}{2} + \sqrt{a - \frac{3}{4}}$ provided $a \geqq 1$, but will not have a root for $a < 1$.

Second Solution. We clear the radicals from the equation in the usual way:

$$a - \sqrt{a + x} = x^2 \,,$$

$$(a - x^2)^2 = a + x \,,$$

$$x^4 - 2ax^2 - x + a^2 - a = 0 \,.$$

We now have an equation of degree 4. However, this equation is quadratic in the letter a; we shall use the device of solving for a in terms of x:

$$a^2 - (2x^2 + 1)a + x^4 - x = 0 \,,$$

$$a = \frac{2x^2 + 1 \pm \sqrt{4x^4 + 4x^2 + 1 - 4x^4 + 4x}}{2}$$

$$= \frac{2x^2 + 1 \pm \sqrt{4x^2 + 4x + 1}}{2} = \frac{2x^2 + 1 \pm (2x + 1)}{2} \,,$$

$$a_1 = x^2 + x + 1 \,, \qquad a_2 = x^2 - x \,.$$

The equation

$$a^2 - (2x^2 + 1)a + x^4 - x = 0$$

has the two roots

$$a_1 = x^2 + x + 1 \,, \qquad a_2 = x^2 - x \,,$$

and so we can write

$$a^2 - (2x^2 + 1)a + x^4 - x = (a - a_1)(a - a_2)$$

$$= (a - x^2 - x - 1)(a - x^2 + x) \,.$$

Therefore, we can write the quartic equation in the form

$$(x^2 - x - a)(x^2 + x - a + 1) = 0 \,.$$

This is readily solved to yield

$$x_{1,2} = \frac{1}{2} \pm \sqrt{\frac{1}{4} + a} = \frac{1}{2} \pm \sqrt{a + \frac{1}{4}} \,,$$

$$x_{3,4} = -\frac{1}{2} \pm \sqrt{\frac{1}{4} + a - 1} = -\frac{1}{2} \pm \sqrt{a - \frac{3}{4}} \,.$$

169. *First Solution.* Let

$$x^2 + 2ax + \frac{1}{16} = y \,,$$

$$-a + \sqrt{a^2 + x - \frac{1}{16}} = y_1 \,.$$

Then the equation takes on the form

$$y = y_1 .$$

We express x in terms of y_1; calculation yields

$$x = y_1^2 + 2ay_1 + \frac{1}{16} .$$

We note that x is expressed in terms of y_1 by a quadratic formula having exactly the same coefficients as that giving y in terms of x. It follows that if we graph the functions

$$y = x^2 + 2ax + \frac{1}{16}$$

$$y_1 = -a \sqrt{a^2 + x - \frac{1}{16}} ,$$

then the two graphs (parabolas) will be symmetrical with respect to the line bisecting the first quadrant (See Figure 10; every point $x = x_0, y = y_0$ of the first curve has an image point $x = y_0, y_1 = x_0$ on the

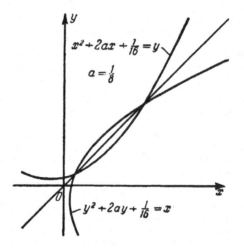

Figure 10

second curve). The points of intersection of these two curves have x-coordinates for which $y = y_1$; these coordinates yield the roots of the equation. These points must lie on the axis of symmetry of the two curves, satisfying the conditions

$$y = x = y_1 .$$

If we solve the equation $y = x$, that is,

$$x^2 + 2ax + \frac{1}{16} = x ,$$

we obtain

$$x_{1,2} = \frac{1 - 2a}{2} \pm \sqrt{\left(\frac{1 - 2a}{2}\right)^2 - \frac{1}{16}} .$$

It is left to the reader to convince himself that for $0 < a < \frac{1}{4}$ both these roots are real and satisfy the given equation.

Second Solution. The problem can be solved in a more conventional way. If we clear radicals in the usual manner, we obtain

$$\left(x^2 + 2ax + a + \frac{1}{16}\right)^2 = a^2 + x - \frac{1}{16} ,$$

or, upon expansion and the collection of terms,

$$x^4 + 4ax^3 + \left(4a^2 + 2a + \frac{1}{8}\right)x^2 + \left(4a^2 + \frac{1}{4}a - 1\right)x$$
$$+ \frac{a}{8} + \frac{1}{16} + \frac{1}{16^2} = 0 .$$

The left member of this equation can be grouped and factored as follows:

$$\left[x^4 + (2a - 1)x^3 + \frac{1}{16}x^2\right]$$
$$+ \left[(2a + 1)x^3 + (4a^2 - 1)x^2 + \left(\frac{a}{8} + \frac{1}{16}\right)x\right]$$
$$+ \left[\left(2a + \frac{17}{16}\right)x^2 + \left(4a^2 + \frac{a}{8} - \frac{17}{16}\right)x + \left(\frac{a}{8} + \frac{1}{16} + \frac{1}{16^2}\right)\right]$$
$$= \left[x^2 + (2a - 1)x + \frac{1}{16}\right]\left[x^2 + (2a + 1)x + \left(2a + \frac{17}{16}\right)\right] .$$

This yields the solutions

$$x^2 + (2a - 1)x + \frac{1}{16} = 0 ,$$

$$x_{1,2} = \frac{1 - 2a}{2} \pm \sqrt{\left(\frac{1 - 2a}{2}\right)^2 - \frac{1}{16}} ;$$

$$x^2 + (2a + 1)x + 2a + \frac{17}{16} = 0 \,,$$

$$x_{3,4} = -\frac{1 + 2a}{2} \pm \sqrt{\left(\frac{1 - 2a}{2}\right)^2 - 2a - \frac{17}{16}} \,.$$

If $0 < a < \frac{1}{4}$, the first two roots are real and satisfy the initial equation; the last two roots are complex numbers.

170. (a) To yield a real number for the left member, for real values of x, the expressions under the radicals in the left member must all be positive. Let us designate these radicands, respectively, starting from the innermost one (from $3x$), by $y_1^2, y_2^2, \cdots, y_{n-1}^2, y_n^2$; we then have

$$3x = x + 2x = y_1^2;$$
$$x + 2y_1 = y_2^2 \,,$$
$$x + 2y_2 = y_3^2 \,,$$
$$\cdots\cdots\cdots\cdots \,,$$
$$x + 2y_{n-2} = y_{n-1}^2 \,,$$
$$x + 2y_{n-1} = y_n^2 \,,$$

where all the numbers $y_1, y_2, \cdots, y_n$ are real and positive. The initial equation takes on the form, in the new designation,

$$y_n = x \,.$$

We shall now prove that $y_1 = x$. Assume that $x > y_1$. Then a comparison of the first and second equations shown above will indicate that $y_1 > y_2$. Similarly, the second and third equations will imply that $y_2 > y_3$; we can continue, in a similar manner, to find

$$y_3 > y_4 > \cdots > y_{n-1} > y_n \,.$$

Hence if $x > y_1$, then $x > y_n$, which contradicts the equation $y_n = x$. The assumption $x < y_1$ will lead, by analogous reasoning, to a similar contradiction. Hence we must have $y_1 = x$.

Since $y_1^2 = 3x$, it follows that

$$3x = x^2 \,,$$

and so we may set down two possible values for x:

$$x_1 = 3 \,, \qquad x_2 = 0 \,.$$

Both values satisfy the given equation.

Remark: We can use another technique to solve this equation. We write it as

$$\underbrace{\sqrt{x + 2\sqrt{x + 2\sqrt{x + \cdots + 2\sqrt{x + 2x}}}}}_{n \text{ radicals}} = x . \qquad (1)$$

If we replace the final x of the left member by the entire expression for x as given by (1), we have

$$x = \underbrace{\sqrt{x + 2\sqrt{x + 2\sqrt{x + \cdots + 2\sqrt{x + 2x}}}}}_{2n \text{ radicals}} .$$

If we again repeat this substitution, we obtain new equations of the same form, except that we have, successively, $3n, 4n, \cdots$ radical signs. Thus we arrive at

$$x = \sqrt{x + 2\sqrt{x + 3\sqrt{x + \cdots}}}$$

$$= \lim_{N \to \infty} \underbrace{\sqrt{x + 2\sqrt{x + 2\sqrt{x + \cdots + 2\sqrt{x + 2x}}}}}_{N \text{ radicals}} \qquad (2)$$

It follows that

$$x = \sqrt{x + 2\sqrt{x + 2\sqrt{x + \cdots}}}$$

$$= \sqrt{x + 2[\sqrt{x + 2\sqrt{x + 2\sqrt{x + \cdots}}}]} = \sqrt{x + 2x} , \qquad (3)$$

which yields $x = \sqrt{3x}, x^2 = 3x$; consequently, $x_1 = 0, x_2 = 3$. This shows that the roots of equation (1) do not depend upon n [since the roots of (2) are independent of n].

The reasoning used here cannot be considered a legitimate solution to the problem, inasmuch as the existence of the limit shown in (2) has not been established and hence cannot be legitimately employed for (3). However, the reasoning can be rigorously justified by a more advanced discussion, which is not undertaken here.

(b) We make successive simplifications of the fraction in the left member:

$$1 + \frac{1}{x} = \frac{x + 1}{x} ;$$

$$1 + \frac{1}{\dfrac{x + 1}{x}} = 1 + \frac{x}{x + 1} = \frac{2x + 1}{x + 1} ;$$

$$1 + \frac{1}{\dfrac{2x + 1}{x + 1}} = 1 + \frac{x + 1}{2x + 1} = \frac{3x + 2}{2x + 1} ;$$

. . . .

We finally arrive at an equation of form

$$\frac{ax + b}{cx + d} = x,$$

where $a, b, c,$ and d are some integers (depending upon n). This equation is in fact a quadratic equation $x(cx + d) = ax + b$, which implies that the given equation can have at most two roots and hence cannot be an identity (since then all values of x would satisfy it; in particular $x = 0$ fails to satisfy the equation).

Without an assignment of value for n we apparently cannot determine these roots. However, let us assume that

$$1 + \frac{1}{x} = x.$$

The successive simplifications of the fraction yield

$$1 + \frac{1}{x} = x;$$

$$1 + \frac{1}{x} = x;$$

$$\cdots\cdots\cdots\cdots,$$

and we finally arrive at the identity $x = x$. Hence, under the assumption, the roots of $1 + \dfrac{1}{x} = x$, or $x^2 - x - 1 = 0$, that is,

$$x_1 = \frac{1 + \sqrt{5}}{2},$$

$$x_2 = \frac{1 - \sqrt{5}}{2},$$

satisfy the given equation. Since the equation has at most two roots, and we have found two roots for it, these represent the complete solution of the problem.

Remark: We display still another method for solving the equation [compare this with the remark following the solution of problem (a)].

We substitute for the final x shown in the "multi-storied" fraction[†] its expression as given by the equation itself. We then have an equation of precisely the same form, except with $2n$ fractional designations.[††] · Continuation of the process leads finally to our writing

---

[†] *Continued fraction* is the terminology usually used for this concept [*Editor*].

[††] Literally, "twice as many stories" [*Editor*].

$$x = \cfrac{1}{1 + \cfrac{1}{1 + \cfrac{1}{1 + \cfrac{1}{1 + \cdot}}}} \left[= \lim_{N \to \infty} \cfrac{1}{1 + \cfrac{1}{1 + \cfrac{1}{1 + \cdot \cdot \cdot + 1 + \cfrac{1}{1}}}} \right], \quad (1)$$

$$\underbrace{}$$
The fraction bar is
repeated N times

where on the left we have an infinite continued fraction.
This yields

$$x = \cfrac{1}{1 + \cfrac{1}{1 + \cfrac{1}{1 + \cfrac{1}{1 + \cdot}}}} = \cfrac{1}{1 + \left[\cfrac{1}{1 + \cfrac{1}{1 + \cfrac{1}{1 + \cdot}}}\right]} = \cfrac{1}{1 + x}, \quad (2)$$

that is, we obtain the quadratic in x,

$$x = \frac{1}{1 + x},$$

which we assumed in the first solution of this problem. This proof now shows that the solution of the equation does not actually depend upon n (a fact we might have adduced at the conclusion of the previous solution).

The reasoning here is not rigorous, inasmuch as the existence of the limit of which we made use has not been proved. However, a rigorous proof can be given by more advanced mathematics.

171. We have

$$x + 3 - 4\sqrt{x - 1} = x - 1 - 4\sqrt{x - 1} + 4$$
$$= (\sqrt{x - 1})^2 - 4\sqrt{x - 1} + 4 = (\sqrt{x - 1} - 2)^2$$

and, analogously,

$$x + 8 - 6\sqrt{x - 1} = x - 1 - 6\sqrt{x - 1} + 9$$
$$= (\sqrt{x - 1} - 3)^2.$$

Hence, the equation can be written in the form

$$(\sqrt{(\sqrt{x - 1} - 2)^2} + \sqrt{(\sqrt{x - 1} - 3)^2} = 1$$

or, since it has been specified that only positive roots are to be considered

$$|\sqrt{x-1}-2| + |\sqrt{x-1}-3| = 1\,,$$

where $|y|$ means the positive numerical value of y.

We consider the several possibilities.

First, if $\sqrt{x-1}-2 \geqq 0$ and $\sqrt{x-1}-3 \geqq 0$, that is, if $\sqrt{x-1} \geqq 3$, $x-1 \geqq 9$, $x \geqq 10$, then $|\sqrt{x-1}-2| = \sqrt{x-1}-2$, $|\sqrt{x-1}-3| = \sqrt{x-1}-3$, and the equation takes on the form

$$\sqrt{x-1}-2 + \sqrt{x-1}-3 = 1\,.$$

Hence,

$$2\sqrt{x-1} = 6$$
$$x-1 = 9\,,$$
$$x = 10\,.$$

If $\sqrt{x-1}-2 \geqq 0$ and $\sqrt{x-1}-3 \leqq 0$, that is, if $\sqrt{x-1} \geqq 2$, $x \geqq 5$, but $\sqrt{x-1} \leqq 3$, $x \leqq 10$, then $|\sqrt{x-1}-2| = \sqrt{x-1}-2$, $|\sqrt{x-1}-3| = -\sqrt{x-1}+3$, and the equation becomes the identity

$$\sqrt{x-1}-2 - \sqrt{x-1}+3 = 1\,.$$

Therefore, the equation is satisfied by *all* values of x between $x=5$ and $x=10$.

If $\sqrt{x-1}-2 \leqq 0$, $\sqrt{x-1}-3 \leqq 0$, that is, if $\sqrt{x-1} \leqq 2$, $x \leqq 5$, then $|\sqrt{x-1}-2| = -\sqrt{x-1}+2$, $|\sqrt{x-1}-3| = -\sqrt{x-1}+3$, and the equation becomes

$$-\sqrt{x-1}+2 - \sqrt{x-1}+3 = 1\,.$$

It follows that

$$2\sqrt{x-1} = 4$$
$$x-1 = 4\,,$$
$$x = 5\,.$$

The case $\sqrt{x-1}-2 \leqq 0$, $\sqrt{x-1}-3 \geqq 0$, is impossible.

In summary, all values of x between 5 and 10, inclusive, that is, $5 \leqq x \leqq 10$, are solutions of the given equation.

172. We shall first look for the real roots lying in the interval 2 to ∞, then in the interval 1 to 2, then 0 to 1, then -1 to 0, and finally $-\infty$ to -1.

Let $x \geqq 2$. Then $x+1 > 0$, $x > 0$, $x-1 > 0$, $x-2 \geqq 0$; hence $|x+1| = x+1$, $|x| = x$, $|x-1| = x-1$; $|x-2| = x-2$, and we have the equation

$$x + 1 - x + 3(x - 1) - 2(x - 2) = x + 2 \,,$$

which is an identity.

Accordingly, all real numbers greater than 2, and 2 itself, are roots of the given equation.

Let $1 \leqq x < 2$. Then $x + 1 > 0$, $x > 0$, $x - 1 \geqq 0$, and $x - 2 < 0$, which implies

$$| x + 1 | = x + 1 \,,$$
$$| x | = x \,,$$
$$| x - 1 | = x - 1 \,,$$
$$| x - 2 | = - (x - 2) \,.$$

We obtain, for this case, the equation

$$x + 1 - x + 3(x - 1) + 2(x - 2) = x + 2 \,.$$

This yields $4x = 8$, or $x = 2$. This value lies in the interval previously considered. Consequently, there is no additional root found between 1 and 2 for the given equation.

Let $0 \leqq x < 1$. Then $| x + 1 | = x + 1$, $| x | = x$, $| x - 1 | = - (x - 1)$, and $| x - 2 | = - (x - 2)$. We have

$$x + 1 - x - 3(x - 1) + 2(x - 2) = x + 2.$$

This yields $x = -1$, but since this lies outside the interval which we used to set up the equation, it must be discarded.

There exists no additional root for the equation in the interval $0 \leqq x < 1$.

Let $-1 \leqq x \leqq 0$. Then $| x + 1 | = x + 1$, $| x | = -x$, $| x - 1 | = -(x - 1)$, and $| x - 2 | = -(x - 2)$. We have

$$x + 1 + x - 3(x - 1) + 2(x - 2) = x + 2 \,.$$

This equation is contradictory; hence there are no roots between -1 and 0, inclusive.

Finally, let $x < -1$. Then $| x + 1 | = -(x + 1)$, $| x | = -x$, $| x - 1 | = -(x - 1)$, and $| x - 2 | = -(x - 2)$; We have

$$-(x + 1) + x - 3(x - 1) + 2(x - 2) = x + 2 \,,$$
$$x = - 2 \,.$$

We obtain the root $x = -2$ from this interval. Therefore, the equation is satisfied by -2, by 2, and by all real numbers exceeding 2.

Remark: The results obtained for this problem become vividly clear if we graph the function

$$y = |x+1| - |x| + 3|x-1| - 2|x-2| - (x+2) .$$

Figure 11 shows in light lines the functions $y_1 = |x+1|$, $y_2 = -|x|$, $y_3 = 3|x-1|$, $y_4 = -2|x-2|$, and $y_5 = -(x+2)$, and in heavy lines the function $y = y_1 + y_2 + y_3 + y_4 + y_5$ (by "composition" of graphs). It is clear from the figure that y crosses the axis at $x = -2$ and at $x = +2$, and thereafter remains on the x axis for $x > 2$.

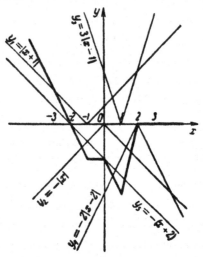

Figure 11

173. From the first equation of the given system we see that

$$y^2 = x^2 , \qquad y = \pm x .$$

If we substitute for y^2 in the second equation, we obtain

$$(x-a)^2 + x^2 = 1 , \tag{1}$$

which, as a quadratic equation, in general yields two possible values for x. Since each value of x can be associated with two values of y, the system will have at most four solutions; This will reduce to at most three solutions if one of the values of x is zero, since this value will go with only one companion value for y, that is, $y = 0$. If we substitute $x = 0$ into equation (1) we obtain

$$a^2 = 1 , \qquad a = \pm 1 .$$

The system can, and will, have precisely three solutions only for these values of a.

The number of solutions of the system reduces to two if the quadratic equation involving x has only one solution (a "double root"). The quadratic equation

$$(x - a)^2 + x^2 = 1 ,$$

or,

$$2x^2 - 2ax + a^2 - 1 = 0$$

will have one root if and only if the discriminant $(B^2 - 4AC)$ vanishes; that is, if

$$a^2 - 2(a^2 - 1) = 0 ,$$

or, $a^2 = 2$; that is, $a = \pm \sqrt{2}$.

174. (a) Formal solution of the system yields

$$x = \frac{a^3 - 1}{a^2 - 1} ,$$

$$y = \frac{-a^2 + a}{a^2 - 1} .$$

If $a + 1 \neq 0$ and $a - 1 \neq 0$, then the system has the single solution

$$x = \frac{a^2 + a + 1}{a + 1} ,$$

$$y = \frac{-a}{a + 1} .$$

If $a = -1$, or if $a = +1$, then the formulas are meaningless; in the first instance we arrive at the system

$$\begin{cases} -x + y = 1 , \\ x - y = 1 , \end{cases}$$

which is a contradictory system. In the second instance we have

$$\begin{cases} x + y = 1 , \\ x + y = 1 , \end{cases}$$

which has an infinite number of solutions (for example, for x arbitrary, $y = 1 - x$).

(b) Solution of the system yields

$$x = \frac{a^4 - 1}{a^2 - 1} ,$$

$$y = \frac{-a^3 + a}{a^2 - 1} .$$

Here, if $a^2 - 1 \neq 0$, the system has the single solution $x = a^2 + 1$, $y = -a$. For $a = -1$ and $a = 1$, we obtain the systems

$$\begin{cases} -x + y = -1, \\ x - y = 1 \end{cases}$$

and

$$\begin{cases} x + y = 1, \\ x + y = 1, \end{cases}$$

both of which have an infinite number of solutions.

(c) We obtain from the first two equations

$$y + z = 1 - ax$$

and

$$ay + z = a - x.$$

If we consider this as a system of two equations in two unknowns, y and z, we obtain

$$y = \frac{a - x - 1 + ax}{a - 1} = \frac{(a - 1)(1 + x)}{a - 1},$$

$$z = \frac{a(1 - ax) - a + x}{a - 1} = \frac{-x(a^2 - 1)}{a - 1} = -(1 + a)x.$$

Hence, if $a \neq 1$, then $y = 1 + x$ and $z = -(1 + a)x$. If these values are substituted into the third equation, we find

$$x + (1 + x) - a(1 + a)x = a^2,$$

$$x(2 - a - a^2) = a^2 - 1$$

$$-x(a + 2)(a - 1) = a^2 - 1.$$

Therefore, if $a - 1 \neq 0$ and $a + 2 \neq 0$, the system has the single solution

$$x = -\frac{a^2 - 1}{(a + 2)(a - 1)} = -\frac{a + 1}{a + 2},$$

$$y = 1 + x = \frac{1}{a + 2},$$

$$z = -(a + 1)x = \frac{(a + 1)^2}{a + 2}.$$

For $a = 1$ and $a = -2$, we obtain the systems

$$\begin{cases} x + y + z = 1 , \\ x + y + z = 1 , \\ x + y + z = 1 \end{cases}$$

and

$$\begin{cases} -2x + y + z = 1 , \\ x - 2y + z = -2 , \\ x + y - 2z = 4 , \end{cases}$$

The first of these systems has an infinite number of solutions, and the second has no solution (from the first two equations we obtain $-x - y + 2z = -1$, which is inconsistent with the third equation).

175. If we subtract the second equation from the first, and the sixth from the fifth, and equate the two expressions for $x_2 - x_3$, we obtain

$$\alpha_1(\alpha_2 - \alpha_3) = \alpha_4(\alpha_2 - \alpha_3) ,$$

or,

$$(\alpha_1 - \alpha_4)(\alpha_2 - \alpha_3) = 0 .$$

Similarly, if we obtain the two expressions related by $x_1 - x_2$ and also by $x_1 - x_3$, we find two more relationships:

$$(\alpha_1 - \alpha_2)(\alpha_3 - \alpha_4) = 0 ,$$
$$(\alpha_1 - \alpha_3)(\alpha_2 - \alpha_4) = 0 .$$

The first of these three relationships implies that either $\alpha_1 = \alpha_4$ or $\alpha_2 = \alpha_3$ (possibly both). Let us suppose that $\alpha_2 = \alpha_3 = \alpha$. Then, from the second relationship, $\alpha_1 = \alpha$ or else $\alpha_4 = \alpha$. Either of these possibilities makes the third equation an identity. Hence, for the system to be consistent, it is necessary that three of the four quantities $\alpha_1, \alpha_2, \alpha_3, \alpha_4$ be equal.

Suppose now that $\alpha_1 = \alpha_2 = \alpha_3 = \alpha$ and that $\alpha_4 = \beta$. Recalling those expressions for the differences $x_1 - x_2$, $x_1 - x_3$, $x_2 - x_3$, with the aid of which we obtained the relationships just exploited between $\alpha_1, \alpha_2, \alpha_3$, and α_4, we find that

$$x_1 = x_2 = x_3 .$$

Designating $x_1 = x_2 = x_3$ by x, and x_4 by y, we find that the six equations in four unknowns reduces to two equations in two unknowns:

$$2x = \alpha^2 ,$$
$$x + y = \alpha\beta ,$$

from which we find that

$$x = \frac{\alpha^2}{2} ;$$

$$y = \alpha \left(\beta - \frac{\alpha}{2} \right) .$$

Remark: Analogous reasoning shows that the more general system

$$\begin{cases} x_1 + x_2 + \cdots + x_{m-1} + x_m = \alpha_1 \alpha_2 \cdots \alpha_{m-1} \alpha_m , \\ x_1 + x_2 + \cdots + x_{m-1} + x_{m+1} = \alpha_1 \alpha_2 \cdots \alpha_{m-1} \alpha_{m+1} , \\ \cdots\cdots\cdots\cdots\cdots\cdots\cdots\cdots\cdots\cdots\cdots\cdots\cdots , \\ x_{n+m-1} + x_{n+m-2} + \cdots + x_{n-1} + x_n = \alpha_{n+m-1} \alpha_{n+m-2} \cdots \alpha_{n-1} \alpha_n , \end{cases}$$

consisting of C_n^m equations in n unknowns $(n > m + 1)$, will be solvable only in the following two cases:

$$\alpha_1 = \alpha_2 = \cdots = \alpha_{n-1} = \alpha, \; \alpha_n = \beta ;$$

here,

$$x_1 = x_2 = \cdots = x_{n-1} = \frac{\alpha^n}{n}, \; x_n = \alpha^{n-1}\left(\beta - \frac{n-1}{n}\alpha\right) .$$

$n - m + 1$, or more, of the quantities $\alpha_1, \alpha_2, \cdots, \alpha_n$ are zero (here, $x_1 = x_2 = \cdots = x_n = 0$).

176. From the first of the given equations,

$$x = 2 - y .$$

Upon substitution into the second, we obtain

$$2y - y^2 - z^2 = 1 ,$$

or

$$z^2 + y^2 - 2y + 1 = 0 ,$$

or

$$z^2 + (y - 1)^2 = 0 .$$

Each of the two terms of the last equation is nonnegative, hence both must vanish. Hence $z = 0$ and $y = 1$, which implies $x = 1$.

Therefore, the system has precisely one real solution.

177. (a) First we note that if x_0 is a root of the given equation, then $-x_0$ is also a root. Consequently, there are as many negative roots as there are positive roots. Moreover, the number 0 is clearly a root of the equation. It suffices then to find how many positive roots there are. Now if $\frac{x}{100} = \sin x$, then

$$|x| = 100 \,|\sin x| \leq 100 \cdot 1 = 100 \,,$$

and so no root can exceed 100 in absolute value.

Let us partition the x-axis from 0 to 100 into segments each of length 2π (except for the final segment, which will be shorter); we shall examine each interval separately to find the roots in it. (See Figure 12).

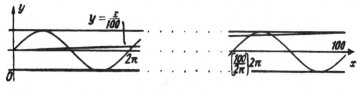

Figure 12

There exists one positive root in the interval from 0 to 2π; in each of the following intervals (excluding the final one) there are two positive roots. To find how many roots may be contributed by the final interval, we examine it separately. Now, $\dfrac{100}{2\pi}$ is a number between 15 and 16 $\left(\dfrac{100}{15} = 6.666 \cdots > 2\pi; \dfrac{100}{16} = 6.25 < 2\pi\right)$; consequently, we have 15 segments each of length 2π and one final segment of length $100 - 15 \cdot 2\pi > 5 > \pi$. This final segment is long enough to contain the complete upper half of the sinusoidal period, and hence it also contributes two roots.

Therefore, in all we have $1 + 14 \cdot 2 + 2 = 31$ positive roots for the given equation, an equal number of negative roots, and, in addition, the root 0; therefore, the equation has 63 roots.

(b) The solution is quite similar to that of problem (a). First, if $\sin x = \log x$, then $x \leq 10$ (inasmuch as $\sin x \leq 1$). Since $2 \cdot 2\pi > 10$, the interval on the x-axis between $x = 0$ and $x = 10$ contains one complete period of the sine curve plus part of a second period. The

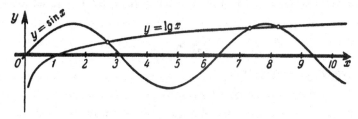

Figure 13

graph of log x intersects the first wave of the sine curve at precisely one point (see Figure 13). Further, since $2\pi + \dfrac{\pi}{2} < 10$, then at the point $x = \dfrac{5\pi}{2}$ we have $\sin x = 1 > \log x$, which means that the graph of log x intersects the first half of the second positive wave of sin x. Since, at $x = 10$, $\log x = 1 > \sin x$, the graph of log x must intersect this second wave another time. Therefore, we conclude that the equation $\sin x = \log x$ has exactly three real roots.

178. It is readily verified that the proposition of the problem is valid for $n = 1$ and $n = 2$:

$$x_1 + x_2 = 6$$
$$x_1^2 + x_2^2 = (x_1 + x_2)^2 - 2x_1x_2 = 6^2 - 2 \cdot 1 = 34 \ .$$

(The sum of the roots of a quadratic equation is equal to the negative of the coefficient of x).

Further, we have

$$x_1^n + x_2^n = (x_1 + x_2)(x_1^{n-1} + x_2^{n-1}) - x_1x_2(x_1^{n-2} + x_2^{n-2})$$
$$= 6(x_1^{n-1} + x_2^{n-1}) - 1 \cdot (x_1^{n-2} + x_2^{n-2}) \ .$$

or

$$x_1^n + x_2^n = 5(x_1^{n-1} + x_2^{n-1})$$
$$+ [(x_1^{n-1} + x_2^{n-1}) - (x_1^{n-2} + x_2^{n-2})] \qquad (1)$$

It follows from this formula, first, that if $x_1^{n-2} + x_2^{n-2}$ and $x_1^{n-1} + x_2^{n-1}$ are integers, then $x_1^n + x_2^n$ is also an integer; thus, by mathematical induction $x_1^n + x_2^n$ is shown to be an integer for all natural numbers n.

Now, let n be the least positive integer such that $x_1^n + x_2^n$ is divisible by 5. It follows from (1) that, in this case, the difference

$$(x_1^{n-1} + x_2^{n-1}) - (x_1^{n-2} + x_2^{n-2})$$

is also divisible by 5. But if we replace n in (1) by $n - 1$, we obtain

$$x_1^{n-1} + x_2^{n-1} = 5(x_1^{n-2} + x_2^{n-2})$$
$$+ (x_1^{n-2} + x_2^{n-2}) - (x_1^{n-3} + x_2^{n-3}) \ ,$$

from which it follows that

$$x_1^{n-3} + x_2^{n-3} = 5(x_1^{n-2} + x_2^{n-2})$$
$$- [(x_1^{n-1} + x_2^{n-1}) - (x_1^{n-2} + x_2^{n-2})]$$

is also divisible by 5. This contradicts the assumption that n is the least integer such that $x_1^m + x_2^m$ is divisible by 5. Hence we must conclude that there cannot exist a positive integer n such that $x_1^n + x_2^n$ is divisible by 5.

179. Let us say that there are n positive numbers (and hence $1000 - n$ negative numbers) among the numbers $a_1, a_2, \cdots, a_{1000}$. Then, in the expansion given, the "mixed products" $a_i a_j$ of the n positive numbers $\left[\text{there will be } \dfrac{n(n+1)}{2} \text{ of these products} \right]$ and the mixed products of the $1000 - n$ negative numbers [there will be

$$\frac{(1000 - n)(1000 - n - 1)}{2}$$

of these products] will be positive terms, and the product of positive by negative numbers [there will be $n(1000 - n)$ of these] will be negative. The condition of the problem requires that

$$\frac{n(n-1)}{2} + \frac{(1000 - n)(1000 - n - 1)}{2} = n(1000 - n) ,$$

or,

$$\frac{n^2 - n + (1000 - n)^2 - (1000 - n)}{2} = 1000n - n^2 ,$$

$$2n^2 - 2000n + \frac{999,000}{2} = 0 ,$$

$$n = \frac{1000 \pm \sqrt{1,000,000 - 999,000}}{2} = \frac{1000 \pm \sqrt{1000}}{2} ,$$

which is impossible.

For the analogous problem posed, we obtain by similar reasoning the requirement

$$n = \frac{10,000 \pm \sqrt{10,000}}{2} = \frac{10,000 \pm 100}{2} .$$

Here it is possible for the expansion to contain an equal number of positive and negative mixed products. For this, it suffices if the initial polynomial contains $\dfrac{10,000 + 100}{2} = 5050$ positive numbers and $\dfrac{10,000 - 100}{2} = 4950$ negative numbers (or vice-versa).

180. First, we have

$$(\sqrt{2} - 1)^1 = \sqrt{2} - 1 ;$$
$$(\sqrt{2} - 1)^2 = 3 - 2\sqrt{2} = \sqrt{9} - \sqrt{8} .$$

The proof will proceed by mathematical induction. Assume that

$$(\sqrt{2} - 1)^{2k-1} = B\sqrt{2} - A = \sqrt{2B^2} - \sqrt{A^2}$$

can be put into the form $\sqrt{N} - \sqrt{N-1}$, that is, that $2B^2 - A^2 = 1$.
We shall show that (replacing k by $k + 1$)

$$(\sqrt{2} - 1)^{2k+1} = B'\sqrt{2} - A'$$

also comes into such a form, that is, that $2B'^2 - A'^2 = 1$. We can write

$$\begin{aligned}
(\sqrt{2} - 1)^{2k+1} &= (\sqrt{2} - 1)^{2k-1}(\sqrt{2} - 1)^2 \\
&= (B\sqrt{2} - A)(3 - 2\sqrt{2}) \\
&= (3B + 2A)\sqrt{2} - (4B + 3A) ;
\end{aligned}$$

consequently,

$$\begin{aligned}
B' &= 3B + 2A , \\
A' &= 4B + 3A ,
\end{aligned}$$

and

$$\begin{aligned}
2B'^2 - A'^2 &= 2(3B + 2A)^2 - (4B + 3A)^2 \\
&= 18B^2 + 24AB + 8A^2 - 16B^2 - 24AB - 9A^2 \\
&= 2B^2 - A^2 = 1 ,
\end{aligned}$$

which is what we wished to show.

Therefore, if the number $(\sqrt{2} - 1)^{2k} = C - D\sqrt{2}$ can be put into the form $\sqrt{N} - \sqrt{N-1}$, then also the number $(\sqrt{2} - 1)^{2k+2} = C' - D'\sqrt{2}$ can be expressed in this form.

The assertion of the problem follows by mathematical induction.

181. If $(A + B\sqrt{3})^2 = C + D\sqrt{3}$, then $C = A^2 + 3B^2$, $D = 2AB$, and

$$(A - B\sqrt{3})^2 = A^2 + 3B^2 - 2AB\sqrt{3} = C - D\sqrt{3} .$$

Consequently, if

$$(A + B\sqrt{3})^2 = 99,999 + 111,111\sqrt{3} ,$$

then also

$$(A - B\sqrt{3})^2 = 99,999 - 111,111\sqrt{3} ,$$

which is an impossibility inasmuch as the left member of this equation is positive and the right member is negative.

182. Assume that $\sqrt[3]{2} = p + q\sqrt{r}$ (p, q, and r rational). If both sides are cubed, we obtain

$$2 = p^3 + 3p^2q\sqrt{r} + 3pq^2r + q^3r\sqrt{r} ,$$

or,

$$2 = p(p^2 + 3q^2r) + q(3p^2 + q^2r)\sqrt{r} .$$

We shall now show that our assumption that $\sqrt[3]{2} = p + q\sqrt{r}$ implies that $\sqrt[3]{2}$ is a rational number. First, if $q = 0$, then $\sqrt[3]{2} = p$ is rational. If $q \neq 0$ and if $3p^2 + q^2r \neq 0$, then from the last of the above equations we obtain

$$\sqrt{r} = \frac{2 - p(p^2 + 3q^2r)}{q(3p^2 + q^2r)} ,$$

from which we find

$$\sqrt[3]{2} = p + q\frac{2 - p(p^2 + 3q^2r)}{q(3p^2 + q^2r)} ,$$

which states that $\sqrt[3]{2}$ is rational. If $3p^2 + q^2r = 0$, then

$$q^2r = -3p^2 ,$$
$$2 = p[p^2 + 3(-3p^2)] = -8p^3 ,$$

and $\sqrt[3]{2} = -2p$ is again a rational number.

It remains to prove that $\sqrt[3]{2}$ is not a rational number. If $\sqrt[3]{2}$ were equal to an irreducible fraction $\frac{m}{n}$ then we would have $2 = \frac{m^3}{n^3}$, or $m^3 = 2n^3$. In this event m^3 (and hence m) would have to be an even number, and therefore would be divisible by 8. We could then write $n^3 = \frac{m^3}{2}$, and, since $\frac{m^3}{2}$ is even, then necessarily n^3 (and therefore n) would have to be even. This contradicts the assumption that $\frac{m}{n}$ is irreducible, and the contradiction proves the statement of the problem.

183. (a) Designate 1.00000000004 by α, and 1.00000000002 by β. Then it is readily seen that the two numbers of the problem can be written as $\frac{1 + \alpha}{1 + \alpha + \alpha^2}$ and $\frac{1 + \beta}{1 + \beta + \beta^2}$. Since $\alpha > \beta$, it is clear that

$$\frac{1 + a}{a^2} = \frac{1}{a^2} + \frac{1}{a} < \frac{1}{\beta^2} + \frac{1}{\beta} = \frac{1 + \beta}{\beta^2} ;$$

$$\frac{a^2}{1 + a} = 1 : \left(\frac{1 + a}{a^2}\right) > 1 : \left(\frac{1 + \beta}{\beta^2}\right) = \frac{\beta^2}{1 + \beta} ;$$

$$\frac{1 + a + a^2}{1 + a} = 1 + \frac{a^2}{1 + a} > 1 + \frac{\beta^2}{1 + \beta^2} = \frac{1 + \beta + \beta^2}{1 + \beta}$$

and, finally,

$$\frac{1+a}{1+a+a^2} = 1 : \left(\frac{1+a+a^2}{1+a}\right) < 1 : \left(\frac{1+\beta+\beta^2}{1+\beta}\right) = \frac{1+\beta}{1+\beta+\beta^2} .$$

Therefore, the second number is greater than the first.

(b) If we designate the two expressions given in the problem by A and B, respectively, we obtain

$$\frac{1}{A} = 1 + \frac{a^n}{1+a+a^2+\cdots+a^{n-1}}$$

$$= 1 + \frac{1}{\dfrac{1+a+a^2+\cdots+a^{n-1}}{a^n}} = 1 + \frac{1}{\dfrac{1}{a^n}+\dfrac{1}{a^{n-1}}+\cdots+\dfrac{1}{a}} ;$$

$$\frac{1}{B} = 1 + \frac{1}{\dfrac{1}{b^n}+\dfrac{1}{b^{n-1}}+\cdots+\dfrac{1}{b}} .$$

It follows that $\dfrac{1}{A} > \dfrac{1}{B}$ or $B > A$.

184. Let X be an arbitrary number. Consider the difference $(X-a)^2 - (x-a)^2$. We note that

$$(X-a)^2 - (x-a)^2 = X^2 - x^2 - 2a(X-x) .$$

We can now write the difference

$$[(X-a_1)^2 + (X-a_2)^2 + \cdots + (X-a_n)^2]$$
$$- [(x-a_1)^2 + (x-a_2)^2 + \cdots + (x-a_n)^2]$$
$$= n(X^2 - x^2) - 2(a_1 + a_2 + \cdots + a_n)(X-x) .$$

If in the right member we substitute $\dfrac{a_1 + a_2 + \cdots + a_n}{n}$ for x, then that member will be nonnegative; in fact, we shall have

$$[(X-a_1)^2 + (X-a_2)^2 + \cdots + (X-a_n)^2]$$
$$- [(x-a_1)^2 + (x-a_2)^2 + \cdots + (x-a_n)^2]$$
$$= n(X^2 - x^2) - 2nx(X-x) = n(X^2 - x^2 - 2Xx + 2x^2)$$
$$= n(X-x)^2 \geq 0 .$$

It follows that the sought-after value of x must be

$$\frac{a_1 + a_2 + \cdots + a_n}{n}$$

185. (a) We have, in all, only three essentially different arrangements insofar as $\emptyset$ is concerned:

(1) a_1, a_2, a_3, a_4:

$$\Phi_1 = (a_1 - a_2)^2 + (a_2 - a_3)^2 + (a_3 - a_4)^2 + (a_4 - a_1)^2 \ .$$

(2) a_1, a_3, a_2, a_4:

$$\Phi_2 = (a_1 - a_3)^2 + (a_3 - a_2)^2 + (a_2 - a_4)^2 + (a_4 - a_1)^2 \ .$$

(3) a_1, a_2, a_4, a_3:

$$\Phi_3 = (a_1 - a_2)^2 + (a_2 - a_4)^2 + (a_4 - a_3)^2 + (a_3 - a_1)^2 \ .$$

Now, it is readily seen that

$$\Phi_3 - \Phi_1 = -2a_2a_4 - 2a_1a_3 + 2a_2a_3 + 2a_1a_4$$
$$= 2(a_2 - a_1)(a_2 - a_4) < 0 \ ;$$
$$\Phi_3 - \Phi_2 = -2a_1a_2 - 2a_3a_4 + 2a_2a_3 + 2a_1a_4$$
$$= 2(a_3 - a_1)(a_2 - a_4) < 0 \ .$$

Consequently, the arrangement we seek is

$$a_1, a_2, a_4, a_3 \ .$$

(b) *First Solution.* Consider the expression

$$\Phi = (a_{i_1} - a_{i_2})^2 + (a_{i_2} - a_{i_3})^2 + \cdots + (a_{i_{n-1}} - a_{i_n})^2 + (a_{i_n} - a_{i_1})^2 \ ,$$

where $a_{i_1}, a_{i_2}, \cdots, a_{i_n}$ are the given numbers in the required order. Consider two of these numbers a_{i_α} and a_{i_β}, where $\alpha < \beta$. We claim that if a_{i_α} is greater than (or, respectively, less than) a_{i_β}, then $a_{i_{\alpha-1}}$ is greater than (or, respectively, less than) $a_{i_{\beta+1}}$. (We assume $\alpha_{i_0} = \alpha_{i_n}$.)

If we assumed the contrary, that is, if

$$(a_{i_\alpha} - a_{i_\beta})(a_{i_{\alpha-1}} - a_{i_{\beta+1}}) < 0 \ ,$$

then the permutation which reverses the order $a_{i_\alpha}, a_{i_{\alpha+1}}, a_{i_{\alpha+2}}, \cdots, a_{i_\beta}$ would decrease the value of the sum Φ, since the difference of the new sum Φ' and the initial sum Φ would then be

$$\Phi' - \Phi = -2a_{i_{\alpha-1}}a_{i_\beta} - 2a_{i_\alpha}a_{i_{\beta+1}} + 2a_{i_{\alpha-1}}a_{i_\alpha} + 2a_{i_\beta}a_{i_{\beta+}}$$
$$= 2(a_{i_\alpha} - a_{i_\beta})(a_{i_{\alpha-1}} - a_{i_{\beta+1}}) \ .$$

This observation enables us to find the full solution of the problem. First, since a cyclic permutation of all the given numbers (the permutation which preserves the relative positions of the numbers—for example, writing them in a circle and merely rotating the circle) does not change the value of Φ, we can assume that a_{i_1} is the smallest of the numbers, that is, $i_1 = 1$. It is then possible to assume that

a_{i_2} and a_{i_n} follow in order of magnitude. In fact, if for example, $a_{i_\beta} < a_{i_2}$ ($\beta \neq n$), then we would have $(a_{i_2} - a_{i_\beta})(a_{i_1} - a_{i_{\beta+1}}) < 0$, and if $a_{i_\beta} < a_{i_n}$ ($\beta \neq 2$), then we would have $(a_{i_1} - a_{i_{\beta-1}})(a_{i_n} - a_{i_\beta}) < 0$. Since it is possible to change the order of the "links" of the chain $a_{i_1}, a_{i_2}, a_{i_3}, \cdots, a_{i_n}, a_{i_1}$ to the reversed order without changing the value of $\emptyset$, we can assume that $a_{i_2} < a_{i_n}$, $i_2 = 2$, and $i_n = 3$.

Further, we assert that the numbers $a_{i_3}, a_{i_{n-1}}$ follow in this order in magnitude, within the pattern of the numbers $a_{i_1}, a_{i_2}, a_{i_n}$ already considered. In fact, if, for example, $a_{i_3} > a_{i_\beta}$ ($\beta \neq 1, 2; n-1, n$), then we would have $(a_{i_3} - a_{i_\beta})(a_{i_2} - a_{i_{\beta+1}}) < 0$. But since we have, moreover, $(a_{i_3} - a_{i_{n-1}})(a_{i_2} - a_{i_n}) > 0$, it follows that $a_{i_3} < a_{i_{n-1}}$, that is, $a_{i_3} = a_4$ and $a_{i_{n-1}} = a_5$.

It is similarly shown that a_{i_4} and $a_{i_{n-3}}$ follow in magnitude after $a_{i_4} < a_{i_{n-2}}$ ($i_4 = 6$, $i_{n-2} = 7$), that the numbers a_{i_5} and $a_{i_{n-3}}$ follow in magnitude after the previously determined $a_{i_5} < a_{i_{n-3}}$ ($i_5 = 8$, $i_{n-3} = 9$), and so on. Finally, we can set down the following scheme.

If $n = 2k$ (even), then

$$a_1 \Bigg/ {\begin{matrix} a_2 - a_4 - a_6 - \cdots - a_{n-2} \\ a_3 - a_5 - a_7 - \cdots - a_{n-1} \end{matrix}} \Bigg\backslash a_n \, ;$$

if $n = 2k + 1$ (odd), then

$$a_1 \Bigg/ {\begin{matrix} a_2 - a_4 - a_6 - \cdots - a_{n-1} \\ a_3 - a_5 - a_7 - \cdots - a_n \end{matrix}} \Bigg| \, .$$

(The schema represents the order of the numbers; for example, for even n we have the order $a_1, a_2, a_4, a_6, \cdots, a_{n-2}, a_n, a_{n-1}, \cdots, a_7, a_5, a_3$).

Second Solution. If the order obtained in the first solution could, in some manner, have been guessed at, the verification could be made by mathematical induction. In fact, for $n = 4$ the proof is quite simple [see the solution of problem (a)]. Assume now that for some even n the sum $\emptyset_n$ in the arrangement given by part (a) for the numbers $a_1 < a_2 < \cdots < a_n$ is less than the sum $\emptyset_n'$ corresponding to any other order. We shall prove that the sum $\emptyset_{n+1}$ corresponding to the scheme shown in part (*a*), as it concerns the $n + 1$ numbers $a_1 < a_2 < \cdots < a_n < a_{n+1}$, is less than the sum $\emptyset_{n+1}'$ corresponding to any other ordering of the $n + 1$ numbers. We have

$$\begin{aligned} \emptyset_{n+1} - \emptyset_n &= (a_n - a_{n+1})^2 + (a_{n+1} - a_{n-1})^2 - (a_n - a_{n-1})^2 \\ &= 2a_{n+1}^2 - 2a_n a_{n+1} - 2a_{n-1}a_{n+1} + 2a_{n-1}a_n \\ &= 2(a_{n+1} - a_n)(a_{n+1} - a_{n-1}) \, . \end{aligned}$$

On the other hand, if in the order answering to a sum $\mathcal{O}'_{n+1}$ the number a_{n+1} were to stand between two numbers a_α and a_β, and if $\mathcal{O}'_n$ answers the array of n numbers obtained from the array of $n + 1$ numbers leading to a sum $\mathcal{O}'_{n+1}$ by striking out the number a_{n+1}, then

$$\mathcal{O}'_{n+1} - \mathcal{O}'_n = (a_\alpha - a_{n+1})^2 + (a_{n+1} - a_\beta)^2 - (a_\alpha - a_\beta)^2$$
$$= 2a^2_{n+1} - 2a_\alpha a_{n+1} - 2a_\beta a_{n+1} + 2a_\alpha a_\beta$$
$$= 2(a_{n+1} - a_\alpha)(a_{n+1} - a_\beta) \geqq \mathcal{O}_{n+1} - \mathcal{O}_n \, .$$

Thus,

$$\mathcal{O}_{n+1} - \mathcal{O}'_{n+1} = [\mathcal{O}_n - \mathcal{O}'_n] + [(\mathcal{O}_{n+1} - \mathcal{O}_n) - (\mathcal{O}'_{n+1} - \mathcal{O}'_n)] \leqq 0$$

(the first pair of brackets parentheses encloses a nonpositive number by the induction hypotheses; the second pair does so by the proof). Here, if the sum $\mathcal{O}'_{n-1}$ differs from $\mathcal{O}_{n+1}$, then either $\mathcal{O}_n - \mathcal{O}'_n < 0$ (and, consequently, $\mathcal{O}_{n+1} - \mathcal{O}'_{n+1} < 0$), or $(\mathcal{O}_{n+1} - \mathcal{O}_n) - (\mathcal{O}'_{n+1} - \mathcal{O}'_n) < 0$ (and, consequently, $\mathcal{O}_{n+1} < \mathcal{O}'_{n+1}$). The transition from n to $n + 1$ is carried out in a similar manner for odd n.

Third Solution. This problem has a less involved geometrical solution. We represent the numbers $a_1 < a_2 < a_3 < \cdots < a_n$ by points $A_1, A_2, A_3, \cdots, A_n$ on a number axis; we designate the intervals $\overline{A_1A_2}, \overline{A_2A_3}, \cdots, \overline{A_{n-1}A_n}$, respectively, by $d_1, d_2, \cdots, d_{n-1}$. Then the sum

$$\mathcal{O} = (a_{i_1} - a_{i_2})^2 + (a_{i_2} - a_{i_3})^2 + \cdots + (a_{i_{n-1}} - a_{i_n})^2 + (a_{i_n} - a_{i_1})^2$$
$$= \overline{A_{i_1}A^2_{i_2}} + \overline{A_{i_2}A^2_{i_3}} + \cdots + \overline{A_{i_n}A^2_{i_1}}$$

is equal to the sum of the squares of interval lengths, or "links" $A_{i_1}A_{i_2}A_{i_3} \cdots A_{i_{n-1}}A_{i_n}A_{i_1}$ [all of which lie on the one straight line; see Figure 14 (a)].

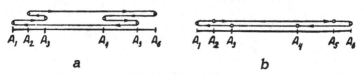

a b

Figure 14

Since the closed overlying curve covers the whole segment $\overline{A_1A_n}$, each of the segments $\overline{A_kA_{k+1}} = d_k$ enters at least twice into its composition (once in the direction from A_k to A_{k+1} and again in the reverse direction). Therefore, regardless of the order in which we take the points, the expanded sum $\varnothing$ when expressed in terms of the lengths $d_1, d_2, \cdots, d_{n-1}$ must contain all the numbers $2d_k^2$; that is, $2d_1^2, 2d_2^2, \cdots, 2d_{n-1}^2$. Further, let $\overline{A_{k-1}A_k} = d_{k-1}$ and $\overline{A_kA_{k+1}} = d_k$ be two adjacent segments. It is clear that if the link of the overlying curve which covers the segment $\overline{A_kA_{k+1}}$ going from A_k to A_{k+1} commences at the point A_k, then the link covering this segment in the reverse direction cannot terminate at the point A_k. Therefore, in all cases there must exist a link which simultaneously covers the segments $\overline{A_{k-1}A_k}$ and $\overline{A_kA_{k+1}}$. It follows that the sum $\varnothing$ must in all cases contain all the numbers $2d_{k-1}d_k$, that is, $2d_1d_2, 2d_2d_3, \cdots, 2d_{n-2}d_{n-1}$.

Now we need only note that if the points are ordered as in the first solution of this problem, then

$$\varnothing = 2d_1^2 + 2d_2^2 + \cdots + 2d_{n-1}^2 + 2d_1d_2 + 2d_2d_3 + \cdots + 2d_{n-2}d_{n-1}$$

[see Figure 14 (b)]. It follows from this, and from what has been said above, that the sum $\varnothing$ will be least for this ordering.

186. (a) First, we may assume that the numbers $a_1, a_2, \cdots, a_n$ and $b_1, b_2, \cdots, b_n$ are all positive, since if some of them are negative it is clear that the inequality will be exaggerated. Consider the broken line $A_0A_1A_2 \cdots A_n$ in Figure 15, where the lengths of the

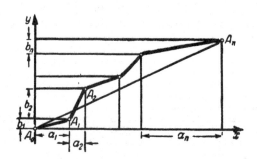

Figure 15

projections of the segments $\overline{A_0A_1}, \overline{A_1A_2} \cdots, \overline{A_{n-1}A_n}$ onto the x-axis are denoted by $a_1, a_2, \cdots, a_n$, and onto the y-axis by $b_1, b_2, \cdots, b_n$. Then, by the Pythagorean theorem, we see that

$$A_0A_1 = \sqrt{a_1^2 + b_1^2},$$

$$A_1A_2 = \sqrt{a_2^2 + b_2^2},$$

$$\cdots\cdots\cdots\cdots\cdots,$$

$$A_{n-1}A_n = \sqrt{a_n^2 + b_n^2},$$

$$A_0A_n = \sqrt{(a_1 + a_2 + \cdots + a_n)^2 + (b_0 + b_1 + \cdots + b_n)^2},$$

from which the inequality given in the problem follows immediately.

The broken line $A_0A_1A_2 \cdots A_n$ can be equal in length to the segment $\overline{A_0A_n}$ only if it is a straight-line segment. This can occur only if $\frac{a_1}{b_1} = \frac{a_2}{b_2} = \cdots = \frac{a_n}{b_n}$, and then we have the equality.

(b) Let h be the height of the pyramid; let $a_1, a_2, \cdots, a_n$ be the lengths of the sides of the base ($a_1 + a_2 + \cdots + a_n = P$, the base perimeter); and let $b_1, b_2, \cdots, b_n$ be the lengths of the perpendiculars from the foot of the altitude (the center of the inscribed circle, for a right pyramid) to each of the base sides, respectively, $\left(\text{then } \frac{1}{2}a_1b_1 + \frac{1}{2}a_2b_2 + \cdots + \frac{1}{2}a_nb_n = S, \text{ the base area}\right)$. Now, the lateral surface area Σ of the pyramid is equal to

$$\frac{1}{2}a_1\sqrt{b_1^2 + h^2} + \frac{1}{2}a_2\sqrt{b_2^2 + h^2} + \cdots + \frac{1}{2}a_n\sqrt{b_n^2 + h^2}.$$

However, by part (a),

$$2\Sigma = \sqrt{(a_1b_1)^2 + (a_1h)^2} + \sqrt{(a_2b_2)^2 + (a_2h)^2} + \cdots + \sqrt{(a_nb_n)^2 + (a_nh)^2}$$
$$\geq \sqrt{(a_1b_1 + a_2b_2 + \cdots + a_na_n)^2 + (a_1h + a_2h + \cdots + a_nh)^2}$$
$$= \sqrt{4S^2 + h^2P^2},$$

The equality here holds only if $a_1b_1 : a_2b_2 : \cdots : a_nb_n = a_1h : a_2h : \cdots : a_nh$—that is, $b_1 = b_2 = \cdots = b_n$.

The statement of the problem follows immediately.

187. We shall investigate separately the cases for which n is even and n is odd.

The integer n is even. Construct the broken line (Figure 16) connecting points $A_1, A_2, \cdots, A_n, A_{n+1}, A_{n+2}$, forming a "step graph," such that the segments $\overline{A_1A_2}, \overline{A_2A_3}, \cdots, \overline{A_{n+1}A_{n+2}}$ are of unit length, and they are successively perpendicular (as shown in Figure 16, where $n = 4$). On each segment $\overline{A_iA_{i+1}}$ ($i = 1, 2, \cdots, n + 1$), or on its extension, we place a point B_i such that the length of the segment $\overline{B_iA_{i+1}}$ is equal to a_i (we shall assume that a_{n+1} is equal to a_1;

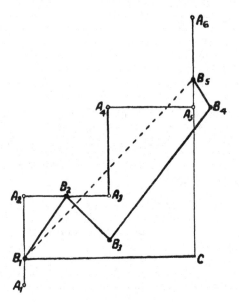

Figure 16

that is, B_{n+1} is selected such that $\overline{B_{n+1}A_{n+2}} = a_1$). In doing this, we place B_i to the left of (or below) the point A_{i+1} if $a_i > 0$, and to the right of (or above) the point A_{i+1} if $a_i < 0$ (in Figure 16, $0 < a_1 < 1$, $0 < a_2 < 1$, $a_3 > 1$, $a_4 < 0$). We now connect the points B_i to form the broken line $B_1B_2 \cdots B_{n+1}$. By the Pythagorean theorem,

$$\overline{B_iB_{i+1}} = \sqrt{B_iA_{i+1}^2 + B_{i+1}A_{i+1}^2} .$$

Now, $\overline{B_iA_{i+1}} = a_i$, and $\overline{B_{i+1}A_{i+1}} = |1 - a_{i+1}|$; consequently,

$$\overline{B_iB_{i+1}} = \sqrt{a_i^2 + (1 - a_{i+1})^2} .$$

Hence the sum in which we are interested,

$$\sqrt{a_1^2 + (1 - a_2)^2} + \sqrt{a_2^2 + (1 - a_3)^2}$$
$$+ \cdots + \sqrt{a_{n-1}^2 + (1 - a_n)^2} + \sqrt{a_n^2 + (1 - a_1)^2}$$

is equal to the length of the broken line $B_1B_2 \cdots B_{n+1}$.

It is obvious that the length of the broken line $B_1B_2 \cdots B_{n+1}$ is not less than the length of the segment $\overline{B_1B_{n+1}}$. We shall now find the length of this segment. We construct the right triangle B_1CB_{n+1} (Figure 16). Then

$$\overline{B_1C} = \overline{A_2A_3} + \overline{A_4A_5} + \cdots + \overline{A_nA_{n+1}} = \frac{n}{2},$$

and

$$\overline{CB_{n+1}} = \overline{A_1A_2} + \overline{A_3A_4} + \cdots + \overline{A_{n-1}A_n} = \frac{n}{2}$$

(for $\overline{A_1B_1} = \overline{A_{n+1}B_{n+1}} = |1 - a_1|$). It follows that

$$\overline{B_1B_{n+1}} = \sqrt{(\overline{B_1C})^2 + (\overline{CB_{n+1}})^2} = \sqrt{\left(\frac{n}{2}\right)^2 + \left(\frac{n}{2}\right)^2} = \frac{n\sqrt{2}}{2}.$$

This proves the inequality sought.

It is not difficult now to determine when the equality holds. In order to arrive at the equality, all the points $B_2, B_3, \cdots, B_n$ must lie on the line segment $\overline{B_1B_{n+1}}$ (that is, B_i must coincide with the points of intersection of the line segments $\overline{B_1B_{n+1}}$ and $\overline{A_iA_{i+1}}$. Because the segment $\overline{B_1B_{n+1}}$ forms a 45° angle with $\overline{B_1C}$ ($\overline{B_1C} = \overline{CB_{n+1}}$), it follows that

$$\overline{B_1A_2} = \overline{A_2B_2} = \overline{B_3A_4} = \overline{A_4B_4} = \cdots = \overline{B_{n-1}A_n} = \overline{A_nB_n},$$

that is, $a_1 = (1 - a_2) = a_3 = (1 - a_4) = \cdots = a_{n-1} = (1 - a_n)$. Thus for n even, the equality holds for

$$a_1 = a_3 = \cdots = a_{n-1} = a,$$
$$a_2 = a_4 = \cdots = a_n = 1 - a,$$

where a is any arbitrary number.

The integer n is odd.[†] Let $a_{n+1} = a_1, a_{n+2} = a_2, \cdots, a_{2n} = a_n$, and consider the sum

$$\sqrt{a_1^2 + (1 - a_2)^2} + \sqrt{a_2^2 + (1 - a_3)^2}$$
$$+ \cdots + \sqrt{a_{2n-1}^2 + (1 - a_{2n})^2} + \sqrt{a_{2n}^2 + (1 - a_1)^2},$$

which is equal to twice the sum

$$\sqrt{a_1^2 + (1 - a_2)^2} + \sqrt{a_2^2 + (1 - a_3)^2}$$
$$+ \cdots + \sqrt{a_{n-1}^{2} + (1 - a_n)^2} + \sqrt{a_n^2 + (1 - a_1)^2}$$

(each term of the last sum is met with twice in the preceding sum-

---

[†] Elucidation for $n = 3$ is left to the reader; the proof for even n does not apply for odd n.

mation). However, it has already been shown that the first sum is less than or equal to $\dfrac{2n \sqrt{2}}{2}$; it follows that

$$\sqrt{a_2^2 + (1 - a_2)^2} + \sqrt{a_3^2 + (1 - a_3)^2}$$
$$+ \cdots + \sqrt{a_{n-1}^2 + (1 - a_n)^2} + \sqrt{a_n^2 + (1 - a_1)^2} \geq \dfrac{n \sqrt{2}}{2};$$

that is, we obtain the required inequality.

The equality sign holds only if

$$a_1 = a_2 = \cdots = a_n = \frac{1}{2}.$$

188. *First Solution.* Both numbers of the inequality are positive; hence upon squaring both sides we have

$$1 - x_1^2 + 1 - x_2^2 + 2 \sqrt{(1 - x_1^2)(1 - x_2^2)} \leq 4 - (x_1^2 + 2x_1x_2 + x_2^2),$$

that is,

$$2 \sqrt{(1 - x_1^2)(1 - x_2^2)} \leq 2 - 2x_1x_2,$$
$$\sqrt{(1 - x_1^2)(1 - x_2^2)} \leq 1 - x_1x_2.$$

If again we square both sides, we have

$$1 - x_1^2 - x_2^2 + x_1^2x_2^2 \leq 1 - 2x_1x_2 + x_1^2x_2^2,$$

and if all terms are transposed to the right side, we have

$$0 \leq (x_1 - x_2)^2.$$

It is clear that the right member of the given inequality dominates the left member; also, the equality can hold only for $x_1 = x_2$.

Second Solution. This problem can also be solved by geometric means; analogous solutions are possible for many more involved problems of this sort. Consider in the Cartesian coordinate system (plane) the unit circle with center at the origin (Figure 17). The coordinates x, y of points on the circle are related by the equation

$$x^2 + y^2 = 1. \tag{1}$$

Select two points M_1 and M_2 on the x-axis having abscissas x_1 and x_2 and where $|x_1| \leq 1$ and $|x_2| \leq 1$ (thus both points are within or else on the circle). Construct perpendiculars from M_1 and M_2 intersecting the upper semicircle in the points N_1 and N_2, respectively. It is clear from equation (1) that $M_1N_1 = \sqrt{1 - x_1^2}$ and $M_2N_2 = \sqrt{1 - x_2^2}$.

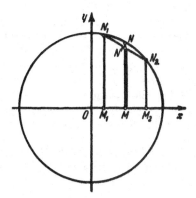

Figure 17

Note now that the abscissa $\dfrac{x_1 + x_2}{2}$ will be the midpoint of the segment $\overline{M_1 M_2}$.[†] We shall designate this point on the x-axis by M, and the point above it on the circle by N.

Clearly, the length of $\overline{MN}$ is $\sqrt{1 - \left(\dfrac{x_1 + x_2}{2}\right)^2}$. Now, the sum $\overline{M_1 N_1} + \overline{M_2 N_2}$ is equal to twice the length of the segment MN', where N' is the point of the trapezoid $M_1 N_1 N_2 M_2$ just above M, and $\overline{MN'}$ is obviously shorter than $\overline{MN}$. The inequality of the problem follows immediately. The equality holds only if the points M_1 and M_2 coincide, that is, if $x_1 = x_2$.

Remark: Many interesting inequalities are suggested by this last proof. For example, consider the unit sphere with center at the origin (Figure 18). Let M_1 and M_2 be any two points in the XY-plane within (or on) the sphere, and let N_1 and N_2 be the points of intersection, with the sphere, of perpendiculars to the plane rising from M_1 and M_2, respectively. Let M be the midpoint of segment $\overline{M_1 M_2}$, and construct the perpendicular $\overline{MN}$ in the plane containing $M_1, M_2, N_1,$ and N_2, where N is the intersection with the sphere. Obviously, MN will intersect the segment from N_1 to N_2, and this intersection point we label N'. If (x_1, y_1) and (x_2, y_2) are the coordinates of points M_1 and M_2, then

---

[†] This is obvious for positive x_1 and x_2; it is easily verified that this will hold even if one of the numbers is (or both are) negative. Also, it suffices to consider only positive x_1 and x_2, since in any other event the inequality is emphasized.

$$M_1N_1 = \sqrt{1 - x_1^2 - y_1^2} \, ;$$

$$M_2N_2 = \sqrt{1 - x_2^2 - y_2^2} \, ;$$

$$MN = \sqrt{1 - \left(\frac{x_1 + x_2}{2}\right)^2 - \left(\frac{y_1 + y_2}{2}\right)^2} \, ;$$

$$MN' = \frac{1}{2}(M_1N_1 + M_2N_2) \, .$$

And since $MN' \leqq MN$, it follows that

$$\sqrt{1 - x_1^2 - y_1^2} + \sqrt{1 - x_2^2 - y_2^2} \leqq 2\sqrt{1 - \left(\frac{x_1 + x_2}{2}\right)^2 - \left(\frac{y_1 + y_2}{2}\right)^2}, \quad (2)$$

provided, however, that all the expressions under the radical are nonnegative. Equality here will hold only for $x_1 = x_2$, $y_1 = y_2$, that is, only when points M_1 and M_2 are coincident.

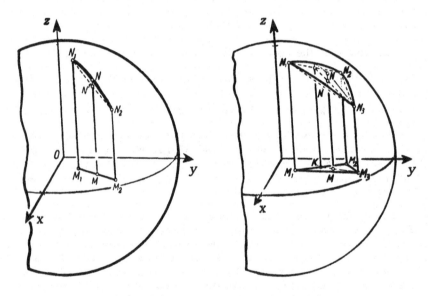

Figure 18   **Figure 19**

A somewhat similar inequality can be produced by using a triangle $M_1M_2M_3$ in the XY-plane and letting M be the point of intersection of the medians of the triangle (Figure 19). We obtain the inequality

$$\sqrt{1 - x_1^2 - y_1^2} + \sqrt{1 - x_2^2 - y_2^2} + \sqrt{1 - x_3^2 - y_3^2}$$

$$\leqq 3\sqrt{1 - \left(\frac{x_1 + x_2 + x_3}{3}\right)^2 - \left(\frac{y_1 + y_2 + y_3}{3}\right)^2}. \quad (3)$$

This results from the fact that segment MN' of the perpendicular from M does not exceed segment MN. Inequality (*3*) is valid only when the expressions under the radicals are nonnegative; the equality holds only for $x_1 = x_2 = x_3$ and $y_1 = y_2 = y_3$; that is, when the points M_1, M_2, M_3 all coincide.

Another inequality arises from the use of a right circular cone having its vertex at the origin and the x-axis as its central axis, and having a vertex angle of 90° (Figure 20):

$$\sqrt{x_1^2 + y_1^2} + \sqrt{x_2^2 + y_2^2} + \sqrt{x_3^2 + y_3^2}$$
$$\geqq 3 \sqrt{\left(\frac{x_1 + x_2 + x_3}{3}\right)^2 + \left(\frac{y_1 + y_2 + y_3}{3}\right)^2}. \tag{4}$$

This is valid for all x_1, x_2, x_3 and y_1, y_2, y_3. Equality holds only when $\frac{x_1}{y_1} = \frac{x_2}{y_2} = \frac{x_3}{y_3}$, that is, when N_1, N_2, and N_3 lie simultaneously on a generator of the cone. Algebraic proofs for inequalities (*2*), (*3*), and (*4*) are very involved.

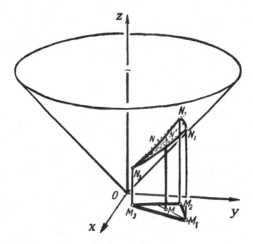

Figure 20

189. Using the trigonometric identity

$$\cos(A + B) = \cos A \cos B - \sin A \sin B,$$

we have

$$\cos\left(\frac{\pi}{2} + \cos x\right) = \cos\frac{\pi}{2}\cos x - \sin\frac{\pi}{2}\sin\cos x$$

$$= -\sin\frac{\pi}{2}\sin\cos x = -\sin\cos x,$$

or,

$$\sin \cos x = -\cos\left(\frac{\pi}{2} + \cos x\right),$$

from which we obtain

$$\cos \sin x - \sin \cos x = \cos \sin x + \cos\left(\frac{\pi}{2} + \cos x\right) \qquad (1)$$

We employ the formula

$$\cos A + \cos B = 2\cos\frac{A+B}{2} \cdot \cos\frac{A-B}{2}$$

in the right member of *(1)* and use the fact that $\cos \alpha = \cos(-\alpha)$ to find

$$\cos \sin x - \sin \cos x$$

$$= 2\cos\frac{\sin x + \frac{\pi}{2} + \cos x}{2} \cdot \cos\frac{-\sin x + \frac{\pi}{2} + \cos x}{2}.$$

Now,

$$|\cos x + \sin x| = \sqrt{\cos^2 x + 2\cos x \sin x + \sin^2 x}$$
$$= \sqrt{1 + \sin 2x} \le \sqrt{2}.$$

(We note that $|\cos x + \sin x| = \sqrt{2}$ only if $\sin 2x = 1$.) In a similar manner we find that

$$|\cos x - \sin x| = \sqrt{\cos^2 x - 2\cos x \sin x + \sin^2 x}$$
$$= \sqrt{1 - \sin 2x} \le \sqrt{2}$$

(and so $|\cos x - \sin x| = \sqrt{2}$ only if $\sin 2x = -1$). Since $\frac{\pi}{2} \approx \frac{3 \cdot 14}{2} = 1 \cdot 57$ and $\sqrt{2} \approx 1 \cdot 41$, we have

$$\frac{\pi}{2} > \frac{\frac{\pi}{2} + \cos x + \sin x}{2} > 0$$

and

$$\frac{\pi}{2} > \frac{\frac{\pi}{2} + \cos x - \sin x}{2} > 0,$$

which means that

$$\cos\frac{\frac{\pi}{2} + \cos x + \sin x}{2}$$

and

$$\cos \frac{\frac{\pi}{2} + \cos x - \sin x}{2}$$

are always positive. Hence the difference $\cos \sin x - \sin \cos x$ is always positive; that is, $\cos \sin x$ exceeds $\sin \cos x$ for all values of x.

190. (a) Write $\log_2 \pi = a$ and $\log_5 \pi = b$. From the equalities $2^a = \pi$ and $5^b = \pi$ we obtain

$$\pi^{1/a} = 2 \,,$$
$$\pi^{1/b} = 5 \,,$$
$$\pi^{1/a} \cdot \pi^{1/b} = 2 \cdot 5 = 10 \,,$$
$$\pi^{1/a + 1/b} = 10 \,.$$

However, $\pi^2 \approx (3.14)^2 < 10$, and therefore we must conclude that $\frac{1}{a} + \frac{1}{b} > 2$, which is what we set out to prove.

(b) Write $\log_2 \pi = a$ and $\log_{\pi^2} = b$. Then we have $2^a = \pi$ and $\pi^b = 2$. Since now $2^{1/b} = \pi$, it follows that $2^a = 2^{1/b}$, or $b = \frac{1}{a}$. The left member of the given inequality now is of form

$$\frac{1}{a} + \frac{1}{1/a} = \frac{1}{a} + a = \frac{a^2 + 1}{a} \,.$$

We are required to show that $\frac{a^2 + 1}{a} > 2$, or, equivalently, that $a^2 + 1 > 2a$ (note that $a > 0$). But $a^2 - 2a + 1 = (a - 1)^2 > 0$, which proves the validity of the given inequality.

191. *First Solution.* We must show that if $\beta > \alpha$, then

(a) $$\sin \beta - \sin \alpha < \beta - \alpha \,.$$

However,

$$\sin \beta - \sin \alpha = 2 \sin \frac{\beta - \alpha}{2} \cos \frac{\beta + \alpha}{2} < 2 \frac{\beta - \alpha}{2} \cdot 1 = \beta - \alpha$$

(for acute angles, $\sin x < x$ and $\cos x < 1$).

(b) We have

$$\tan \beta - \tan \alpha > \beta - \alpha \,.$$

Clearly,

$$\beta - \alpha < \tan (\beta - \alpha) = \frac{\tan \beta - \tan \alpha}{1 + \tan \beta \tan \alpha} < \tan \beta - \tan \alpha$$

(for acute angles, tan $x > x$).

Second Solution. We consider only part (a) since part (b) is analogous.

In Figure 21, let the radius of the circle be unity; then the chord $\overline{AE}$ is equal to the radian measure of α, and $\overline{AF} = \beta$. If $\overline{EM}$ and $\overline{FP}$ are perpendiculars from E and F to OA, then, designating by $S(OEA)$ the area of the triangle OEA, and so on, we have

$$S(OEA) = \frac{1}{2}\sin\alpha,$$

$$S(OFA) = \frac{1}{2}\sin\beta,$$

and if $S_c(OEA)$ is the area of the sector OEA, and so on, then

$$S_c(OEA) = \frac{1}{2}\alpha,$$

$$S_c(OFA) = \frac{1}{2}\beta.$$

We readily read from the figure that $\alpha - \sin\alpha < \beta - \sin\beta$.

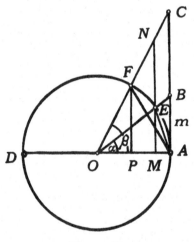

Figure 21

192. Reference is made to Figure 21, and the same terminology is used as in the second solution of problem 191. The perpendicular

AC is tangent to the circle at A. We read from the figure that

$$S(OAB) = \frac{1}{2} \tan \alpha \cdot , \qquad S(OAC) = \frac{1}{2} \tan \beta ,$$

$$S_c(OAE) = \frac{1}{2} \alpha , \qquad S_c(OAF) = \frac{1}{2} \beta .$$

Consequently,

$$\frac{\tan \alpha}{\alpha} = \frac{S(OAB)}{S_c(OAE)} ,$$

$$\frac{\tan \beta}{\beta} = \frac{S(OAC)}{S_c(OAF)} .$$

Also, it is readily seen that

$$\frac{S(OAB)}{S_c(OAE)} < \frac{S(OAB)}{S(OEM)} ,$$

$$\frac{S(OBC)}{S_c(OEF)} > \frac{S(OEC)}{S(OEN)} ,$$

and

$$\frac{S(OAB)}{S(OEM)} = \frac{S(OBC)}{S(OEN)} .$$

Thus

$$\frac{S(OAB)}{S_c(OAE)} < \frac{S(OBC)}{S_c(OEF)} .$$

From the fact that $\dfrac{S(OBC)}{S_c(OEF)} > \dfrac{S(OAB)}{S_c(OAE)}$, it follows that

$$\frac{S(OAB) + S(OBC)}{S_c(OAE) + S_c(OEF)} > \frac{S(OAB)}{S_c(OAE)} ,$$

that is, that

$$\frac{S(OAC)}{S_c(OAF)} > \frac{S(OAB)}{S_c(OAE)} ,$$

which is what we wished to prove.

193. Let arc sin cos arc sin $x = \alpha$. The angle α is bounded between 0 and $\frac{\pi}{2}$; that is, $0 \leq \alpha \leq \frac{\pi}{2}$, inasmuch as $0 \leq \cos \arcsin x \leq 1$ $\left(-\frac{\pi}{2} \leq \arcsin x \leq \frac{\pi}{2} \right)$. Further,

$$\sin \alpha = \cos \text{arc} \sin x \, .$$

Consequently,

$$\text{arc} \sin x = \pm \left(\frac{\pi}{2} - \alpha \right) \, ,$$

and

$$x = \sin \left[\pm \left(\frac{\pi}{2} - \alpha \right) \right] = \pm \cos \alpha \, .$$

Similarly, if $\text{arc} \cos \sin \text{arc} \cos x = \beta$, then $0 \leqq \beta \leqq \frac{\pi}{2}$ (for $0 \leqq \sin \text{arc} \cos x \leqq 1$, since $0 \leqq \text{arc} \cos x \leqq \pi$), and

$$\cos \beta = \sin \text{arc} \cos x \, .$$

Consequently,

$$\text{arc} \cos x = \frac{\pi}{2} \mp \beta \, ,$$

and

$$x = \cos \left(\frac{\pi}{2} \mp \beta \right) = \pm \sin \beta \, .$$

Since $\cos \alpha = \sin \beta (= \pm x)$, conclude that

$$\alpha + \beta = \text{arc} \sin \cos \text{arc} \sin x + \text{arc} \cos \sin \text{arc} \cos x = \frac{\pi}{2} \, .$$

194. Assume that the series

$$\cos 32x + a_{31} \cos 31x + a_{30} \cos 30x + a_{29} \cos 29x$$
$$+ \cdots + a_2 \cos 2x + a_1 \cos x \qquad (1)$$

is always positive for all values of x. Substituting $x + \pi$ for x in this series, we obtain

$$\cos 32(x + \pi) + a_{31} \cos 31(x + \pi) + a_{30} \cos 30(x + \pi)$$
$$+ a_{29} \cos 29(x + \pi) + \cdots + a_2 \cos 2(x + \pi) + a_1 \cos (x + \pi)$$
$$= \cos 32x - a_{31} \cos 31x + a_{30} \cos 30x - a_{29} \cos 29x$$
$$+ \cdots + a_2 \cos 2x - a_1 \cos x \, , \qquad (2)$$

which must also be positive for all x. Now if the two series (1) and (2) are added, we obtain

$$\cos 32x + a_{30} \cos 30x + \cdots + a_4 \cos 4x + a_2 \cos 2x \, , \qquad (3)$$

which also can take on only positive value for any x.

In (3) we now substitute $x + \frac{\pi}{2}$ for x to obtain

$$\cos 32\left(x + \frac{\pi}{2}\right) + a_{30}\cos 30\left(x + \frac{\pi}{2}\right) + a_{28}\cos 28\left(x + \frac{\pi}{2}\right)$$

$$+ \cdots + a_4\cos 4\left(x + \frac{\pi}{2}\right) + a_2\cos 2\left(x + \frac{\pi}{2}\right)$$

$$= \cos 32x - a_{30}\cos 30x + a_{28}\cos 28x - \cdots + a_4\cos 4x - a_2\cos 2x .$$

The sum of (3) and the final series yields the new series

$$\cos 32 + a_{28}\cos 28x + a_{24}\cos 24x + \cdots + a_8\cos 8x + a_4\cos 4x ,$$

which will have to be positive for all x.

Replacing x by $x + \frac{\pi}{4}$ in the last sum obtained, and adding the resulting series to the previous one, we obtain

$$\cos 32x + a_{24}\cos 24x + a_{16}\cos 16x + a_8\cos 8x .$$

Replacing in this series, x by $x + \frac{\pi}{8}$, and adding the resulting expression to this one, we obtain

$$\cos 32x + a_{16}\cos 16x .$$

Finally, in another step, we find that $\cos 32x$ can take on only positive values for all x. But this is a contradiction, since if $x = \frac{\pi}{32}$, then $\cos 32x = \cos \pi = -1$. This contradiction proves the assertion of the problem.

195. The well-known half-angle formula of trigonometry can be written as

$$2\sin\frac{\alpha}{2} = \pm\sqrt{2 - 2\cos\alpha} ,$$

where the plus-or-minus sign is determined by the quadrant in which $\frac{\alpha}{2}$ lies. We shall use this formula to find the sines of the angles

$$a_1 \cdot 45° ,$$

$$\left(a_1 + \frac{a_1 a_2}{2}\right) \cdot 45° ,$$

$$\left(a_1 + \frac{a_1 a_2}{2} + \frac{a_1 a_2 a_3}{4}\right) \cdot 45° ,$$

$$\cdots\cdots\cdots\cdots\cdots\cdots\cdots\cdots\cdots\cdots ,$$

$$\left(a_1 + \frac{a_1 a_2}{2} + \frac{a_1 a_2 a_3}{4} + \cdots + \frac{a_1 a_2 a_3 \cdots a_n}{2^{n-1}}\right) \cdot 45° .$$

Assume that we have already found the sine of the angle

$$\left(a_1 + \frac{a_1 a_2}{2} + \frac{a_1 a_2 a_3}{4} + \cdots + \frac{a_1 a_2 \cdots a_k}{2^{k-1}}\right) \cdot 45°,$$

where $a_1, a_2, a_3, \cdots, a_k$ have the individual values $+1$ or -1. Since

$$2\left(a_1 + \frac{a_1 a_2}{2} + \frac{a_1 a_2 a_3}{4} + \cdots + \frac{a_1 a_2 \cdots a_k}{2^{k-1}} + \frac{a_1 a_2 \cdots a_k a_{k+1}}{2^k}\right) \cdot 45°$$

$$= \left[\pm 90° \pm \left(a_2 + \frac{a_2 a_3}{2} + \cdots + \frac{a_2 a_3 \cdots a_k a_{k+1}}{2^{k-1}}\right) \cdot 45°\right],$$

(where the plus sign refers to $a_1 = +1$ and the minus sign refers to $a_1 = -1$), and since

$$\cos\left[\pm 90° \pm \left(a_2 + \frac{a_2 a_3}{2} + \cdots + \frac{a_2 a_3 \cdots a_{k+1}}{2^{k-1}}\right) \cdot 45°\right]$$

$$= -\sin\left(a_2 + \frac{a_2 a_3}{2} + \cdots + \frac{a_2 a_3 \cdots a_{k+1}}{2^{k-1}}\right) \cdot 45°,$$

we may determine the following:

$$2\sin\left(a_1 + \frac{a_1 a_2}{2} + \cdots + \frac{a_1 a_2 \cdots a_k}{2^{k-1}} + \frac{a_1 a_2 \cdots a_k a_{k+1}}{2^k}\right) \cdot 45°$$

$$= \pm\sqrt{2 + 2\sin\left(a_2 + \frac{a_2 a_3}{2} + \cdots + \frac{a_2 a_3 \cdots a_{k+1}}{2^{k-1}}\right) \cdot 45°} \,.$$

Keeping in mind that all angles are (positively or negatively) acute, we see that even

$$\left(1 + \frac{1}{2} + \frac{1}{4} + \cdots + \frac{1}{2^{n-1}}\right) \cdot 45° = 90° - \frac{1}{2^{n-1}} 90°$$

is less than $90°$, that the sign of these angles is determined by the sign of a_1, and that the square root in the final formula must be taken with a plus or a minus sign in accordance with the sign of a_1. In brief, we can write

$$2\sin\left(a_1 + \frac{a_1 a_2}{2} + \cdots + \frac{a_1 a_2 \cdots a_k}{2^{k-1}} + \frac{a_1 a_2 \cdots a_k a_{k+1}}{2^k}\right) \cdot 45°$$

$$= a_1\sqrt{2 + 2\sin\left(a_2 + \frac{a_2 a_3}{2} + \cdots + \frac{a_2 a_3 \cdots a_k}{2^{k-1}}\right) \cdot 45°} \,.$$

Now, it is clear that

$$2\sin a_1 45° = a_1\sqrt{2} \,.$$

From this we obtain

$$2 \sin \left(a_1 + \frac{a_1 a_2}{2} \right) \cdot 45° = a_1 \sqrt{2 + a_2 \sqrt{2}} \, ,$$

$$2 \sin \left(a_1 + \frac{a_1 a_2}{2} + \frac{a_1 a_2 a_3}{4} \right) \cdot 45° = a_1 \sqrt{2 + a_2 \sqrt{2 + a_3 \sqrt{2}}} \, ,$$

$$2 \sin \left(a_1 + \frac{a_1 a_2}{2} + \frac{a_1 a_2 a_3}{4} + \frac{a_1 a_2 a_3 a_4}{8} \right) \cdot 45°$$
$$= a_1 \sqrt{2 + a_2 \sqrt{2 + a_3 \sqrt{2 + a_4 \sqrt{2}}}} \, ,$$

$$\cdots\cdots\cdots\cdots\cdots\cdots\cdots\cdots\cdots\cdots\cdots\cdots\cdots\cdots\cdots\cdots\cdots , $$

$$2 \sin \left(a_1 + \frac{a_1 a_2}{2} + \frac{a_1 a_2 a_3}{4} + \cdots + \frac{a_1 a_2 a_3 \cdots a_n}{2^{n-1}} \right) \cdot 45°$$
$$= a_1 \sqrt{2 + a_2 \sqrt{2 + a_3 \sqrt{2 + \cdots + a_n \sqrt{2}}}} \, ,$$

which is what we wished to show.

196. The expansion of the given expression will take on the form

$$(1 - 3x + 3x^2)^{743}(1 + 3x - 3x^2)^{744}$$
$$= A_0 + A_1 x + A_2 x^2 + \cdots + A_n x^n \, , \qquad (1)$$

where $A_0, A_1, A_2, \cdots, A_n$ are the coefficients whose sum we wish to find, and the degree n of this polynomial is $743 \cdot 2 + 744 \cdot 2 = 2974$. In equation (*1*) let $x = 1$; we then have

$$1^{743} \cdot 1^{744} = A_0 + A_1 + A_2 + \cdots + A_n \, .$$

The sum we seek is equal to 1.

197. Assume that we have expanded the two expressions, obtaining two polynomials in x. Now let us replace x by $-x$ in each polynomial and rewrite them: the coefficients of odd powers of x change sign, and the coefficients of even powers do not. In particular, the coefficient of x^{20} in each of the expansions remains unchanged. Hence, insofar as the coefficient of x^{20} is concerned, we may as well compare the coefficients of x^{20} in the two expressions $(1 + x^2 + x^3)^{1000}$ and $(1 - x^2 - x^3)^{1000}$, which are obtained, respectively, from the given expressions, by replacing x by $-x$.

Now it is easily shown that the first of these new polynomials has the larger coefficient for x^{20}. In fact, the expansion of $(1 - x^2 + x^3)^{1000}$ contributes only positive terms to the sum which makes up the coefficient of x^{20}; the expansion of $(1 - x^2 - x^3)^{1000}$ cannot produce as large a coefficient for x^{20}, since the sum making up that coefficient is comprised of like terms having coefficients of the same absolute values as those of the first expansion, but some of them are negative.

Therefore, the coefficient of x^{20} in $(1 + x^2 - x^3)^{1000}$ is greater than that in $(1 - x^2 + x^3)^{1000}$.

198. The proof of the problem follows from the following:

$$(1 - x + x^2 - x^3 + \cdots - x^{99} + x^{100})(1 + x + x^2 + x^3 + \cdots + x^{99} + x^{100})$$

$$= [(1 + x^2 + x^4 + \cdots + x^{100}) - x(1 + x^2 + x^4 + \cdots + x^{98})]$$

$$\times [(1 + x^2 + x^4 + \cdots + x^{100}) + x(1 + x^2 + x^4 + \cdots + x^{98})]$$

$$= (1 + x^2 + x^4 + \cdots + x^{100})^2 - x^2 (1 + x^2 + x^4 + \cdots + x^{98})^2 .$$

199. (a) Using the formulas for the sum of a geometric progression, and the binomial theorem, we obtain

$$(1 + x)^{1000} + x(1 + x)^{999} + x^2(1 + x)^{998} + \cdots + x^{1000}$$

$$= \frac{\dfrac{x^{1001}}{1 + x} - (1 + x)^{1000}}{\dfrac{x}{1 + x} - 1} = \frac{x^{1001} - (1 + x)^{1001}}{x - 1 - x} = (1 + x)^{1001} - x^{1001}$$

$$= 1 + 1001x + C_{1001}^2 x^2 + C_{1001}^3 x^3 + \cdots + 1001x^{1000} .$$

Therefore, the coefficient we seek is equal to

$$C_{1001}^{50} = \frac{1001!}{50!\,951!} .$$

(b) Designate the given series by $P(x)$. Then we can write

$$(1 + x)P(x) - P(x)$$

$$= [(1 + x)^2 + 2(1 + x)^3 + \cdots + 999(1 + x)^{1000} + 1000(1 + x)^{1001}]$$

$$- [(1 + x) + 2(1 + x^2) + 3(1 + x^3) + \cdots + 1000(1 + x)^{1000}]$$

$$= 1000(1 + x)^{1001} - [(1 + x) + (1 + x)^2 + (1 + x)^3 + \cdots + (1 + x)^{1000}]$$

$$= 1000(1 + x)^{1001} - \frac{(1 + x)^{1001} - (1 + x)}{1 + x - 1} = 1000(1 + x)^{1001}$$

$$- \frac{(1 + x)^{1001} - (1 + x)}{x} .$$

It follows that

$$P(x) = \frac{1000(1 + x)^{1001}}{x} - \frac{(1 + x)^{1001} - (1 + x)}{x^2}$$

$$= 1000[1001 + C_{1001}^2 x + C_{1001}^3 x^2 + \cdots + 1001x^{999} + x^{1000}]$$

$$- [C_{1001}^2 + C_{1001}^3 x + C_{1001}^4 x^2 + \cdots + 1001x^{998} + x^{999}] .$$

Therefore, the coefficient sought is equal to

$$1000C_{1001}^{51} - C_{1001}^{52} = \frac{1000 \cdot 1001!}{51! \cdot 950!} - \frac{1001!}{52! \cdot 950!}$$

$$= \frac{1001!}{52! \cdot 950!} \,[52 \cdot 1000 - 950] = \frac{51,050 \cdot 1001!}{52! \cdot 950!} \,.$$

200. We shall first determine the constant term obtained by expanding

$$\underbrace{(\cdots((x-2)^2 - 2)^2 - \cdots - 2)^2}_{k \text{ times}}$$

and collecting like terms. Clearly, this term is equal to what is obtained if we set $x = 0$; that is,

$$\underbrace{(\cdots(((-2)^2 - 2)^2 - \cdots - 2)^2}_{k \text{ times}} = \underbrace{(\cdots((4 - 2)^2 - 2)^2 - \cdots - 2)^2}_{k - 1 \text{ times}}$$

$$= \underbrace{(\cdots((4 - 2)^2 - 2)^2 - \cdots - 2)^2}_{k - 2 \text{ times}}$$

$$= \cdots + ((4 - 2)^2 - 2)^2 = (4 - 2)^2 = 4\,.$$

Designate by A_k the coefficient of x, by B_k the coefficient of x^2, and by $P_k x^3$ the sum of all the terms containing higher powers of x (this is x^3 times a polynomial in x). We then have

$$\underbrace{(\cdots((x-2)^2 - 2)^2 - \cdots - 2)^2}_{k \text{ times}} = P_k x^3 + B_k x^2 + A_k x + 4\,.$$

However,

$$\underbrace{(\cdots((x - 2)^2 - 2)^2 - 2)^2 - \cdots - 2)^2}_{k \text{ times}}$$

$$= [\underbrace{(\cdots((x - 2)^2 - 2)^2 - \cdots - 2)^2}_{k - 1 \text{ times}} - 2]^2$$

$$= [(P_{k-1} x^3 + B_{k-1} x^2 + A_{k-1} x + 4) - 2]^2$$

$$= (P_{k-1} x^3 + B_{k-1} x^2 + A_{k-1} x + 2)^2$$

$$= P_{k-1}^2 x^6 + 2 P_{k-1} B_{k-1} x^5 + (2 P_{k-1} A_{k-1} B_{k-1}^2) x^4$$

$$\quad + (4 P_{k-1} + 2 B_{k-1} A_{k-1}) x^3 + (4 B_{k-1} + A_{k-1}^2) x^2 + 4 A_{k-1} x + 4$$

$$= [P_{k-1}^2 x^3 + 2 P_{k-1} B_{k-1} x^2 + (2 P_{k-1} A_{k-1} + B_{k-1}^2) x$$

$$\quad + (4 P_{k-1} + 2 B_{k-1} A_{k-1})] x^3 + (4 B_{k-1} + A_{k-1}^2) x^2 + 4 A_{k-1} x + 4\,.$$

From this we obtain

$$A_k = 4A_{k-1} ,$$
$$B_k = A_{k-1}^2 + 4B_{k-1} .$$

Since $(x - 2)^2 = x^2 - 4x + 4$, we have $A_1 = -4$. Consequently, $A_2 = -4 \cdot 4 = -4^2$, $A_3 = -4^3$, $\cdots$, and, in general, $A_k = -4^k$. We shall now find B_k:

$$
\begin{aligned}
B_k &= A_{k-1}^2 + 4B_{k-1} = A_{k-1}^2 + 4(A_{k-2}^2 + 4B_{k-2}) \\
&= A_{k-1}^2 + 4A_{k-2}^2 + 4^2(A_{k-3}^2 + 4B_{k-3}) \\
&= A_{k-1}^2 + 4A_{k-2}^2 + 4^2(A_{k-3}^2 + 4^3(A_{k-4}^2 + 4B_{k-4})) \\
&= \cdots = A_{k-1}^2 + 4A_{k-2}^2 + 4^2 A_{k-3}^2 \\
&\quad + \cdots + 4^{k-3}A_2^2 + 4^{k-2}A_1^2 + 4^{k-1}B_1 .
\end{aligned}
$$

If now we substitute

$$
\begin{array}{ll}
B_1 = 1 , & A_3 = -4^3 , \\
A_1 = 4 , & \cdots\cdots\cdots , \\
A_2 = -4^2 , & A_{k-1} = -4^{k-1}
\end{array}
$$

we arrive at

$$
\begin{aligned}
B_k &= 4^{2k-2} + 4 \cdot 4^{2k-4} + 4^2 \cdot 4^{2k-6} + \cdots + 4^{k-2} \cdot 4^2 + 4^{k-1} \cdot 1 \\
&= 4^{2k-2} + 4^{2k-3} + 4^{2k-4} + \cdots + 4^{k+1} + 4^k + 4^{k-1} \\
&= 4^{k-1}(1 + 4 + 4^2 + 4^3 + \cdots + 4^{k-2} + 4^{k-1}) \\
&= 4^{k-1} \frac{4^k - 1}{4 - 1} = \frac{4^{2k-1} - 4^{k-1}}{3} .
\end{aligned}
$$

201. (a) *First solution.* Since $x^k - 1$ is divisible by $x - 1$, for all natural numbers k, and since we can write

$$
\begin{aligned}
x + x^3 + x^9 + x^{27} + x^{81} + x^{243} &= (x - 1) + (x^3 - 1) \\
&\quad + (x^9 - 1) + (x^{27} - 1) + (x^{81} - 1) + (x^{243} - 1) + 6 ,
\end{aligned}
$$

we see that the given polynomial gives a remainder of 6 upon division by $x - 1$.

Second Solution. Let $q(x)$ be the quotient resulting from division of the given polynomial by $x - 1$, and let r be the remainder. Then

$$x + x^3 + x^9 + x^{27} + x^{81} + x^{243} = q(x)(x - 1) + r .$$

The substitution $x = 1$ into this identity yields $r = 6$.

(b) Let $q(x)$ be the quotient and let $r_1 x + r_2$ be the remainder obtained by dividing the given polynomial by $x^2 - 1$. Then

$$x + x^3 + x^9 + x^{27} + x^{81} + x^{243} = q(x)(x^2 - 1) + r_1 x + r_2 \,.$$

If we substitute, first, $x = 1$ into this identity, and then $x = -1$, we derive the two equations (both of which must hold):

$$6 = r_1 + r_2$$

and

$$-6 = -r_1 + r_2 \,.$$

Solution of this system yields $r_1 = 6$ and $r_2 = 0$.

Therefore, the remainder we seek is $6x$.

202. Designate the unknown polynomial by $p(x)$, and let $q(x)$ designate the quotient and $r(x) = ax + b$ the remainder resulting from division $P(x)$ by $(x - 1)(x - 2)$. Then

$$P(x) = (x - 1)(x - 2)q(x) + ax + b \,. \qquad (1)$$

By the conditions of the problem,

$$p(x) = (x - 1)q_1(x) + 2 \,,$$

whence $p(1) = 2$;

$$p(x) = (x - 2)q_2(x) + 1 \,,$$

whence $p(2) = 1$.

If we substitute $x = 1$ and $x = 2$, successively, in (1), we obtain the two equations

$$2 = p(1) = a + b \,,$$

and

$$1 = p(2) = 2a + b \,,$$

from which we obtain

$$a = -1 \,,$$
$$b = 3 \,.$$

Therefore, the remainder sought is $-x + 3$.

203. The polynomial $x^4 + x^3 + 2x^2 + x + 1$ is factorable into $(x^2 + 1)(x^2 + x + 1)$. It follows that this polynomial is a divisor of

$$x^{12} - 1 = (x^6 - 1)(x^6 + 1)$$
$$= (x^3 - 1)(x^3 + 1)(x^2 + 1)(x^4 - x^2 + 1) \,,$$

and, specifically, that

$$x^4 + x^3 + 2x^2 + x + 1 = \frac{x^{12} - 1}{(x-1)(x^3+1)(x^4-x^2+1)}$$

$$= \frac{x^{12} - 1}{x^9 - x^7 - x^6 + 2x^5 - 2x^3 + x^2 + x - 1} .$$

Dividing $x^{1951} - 1$ by $x^4 + x^3 + 2x^2 + x + 1$ is equivalent to dividing $x^{1951} - 1$ first by $x^{12} - 1$ and then multiplying the result by

$$x^9 - x^7 - x^6 + 2x^5 - 2x^3 + x^2 + x - 1 .$$

However, it is readily found that

$$\frac{x^{1951} - 1}{x^{12} - 1} = x^{1939} + x^{1927} + x^{1915} + x^{1903} + \cdots + x^{19} + x^7 + \frac{x^7 - 1}{x^{12} - 1}$$

(this is conveniently found if we note that

$$x^{1951} - 1 = x^7[(x^{12})^{162} - 1] + x^7 - 1$$

and use the well-known formula for dividing the difference of two even powers by the difference of the bases). It follows that the coefficient we seek coincides with the coefficient of x^{15} in the product

$$\left(x^{1939} + x^{1927} + \cdots + x^{31} + x^{19} + x^7 + \frac{x^7 - 1}{x^{12} - 1} \right)$$

$$\times (x^9 - x^7 - x^6 + 2x^5 - 2x^3 + x^2 + x - 1) ,$$

and this coefficient is equal to 1.

204. (a) Write $x = \sqrt{2} + \sqrt{3}$. Then

$$x^2 = 2 + 2\sqrt{2} \cdot \sqrt{3} + 3 = 5 + 2\sqrt{6} ,$$

from which we obtain

$$x^2 - 5 = 2\sqrt{6} ,$$
$$x^4 - 10x^2 + 25 = 24 ,$$
$$x^4 - 10x^2 + 1 = 0 .$$

This equation satisfies the condition of the problem.

   (b) Let $x = \sqrt{2} + \sqrt[3]{3}$. Then we can write

$$x = \sqrt{2} + \sqrt[3]{3} ,$$
$$x^2 = 2 + 2\sqrt{2} \cdot \sqrt[3]{3} + \sqrt[3]{9} ,$$
$$x^3 = 2\sqrt{2} + 3 \cdot 2 \cdot \sqrt[3]{3} + 3 \cdot \sqrt{2} \cdot \sqrt[3]{9} + 3 .$$

Two of the three irrational numbers can be readily eliminated from these three equations, namely, $\sqrt{2}$, $\sqrt[3]{3}$, and $\sqrt[3]{9}$, by finding their corresponding expressions in terms of x, x^2, and x^3. In fact, from the first and second equations we obtain

$$\sqrt[3]{3} = x - \sqrt{2},$$
$$\sqrt[3]{9} = x^2 - 2 - 2\sqrt{2} \cdot \sqrt[3]{3}$$
$$= x^2 - 2 - 2\sqrt{2}(x - \sqrt{2}) = x^2 + 2 - 2\sqrt{2x}.$$

If these substitutions are made in the third equation, we obtain

$$x^3 = 2\sqrt{2} + 6(x - \sqrt{2}) + 3\sqrt{2}(x^2 + 2 - 2\sqrt{2x}) + 3,$$

from which we find

$$x^3 + 6x - 3 = \sqrt{2}(3x^2 + 2).$$

Now we may eliminate the radical by squaring and transposing all terms to the left side:

$$x^6 + 36x^2 + 9 + 12x^4 - 6x^3 - 36x = 18x^4 + 24x^2 + 8;$$
$$x^6 - 6x^4 - 6x^3 + 12x^2 - 36x + 1 = 0.$$

This equation satisfies the condition of the problem.

205. Using the well-known relations between roots and coefficients of a quadratic equation, we obtain

$$\alpha + \beta = -p, \qquad \alpha\beta = 1;$$
$$\gamma + \delta = -q, \qquad \gamma\delta = 1.$$

Therefore,

$$(\alpha - \gamma)(\beta - \gamma)(\alpha + \delta)(\beta + \delta)$$
$$= [(\alpha - \gamma)(\beta + \delta)][(\beta - \gamma)(\alpha + \delta)]$$
$$= (\alpha\beta + \alpha\delta - \beta\gamma - \gamma\delta)(\alpha\beta + \beta\delta - \alpha\gamma - \gamma\delta)$$
$$= (\alpha\delta - \beta\gamma)(\beta\delta - \alpha\gamma) = \alpha\beta\delta^2 - \alpha^2\gamma\delta - \beta^2\gamma\delta + \alpha\beta\gamma^2$$
$$= \delta^2 - \alpha^2 - \beta^2 + \gamma^2 = [(\delta + \gamma)^2 + 2\delta\gamma] - [(\alpha + \beta)^2 - 2\alpha\beta]$$
$$= (q^2 - 2) - (p^2 - 2) = q^2 - p^2.$$

206. If α and β are roots of the equation

$$x^2 + px + q = 0,$$

then $(x - \alpha)(x - \beta) = x^2 + px + q$. Consequently,

$$(\alpha - \gamma)(\beta - \gamma)(\alpha - \delta)(\beta - \delta)$$
$$= [(\gamma - \alpha)(\gamma - \beta)][(\delta - \alpha)(\delta - \beta)]$$
$$= (\gamma^2 + p\gamma + q)(\delta^2 + p\delta + q) .$$

However,

$$\gamma + \delta = -P ,$$
$$\gamma\delta = Q ,$$

which means that

$$(\alpha - \gamma)(\beta - \gamma)(\alpha - \delta)(\beta - \delta) = (\gamma^2 + p\gamma + q)(\delta^2 + p\delta + q)$$
$$= \gamma^2\delta^2 + p\gamma^2\delta + q\gamma^2 + p\gamma\delta^2 + p^2\gamma\delta + pq\gamma + q\delta^2 + pq\delta + q^2$$
$$= (\gamma\delta)^2 + p\gamma\delta(\gamma + \delta) + q[(\gamma + \delta)^2 - 2\gamma\delta] + p^2\gamma\delta + pq(\gamma + \delta) + q^2$$
$$= Q^2 - pPQ + q(P^2 - 2Q) + p^2Q - pqP + q^2$$
$$= Q^2 + q^2 - pP(Q + q) + qP^2 + p^2Q - 2qQ .$$

207. *First Solution.* We solve for the constant a in the second equation and substitute into the first:

$$a = -(x^2 + x) ,$$
$$x^2 - (x^2 + x)x + 1 = 0 ,$$
$$x^3 - 1 = 0 ,$$
$$(x - 1)(x^2 + x + 1) = 0 .$$

We obtain the roots

$$x_1 = 1 ,$$

and

$$x_{2,3} = \frac{-1 \pm i\sqrt{3}}{2} ,$$

and, consequently, since $a = -(x^2 + x)$,

$$a_1 = -2 ,$$

and

$$a_{2,3} = 1 .$$

Second Solution. Using the results of problem 206, we can assert that the necessary and sufficient condition for the given equations to have a common root is the vanishing of the following expression:

$$a^2 + 1 - a \cdot 1(a + 1) + a^3 - 2a = a^3 - 3a + 2$$
$$= (a - 1)(a^2 + a - 2) = (a - 1)^2(a + 2) .$$

This implies that

$$a_1 = -2 \, ,$$

and

$$a_{2,3} = 1 \, .$$

208. Let $(x - a)(x - 10) + 1 = (x + b)(x + c)$. If we make the substitution $x = -b$ in this identity we obtain

$$(-b - a)(-b - 10) + 1 = (-b + b)(-b + c) = 0 \, .$$

Thus, necessarily,

$$(b + a)(b + 10) = -1 \, .$$

Since a and b are to be integers, $b + a$ and $b + 10$ must also be integers. However, -1 can be the product of two integers only if one of the integers is $+1$ and the other is -1. Therefore, there are only two possibilities:

(1) $b + 10 = 1$, $b = -9$; then $b + a = -9 + a = -1$, that is, $a = 8$. In this case,

$$(x - 8)(x - 10) + 1 = (x - 9)^2 \, .$$

(2) $b + 10 = -1$, $b = -11$; then $b + a = -11 + a = 1$, that is, $a = 12$. In this case,

$$(x - 12)(x - 10) + 1 = (x - 11)^2 \, .$$

209. A polynomial of degree four can be represented as a product of two polynomial factors in two ways: a first-degree and a third-degree polynomial, or two quadratic polynomials. We investigate each case separately.

For the first case we have

$$x(x - a)(x - b)(x - c) + 1 = (x + p)(x^3 + qx^2 + rx + s) \, . \qquad (1)$$

[The coefficient of x in the first factor on the right side and of x^3 in the second factor are both either 1 or -1, since the coefficient of x^4 on the left is 1. However, the equation $x(x - a)(x - b)(x - c) + 1 = (-x + p_1)(-x^3 + q_1x^2 + r_1x + s_1)$ can be written in form (1) by using -1 as a factor on both sides.]

If in identity (1) we substitute, in turn, $x = 0$, $x = a$, $x = b$, and $x = c$, and note that 1 can be expressed as the product of two factors in only two ways, $1 = 1 \cdot 1$ or $1 = (-1) \cdot (-1)$, we see that the four distinct numbers

$$0 + p = p \,,$$
$$a + p \,,$$
$$b + p \,,$$
$$c + p$$

(distinct since, by hypothesis, a, b, and c are distinct) can have only two values, $+1$ or -1. This is a contradiction.

For the second case we have

$$x(x - a)(x - b)(x - c) + 1 = (x^2 + px + q)(x^2 + rx + s) \,.$$

Substituting, successively, $x = 0$, $x = a$, $x = b$, and $x = c$, we find that both the polynomials $x^2 + px + q$ and $x^2 + rx + s$ can take on only values of $+1$ or -1. Now, the quadratic trinomial $x^2 + px + q$ cannot have the same value a for three distinct values of x (otherwise the quadradic equation $x^2 + px + q - a = 0$ would have three distinct roots), and so for two of the four distinct values $x = 0$, $x = a$, $x = b$, and $x = c$ this trinomial has value 1, and for the other two it has value -1. Supppse that $0^2 + p \cdot 0 + q = q = 1$, and let $x = a$ be the other value among the numbers $x = a$, $x = b$, and $x = c$ for which this trinomial has value 1. Then for $x = b$ and for $x = c$ this trinomial has value -1. This yields the equations

$$a^2 + pa + 1 = 1 \,,$$
$$b^2 + pb + 1 = -1 \,,$$
$$c^2 + pc + 1 = -1 \,.$$

From $a^2 + pa = a(a + p) = 0$ we find that $a + p = 0$ and $p = -a$ (by hypothesis $a \neq 0$). Then the last two equations take on the form

$$b^2 - ab = b(b - a) = -2 \,,$$
$$c^2 - ac = c(c - a) = -2 \,.$$

If we subtract the first equation from the second, we obtain

$$b^2 - ab - c^2 + ac = (b - c)(b + c) - a(b - c)$$
$$= (b - c)(b + c - a) = 0 \,,$$

which yields (since $b \neq c$)

$$b + c - a = 0 \,,$$
$$a = b + c \,,$$
$$b - a = -c \,,$$
$$c - a = -b \,.$$

Now, from the equation

$$b(b - a) = -bc = -2$$

we find the following values for $b, c,$ and a:

$b = 1$,
$c = 2$,
$a = b + c = 3$,
$x(x - a)(x - b)(x - c) + 1 = x(x - 3)(x - 1)(x - 2) + 1$
$$= (x^2 - 3x + 1)^2 ,$$

and

$b = -1$,
$c = -2$,
$a = b + c = -3$,
$x(x - a)(x - b)(x - c) + 1 = x(x + 3)(x + 1)(x + 2) + 1$
$$= (x^2 - 3x + 1)^2 .$$

Analogously, if $x^2 + px + q$ assumes the value -1 for $x = 0$ and $x = a$, and hence the value $+1$ for $x = b$ and $x = c$, we have

$$q = -1 ,$$
$$a^2 + pa - 1 = -1 ,$$
$$b^2 + pc - 1 = 1 ,$$
$$c^2 + pc - 1 = 1 ,$$

from which we obtain

$p = -a$, $\qquad\qquad$ $a = b + c$,
$b(b - a) = c(c - a) = 2$, $\qquad$ $b - a = -c$,
$b^2 - ab - c^2 + ac = 0$, $\qquad$ $-bc = 2$.
$(b - c)(b + c - a) = 0$,

In this way we obtain two additional systems from which to obtain values for $a, b,$ and c:

$b = 2$,
$c = -1$,
$a = b + c = 1$,
$x(x - a)(x - b)(x - c) = x(x - 1)(x - 2)(x + 1) + 1$
$$= (x^2 - x - 1)^2 ;$$

$$b = 1 \,,$$
$$c = -2 \,,$$
$$a = b + c = -1 \,,$$
$$x(x - a)(x - b)(x - c) = x(x + 1)(x - 1)(x + 2) + 1$$
$$= (x^2 + x - 1)^2 \,.$$

Remark: Another solution of this problem is given in the final part of the solution of problem 210 (b).

210. (a) Assume that

$$(x - a_1)(x - a_2)(x - a_3) \cdots (x - a_n) - 1 = p(x)q(x) \,,$$

where $p(x)$ and $q(x)$ are both polynomials of degree $\geqq 1$ with integral coefficients, and that the sum of their degrees is n. We may assume that the leading coefficient in each polynomial is 1 (compare with the preceding problem). If we make the successive substitutions $x = a_1, x = a_2, x = a_3, \cdots, x = a_n$, and take into consideration that -1 is essentially factorable only as $-1 = 1 \cdot (-1)$, we see that, for each of these values of x, either $p(x) = 1$ and $q(x) = -1$, or vice-versa. Therefore, the sum $p(x) + q(x)$ vanishes for $x = a_1, a_2, \cdots, a_n$, and the equation $p(x) + q(x) = 0$ has as its roots $x_1 = a_1, x_2 = a_2, \cdots, x_n = a_n$. It follows that $p(x) + q(x)$ is divisible by $x - a_1 \; x - a_2, \cdots, \; x - a_n$, and hence by the product $(x - a_1)(x - a_2) \cdots (x - a_n)$. However, the degree of the polynomial $p(x) + q(x)$ is only the larger of the degrees of $p(x)$ or $q(x)$, which is less than n [n is the degree of $(x - a_1)(x - a_2) \cdots (x - a_n) - 1$]. Therefore, $p(x) + q(x)$ cannot be divisible by the product $(x - a_1)(x - a_2) \cdots (x - a_n)$, and because of this contradiction we must conclude that the factorization assumed at the beginning of this proof is not possible.

(b) Assume that

$$(x - a_1)(x - a_2)(x - a_3) \cdots (x - a_n) - 1 = p(x)q(x) \,,$$

where $p(x)$ and $q(x)$ are polynomials of degree $\geqq 1$, and whose leading coefficients are each 1. If in this identity we substitute, successively, $x = a_1, x = a_2, x = a_3, \cdots, x = a_n$, then we find that for each of these values of x, $p(x) = 1$, $q(x) = 1$, or else $p(x) = -1$, $q(x) = -1$.

Since $p(x) - q(x)$ vanishes for n distinct values of x, we must conclude that $p(x) - q(x) \equiv 0$, or $p(x) \equiv q(x)$ [compare with the solution of problem (a)]; also, the integer n must be even, say $n = 2k$, where k is the common degree of the identical polynomials $p(x)$ and $q(x)$. We can rewrite the identity in the form

$$(x - a_1)(x - a_2)(x - a_3) \cdots (x - a_{2k}) = p^2(x) - 1 ,$$

or,

$$(x - x_1)(x - a_2)(x - a_3) \cdots (x - a_{2k}) = [p(x) + 1][p(x) - 1] .$$

Now, the product of the two polynomials $p(x) + 1$ and $p(x) - 1$ vanishes for $x = a_1$, $x = a_2$, $\cdots$, $x = a_{2k}$. Consequently, for each of these values of x at least one of these polynomials vanishes, which means that either $p(x) + 1$ or else $p(x) - 1$ (or both) is divisible by $x - a_1$ and this conclusion holds also for divisibility by $x - a_2$ $x - a_3$, and so on. Since a polynomial of degree k cannot be divisible by the product of more than k factors of form $x - a_i$, and since the polynomial $p(x) + 1$ of degree k, with leading coefficient 1, is divisible by k factors of form $x - a_i$, it follows that this polynomial is identically equal to this product. That is, $p(x) + 1$ is the product of k of the factors $x - a_1$, $x - a_2$, $\cdots$, $x - a_{2k}$, and $p(x) - 1$ is the product of the remaining k factors.

We may assume, with generality, that

$$p(x) + 1 = (x - a_1)(x - a_3) \cdots (x - a_{2k-1}) ,$$
$$p(x) - 1 = (x - a_2)(x - a_4) \cdots (x - a_{2k}) .$$

If the second equation is subtracted from the first, we obtain

$$2 = (x - a_1)(x - a_3) \cdots (x - a_{2k-1}) - (x - a_2)(x - a_4) \cdots (x - a_{2k}) .$$

Now, to consider a specific case, if we let $x = a_2$, we obtain the factorization of the integer 2 as the product of k integral factors:

$$2 = (a_2 - a_1)(a_2 - a_3) \cdots (a_2 - a_{2k-1}) . \tag{1}$$

Since the integer 2 cannot be expressed as a product of more than three distinct integers [for example, $2 = 1 \cdot (-1) \cdot (-2)$] it follows that $k \leqq 3$. But the condition $k = 3$ is impossible in (1). Assume, for example, that $k = 3$ and that $a_1 < a_3 < a_5$; then $2 = (a_2 - a_1)(a_2 - a_3)(a_2 - a_5)$, where $a_2 - a_1 > a_2 - a_3 > a_2 - a_5$, and hence $a_2 - a_1 = 1$, $a_2 - a_3 = -1$, and $a_2 - a_5 = -2$. If we substitute $x = a_4$ in the formula

$$2 = (x - a_1)(x - a_3)(x - a_5) - (x - a_2)(x - a_4)(x - a_6) ,$$

we arrive at the other representation of the integer 2 as a product of three "polynomials":

$$2 = (a_4 - a_1)(a_4 - a_3)(a_4 - a_5) ,$$

where also $a_4 - a_1 > a_4 - a_3 > a_4 - a_5$. It follows that $a_4 - a_1 = 1$, $a_4 - a_3 = -1$, and $a_4 - a_5 = -2$, which implies that $a_4 = a_2$, which contradicts the conditions of the problem.

Hence only two cases are possible: $k = 2$ and $k = 1$.

If $k = 1$, we have

$$2 = (x - a_1) - (x - a_2) \, ,$$

which implies that $a_2 = a_1 + 2$, and if we designate a_2 simply by a, we have

$$(x - a_1)(x - a_2) + 1 = (x - a)(x - a - 2) + 1$$
$$= (x - a - 1)^2$$

(compare with the solution of problem 208).

If $k = 2$, we have

$$2 = (x - a_1)(x - a_3) - (x - a_2)(x - a_4) \, ,$$

where we will consider $a_1 < a_3$ and $a_2 < a_4$. If we substitute $x = a_2$ in this equation, and then substitute $x = a_4$, (thus obtaining two equations) we have

$$2 = (a_2 - a_1)(a_2 - a_3) \, , \qquad a_2 - a_1 > a_2 - a_3 \, ,$$
$$2 = (a_4 - a_1)(a_4 - a_3) \, , \qquad a_4 - a_1 > a_4 - a_3 \, .$$

However, the integer 2 can be expressed (in diminishing order) only in two ways: $2 = 2 \cdot 1$ and $2 = (-1) \cdot (-2)$. Since, moreover, $a_2 - a_1 < a_4 - a_1$, we obtain

$$a_2 - a_1 = -1 \, , \qquad a_2 - a_3 = -2 \, ,$$
$$a_4 - a_1 = 2 \, , \qquad a_4 - a_3 = 1 \, ,$$

from which, replacing a_1 by a, we find

$$a_2 = a - 1 \, ,$$
$$a_3 = a + 1 \, ,$$
$$a_4 = a + 2 \, ,$$

and

$$(x - a_1)(x - a_2)(x - a_3)(x - a_4) + 1$$
$$= (x - a)(x - a + 1)(x - a - 1)(x - a - 2) + 1$$
$$= [x^2 - (2a - 1)x + a^2 + a - 1]^2$$

(see the solution of problem 209).

211. As in the solution of the preceding problem, assume that

$$(x - a_1)^2(x - a_2)^2(x - a_3)^2 \cdots (x - a_n)^2 + 1 = p(x)q(x) \, , \qquad (1)$$

where $p(x)$ and $q(x)$ are certain polynomials with integral coefficients (each with leading coefficient 1). In this event, either $p(x) = 1$ and $q(x) = 1$, or else $p(x) = -1$ and $q(x) = -1$, for each of the values $x = a_1$, $x = a_2$, $x = a_3$, $\cdots$, $x = a_n$. We shall show that the polynomial $p(x)$ [and, of course, $q(x)$] will be, for all values $x = a_1$, $x = a_2$, $\cdots$, $x = a_n$, either equal to 1, or for all these values of x equal to -1.

In fact, if (for example) the polynomial $p(x)$ takes on the value 1 for $x = a_i$ but assumes the value -1 for $x = a_j$ ($i \neq j$), then for some intermediate value of x between a_i and a_j it must become zero [the polynomial $p(x)$ is continuous; its graph will go from the -1 value for $x = a_i$ to the value $+1$ for $x = a_j$, thus crossing the x-axis for some x such that $a_i < x < a_j$]. However, this is impossible, since the left side of equation (1) is always greater than 1 and therefore cannot become zero.

Let us assume that $p(x)$ and $q(x)$ both take on the value 1 for $x = a_1$ $x = a_2$, $\cdots$, $x = a_n$. In this case, both $p(x) - 1$, and $q(x) - 1$ become zero for $x = a_1$, $x = a_2$, $\cdots$, $x = a_n$, and, consequently, $p(x) - 1$ and $q(x) - 1$ are divisible by the product $(x - a_1) \cdot (x - a_2) \cdots (x - a_n)$. Since the sum of the degrees of the polynomials $p(x)$ and $q(x)$ must be equal to the degree of $(x - a_1)^2 (x - a_2)^2 \cdots (x - a_n)^2 + 1$, that is, $2n$, it follows that $p(x) - 1 = (x - a_1) \cdots (x - a_n)$ and $q(x) - 1 = (x - a_1) \cdots (x - a_n)$ (compare with the solution of the preceding problem). We also have the identity

$$(x - a_1)^2(x - a_2)^2 \cdots (x - a_n)^2 + 1 = p(x)q(x)$$
$$= [(x - a_1) \cdots (x - a_n) + 1][(x - a_1) \cdots (x - a_n) + 1]$$
$$= (x - a_1)^2(x - a_2)^2 \cdots (x - a_n)^2$$
$$+ 2(x - a_1)(x - a_2) \cdots (x - a_n) + 1 ,$$

from which we must conclude that

$$(x - a_1)(x - a_2) \cdots (x - a_n) \equiv 0 .$$

This is impossible; hence we must conclude that neither $p(x)$ nor $q(x)$ can take on the value 1 for all the $x = a_i$. It can be shown in exactly the same way that neither $p(x)$ nor $q(x)$ can assume the value -1 at the points $x = a_1$, $x = a_2$, $\cdots$, $x = a_n$ (if they could, we would obtain $p(x) = q(x) = (x - a_1)(x - a_2) \cdots (x - a_n) - 1$).

Therefore, the proposed factorization of

$$(x - a_1)^2(x - a_2)^2 \cdots (x - a_n)^2 + 1$$

as the product of two polynomials with integral coefficients is impossible.

212. Let the polynomial $P(x)$ be equal to 7 for $x = a$, $x = b$, $x = c$, and $x = d$, where a, b, c, and d are integers. Then the equation $P(x) - 7$ has four integral roots a, b, c, and d. This means that the polynomial $P(x) - 7$ is divisible by $x - a$, $x - b$, $x - c$, and $x - d$,[†] that is,

$$P(x) - 7 = (x - a)(x - b)(x - c)(x - d)p(x) ,$$

where $p(x)$ is the remaining factor (it may be constant).

Now let us suppose that the polynomial $P(x)$ takes the value 14 for the integral value $x = A$. Upon substituting $x = A$ into the preceding identity, we obtain, since $P(A) = 14$,

$$7 = (A - a)(A - b)(A - c)(A - d)p(A) ,$$

which is impossible, since the integers $A - a$, $A - b$, $A - c$, and $A - d$ are all distinct, and 7 cannot be expressed as a product of five factors of which at least four are distinct.

213. If a polynomial of seventh degree is factorable as the product of two polynomials $p(x)$ and $q(x)$ with integral coefficients, then the degree of one of these factors does not exceed 3. Let us assume that $p(x)$ is such a factor. If $P(x)$ has the value $+1$ or else -1 for seven integral values of x, then $p(x)$ has the value $+1$ or -1, for those same values of x [since all coefficients are assumed to be integers, and since $p(x)q(x) = P(x)$]. Among the seven integral values of x for which $p(x)$ has the value $+1$ or -1, there must be at least four for which $p(x)$ will have the value 1 or else four for which $p(x)$ will have the value -1. In the first instance the third-degree equation $p(x) - 1 = 0$ has four roots, and in the second instance the equation $p(x) + 1 = 0$ has four roots. No such polynomial $p(x)$ of third degree can exist, since neither $p(x) - 1 = 0$ nor $p(x) + 1 = 0$ can have four roots [They would have to be divisible by a polynomial of fourth degree. Compare with the solution of problem 210 (a).]

214. Let p and q be two integers which are either both even or both odd. Then the difference $P(p) - P(q)$ is even, since the value

---

† Assume that $P(x) - 7$ has a remainder r when divided by $x - a$. Then

$$P(x) - 7 = (x - a)Q(x) + r .$$

If we set $x = a$ in this equality, we obtain $7 - 7 = 0 + r$ (that is, $r = 0$), and this means that $P(x) - 7 = (x - a)Q(x)$ is divisible by $x - a$.

$$P(p) - P(q) = a_0(p^n - q^n) + a_1(p^{n-1} - q^{n-1})$$
$$+ \cdots + a_{n-2}(p^2 - q^2) + a_{n-1}(p - q)$$

is divisible by the even number $p - q$.

In particular, for even p the difference $P(p) - P(0)$ is even. But by hypothesis $P(0)$ is odd; consequently, $P(p)$ must also be odd, and therefore $P(p) \neq 0$. Analogously, for odd p the difference $P(p) - P(1)$ is even. Since by hypothesis $P(1)$ is odd, it must follow, by the same reasoning used above, that $P(p) \neq 0$.

Consequently, $P(x)$ cannot become zero for any integral value of x (either even or odd); that is, the polynomial $P(x)$ does not have integral roots.

215. Let us assume that the equation $P(x) = 0$ has the rational root $x = \dfrac{k}{l}$, that is, $P\left(\dfrac{k}{l}\right) = 0$. Let us write the polynomial $P(x)$ in powers of $x - p$; we shall express it as

$$P(x) = c_0(x - p)^n + c_1(x - p)^{n-1}$$
$$+ c_2(x - p)^{n-2} + \cdots + c_{n-1}(x - p) + c_n ,$$

where $c_0, c_1, c_2, \cdots, c_n$ are certain integers which are readily found in terms of a_i [c_0 is equal to the leading coefficient a_0 of the polynomial $P(x)$, c_1 is equal to the leading coefficient of the polynomial $P(x) - c_0(x - p)^n$ of degree $n - 1$, c_2 is equal to the leading coefficient of the polynomial $P(x) - c_0(x - p)^n - c_1(x - p)^{n-1}$ of degree $n - 2$, and so on.] If in the last expression for $P(x)$ we set $x = p$, we obtain $c_n = P(p) = \pm 1$.

If in the same expression we set $x = \dfrac{k}{l}$ and multiply the result by l^n, we obtain

$$l^n P\left(\frac{k}{l}\right) = c_0(k - pl)^n + c_1 l(k - pl)^{n-1}$$
$$+ c_2 l^2(k - pl)^{n-2} + \cdots + c_{n-1} l^{n-1}(k - pl) + c_n l^n = 0 ,$$

from which it follows that if $P\left(\dfrac{k}{l}\right) = 0$, then

$$\frac{c_n l^n}{k - pl} = \frac{\pm l^n}{k - pl}$$
$$= -c_0(k - pl)^{n-1} - c_1 l(k - pl)^{n-2} - \cdots$$
$$- c_{n-2} l^{n-2}(k - pl) - c_{n-1} l^{n-1}$$

is an integer. But since pl is divisible by l, and k is relatively prime to l (we assume, of course, that the fraction $\dfrac{k}{l}$ is in lowest terms), then $k - pl$ is relatively prime to l, and, consequently, $k - pl$ is also relatively prime to l^n. It follows that $\dfrac{\pm l^n}{k - pl}$ can be an integer only if $k - pl = \pm 1$. We may show, in exactly the same manner, that $k - ql = \pm 1$.

Subtracting the equality $k - pl = \pm 1$ from the equality $k - ql = \pm 1$, we obtain $(p - q)l = 0$ or $(p - q)l = \pm 2$. But $(p - q)l > 0$, since $p > q$ and $l > 0$, and, consequently, $(p - q)l = 2$, $k - pl = -1$, and $k - ql = 1$.

Hence, if $p - q > 2$, then the equation $P(x) = 0$ cannot have any rational root. If, however, $p - q = 2$ or $p - q = 1$, then a rational root $\dfrac{k}{l}$ may exist.

Upon adding the equations

$$k - pl = -1 \, ,$$
$$k - ql = 1 \, ,$$

we obtain

$$2k - (p + q)l = 0 \, ,$$
$$\frac{k}{l} = \frac{p + q}{2} \, ,$$

which is what we sought to prove.

216. (a) Let us assume that the polynomial can be expressed as a product of two polynomial factors having integral coefficients:

$$x^{2222} + 2x^{2220} + 4x^{2218} + \cdots + 2220x^2 + 2222$$
$$= (a_n x^n + a_{n-1} x^{n-1} + a_{n-2} x^{n-2} + \cdots + a_0)$$
$$\times (b_m x^m + b_{m-1} x^{m-1} + b_{m-2} x^{m-2} + \cdots + b_0) \, ,$$

where $m + n = 2222$. Here, $a_0 b_0 = 2222$, and therefore one of the two integers a_0 and b_0 will be even and the other will be odd. Let us assume that a_0 is the even integer and b_0 is odd. We shall show that all the coefficients of the polynomial $a_n x^n + a_{n-1} x^{n-1} + \cdots + a_0$ must be even. Indeed, suppose that a_k is the first odd coefficient of this polynomial to appear, reading from right to left. Then the coefficient of x^k in the product

$$(a_n x^n + a_{n-1} x^{n-1} + \cdots + a_0)(b_m x^m + b_{m-1} x^{m-1} + \cdots + b_0)$$

will be equal to

$$a_k b_0 + a_{k-1} b_1 + a_{k-2} b_2 + \cdots + a_0 b_k \qquad (1)$$

(for $k > m$, this sum ends with the term $a_{k-m}b_m$). This coefficient is equal to the corresponding coefficient of x^k in the initial polynomial. That is, it is equal to zero if k is odd, and even if k is even (since all the coefficients of the polynomial, except the first, by hypothesis, are even, and $k \leq n < 2222$). But since, by assumption, all the numbers $a_{k-1}, a_{k-2}, a_{k-3}, \cdots, a_0$ are even, then in sum (1) all the terms other than the first must be even, and, therefore, the product $a_k b_0$ must also be even, which cannot be since a_k and b_0 are both odd.

Therefore, all the coefficients of the polynomial

$$a_n x^n + a_{n-1} x^{n-1} + \cdots + a_0$$

must be even, which contradicts the fact that $a_n b_n$ must be equal to 1. Our assumption that it is possible to write the given polynomial as the product of two polynomial with integral coefficients is therefore untenable.

(b) Let us set $x = y + 1$. We then have

$$x^{250} + x^{249} + x^{248} + \cdots + x + 1$$
$$= (y + 1)^{250} + (y + 1)^{249} + \cdots + (y + 1) + 1$$
$$= \frac{(y + 1)^{251} - 1}{(y + 1) - 1} = \frac{1}{y} [(y + 1)^{251} - 1]$$
$$= y^{250} + 251 y^{249} + C_{251}^2 y^{248} + C_{251}^3 y^{247} + \cdots + C_{251}^2 y + 251 .$$

Further, since all the coefficients of the last-written polynomial, except the first, are divisible by the prime number 251 $\Big[$inasmuch as $C_{251}^k = \dfrac{251 \cdot 250 \cdot 249 \cdots (251 - k + 1)}{1 \cdot 2 \cdot 3 \cdots k} \Big]$, and since the constant term of the polynomial is 251, which is not divisible by 251^2, we can, by using reasoning almost identical to that used in problem (a) and merely replacing eveness and oddness of coefficients with divisibility by 251, conclude that a necessary condition for the given polyomial to be expressed as the product of two factors is that all the coefficients of one of the factors be divisible by 251. However, this is impossible, since the first coefficient of the given polynomial is 1.

217. Let us write the polynomials in the forms

$$A = a_0 + a_1 x + a_2 x^2 + \cdots + a_n x^n ,$$
$$B = b_0 + b_1 x + b_2 x^2 + \cdots + b_m x^m .$$

Since by hypothesis not all the coefficients in the product are divisible by 4, not all the coefficients in both polynomials can be even. Consequently, in one of them (say, polynomial B) not all the coefficients will be even. Now assume that polynomial A also contains some odd coefficient. Let us examine the first of these to appear (that with the smallest subscript), and let us assume that this is the coefficient a_s. Further, let the first odd coefficient of polynomial B be b_k, and consider the coefficient of x^{k+s} in the product of the polynomials A and B. The term x^{k+s} in this product can be obtained only from those powers of x the sum of whose exponents is equal to $k + s$; consequently, this coefficient is equal to

$$a_0 b_{k+s} + a_1 b_{k+s-1} + \cdots + a_{s-1} b_{k+1} + a_s b_k + a_{s+1} b_{k-1} + \cdots + a_{s+k} b_0 .$$

In this sum all the products appearing before $a_s b_k$ are even, since all the numbers $a_0, a_1, \cdots, a_{s-1}$ are even. All the products appearing after $a_s b_k$ are also even, since all the numbers $b_{k-1}, b_{k-2}, \cdots, b_0$ are even. But the product $a_s b_k$ is odd, since both a_s and b_k are odd numbers. Consequently, the sum is also odd, and this contradicts the requirement that all the product coefficients be even. Therefore, the assumption that polynomial A has odd coefficients is untenable. Therefore, all the coefficients of A must be even, as was to be proved.

218. We shall prove that for an arbitrary rational, but not integral, value of x, the polynomial $P(x)$ cannot be an integer, (nor zero, which we consider an integer).

Let $x = \dfrac{p}{q}$, where p and q are relatively prime (that is, this fraction is in lowest terms). Then

$$\begin{aligned}
P(x) &= x^n + a_1 x^{n-1} + a_2 x^{n-2} + \cdots + a_{n-1} x + a_n \\
&= \frac{p^n}{q^n} + a_1 \frac{p^{n-1}}{q^{n-1}} + a_2 \frac{p^{n-2}}{q^{n-2}} + \cdots + a_{n-1} \frac{p}{q} + a_n \\
&= \frac{p^n + a_1 p^{n-1} q + a_2 p^{n-2} q^2 + \cdots + a_{n-1} p q^{n-1} + a_n q^n}{q^n} \\
&= \frac{p^n + q(a_1 p^{n-1} + a_2 p^{n-2} q + \cdots + a_{n-1} p q^{n-2} + a_n q^{n-1})}{q^n} .
\end{aligned}$$

The number p^n, as well as p, is relatively prime to q; consequently, $p^n + q(a_1 p^{n-1} + \cdots + a_n q^{n-1})$ is also relatively prime to q, hence also to q^n. Therefore $P(x)$ becomes an irreducible fraction which cannot be an integer.[†]

[†] It is as easy to prove the more general theorem that if $\dfrac{p}{q}$ (in lowest terms) is a zero of $P(x)$, then p divides a_n and q divides the leading coefficient [*Editor*].

219. Let N be a certain integer and let $P(N) = M$. For any integer k,

$$P(N + kM) - P(N) = a_0[(N + kM)^n - N^n]$$
$$+ a_1[(N + kM)^{n-1} - N^{n-1}] + \cdots + a_{n-1}[(N + kM) - N]$$

is divisible by kM—since $(N + kM)^l - N^l$ is divisible by $[(N + kM) - N = kM]$—and hence also by M. Therefore, for any integer k, $P(N + kM)$ is divisible by M.

Thus, if we prove that among the values $P(N + kM)$ $(k = 0, 1, 2, \cdots)$ there are integers distinct from $\pm M$, then this will prove that not all of them can be prime. But the polynomial $P(x)$ of nth degree assumes a given value A for at most n distinct values of x (since otherwise the nth degree equation $P(x) - A = 0$ would have more than n roots). Hence, among the first $2n + 1$ values of $P(N + kM)$ $(k = 0, 1, 2, \cdots, 2n)$ there must be at least one which is distinct from M or $-M$.

220. First we show that every polynomial $P(x)$ of degree n can be expressed in terms of (as "a linear combination" of) polynomials of the form

$$P_0(x) = 1 ,$$
$$P_1(x) = x ,$$
$$P_2(x) = \frac{x(x-1)}{1 \cdot 2} ,$$
$$\cdots\cdots\cdots\cdots\cdots ,$$
$$P_n(x) = \frac{x(x-1)(x-2)\cdots(x-n+1)}{1 \cdot 2 \cdot 3 \cdots n} ,$$

each supplied with suitable numerical coefficients b_i, namely,

$$P(x) = b_n P_n(x) + b_{n-1} P_{n-1}(x) + \cdots + b_1 P_1(x) + b_0 P_0(x) .$$

To prove this, we note that if b_n is chosen such that the number $\frac{b_n}{n!}$ is equal to the leading coefficient of the polynomial $P(x)$, then $P(x)$ and $b_n P_n(x) + b_{n-1} P_{n-1}(x) + \cdots + b_0 P_0(x)$ have identical coefficients for x^n. If b_{n-1} is chosen such that $\frac{b_{n-1}}{(n-1)!}$ is equal to the leading coefficient of the new polynomial $P(x) - b_n P_n(x)$, then $P(x)$ and $b_n P_n(x) + b_{n-1} P_{n-1}(x) + \cdots + b_0 P_0(x)$ have identical coefficients for both x^n and x^{n-1}. If, in addition, $\frac{b_{n-2}}{(n-2)!}$ is equal to the leading coefficient of the polynomial $P(x) - b_n P_n(x) - b_{n-1} P_{n-1}(x)$, then $P(x)$

and $b_n P_n(x) + b_{n-1} P_{n-1}(x) + b_{n-2} P_{n-2}(x) + \cdots + b_0 P_0(x)$ have identical coefficients for x^n, x^{n-1}, x^{n-2}, and so on. Thus, we can determine $b_n, b_{n-1}, \cdots, b_1, b_0$ in such a way that the polynomials $P(x)$ and $b_n P_n(x) + b_{n-1} P_{n-1}(x) + \cdots + b_1 P_1(x) + b_0 P_0(x)$ completely coincide.

Express the given polynomial $P(x)$ of degree n, which has the property that $P(0), P(1), \cdots, P(n)$ are integers, in the form (as outlined above)

$$P(x) = b_0 P_0(x) + b_1 P_1(x) + b_2 P_2(x) + \cdots + b_n P_n(x) \ .$$

We see that

$$P_1(0) = P_2(0) = \cdots = P_n(0)$$
$$= P_2(1) = P_3(1) = \cdots = P_n(1) = P_3(2) = \cdots = P_n(2)$$
$$= \cdots = P_{n-1}(n-2) = P_n(n-2) = P_n(n-1) = 0 \ ,$$
$$P_0(0) = P_1(1) = P_2(2) = \cdots = P_{n-1}(n-1) = P_n(n) = 1 \ .$$

Therefore,

$$P(0) = b_0 P_0(0) \ ,$$

whence $b_0 = P(0)$;

$$P(1) = b_0 P_0(1) + b_1 P_1(1) \ ,$$

whence $b_1 = P(1) - b_0 P_0(1)$;

$$P(2) = b_0 P_0(2) + b_1 P_1(2) + b_2 P_2(2) \ ,$$

from which it follows that

$$b_2 = P(2) - b_0 P_0(2) - b_1 P_1(2) \ ,$$
$$\cdots\cdots\cdots\cdots\cdots\cdots\cdots\cdots\cdots\cdots\cdots\cdots , $$
$$P(n) = b_0 P_0(n) + b_1 P_1(n) + \cdots + b_{n-1} P_{n-1}(n) + b_n P_n(n) \ ,$$

and so

$$b_n = P(n) - b_0 P_0(n) - b_1 P_1(n) - \cdots - b_{n-1} P_{n-1}(n) \ .$$

Thus, all the coefficients $b_0, b_1, b_2, \cdots, b_n$ are integers.

221. (a) It was shown in the solution of problem 220 that a polynomial of degree n can be expressed as a linear combination of polynomials of form $P_0(x), P_1(x), \cdots, P_n(x)$ (see problem 220), where the coefficients of P_i in the linear combination (series) are integers, and provided that $P(k)$ is an integer for all integers k. The proof needed only the fact that $P(x)$ had integral values for $k = 0, 1, 2, \cdots, n$. Hence the polynomial of the present problem can, under the given conditions, be so represented. It is clear that the linear combination $b_n P_n(x) + \cdots b_0 P_0(x)$ must have integral value for *all* integers, hence, so must $P(x)$.

(b) If the polynomial $P(x) = a_n x^n + a_{n-1} x^{n-1} + a_{n-2} x^{n-2} + \cdots + a_1 x + a_0$ has integral values for $x = k, k+1, k+2, \cdots, k+n$, then the polynomial

$$Q(x) = P(x - k) = a_n(x - k)^n + a_{n-1}(x - k)^{n-1} + \cdots + a_1(x - k) + a_0$$

has integral values for $x = 0, 1, 2, 3, \cdots, n$. It follows from problem (a) that $Q(x)$ has integral values for every integer x. Therefore we must conclude that the polynomial $P(x) = Q(x - k)$ also has integral values for every integer x.

(c) Let the polynomial $P(x) = a_n x^n + a_{n-1} x^{n-1} + \cdots + a_1 x + a_0$ have integral values for $x = 0, 1, 4, 9, \cdots, n^2$. Then the polynomial $Q(x) = P(x^2) = a_n(x^2)^n + a_{n+1}(x^2)^{n-1} + \cdots + a_1 x^2 + a_0$ of degree $2n$ has integral values for $2n + 1$ consecutive values of x, that is, for $x = -n, -(n-1), -(n-2), \cdots, -1, 0, 1, \cdots, n-1, n$. In fact, it is obvious that

$$Q(0) = P(0) , \qquad\qquad Q(3) = Q(-3) = P(9) ,$$
$$Q(1) = Q(-1) = P(1) , \qquad \cdots\cdots\cdots\cdots\cdots\cdots ,$$
$$Q(2) = Q(-2) = P(4) , \qquad Q(n) = Q(-n) = P(n^2) ,$$

and all these numbers, by hypothesis, are integers. Consequently, basing our reasoning on problem (b), we may say that the polynomial $Q(x)$ has integral values for every integral value of x. This also means that $P(k^2) = Q(k)$ is, for any integer k, an integer. As an example, we may use $P(x) = \dfrac{x(x-1)}{12}$, for which

$$Q(x) = P(x^2) = \frac{x^2(x^2 - 1)}{12} = \frac{x^2(x - 1)(x + 1)}{12}$$

$$= 2\,\frac{(x + 2)(x + 1)x(x - 1)}{1\cdot2\cdot3\cdot4} - \frac{(x + 1)x(x - 1)}{1\cdot2\cdot3} .$$

222. (a) Using De Moivre's formula and the binomial theorem, we have

$$\cos 5a + i \sin 5a = (\cos a + i \sin a)^5$$
$$= \cos^5 a + 5\cos^4 a \cdot i \sin a + 10 \cos^3 a \cdot (i \sin a)^2$$
$$\quad + 10 \cos^2 a \cdot (i \sin a)^3 + 5 \cos a \cdot (i \sin a)^4 + (i \sin a)^5$$
$$= (\cos^5 a - 10 \cos^3 a \cdot \sin^2 a + 5 \cos a \cdot \sin^4 a)$$
$$\quad + i(5 \cos^4 a \cdot \sin a - 10 \cos^2 a \cdot \sin^3 a + \sin^5 a) .$$

Equating the real and imaginary parts of the left and right sides, we obtain the required formulas.

(b) As in the solution of problem (a), we have

$$\cos na + i\sin na = (\cos a + i\sin a)^n$$
$$= \cos^n a + C_n^1 \cos^{n-1} a \cdot i\sin a + C_n^2 \cos^{n-2} a \cdot (i\sin a)^2$$
$$+ C_n^3 \cos^{n-3} a \cdot (i\sin a)^3 + C_n^4 \cos^{n-4} a \cdot (i\sin a)^4 + \cdots$$
$$= (\cos^n a - C_n^2 \cos^{n-2} a \cdot \sin^2 a + C_n^4 \cos^{n-4} a \cdot \sin^4 a - \cdots)$$
$$+ i(C_n^1 \cos^{n-1} a \cdot \sin a - C_n^3 \cos^{n-3} a \cdot \sin^3 a + \cdots).$$

Verification of the desired identities is immediate.

223. Using the formulas of problem 222 (b) we have

$$\tan 6a = \frac{\sin 6a}{\cos 6a} = \frac{6\cos^5 a \sin a - 20\cos^3 a \sin^3 a + 6\cos a \sin^5 a}{\cos^6 a - 15\cos^4 a \sin^2 a + 15\cos^2 a \sin^4 a - \sin^6 a}$$

Dividing the numerator and denominator of the last fraction by $\cos^6 \alpha$, we obtain the desired formula:

$$\tan 6a = \frac{6\tan a - 20\tan^3 a + 6\tan^5 a}{1 - 15\tan^2 a + 15\tan^4 a - \tan^6 a}.$$

224. We may rewrite the equation $x + \dfrac{1}{x} = 2\cos \alpha$ in the form

$$x^2 + 1 = 2x\cos a$$

or

$$x^2 - 2x\cos a + 1 = 0.$$

Thus,

$$x = \cos a \pm \sqrt{\cos^2 a - 1} = \cos a \pm i\sin a.$$

It follows from De Moivre's theorem that

$$x^n = \cos na \pm i\sin na;$$
$$\frac{1}{x^n} = \frac{1}{\cos na \pm i\sin na} = \cos na \mp i\sin na.$$

By addition we obtain

$$x^n + \frac{1}{x^n} = 2\cos na.$$

225. Let us consider the sum

$$[\cos \varphi + i\sin \varphi] + [\cos(\varphi + \alpha) + i\sin(\varphi + \alpha)]$$
$$+ [\cos(\varphi + 2\alpha) + i\sin(\varphi + 2\alpha)] + \cdots$$
$$+ [\cos(\varphi + n\alpha) + i\sin)\varphi + n\alpha)].$$

We have now only to compute the coefficients for the imaginary and real parts of this sum. By designating $\cos \varphi + i \sin \varphi$ as a and $\cos \alpha + i \sin \alpha$ as x, and applying the formula for the multiplication of complex numbers, and De Moivre's formula, we find that the sum under consideration is equal to

$$a + ax + ax^2 + \cdots + ax^n = \frac{ax^{n+1} - a}{x - 1}$$

$$= (\cos \varphi + i \sin \varphi) \frac{\cos (n + 1)\alpha + i \sin (n + 1)\alpha - 1}{\cos \alpha + i \sin \alpha - 1}$$

$$= (\cos \varphi + i \sin \varphi) \frac{[(\cos (n + 1)\alpha - 1)] + i[\sin (n + 1)\alpha]}{[(\cos \alpha - 1) + i \sin \alpha]}$$

$$= (\cos \varphi + i \sin \varphi) \frac{-2 \sin^2 \dfrac{n + 1}{2} \alpha + 2i \sin \dfrac{n + 1}{2} \alpha \cos \dfrac{n + 1}{2} \alpha}{-2 \sin^2 \dfrac{\alpha}{2} + 2i \sin \dfrac{\alpha}{2} \cos \dfrac{\alpha}{2}}$$

$$= (\cos \varphi + i \sin \varphi) \frac{2i \sin \dfrac{n + 1}{2} \alpha \left[\cos \dfrac{n + 1}{2} \alpha + i \sin \dfrac{n + 1}{2} \alpha \right]}{2i \sin \dfrac{\alpha}{2} \left[\cos \dfrac{\alpha}{2} + i \sin \dfrac{\alpha}{2} \right]}$$

$$= \frac{\sin \dfrac{n + 1}{2} \alpha}{\sin \dfrac{\alpha}{2}} (\cos \varphi + i \sin \varphi)$$

$$\times \frac{\left(\cos \dfrac{n + 1}{2} \alpha + i \sin \dfrac{n + 1}{2} \alpha \right) \left(\cos \dfrac{\alpha}{2} - i \sin \dfrac{\alpha}{2} \right)}{\cos^2 \dfrac{\alpha}{2} + \sin^2 \dfrac{\alpha}{2}}$$

$$= \frac{\sin \dfrac{n + 1}{2} \alpha}{\sin \dfrac{\alpha}{2}} \left[\cos \left(\varphi + \frac{n}{2} \alpha \right) + i \sin \left(\varphi + \frac{n}{2} \alpha \right) \right].$$

$\left[\vphantom{\frac{\alpha}{2}}\right.$ Here we have again used the formula for the multiplication of complex numbers and also the fact that $\cos \dfrac{\alpha}{2} - i \sin \dfrac{\alpha}{2} = \cos \left(-\dfrac{\alpha}{2} \right) + i \sin \left(-\dfrac{\alpha}{2} \right).\left.\vphantom{\frac{\alpha}{2}}\right]$ The required identities follow immediately.

226. Employing the identity $\cos^2 x = \dfrac{1 + \cos 2x}{2}$ and the result of the preceding problem, we obtain

$$\cos^2 \alpha + \cos^2 2\alpha + \cdots + \cos^2 n\alpha$$

$$= \frac{1}{2}[\cos 2\alpha + \cos 4\alpha + \cdots + \cos 2n\alpha + n]$$

$$= \frac{1}{2}\left[\frac{\sin(n+1)\alpha \cos n\alpha}{\sin \alpha} - 1\right] + \frac{n}{2}$$

$$= \frac{\sin(n+1)\alpha \cos n\alpha}{2 \sin \alpha} + \frac{n-1}{2}.$$

But since $\sin^2 x = 1 - \cos^2 x$, we have

$$\sin^2 \alpha + \sin^2 2\alpha + \cdots + \sin^2 n\alpha$$

$$= n - \frac{\sin(n+1)\alpha \cos n\alpha}{2 \sin \alpha} - \frac{n-1}{2}$$

$$= \frac{n+1}{2} - \frac{\sin(n+1)\alpha \cos n\alpha}{2 \sin \alpha}.$$

227. We must compute the real part and the coefficient for the imaginary part of the sum

$$(\cos a + i \sin a) + C_n^1(\cos 2a + i \sin 2a)$$
$$+ C_n^2(\cos 3a + i \sin 3a) + \cdots + [\cos(n+1)a + i \sin(n+1)a].$$

Designating $\cos a + i \sin a$ as x, and using De Moivre's formula and the binomial formula, we can transform the sum into the following form:

$$x + C_n^1 x^2 + {}_n^2 x^3 + \cdots + x^{n+1} = x(x+1)^n$$

$$= (\cos a + i \sin a)(\cos a + 1 + i \sin a)^n$$

$$= (\cos a + i \sin a)\left(2\cos^2 \frac{a}{2} + 2i \cos \frac{a}{2} \sin \frac{a}{2}\right)^n$$

$$= 2^n \cos^n \frac{a}{2}(\cos a + i \sin a)\left(\cos \frac{na}{2} + i \sin \frac{na}{2}\right)$$

$$= 2^n \cos^n \frac{a}{2}\left(\cos \frac{n+2}{2}a + i \sin \frac{n+2}{2}a\right).$$

It follows that

$$\cos a + C_n^1 \cos 2a + C_n^2 \cos 3a + \cdots + \cos(n+1)a$$

$$= 2^n \cos^n \frac{a}{2} \cos \frac{n+2}{2}a,$$

$$\sin a + C_n^1 \sin 2a + C_n^2 \sin 3a + \cdots + \sin(n+1)a$$

$$= 2^n \cos^n \frac{a}{2} \sin \frac{n+2}{2}a.$$

228. We use the trigonometric identity

$$\sin A \sin B = \frac{1}{2} [\cos (A - B) - \cos (A + B)]$$

for the terms of the sum, obtaining

$$\frac{1}{2} \left[\cos \frac{(m - n)\pi}{p} + \cos \frac{2(m - n)\pi}{p} + \cos \frac{3(m - n)\pi}{p} \right.$$
$$\left. + \cdots + \cos \frac{(p - 1)(m - n)\pi}{p} \right]$$
$$- \frac{1}{2} \left[\cos \frac{(m + n)\pi}{p} + \cos \frac{2(m + n)\pi}{p} + \cos \frac{3(m + n)\pi}{p} \right.$$
$$\left. + \cdots + \cos \frac{(p - 1)(m + n)\pi}{p} \right].$$

The sum

$$\cos \frac{k\pi}{p} + \cos \frac{2k\pi}{p} + \cos \frac{3k\pi}{p} + \cdots + \cos \frac{(p - 1)k\pi}{p}$$

is equal to $p - 1$ if k is divisible by $2p$ (here, every summand of the sum is equal to 1). In the event k is not divisible by $2p$, however, this sum, according to problem 225, is equal to

$$\frac{\sin \dfrac{pk\pi}{2p} \cos \dfrac{(p - 1)k\pi}{2p}}{\sin \dfrac{k\pi}{2p}} - 1 = \sin k \frac{\pi}{2} \cdot \frac{\cos \left(k \dfrac{\pi}{2} - \dfrac{k\pi}{2p} \right)}{\sin \dfrac{k\pi}{2p}} - 1$$

$$= \begin{cases} 0, & \text{if } k \text{ is odd}, \\ -1, & \text{if } k \text{ is even}. \end{cases}$$

Both of the numbers $m + n$ and $m - n$ will be simultaneously even or odd; in particular, if either of $m + n$ or $m - n$ is divisible by $2p$, then both $m + n$ and $m - n$ are even. The equation sought follows immediately.

229. Consider the equation $x^{2n+1} - 1 = 0$, which has roots

$$1,$$
$$\cos \frac{2\pi}{2n + 1} + i \sin \frac{2\pi}{2n + 1},$$
$$\cos \frac{4\pi}{2n + 1} + i \sin \frac{4\pi}{2n + 1},$$
$$\cdots\cdots\cdots\cdots\cdots\cdots\cdots,$$
$$\cos \frac{4n\pi}{2n + 1} + i \sin \frac{4n\pi}{2n + 1}.$$

Since the coefficient of the absent term x^{2n} in this equation may be taken as 0, the sum of all these roots is equal to zero:

$$\left(1 + \cos\frac{2\pi}{2n+1} + \cos\frac{4\pi}{2n+1} + \cdots + \cos\frac{4n\pi}{2n+1}\right)$$
$$+ i\left(\sin\frac{2\pi}{2n+1} + \sin\frac{4\pi}{2n+1} + \cdots + \sin\frac{4n\pi}{2n+1}\right) = 0.$$

Consequently, the expression inside each set of parenthesis is equal to zero; in particular,

$$\cos\frac{2\pi}{2n+1} + \cos\frac{4\pi}{2n+1} + \cdots + \cos\frac{4n\pi}{2n+1} = -1.$$

However,

$$\cos\frac{2\pi}{2n+1} = \cos\frac{4n\pi}{2n+1},$$
$$\cos\frac{4\pi}{2n+1} = \cos\frac{(4n-2)\pi}{2n+1},$$

and so on, which implies that

$$2\left(\cos\frac{2\pi}{2n+1} + \cos\frac{4\pi}{2n+1} + \cdots + \cos\frac{2n\pi}{2n+1}\right) = -1.$$

That is,

$$\cos\frac{2\pi}{2n+1} + \cos\frac{4\pi}{2n+1} + \cdots + \cos\frac{2n\pi}{2n+1} = -\frac{1}{2}.$$

Remark: It is also possible to prove this result by using the results of problem 225.

230. (a) From the result of problem 222 (b) we have

$$\sin(2n+1)a = C^1_{2n+1}(1-\sin^2 a)^n \sin a$$
$$- C^3_{2n+1}(1-\sin^2 a)^{n-1}\sin^3 a + \cdots + (-1)^n \sin^{2n+1}a,$$

whence it follows that the numbers

$$0,$$
$$\sin\frac{\pi}{2n+1},$$
$$\sin\frac{2\pi}{2n+1},$$
$$\cdots\cdots\cdots,$$

$$\sin \frac{n\pi}{2n+1},$$

$$\sin\left(-\frac{\pi}{2n+1}\right) = \sin \frac{\pi}{2n+1},$$

$$\sin\left(-\frac{2\pi}{2n+1}\right) = -\sin \frac{2\pi}{2n+1},$$

$$\dots\dots\dots\dots\dots\dots\dots\dots\dots,$$

$$\sin\left(-\frac{n\pi}{2n+1}\right) = -\sin \frac{n\pi}{2n+1}$$

are the roots of the following equation of degree $(2n+1)$:

$$C_{2n+1}^1(1-x^2)^n x - C_{2n+1}^3(1-x^2)^{n-1}x^3 + \cdots + (-1)^n x^{2n+1} = 0.$$

Consequently, the numbers

$$\sin^2 \frac{\pi}{2n+1},$$

$$\sin^2 \frac{2\pi}{2n+1},$$

$$\dots\dots\dots\dots,$$

$$\sin^2 \frac{n\pi}{2n+1}$$

are roots of an equation of degree n, such as

$$C_{2n+1}^1(1-x)^n - C_{2n+1}^3(1-x)^{n-1}x + \cdots + (-1)^n x^n = 0.$$

(b) Let us replace n with $2n+1$ in the formula of problem 222 (b) and write it in the following form:

$$\sin(2n+1)a = \sin^{2n+1}a(C_{2n+1}^1 \cot^{2n}a - C_{2n+1}^3 \cot^{2n-2}a$$
$$+ C_{2n+1}^5 \cot^{2n-4}a - \cdots).$$

Whence it follows that for

$$a = \frac{\pi}{2n+1},$$

$$\frac{2\pi}{2n+1},$$

$$\frac{3\pi}{2n+1},$$

$$\dots\dots\dots,$$

$$\frac{n\pi}{2n+1}$$

the following equation holds:

$$C^1_{2n+1} \cot^{2n}a - C^3_{2n+1} \cot^{2n-2}a + C^5_{2n+1} \cot^{2n-4}a - \cdots = 0 .$$

Therefore, the numbers

$$\cot^2 \frac{\pi}{2n+1} ,$$

$$\cot^2 \frac{2\pi}{2n+1} ,$$

$$\cdots\cdots\cdots ,$$

$$\cot^2 \frac{n\pi}{2n+1}$$

are roots of the following equation of degree n:

$$C^1_{2n+1} x^n - C^3_{2n+1} x^{n-1} + C^5_{2n+1} x^{n-2} - \cdots = 0 .$$

231. (a) The sum of the roots of the nth degree equation

$$x^n - \frac{C^3_{2n+1}}{C^1_{2n+1}} x^{n-1} + \frac{C^5_{2n+1}}{C^1_{2n+1}} x^{n-2} - \cdots = 0$$

[see the solution of problem 230 (b)] is equal to the coefficient of x^{n-1} taken with the opposite sign (see the remarks preceding problem 222); that is,

$$\cot^2 \frac{\pi}{2n+1} + \cot^2 \frac{2\pi}{2n+1} + \cot^2 \frac{3\pi}{2n+1} + \cdots$$

$$+ \cot^2 \frac{n\pi}{2n+1} = \frac{C^3_{2n+1}}{C^1_{2n+1}} = \frac{n(2n-1)}{3} .$$

(b) Since $\csc^2 \alpha = \cot^2 \alpha + 1$, the formula of part (a) implies

$$\csc^2 \frac{\pi}{2n+1} + \csc^2 \frac{2\pi}{2n+1} + \csc^2 \frac{3\pi}{2n+1}$$

$$+ \cdots + \csc^2 \frac{n\pi}{2n+1} = \frac{n(2n-1)}{3} + n = \frac{2n(n+1)}{3} .$$

232. (a) *First Solution.* The numbers

$$\sin^2 \frac{\pi}{2n+1} ,$$

$$\sin^2 \frac{2\pi}{2n+1} ,$$

$$\cdots\cdots\cdots ,$$

$$\sin^2 \frac{n\pi}{2n+1}$$

are the roots of the nth degree equation obtained in the solution of problem 230 (a). The coefficient of the highest-degree term x^n, of this equation is equal to

$$(-1)^n(C_{2n+1}^1 + C_{2n+1}^3 + \cdots + C_{2n+1}^{2n-1} + 1) \,.$$

But the sum in the parentheses is half the sum of the binomial coefficients

$$1 + C_{2n+1}^1 + C_{2n+1}^2 + \cdots + C_{2n+1}^{2n} + 1 \,,$$

which is equal to $(1 + 1)^{2n+1} = 2^{2n+1}$. Consequently, the coefficient of x^n in the equation is equal to $(-1)^n 2^{2n}$. Furthermore, the constant term of this equation is

$$C_{2n+1}^1 = 2n + 1 \,.$$

Now, the product of the roots of a polynomial equation of degree n with leading coefficient 1 is equal to $(-1)^n$ times the constant term [if the polynomial has leading coefficient $a_0 \neq 1$, then the constant term a_n is equal to $(-1)^n$ times the product divided by a_0]. Therefore, we have

$$(-1)^n \sin^2 \frac{\pi}{2n+1} \sin^2 \frac{2\pi}{2n+1} \cdots \sin^2 \frac{n\pi}{2n+1} = (-1)^n \frac{2n+1}{2^{2n}} \,,$$

and, consequently,

$$\sin \frac{\pi}{2n+1} \sin \frac{2\pi}{2n+1} \cdots \sin \frac{n\pi}{2n+1} = \frac{\sqrt{2n+1}}{2^n} \,.$$

It can be proved, in an analogous manner, that

$$\sin \frac{\pi}{2n} \sin \frac{2\pi}{2n} \cdots \sin \frac{(n-1)\pi}{2n} = \frac{\sqrt{n}}{2^{n-1}} \,.$$

Second Solution. The roots of the equation $x^{2n} - 1 = 0$ are 1, -1, $\cos \frac{\pi}{n} + i \sin \frac{\pi}{n}$, $\cos \frac{2\pi}{n} + i \sin \frac{2\pi}{n}$, $\cos \frac{3\pi}{n} + i \sin \frac{3\pi}{n}$, $\cdots$, $\cos \frac{(2n-1)\pi}{n} + i \sin \frac{(2n-1)\pi}{n}$. Therefore, we can write

$$x^{2n} - 1 = (x - 1)(x + 1)\left(x - \cos \frac{\pi}{n} - i \sin \frac{\pi}{n}\right)$$

$$\times \left(x - \cos \frac{2\pi}{n} - i \sin \frac{2\pi}{n}\right) \cdots \left[x - \cos \frac{(n-1)\pi}{n}\right.$$

$$\left. - i \sin \frac{(n-1)\pi}{n}\right]\left[x - \cos \frac{(n+1)\pi}{n} - i \sin \frac{(n+1)\pi}{n}\right] \times \cdots$$

$$\times \left[x - \cos \frac{(2n-1)\pi}{n} - i \sin \frac{(2n-1)\pi}{n}\right].$$

However,

$$\cos \frac{(2n - k)\pi}{n} = \cos \frac{k\pi}{n} \, ,$$

$$\sin \frac{(2n - k)\pi}{n} = - \sin \frac{k\pi}{n} \, ;$$

whence it follows that

$$\left(x - \cos \frac{\pi}{n} - i \sin \frac{\pi}{n}\right)\left[x - \cos \frac{(2n - 1)\pi}{n} - i \sin \frac{(2n - 1)\pi}{n}\right]$$

$$= x^2 - 2x \cos \frac{\pi}{n} + 1 \, ,$$

$$\left(x - \cos \frac{2\pi}{n} - i \sin \frac{2\pi}{n}\right)\left[x - \cos \frac{(2n - 2)\pi}{n} - i \sin \frac{(2n - 2)\pi}{n}\right]$$

$$= x^2 - 2x \cos \frac{2\pi}{n} + 1 \, ,$$

$$\dotsb \, ,$$

$$\left[x - \cos \frac{(n - 1)\pi}{n} - i \sin \frac{(n - 1)\pi}{n}\right]$$

$$\times \left[x - \cos \frac{(n + 1)\pi}{n} - i \sin \frac{(n + 1)\pi}{n}\right]$$

$$= x^2 - 2x \cos \frac{(n - 1)\pi}{n} + 1 \, .$$

Therefore, the decomposition of the polynomial $x^{2n} - 1$ into factors can be written:

$$x^{2n} - 1 = (x^2 - 1)\left(x^2 - 2x \cos \frac{\pi}{n} + 1\right)\left(x^2 - 2x \cos \frac{2\pi}{n} + 1\right) \times \cdots$$

$$\times \left(x^2 - 2x \cos \frac{(n - 1)\pi}{n} + 1\right) .$$

It follows that

$$\frac{x^{2n} - 1}{x^2 - 1} = x^{2n-2} + x^{2n-4} + \cdots + x^2 + 1$$

$$= \left(x^2 - 2x \cos \frac{\pi}{n} + 1\right)\left(x^2 - 2x \cos \frac{2\pi}{n} + 1\right) \times \cdots$$

$$\times \left[x^2 - 2x \cos \frac{(n - 1)\pi}{n} + 1\right] .$$

If here we set $x = 1$ and use the identity

$$2 - 2 \cos a = 4 \sin^2 \frac{a}{2},$$

we obtain

$$n = 4^{n-1} \sin^2 \frac{\pi}{2n} \sin^2 \frac{2\pi}{2n} \cdots \sin^2 \frac{(n-1)\pi}{2n},$$

from which it follows that

$$\sin \frac{\pi}{2n} \sin \frac{2\pi}{2n} \cdots \sin \frac{(n-1)\pi}{2n} = \frac{\sqrt{n}}{2^{n-1}}.$$

It is proved in the same manner that

$$\sin \frac{\pi}{2n+1} \sin \frac{2\pi}{2n+1} \cdots \sin \frac{n\pi}{2n+1} = \frac{\sqrt{2n+1}}{2^n}.$$

(b) We can obtain the required result by reference to either the first or the second solution of problem (a). We shall not repeat those solutions here, but we shall derive the formulas we need from the formulas of problem (a).

Since

$$\sin \frac{\pi}{2n+1} = \sin \frac{2n\pi}{2n+1},$$

$$\sin \frac{3\pi}{2n+1} = \sin \frac{(2n-2)\pi}{2n+1},$$

$$\sin \frac{5\pi}{2n+1} = \sin \frac{(2n-4)\pi}{2n+1},$$

$$\cdots\cdots\cdots\cdots\cdots,$$

it follows that

$$\sin \frac{2\pi}{2n+1} \sin \frac{4\pi}{2n+1} \sin \frac{6\pi}{2n+1} \cdots \sin \frac{2n\pi}{2n+1}$$

$$= \sin \frac{\pi}{2n+1} \sin \frac{2\pi}{2n+1} \sin \frac{3\pi}{2n+1} \cdots \sin \frac{n\pi}{2n+1} = \frac{\sqrt{2n+1}}{2^n}$$

[see problem (a)]. If we divide this formula by

$$\sin \frac{\pi}{2n+1} \sin \frac{2\pi}{2n+1} \cdots \sin \frac{n\pi}{2n+1} = \frac{\sqrt{2n+1}}{2^n}$$

and use the identities

$$\sin \frac{2\pi}{2n+1} = 2 \sin \frac{\pi}{2n+1} \cos \frac{\pi}{2n+1} \,,$$

$$\sin \frac{4\pi}{2n+1} = 2 \sin \frac{2\pi}{2n+1} \cos \frac{2\pi}{2n+1} \,,$$

$$\cdots\cdots\cdots\cdots\cdots\cdots\cdots\cdots\cdots\cdots\cdots\cdots\cdots\cdots\cdots\cdots\cdots\cdots\cdots \,,$$

$$\sin \frac{2n\pi}{2n+1} = 2 \sin \frac{n\pi}{2n+1} \cos \frac{n\pi}{2n+1} \,,$$

we obtain

$$\cos \frac{\pi}{2n+1} \cos \frac{2\pi}{2n+1} \cos \frac{3\pi}{2n+1} \cdots \cos \frac{n\pi}{2n+1} = \frac{1}{2^n}.$$

Similarly, we have

$$\left[\cos \frac{\pi}{2n} \cos \frac{2\pi}{2n} \cdots \cos \frac{(n-1)\pi}{2n} \right]\left[\sin \frac{\pi}{2n} \sin \frac{2\pi}{2n} \cdots \sin \frac{(n-1)\pi}{2n} \right]$$

$$= \frac{1}{2^{n-1}} \sin \frac{\pi}{n} \sin \frac{2\pi}{n} \cdots \sin \frac{(n-1)\pi}{n} \,.$$

But

$$\sin \frac{\pi}{n} = \sin \frac{(n-1)\pi}{n} \,,$$

$$\sin \frac{2\pi}{n} = \sin \frac{(n-2)\pi}{n} \,,$$

$$\cdots\cdots\cdots\cdots\cdots\cdots\cdots\cdots \,,$$

$$\sin \frac{\pi}{2} = 1 \,.$$

Therefore, for odd n $(n = 2k + 1)$:

$$\sin \frac{\pi}{n} \sin \frac{2\pi}{n} \cdots \sin \frac{(n-1)\pi}{n} \cdot$$

$$= \left(\sin \frac{\pi}{2k+1} \sin \frac{2\pi}{2k+1} \cdots \sin \frac{k\pi}{2k+1} \right)^2 = \left(\frac{\sqrt{2k+1}}{2^k} \right)^2 = \frac{n}{2^{n-1}} \,.$$

For even n $(n = 2k)$:

$$\sin \frac{\pi}{n} \sin \frac{2\pi}{n} \cdots \sin \frac{(n-1)\pi}{n}$$

$$= \left[\sin \frac{\pi}{2k} \sin \frac{2\pi}{2k} \cdots \sin \frac{(k-1)\pi}{2k} \right]^2 = \left(\frac{\sqrt{k}}{2^{k-1}} \right)^2 = \frac{n}{2^{n-1}}$$

[see solution (a)]. Whence we obtain

$$\cos \frac{\pi}{2n} \cos \frac{2\pi}{2n} \cdots \cos \frac{(n-1)\pi}{2n} = \frac{1}{2^{n-1}} \frac{2^{n/n-1}}{\sqrt{n}/2^{n-1}} = \frac{\sqrt{n}}{2^{n-1}} .$$

Remark: If we divide the formulas of problem (a) by the corresponding formulas of problem (b), we arrive at

$$\tan \frac{\pi}{2n+1} \tan \frac{2\pi}{2n+1} \cdots \tan \frac{n\pi}{2n+1} = \sqrt{2n+1} ;$$

$$\tan \frac{\pi}{2n} \tan \frac{2\pi}{2n} \cdots \tan \frac{(n-1)\pi}{2n} = 1 .$$

Moreover, the second of these equalities is obvious, since

$$\tan \frac{\pi}{2n} \tan \frac{(n-1)\pi}{2n} = \tan \frac{\pi}{2n} \cot \frac{\pi}{2n} = 1 ,$$

$$\tan \frac{2\pi}{2n} \tan \frac{(n-2)\pi}{2n} = \cdots = \tan \frac{(n-1)\pi}{4n} \tan \frac{(n+1)\pi}{4n} = 1 ,$$

$$\tan \frac{n\pi}{4n} = 1 .$$

From this equality and from the second formula of problem 232 (a) we may readily derive the formula

$$\cos \frac{\pi}{2n} \cos \frac{2\pi}{2n} \times \cdots \times \cos \frac{(n-1)\pi}{2n} = \frac{\sqrt{\pi}}{2^{n-1}} .$$

These solutions can also be obtained in a manner analogous to that used in the first solution of problem 232 (a).

233. We first show that if α (in radians) is an acute angle, then

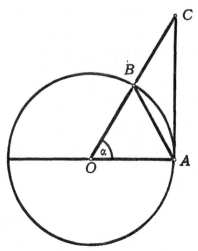

Figure 22

$$\sin \alpha < \alpha < \tan \alpha .$$

We have (see Figure 22; S is area and the circle has radius 1)

$$S_{\triangle AOB} = \frac{1}{2} \sin \alpha ,$$

$$S_{\text{Sector } AOB} = \frac{1}{2} \alpha ,$$

$$S_{\triangle AOC} = \frac{1}{2} \tan \alpha .$$

But since

$$S_{\triangle AOB} < S_{\text{Sector } AOB} < s_{AOC} ,$$

We have

$$\sin \alpha < \alpha < \tan \alpha .$$

This double inequality implies that

$$\cot a < \frac{1}{2} < \csc a .$$

Therefore, it follows from the formulas of problems 231 (a) and (b) that

$$\frac{n(2n-1)}{3} = \cot^2 \frac{\pi}{2n+1} + \cot^2 \frac{2\pi}{2n+1}$$

$$+ \cot^2 \frac{3\pi}{2n+1} + \cdots + \cot^2 \frac{n\pi}{2n+1}$$

$$< \left(\frac{2n+1}{\pi} \right)^2 + \left(\frac{2n+1}{2\pi} \right)^2 + \left(\frac{2n+1}{3\pi} \right)^2 + \cdots + \left(\frac{2n+1}{n\pi} \right)^2$$

$$< \csc^2 \frac{\pi}{2n+1} + \csc^2 \frac{2\pi}{2n+1}$$

$$+ \csc^2 \frac{3\pi}{2n+1} + \cdots + \csc^2 \frac{n\pi}{2n+1} = \frac{2n(n+1)}{3} .$$

If we divide all the terms of the last double inequality by $\frac{(2n+1)^2}{\pi^2}$, we obtain

$$\frac{2n}{2n+1} \cdot \frac{2n-1}{2n+1} \cdot \frac{\pi^2}{6} = \left(1 - \frac{1}{2n+1} \right)\left(1 - \frac{2}{2n+2} \right) \cdot \frac{\pi^2}{6}$$

$$< 1 + \frac{1}{2^2} + \frac{1}{3^2} + \cdots + \frac{1}{n^2}$$

$$< \frac{2n}{2n+1} \cdot \frac{2n+2}{2n+1} \cdot \frac{\pi^2}{6}$$

$$= \left(1 - \frac{1}{2n+1}\right)\left(1 + \frac{1}{2n+1}\right) \cdot \frac{\pi^2}{6},$$

as was to be proved.

234. (a) Assume that the point M is on the arc A_1A_n of the circle (Figure 23). Designate arc MA_1 as a; then the arcs MA_2, MA_3, $\cdots$, MA_n are equal, respectively, to

$$a + \frac{2\pi}{n},$$

$$a + \frac{4\pi}{n},$$

$$\cdots \cdots,$$

$$a + \frac{2(n-1)\pi}{n}.$$

But the length of a chord AB of a circle with radius R is equal to $2R \sin \frac{AB}{2}$. (This is readily discerned from the equilateral triangle AOB, where O is the center of the circle.) It is clear, then, that the sum which interests us is equal to

$$4R^2\left\{\sin^2 \frac{a}{2} + \sin^2 \left(\frac{a}{2} + \frac{\pi}{n}\right) + \sin^2 \left(\frac{a}{2} + \frac{2\pi}{n}\right) + \cdots \right.$$
$$\left. + \sin^2 \left(\frac{a}{2} + \frac{(n-1)\pi}{n}\right)\right\}.$$

We shall now evaluate the expression in the brackets. Using the half-angle formula, $\sin^2 x = \dfrac{1 - \cos 2x}{2}$, we find that the bracketed expression is equal to

$$S = \frac{n}{2} - \left\{\cos a + \cos \left(a + \frac{2\pi}{n}\right) + \cos \left(a + \frac{4\pi}{n}\right) + \cdots \right.$$
$$\left. + \cos \left(a + \frac{2(n-1)\pi}{n}\right)\right\}.$$

But, by the identity in problem 225, we have

$$\cos a + \cos \left(a + \frac{2\pi}{n}\right) + \cdots + \cos \left[a + \frac{2(n-1)\pi}{n}\right]$$

$$= \frac{\sin \pi \cos\left[a + \frac{(n-1)\pi}{n} \right]}{\sin \frac{\pi}{n}} = 0 ,$$

consequently, $S = \frac{n}{2}$. The assertion of the problem follows.

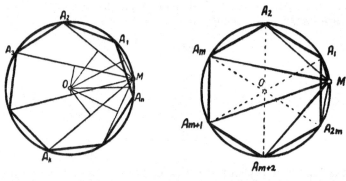

Figure 23   **Figure 24**

Remark: For even $n = 2m$ (Figure 24) the assertion of the problem is obvious, since, by the Pythagorean theorem,

$$MA_1^2 + MA_{m+1}^2 = MA_2^2 + MA_{m+2}^2 = \cdots = MA_m^2 + MA_{2m}^2 = 4R^2 .$$

(b) Let $A_1B_1, A_2B_2, \cdots, A_nB_n$ be perpendiculars drawn from points $A_1, A_2, \cdots, A_n$ to the line OM [Figure 25 (a)]. Then it is readily found (from the law of cosines and consideration of the triangle OA_kB_k) that

$$MA_k^2 = MO^2 + OA_k^2 - 2MO \cdot OB_k = l^2 + R^2 - 2l \cdot OB_k$$
$$(k = 1, 2, \cdots, n) ,$$

where the segments OB_k are taken with the plus or minus sign depending upon whether the point B_k is located on the segment OM or on its extension.

Consequently,

$$MA_1^2 + MA_2^2 + \cdots + MA_n^2 = n(l^2 + R^2) - 2l(OB_1 + OB_2 + \cdots + OB_n) .$$

But if $\angle MOA_1 = a$, then

$$OB_1 = OA_1 \cos \angle A_1OM = R \cos a ,$$

$$OB_2 = R \cos \left(a + \frac{2\pi}{n} \right),$$

$$OB_3 = R \cos \left(a + \frac{4\pi}{n} \right),$$

$$\cdots,$$

$$OB_n = R \cos \left[a + \frac{2(n-1)\pi}{n} \right].$$

However, it was shown in the solution of problem (a) that

$$\cos a + \cos \left(a + \frac{2\pi}{n} \right) + \cdots + \cos \left[a + \frac{2(n-1)\pi}{n} \right] = 0,$$

and so $OB_1 + OB_2 + \cdots + OB_n = 0$.

The assertion of the problem follows.

Remark: For even $n = 2m$ the assertion of the problem is obvious from Figure 25 (b). In this case

$$OB_1 + OB_{m+1} = OB_2 + OB_{m+2} = \cdots = OB_m + OB_{2m} = 0.$$

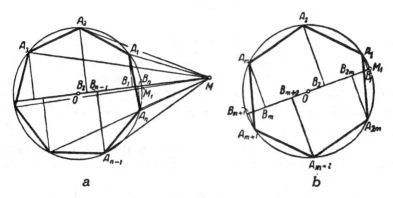

a   b

Figure 25

(c) Let M_1 be the projection of point M onto the plane of an m-sided polygon (Figure 26). We have

$$MA_k^2 = M_1 A_k^2 + MM_1^2 \qquad (k = 1, 2, \cdots, n),$$

and, consequently,

$$MA_1^2 + MA_2^2 + \cdots + MA_n^2 = M_1 A_1^2 + M_1 A_2^2 + \cdots + M_1 A_n^2 + n \cdot MM_1^2.$$

But

$$M_1A_1^2 + M_1A_2^2 + \cdots + MA_n^2 = n(R^2 + OM_1^2)$$

[see problem (b)], and

$$l^2 = OM^2 = OM_1^2 + M_1M^2 ,$$

whence follows the assertion of the problem.

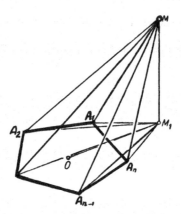

Figure 26

235. (a) The solution of this problem follows directly from the theorem of problem 234 (a), if we take into account that for an even n, the even (and odd) vertices of the n-sided polygon are also the vertices of those regular $\dfrac{n}{2}$-gons $\left(\text{abbreviation for polygons with } \dfrac{n}{2} \text{ sides}\right)$ inscribed in the circle.

(b) Let $n = 2m + 1$. We deduce from the solution of problem 234 (a) that it is sufficient to prove equality of the following two sums:

$$S_1 = \sin\frac{a}{2} + \sin\left(\frac{a}{2} + \frac{2\pi}{2m+1}\right) + \sin\left(\frac{a}{2} + \frac{4\pi}{2m+1}\right)$$

$$+ \cdots + \sin\left(\frac{a}{2} + \frac{2m\pi}{2m+1}\right) ,$$

and

$$S_2 = \sin\left(\frac{a}{2} + \frac{\pi}{2m+1}\right) + \sin\left(\frac{a}{2} + \frac{3\pi}{2m+1}\right)$$

$$+ \cdots + \sin\left(\frac{a}{2} + \frac{(2m-1)\pi}{2m+1}\right) .$$

But, by problem 225, we have

$$S_1 = \frac{\sin \dfrac{(m+1)\pi}{2m+1} \sin \left(\dfrac{a}{2} + \dfrac{m\pi}{2m+1} \right)}{\sin \dfrac{\pi}{2m+1}},$$

$$S_2 = \frac{\sin \dfrac{m\pi}{2m+1} \sin \left[\dfrac{a}{2} + \dfrac{\pi}{2m+1} + \dfrac{(m-1)\pi}{2m+1} \right]}{\sin \dfrac{\pi}{2m+1}} = S_1 .$$

236. In problem 234 (a) we found that the sum of the squares of the distances from a point on a circle circumscribed about a regular n-sided polygon to all of its vertices is equal to $2nR^2$. If we assume that M coincides with A_1, then the sum of all sides and diagonals of the n-gon emanating from one vertex is equal to $2nR^2$. If we multiply this sum by n (the number of vertices of the n-gon) then we obtain double the sum of all the sides and diagonals of the n-gon (since every side or diagonal has two ends, it counts twice in the sum). The sum we seek is therefore equal to $\dfrac{n}{2} \cdot 2nR^2 = n^2R^2$.

(b) For a regular polygon, the sum of all the sides and diagonals which emanate from one vertex A_1 is equal to

$$2R \left[\sin \frac{\pi}{n} + \sin \frac{2\pi}{n} + \cdots + \sin \frac{(n-1)\pi}{n} \right]$$

$$= 2R \frac{\sin \dfrac{\pi}{2} \sin \dfrac{(n-1)\pi}{2n}}{\sin \dfrac{\pi}{2n}} = 2R \cot \frac{\pi}{2n}$$

[compare with the solution of problem 235 (b)]. When we multiply this sum by n and halve it we obtain the required result: $Rn \cot \dfrac{\pi}{2n}$.

(c) The product of all the sides and all the diagonals of the n-gon emanating from one vertex is equal to

$$2^{n-1}R^{n-1} \sin \frac{\pi}{n} \sin \frac{2\pi}{n} \cdots \sin \frac{(n-1)\pi}{n} = 2^{n-1}R^{n-1} \frac{n}{2^{n-1}}$$

[see the solution of problem 232 (a)]. By raising this product to the nth power and extracting the square root, we obtain the required result.

237. Let us compute the sum of the 50th powers of the sides and diagonals emanating from one of the vertices, say A_1. This problem reduces to finding the sum

$$\Sigma = \left(2R \sin \frac{\pi}{100}\right)^{50} + \left(2R \sin \frac{2\pi}{100}\right)^{50} + \cdots + \left(2R \sin \frac{99\pi}{100}\right)^{50}$$

[compare with the solution of problem 234 (a)]. Thus we must make a summation of the 50th powers of the sines of several angles. But

$$\sin^{50} a = \left[\frac{(\cos a + i \sin a) - (\cos a - i \sin a)}{2i}\right]^{50}$$

$$= \frac{\left(x - \dfrac{1}{x}\right)^{50}}{-2^{50}} = -\frac{1}{2^{50}}\left(x - \frac{1}{x}\right)^{50},$$

where we write $\cos a + i \sin a = x$; in this case, $\cos a - i \sin a = \dfrac{1}{x}$. Therefore,

$$\sin^{50} \alpha = \frac{-1}{2^{50}}\left(x^{50} - C_{50}^1 x^{49}\frac{1}{x} + C_{50}^2 x^{48}\frac{1}{x^2} - \cdots\right.$$

$$+ C_{50}^{24} x^{26}\frac{1}{x^{24}} - C_{50}^{24} x^{25}\frac{1}{x^{25}} + C_{50}^{26} x^{24}\frac{1}{x^{26}} - \cdots$$

$$\left. + C_{50}^{48} x^2\frac{1}{x^{48}} - C_{50}^{49} x\frac{1}{x^{49}} + \frac{1}{x^{50}}\right)$$

$$= -\frac{1}{2^{50}}\left[\left(x^{50} + \frac{1}{x^{50}}\right) - C_{50}^1\left(x^{48} + \frac{1}{x^{48}}\right) + C_{50}^2\left(x^{46} + \frac{1}{x^{46}}\right) - \cdots\right.$$

$$\left. + C_{50}^{24}\left(x^2 + \frac{1}{x^2}\right) + C_{50}^{25}\right]$$

$$= -\frac{1}{2^{50}}(2 \cos 50\alpha - 2C_{50}^1 \cos 48\alpha$$

$$+ 2C_{50}^2 \cos 46\alpha - \cdots + 2C_{50}^{24} \cos 2\alpha + C_{50}^{25})$$

$\left[\text{here we make use of the fact that } x^k + \dfrac{1}{x^k} = (\cos ka + i \sin ka) + \right.$ $\left.(\cos ka - i \sin ka) = 2 \cos ka\right]$.

Hence the sum Σ may be expressed in the following way.

$$\Sigma = -R^{50}\left[2\left(\cos 50\frac{\pi}{100} + \cos 50\frac{2\pi}{100} + \cdots + \cos 50\frac{99\pi}{100}\right)\right.$$

$$\left. - 2C_{50}^1\left(\cos 48\frac{\pi}{100} + \cos 48\frac{2\pi}{100} + \cdots + \cos 48\frac{99\pi}{100}\right)\right.$$

$$+ 2C_{50}^2\left(\cos 46\frac{\pi}{100} + \cos 46\frac{2\pi}{100} + \cdots + \cos 46\frac{99\pi}{100}\right)$$

$$\cdots\cdots\cdots\cdots\cdots\cdots\cdots\cdots\cdots\cdots\cdots\cdots\cdots\cdots\cdots$$

$$+ 2C_{50}^{24}\left(\cos 2\frac{\pi}{100} + \cos 2\frac{2\pi}{100} + \cdots + \cos 2\frac{99\pi}{100}\right) - 99C_{50}^{25}\Bigr]$$

$$= -R^{50}[2s_1 - 2C_{50}^1 s_2 + 2C_{50}^2 s_3 - \cdots + 2C_{50}^{24} s_{25} - 99C_{50}^{25}] \,,$$

where $s_1, s_2, \cdots, s_{25}$ replace the sums in the parentheses. But from the formula in problem 225 it follows that $s_1 = s_2 = \cdots = s_{25} = -1$. Therefore we have

$$\Sigma = R^{50}(2 - 2C_{50}^1 + 2C_{50}^2 - \cdots + 2C_{50}^{24} + 99C_{50}^{25})$$
$$= R^{50}(1 - C_{50}^1 + C_{50}^2 - C_{50}^3 + \cdots + C_{50}^{24} - C_{50}^{25}$$
$$\qquad + C_{50}^{26} - \cdots + C_{50}^{48} - C_{50}^{49} + 1 - 100C_{50}^{25})$$
$$= R^{50}[(1 - 1)^{50} + 100C_{50}^{25}] = 100C_{50}^{25}R^{50} \,.$$

From this we immediately obtain the result that the sum of the 50th powers of all the sides and diagonals of a 100-sided polygon is equal to

$$\frac{100\Sigma}{2} = 5000C_{50}^{25}R^{50} = \frac{5000\cdot 50!}{(25!)^2}R^{50} \,.$$

238. We make use of the fact that in a triangle with integral sides $a, b,$ and c the number $2\cos A = \dfrac{b^2 + c^2 - a^2}{bc}$ (law of cosines, where A is the angle between b and c) is rational. Further, if $A = \dfrac{m}{n}\cdot 180°$, where m and n are integers, then $\cos nA = \cos(m\cdot 180°) = (-1)^m$.

We shall now show that $2\cos nA$ for an arbitrary integer n can be expressed as a polynomial of degree n in terms of $2\cos A$ with integral coefficients and having leading coefficient 1; that is, that

$$2\cos nA = (2\cos A)^n + a_1(2\cos A)^{n-2} + a_2(2\cos A)^{n-4} + \cdots \,,$$

where $a_1, a_2, \cdots$ are integers. This assertion may by deduced from the first formula of problem 222 (a), or else proven by mathematical induction. Indeed, it is true for $n = 1$ and $n = 2$, since

$$2\cos A = 2\cos A \,,$$
$$2\cos 2A = (2\cos A)^2 - 2.$$

But from the known product-to-sum identity we have

$$\cos(n + 2)A + \cos nA = 2\cos A\cos(n + 1)A \,,$$

and so

$$\cos(n+2)A = 2\cos A\cos(n+1)A - \cos nA ,$$

from which it immediately follows that if the assertion is true for the values n and $n+1$, then it is also true for $n+2$. Since the assertion is valid for $n=1$ and $n=2$, it follows that it is true for all values of n.

By setting $2\cos A = x$ and $\cos mA = (-1)^m$ in the resulting formula we obtain the following equation with respect to the unknown x:

$$x^n + a_1 x^{n-2} + a_2 x^{n-4} + \cdots - 2(-1)^n = 0 .$$

Thus x is a rational root (since, here, $x = 2\cos A$ is rational) of an equation with integral coefficients whose highest-degree term has coefficient unity. But all the rational roots of such an equation must be integers (see problem 218); therefore, $x = 2\cos A$ must be an integer. Since $0 \le \cos A \le 1$, the only possibilities are $\cos A = 0$ or else $\cos A = \pm\frac{1}{2}$; that is, A may be $60°, 90°$, or $120°$ (these are the only angles greater than zero but less than $\pi = 180°$ for which $2\cos A$ is an integer).

The existence of $60°, 90°$, and $120°$ angles in triangles with integral sides may be shown by simple examples. A triangle with sides $(3,4,5)$ is a right triangle—it has a $90°$ angle. In a triangle with sides $(1,1,1)$ all the angles are $60°$. Finally, we cen verify (by the law of cosines, for example) that a triangle with sides $(3,8,7)$ will have a $60°$ angle (between side 3 and side 8) and a $(3,5,7)$ triangle will have a $120°$ angle (see also problem 128).

239. (a) Let us assume that the ratio of $\theta = \arccos\frac{1}{p}$ to $180°$ is a rational number, that is, that $\theta = \frac{m}{n}\cdot 180°$, where m and n are integers (which will be considered relatively prime). From the second formula of problem 222 (b) we have

$$\sin n\theta = C_n^1\cos^{n-1}\theta\sin\theta - C_n^3\cos^{n-3}\theta\sin^3\theta + C_n^5\cos^{n-5}\theta\sin^5\theta - \cdots .$$

In the case of interest to us we have

$$\cos\theta = \frac{1}{p} ,$$
$$\sin\theta = \sqrt{1-\cos^2\theta} = \frac{\sqrt{p^2-1}}{p} ,$$

and $\sin m\theta = \sin 180°m = 0$. Substituting these values into the last equality, we obtain

$$0 = \frac{\sqrt{p^2-1}}{p^n}\left[n - \frac{n(n-1)(n-2)}{3!}(p^2-1) \right.$$
$$\left. + \frac{n(n-1)(n-2)(n-3)(n-4)}{5!}(p^2-1)^2 - \cdots \right].$$

Since $p \neq 1$, we have $\dfrac{\sqrt{p^2-1}}{p^n}$, $\neq 0$, and therefore the sum within brackets must be equal to zero. Since all the terms of this sum, except the first, are even integers (p^2-1 is even, since p is odd), the first term must also be even. Hence n is even, $n = 2k$, and therefore m is odd.

Since m is odd, it follows that

$$\cos k\theta = \cos \frac{m\theta}{2} = \cos \frac{m}{2}\pi = 0 .$$

From the first formula of problem 222 (b) we have

$$\cos k\theta = \cos^k \theta - C_k^2 \cos^{k-2} \theta \sin^2 \theta + C_k^4 \cos^{k-4} \theta \sin^4 \theta - \cdots .$$

By setting $\cos\theta = \dfrac{1}{p}$, $\sin\theta = \dfrac{\sqrt{p^2-1}}{p}$, and $\cos k\theta = 0$ we obtain

$$0 = \frac{1}{p^k}[1 - C_k^2(p^2-1) + C_k^4(p^2-1)^2 - \cdots] .$$

Here all the summands in the brackets, except the first, are even integers, and the first is equal to unity, that is, is odd. Therefore, the resulting equality is impossible. This contradiction proves that the ratio of $\theta = \arccos \dfrac{1}{p}$ to $180°$ cannot be a rational number.

(b) Let us assume that $\theta = \arctan \dfrac{p}{q}$ contains a rational number of degrees, that is, that θ can be expressed as $\theta = \dfrac{m}{n}180°$, where m and n are integers. We shall consider p and q relatively prime, which is permissible, since we are interested only in the ratio $\dfrac{p}{q}$. We shall use the DeMoivre formulas

$$(\cos\theta + i\sin\theta)^n = \cos n\theta + i\sin n\theta$$

and

$$(\cos\theta - i\sin\theta)^n = \cos n\theta - i\sin n\theta .$$

Since $\theta = \dfrac{m}{n}180°$, we have $\sin n\theta = \sin 180°m = 0$, and, consequently,

$$(\cos\theta + i\sin\theta)^n = (\cos\theta - i\sin\theta)^n .$$

Dividing both sides of this equality by $\cos^n \theta$ $\left(\cos \theta \neq 0, \text{ since } \tan \theta = \dfrac{\sin \theta}{\cos \theta} = \dfrac{p}{q}, \text{ where } q \neq 0\right)$, we obtain

$$(1 + i \tan \theta)^n = (1 - i \tan \theta)^n ,$$

that is,

$$\left(1 + i \frac{p}{q}\right)^n = \left(1 - i \frac{p}{q}\right)^n ,$$

or, after multiplying by q^n,

$$(q + ip)^n = (q - ip)^n .$$

We shall now show that this equality is impossible when p and q are integers, relatively prime, $p \neq 0$, $q \neq 0$, and p and q are not simultaneously equal to ± 1. To this end we express the equality as

$$(q - ip)^n = [(q - ip) + 2ip]^n$$
$$= (q - ip)^n + C_n^1(q - ip)^{n-1}2ip + C_n^2(q - ip)^{n-2}(2ip)^2 + \cdots$$
$$+ C_n^{n-1}(q - ip)(2ip)^{n-1} + (2ip)^n .$$

Eliminating the term $(q - ip)^n$ from the left and the right sides, dividing by $2ip$, and transposing $(2ip)^{n-1}$ to the left side, we obtain

$$-(2ip)^{n-1} = (q - ip)[C_n^1(q - ip)^{n-2} +$$
$$C_n^2(q - ip)^{n-3}2ip + \cdots + C_n^{n-1}(2ip)^{n-2}] .$$

Each side of this equality is a complex number; by equating the moduli of these complex numbers and taking into account the fact that the modulus of the product equals the product of the moduli of the factors, we find

$$(2p)^{2n-2} = (q^2 + p^2)B ,$$

where B is the modulus of the expression within the brackets on the right side of the preceding equality. Thus B is equal to the sum of the squares of the real and imaginary parts of the number,

$$C_n^1(q - ip)^{n-2} + C_n^2(q - ip)^{n-3}2ip + \cdots + C_n^{n-1}(2ip)^{n-2} ,$$

and is therefore an integer. Thus $(2p)^{2n-2}$ is divisible by $p^2 + q^2$. But since p and q are relatively prime, $p^2 + q^2$ does not have common factors with p and q; that is, 2^{2n-2} must be divisible by $p^2 + q^2$. The numbers p and q are either both odd, or one is even and the other odd. If one is even and the other odd, then $p^2 + q^2$ will be odd, but if $p = 2r + 1$ and $q = 2s + 1$ (both odd), then

$$p^2 + q^2 = 2(2r^2 + 2r + 2s^2 + 2s + 1)$$

will be even, but it will contain the odd factor $2(r^2 + r + s^2 + s) + 1$. This odd factor becomes unity only for $p = \pm 1$, $q = \pm 1$. Therefore for all remaining cases 2^{2n-2} cannot be divisible by $p^2 + q^2$. That is, $\theta = \arctan \dfrac{p}{q}$ cannot contain a rational number of degrees.

240. *First Solution.* Assume that a is not divisible by p. Then the integers $a, 2a, 3a, \cdots, (p-1)a$ will also fail to be divisible by p, and they will yield different remainders when divided by p [if ka and la $(p - 1 \geqq k > l)$ produced identical remainders when divided by p, then the difference $ka - la = (k - l)a$ would be divisible by p, which is impossible, since p is prime, a is not divisible by p, and $k - l$ is less than p.] But since, upon division by p, all possible remainders are exhausted by the $p - 1$ numbers $1, 2, 3, \cdots, p - 1$, it necessarily follows that

$$a = q_1 p + a_1 ,$$
$$2a = q_2 p + a_2 ,$$
$$3a = q_3 p + a_3 ,$$
$$\cdots\cdots\cdots\cdots ,$$
$$(p - 1)a = q_{p-1} p + a_{p-1} ,$$

where $a_1, a_2, \cdots, a_{p-1}$ are the positive integers $1, 2, \cdots, p - 1$, taken in some order. Multiplying together all these equalities, we obtain

$$[1\cdot2\cdots(p - 1)]a^{p-1} = Np + a_1 a_2 \cdots a_{p-1} ,$$
$$[1\cdot2\cdots(p - 1)](a^{p-1} - 1) = Np .$$

Therefore it follows that $a^{p-1} - 1$ is divisible by p, and $a^p - a$ is also divisible by p. (If a is initially divisible by p, then the assertion of Fermat's theorem is obvious.)

Second Solution. The theorem is obvious for $a = 1$, since in that case $a^p - a = 1 - 1 = 0$. We shall now apply mathematical induction to the positive integer a, and show that if $a^p - a$ is divisible by p, then it will follow that $(a + 1)^p - (a + 1)$ is divisible by p.

By use of the binomial theorem we find that

$$(a + 1)^p - (a + 1) = a^p + pa^{p-1} + C_p^2 a^{p-2} + C_p^3 a^{p-3} + \cdots + pa + 1 - a - 1$$
$$= (a^p - a) + pa^{p-1} + C_p^2 a^{p-2} + \cdots + C_p^{p-2} a^2 + pa .$$

But all the binomial coefficients

$$C_p^k = \frac{p(p - 1)(p - 2)\cdots(p - k + 1)}{1\cdot2\cdot3\cdots k}$$

are divisible by the prime number p, since C_p^k is an integer and since the numerator of the above expression for C_p^k contains the factor p whereas the denominator does not contain this factor. But since, by the induction hypothesis, $a^p - a$ is also divisible by p, it follows that $(a + 1)^p - (a + 1)$ is divisible by p.

We shall detail a more elegant variation of the same proof. Because all the binomial coefficients C_p^k are divisible by p, the difference

$$(A + B)^p - A^p - B^p$$
$$= pA^{p-1}B + C_p^2 A^{p-2}B^2 + \cdots + C_p^{p-2}A^2B^{p-2} + pAB^{p-2}$$

where A and B are any integers, is always divisible by p. If we apply this result, we find that

$$(A + B + C)^p - A^p - B^p - C^p$$
$$= \{[(A + B) + C]^p - (A + B)^p - C^p\} + (A + B)^p - A^p - B^p$$

is always divisible by p; that

$$(A + B + C + D)^p - A^p - B^p - C^p - D^p$$
$$= \{[(A + B + C) + D]^p - (A + B + C)^p - D^p\}$$
$$+ (A + B + C)^p - A^p - B^p - C^p$$

is always divisible by p; and that, in general,

$$(A + B + C + \cdots + K)^p - A^p - B^p - C^p - \cdots - K^p$$

is always divisible by p.

If in the last expression we now set $A = B = C = \cdots = K = 1$, and if we let the number of these terms be equal to a, we arrive at Fermat's theorem: $a^p - a$ is divisible by p.

241. The proof of Euler's theorem is completely analogous to the first proof of Fermat's theorem. Let $k_1, k_2, \cdots, k_r$ be the set of natural numbers which are less than, and relatively prime to, the integer N. We consider the r numbers $k_1 a, k_2 a, \cdots, k_r a$. All of these are relatively prime to N (since a is by hypothesis also relatively prime to N), and all of them, when divided by N, will yield different remainders (this is proven exactly as in problem 240). We may write

$$k_1 a = q_1 N + a_1 ,$$
$$k_2 a = q_2 N + a_2 ,$$
$$\cdots\cdots\cdots\cdots ,$$
$$k_r a = q_r N + a_r ,$$

where $a_1, a_2, \cdots, a_r$ must be the same numbers as $k_1, k_2, \cdots, k_r$, though in different order (since clearly the a_i are distinct, all less than N, and if a_i were not relatively prime to N, then neither would be k_i).

If we multiply together all the equalities, we obtain

$$k_1 k_2 \cdots k_r a^r = NM + a_1 a_2 \cdots a_r \,,$$
$$k_1 k_2 \cdots k_r (a^r - 1) = NM \,,$$

whence it follows that the integer $a^r - 1$ is divisible by N.

242. We shall prove this by mathematical induction. First, it is evident that for $m = 1$ the proposition of the problem is correct:

$$2^1 - 1 = 1 \,,$$
$$2^2 - 1 = 3 \,,$$
$$2^3 - 1 = 7$$

are not divisible by 5. We shall show that the proposition holds also for $n = 2$. Let 2^k be the smallest power of the number 2 which will produce a remainder of 1 when divided by $5^2 = 25$. (That is, k is least such that $2^k - 1$ is divisible by 25.) Now, assume that $k < 5^2 - 5 = 25 - 5 = 20$. If 20 is not divisible by k (that is, $20 = qk + r$, where $0 < r < k$), then we obtain

$$2^{20} - 1 = 2^{qk+r} - 1 = 2^r(2^{qk} - 1) + (2^r - 1) \,.$$

But $2^{20} - 1$ is divisible by 25 (by Euler's theorem), and $2^{qk} - 1 = (2^k)^q - 1^q$ is divisible by $2^k - 1$, which, by assumption, is also divisible by 25; therefore $2^r - 1$ must be divisible by 25, which is a contradiction of the assumption that k is the smallest number for which $2^k - 1$ is divisible by 25. Therefore, k must be a divisor of the number 20; that is, k can be equal to only 2, 4, 5, or 10. But $2^2 - 1 = 3$, $2^5 - 1 = 31$, and $2^{10} - 1 = 1023$ are not divisible by 5, whereas $2^4 - 1 = 15$ is divisible by 5, but is not divisible by 25. Thus, the proposition of the problem holds also for $n = 2$.

Let us now assume that the proposition of the problem holds for some n, but is invalid for $n + 1$, that is, that the least k such that $2^k - 1$ is divisible by 5^{n+1} is less than $5^{n+1} - 5^n = 4 \cdot 5^n$. We can show, exactly as above (for $n = 2$), that k must be a divisor of the number $4 \cdot 5^n$. But, moreover, we can show analogously that the number $5^n - 5^{n-1} = 4 \cdot 5^{n-1}$ must be the divisor of the number k. If it were true that $k = q \cdot 4 \cdot 5^{n-1} + r$, where $0 < r < 4 \cdot 5^{n-1}$, then $5^r - 1$ would be divisible by 5^n, which contradicts the hypothesis that the proposition of the problem is valid for the number n. Thus k has only one possible value: $k = 4 \cdot 5^{n-1}$.

Since the number

$$2^{5^{n-1}-5^{n-2}} - 1 = 2^{4 \cdot 5^{n-2}} - 1$$

is divisible by 5^{n-1} (from Euler's theorem) and is not divisible by 5^n (otherwise the hypothesis would not be true for n), then

$$2^{4 \cdot 5^{n-2}} = q \cdot 5^{n-1} + 1 \, ,$$

where q is not divisible by 5. From the expansion

$$(a + b)^5 = a^5 + 5a^4b + 10a^2b^2 + 10a^2b^3 + 5ab^4 + b^5$$

we obtain

$$2^{4 \cdot 5^{n-1}} - 1 = (2^{4 \cdot 5^{n-2}})^5 - 1 = (q \cdot 5^{n-1} + 1)^5 - 1$$
$$= 5^{n+1}(q^5 \cdot 5^{4n-6} + q^4 \cdot 5^{3n-5} + 2q^3 \cdot 5^{2n-4} + 2q^2 \cdot 5^{n-3}) + q \cdot 5^n \, ,$$

whence it is clear that $2^{4 \cdot 5^{n-1}} - 1$ is not divisible by 5^{n+1}. Hence, the truth of the proposition for n implies that it is also true for $n + 1$, whence the statement holds for all integers.

243. According to Euler's theorem (see problem 241), the number

$$2^{5^{10}-5^9} - 1 = 2^{4 \cdot 5^9} - 1 = 2^{7,812,500} - 1$$

is divisible by 5^{10}; therefore, for $n \geqq 10$ the difference

$$2^{7,812,500+n} - 2^n = 2^n(2^{7,812,500} - 1)$$

is divisible by 10^{10}. That is, the last ten digits of the number $2^{7,812,500+n}$ coincide with 2^n. This means that the last ten digits of the numbers of the sequence $2^1, 2^2, 2^3, \cdots, 2^n, \cdots$ will repeat after every 7,812,500 numbers. Moreover, this periodicity begins with the tenth number of this sequence, that is, with 2^{10}.

It follows from the result of problem 242 that the period is, actually not less than 7,812,500.

Remark: It can be proven analogously that the last n digits of the numbers in the sequence under consideration will repeat every $4 \cdot 5^{n-1}$ numbers, beginning with the nth number of this sequence (for example, the last two digits would repeat, beginning with the second number, every 20 numbers).

244. We shall prove an even more general theorem, namely, that for any integer N there is always a power of 2 whose last N digits will always be ones and twos. Since $2^5 = 32$ and $2^9 = 512$, the proposition is valid for $N = 1$ and $N = 2$. We shall carry out the proof by mathematical induction.

Assume that for some natural number N the final N digits of the number 2^n are ones and twos. We are to show that there is a power of 2 whose last $N + 1$ digits will be ones and twos.

According to the induction hypothesis, $2^n = a \cdot 10^N + b$, where b is an N-digit integer whose digits are all ones or twos. Let us designate the number $5^N - 5^{N-1} = 4 \cdot 5^{N-1}$ by the letter r; then by Euler's theorem (problem 241), the difference $2^r - 1$ will be divisible by 5^N. It follows that if the integer k is divisible by 2^{N+1}, then the difference $2^r k - k = k(2^r - 1)$ will be divisible by $2 \cdot 10^N$. That is, the final N digits of $2^r k$ and k will coincide, and the $(N + 1)$st digit from the end of each will be both odd or both even.

Let us now consider the following five powers of 2:

$$2^n ,$$
$$2^{n+r} = 2^r \cdot 2^n ,$$
$$2^{n+2r} = 2^r \cdot 2^{n+r} ,$$
$$2^{n+3r} = 2^r \cdot 2^{n+2r} ,$$
$$2^{n+4r} = 2^r \cdot 2^{n+3r} .$$

From what we have shown, the final N digits of all of these numbers will coincide (each of the numbers, as well as 2^n, will terminate in the same number, b, which consists entirely of one and twos), but the digits in the $(N + 1)$st place from the end of all of them will be simultaneously even or odd. We shall now prove that the digits in the $(N + 1)$st place from the end cannot be identical for any two of the five numbers. In fact, the difference of any two of the numbers can be expressed as $2^{n+m_1 r} (2^{m_2 r} - 1)$, where $m_1 = 0, 1, 2,$ or 3, but $m_2 = 1, 2, 3,$ or 4. If this difference is divisible by 10^{N+1}, then $2^{m_2 r} - 1$ must be divisible by 5^{N+1}; but since

$$m_2 r = m_2 \cdot (5^N - 5^{N-1}) < 5 \cdot (5^N - 5^{N-1}) = 5^{N+1} - 5^N ,$$

this contradicts the result of problem 242.

Therefore, the digits standing in the $(N + 1)$st place from the end of the above five integers must be either $1, 3, 5, 7,$ and 9 (all appearing) or else $0, 2, 4, 6,$ and 8 (in what order, we do not know). In either event, in one of these integers the $(N + 1)$st digit from the end must be 1 or else 2. This means that, in any event, there exists a power of 2 whose final $N + 1$ digits can comprise only the digits 1 and 2. This induction proves the proposition of the problem.

245. Let a be one of the numbers of the sequence $2, 3, \cdots, p - 2$. Consider the integers

$$a, 2a, \cdots, (p-1)a\,.$$

Clearly, no two of these integers can yield the same remainder upon division by p; therefore those remainders will be the positive integers $1, 2, 3, \cdots, p-1$ (all appearing, but in what order is unknown and not important). (Compare with the solution of problem 240.) In particular, there will be an integer b in the sequence $1, 2, \cdots, p-1$ such that ba, when divided by p, will have a remainder of 1. Now, $b \neq 1$ and $b \neq p-1$, since $2 \leqq a \leqq p-2$, and were $b = 1$ then the number $ba = a$ when divided by p would yield remainder $a \neq 1$; and were $b = p-1$ the number $ba = (p-1)a = pa - a$ when divided by p would yield the remainder $p - a \neq 1$. Moreover, $b \neq a$, since if a^2 yielded a remainder of 1 when divided by p, then $a^2 - 1 = (a+1)(a-1)$ would be divisible by p, which is possible only for $a = 1$ and $a = p-1$. Therefore, $2 \leqq b \leqq p-2$, and $b \neq a$, which means that each of the numbers $2, 3, \cdots, p-2$ can be paired with one other distinct integer of this set such that the product of this pair yields a remainder of 1 upon division by p. Accordingly, the product $2 \cdot 3 \cdot 4 \cdots p - 2$ itself yields a remainder of 1 when divided by p.

Now the number $p - 1$ may be thought of as yielding the remainder -1 upon division by p. It follows that

$$(p-1)! = 1 \cdot 2 \cdot 3 \cdots (p-2)(p-1) = [2 \cdot 3 \cdots (p-2)] \cdot (p-1)$$

has the remainder -1 when divided by p. That is,

$$(p-1)! = kp - 1\,;$$
$$(p-1)! + 1 = kp\,,$$

which says that $(p-1)! + 1$ is divisible by p.

If p is not prime, it must have a prime divisor $q < p$. Then $(p-1)!$ is divisible by q, since q is one of the factors in $(p-1)!$. But then q cannot divide $(p-1)! + 1$; hence neither can p.

246. Let $p = 4n + 1$ be a prime number. By Wilson's theorem (problem 245), the number $(p-1)! + 1 = (1 \cdot 2 \cdot 3 \cdots 4n) + 1$ is divisible by p. Now in $(4n)!$ we shall replace each factor exceeding $2n$ by the identical number expressed in terms of p and n. For example, since $p = 4n + 1$, we may write $2n + 1 = p - 2n$, $2n + 2 = p - (2n-1)$, and so on, until, finally, $4n = p - 1$. Then $(p-1)! = (4n)! = 1 \cdot 2 \cdot 3 \cdot \cdots (2n-1)2n(p-2n)[p - (2n-1)] \cdots (p-1)$.

It is readily seen that if the right side is expanded, then every term will have p as a factor, except a final term which will be equal to $[(2n)!]^2$. Therefore, $(p-1)! + 1 = Ap + [(2n)!]^2 + 1$, where A represents an expression unimportant to us. Since this number is divisible by p, and the term Ap is divisible by p, it follows that $[(2n)!]^2 + 1$ is also divisible by p. Therefore, the number $x = (2n)! = \left(\dfrac{p-1}{2}\right)!$ satisfies the conditions of the problem.

Remark: We note that if the integer x has the remainder x_1 when divided by p, then since

$$x^2 + 1 = (kp + x_1)^2 + 1 = (k^2 p + 2k x_1)p + x_1^2 + 1 ,$$

it follows that $x_1^2 + 1$ is divisible by p. Therefore we might stipulate, as an additional condition of the problem, that x < p, since such an $x = x_1$ does exist.

247. (a) The assertion of the problem follows immediately from the identity
$$(a^2 + b^2)(a_1^2 + b_1^2) = (aa_1 + bb_1)^2 + (ab_1 - ba_1)^2 .$$

(b) First, it is easily shown that no number of the form $4n + 3$ can be expressed as the sum of two squares. In fact, the square of every even number may be expressed as $4k$, and the square of an odd number may be expressed as $4k + 1$, since

$$(2a + 1)^2 = 4(a^2 + a) + 1 .$$

Accordingly, the sum of the squares of two even numbers may have the form $4n$; the sum of the squares of two odd numbers may have the form $4n + 2$; and finally, the sum of the squares of an even and an odd number may have the form $4n + 1$. Thus, an integer which can be written as $4n + 3$ cannot be the sum of two squares.

It is a more involved matter to show that every prime number of the form $4n + 1$ may be expressed as the sum of the squares of two integers. Let p be a prime of form $4n + 1$. We know from problem 246 that some multiple of p can be expressed as the sum of two squares; in fact, since there exists an integer x such that p divides $x^2 + 1$, there is an m such that

$$mp = x^2 + 1 . \tag{1}$$

From the remark following problem 246 we may find $x < p$, whence $x^2 + 1 < p^2$, and so in (1) we may assume that, because of the suitable choice of x, $m < p$. If $m = 1$, the proof is completed. Hence we shall assume that $m \neq 1$.

Now, if m is even, then $x^2 + 1$ is even, whence x must be odd. Then we can write

$$\frac{m}{2}p = \left(\frac{x+1}{2}\right)^2 + \left(\frac{x-1}{2}\right)^2 .$$

That is, there exists an $m^1 = \dfrac{m}{2}$ such that

$$m^1 p = x_1^2 + y_1^2 .$$

If m^1 is again even, then either x_1 and y_1 are both even, or else both are odd. In either event, we can easily determine (reasoning as above) that there exists an integer $m^{11}\left(= \dfrac{m^1}{2} \right)$ such that

$$m^{11} p = x_2^2 + y_2^2 .$$

Now, if m is a power of 2, then the proof concludes in an obvious way. Hence we need consider only the case in which m is odd (to which the problem reduces if m fails to be a power of 2) and that we have

$$mp = x^2 + y^2$$

(whether $y^2 = 1$ or not is not important in the sequel).

Let x_1 and y_1 be the least remainders in absolute value which can result when x and y, respectively, are divided by m:

$$x = mr + x_1 ,$$
$$y = ms + y_1$$

(either of x_1 or y_1, or both, may be negative integers). Then $|x_1|$ and $|y_1|$ are both less than $\dfrac{m}{2}$ (equality cannot hold, since m is odd), and we can write

$$mp = x^2 + y^2 = (m^2 r^2 + 2mr x_1 + x_1^2) + (m^2 s^2 + 2ms y_1 + y_1^2) .$$

It is clear that $x_1^2 + y_1^2$ is divisible by m:

$$x_1^2 + y_1^2 = mn$$

(it is readily found that $n = p - mr^2 - 2rx_1 - ms_1 - 2sy_1$).

We note that $n < \dfrac{m}{2}$; indeed, since $x_1 < \dfrac{m}{2}$ and $y_1 < \dfrac{m}{2}$, it follows that

$$mn = x_1^2 + y_1^2 < \left(\frac{m}{2}\right)^2 + \left(\frac{m}{2}\right)^2 = \frac{m^2}{2} .$$

In addition, $n \neq 0$, since otherwise x and y would both be divisible by m and $mp = x^2 + y^2$ would be divisible by m^2, which is impossible since p is prime and m is distinct from 1 and less than p.

We shall now show that np may be expressed as the sum of the squares of two integers. From the identity in problem (a) we have

$$mn \cdot mp = m^2np = (x^2 + y^2)(x_1^2 + y_1^2) = (xx_1 + yy_1)^2 + (xy_1 - yx_1)^2 .$$

But since $x = mr + x_1$ and $y = ms + y_1$, the numbers

$$xx_1 + yy_1 = mrx_1 + x_1^2 + msy_1 + y_1^2$$
$$= mrx_1 + msy_1 + (x_1^2 + y_1^2) = m(rx_1 + sy_1 + n) ,$$
$$xy_1 - yx_1 = mry_1 + x_1y_1 - msx_1 - x_1y_1 = m(ry_1 - sx_1)$$

are divisible by m. Thus we obtain

$$np = \left(\frac{xx_1 + yy_1}{m} \right)^2 + \left(\frac{xy_1 - x_1y}{m} \right)^2 ,$$

which displays np as the sum of squares.

If, now, $n = 1$, we are finished. If $n \neq 1$, we can, by using exactly the same method, decrease this number, that is, find an $n_1 < n$ such that $n_1 p$ can be expressed as the sum of the squares of two integers. If n_1 fails to be 1, we can find an $n_2 < n_1$ such that $n_2 p$ may be expressed as the sum of the squares of two integers. Continuation of this process produces a strictly decreasing sequence of positive integers for n_i, which must terminate with 1. , Therefore, the number $1 \cdot p$ can be expressed as the sum of the squares of two integers:

$$p = X^2 + Y^2 ,$$

as was to be shown.

(c) First, it follows almost immediately from the theorems in problems (a) and (b) that if a composite number N contains prime factors of the form $4n + 3$, but only even powers of them, then N can be expressed as the sum of the squares of two integers. Indeed, in that case the number N can be expressed as a product $P^2 \cdot Q$, where all the prime factors of P are of form $4n + 3$, whereas all the odd prime factors of Q are of form $4n + 1$. Since $2 = 1^2 + 1^2$, then, from the theorems of problem (b), all the prime factors of Q can be expressed as the sum of the squares of two integers. In that event, it follows from the theorem in problem (a) that even Q may be expressed as $Q = x^2 + y^2$. But since this is so,

$$N = p^2 \cdot Q = (px)^2 + (py)^2$$

may also be expressed as the sum of the squares of two integers. This proves one part of problem (c).

Let the composite number N, now, contain a prime factor p of the form $4n + 3$ to an odd power: $N = p^{2k+1} \cdot m$ (where m is not now divisible by p). We shall prove that N cannot be expressed as the sum of the squares of two integers. Indeed, assume that

$$N = x^2 + y^2 \, ,$$

where x and y are integers. Then upon dividing x^2, y^2, and N by the square of the greatest common factor of x and y, we must arrive at the equality

$$M = X_1^2 + Y_1^2 \, ,$$

where M is still divisible by p: $M = M_1 p$. By substituting for X_1 and Y_1 their remainders x_1 and y_1 upon division by p, we obtain the equality $mp = x_1^2 + y_1^2$, where $m < p$ [compare with the remarks after problem (246)]. But here, as in solution (b), p can be written as the sum of the squares of two integers, which is impossible [see the beginning of the solution of part (b)]. This completes the proof.

248. For $p = 2$ we have, trivially, $2 = 1^2 + 0^2 + 1$. For an odd prime p we shall give a constructive method for finding two numbers x and y, both less than $\dfrac{p}{2}$, which satisfy the condition of the problem.

Consider the $\dfrac{p+1}{2}$ integers $0, 1, 2, \cdots, \dfrac{p-1}{2}$. The squares of any two of these numbers, when divided by p, will yield different remainders. In fact, the equations

$$x_1^2 = k_1 p + r \, ,$$
$$x_2^2 = k_2 p + r$$

would imply that

$$x_1^2 - x_2^2 = (x_1 - x_2)(x_1 + x_2) = (k_1 - k_2)p \, .$$

That is,

$$(x_1 - x_2)(x_1 + x_2)$$

would be divisible by p, which is impossible since $x_1 < \dfrac{p}{2}$ and $x_2 < \dfrac{p}{2}$, and so

$$x_1 + x_2 < p \, ,$$
$$|x_1 - x_2| < p$$

(remember that p is a prime number). Hence the numbers of the set $0^2, 1^2, 2^2, \cdots, \left(\dfrac{p-1}{2}\right)^2$ yield $\dfrac{p+1}{2}$ distinct (nonnegative) remainders when divided by p. This implies that the $\dfrac{p+1}{2}$ (negative) numbers $-1, -1^2 - 1, -2^2 - 1, \cdots, -\left(\dfrac{p-1}{2}\right)^2 - 1$ when divided by p also yield $\dfrac{p+1}{2}$ different (nonnegative) remainders (if $-x_1^2 - 1$ and $-x_2^2 - 1$

yield identical remainders, then x_1^2 and x_2^2 also yield identical remainders[†]. But since there are only p distinct (nonnegative) remainders possible after division by p (namely, $0, 1, 2, \cdots, p-1$), it is clear that of the $p+1$ numbers $0^2, 1^2, 2^2, \cdots, \left(\dfrac{p-1}{2}\right)^2, -1, -1^2-1,$ $-2^2-1, \cdots, -\left(\dfrac{p-1}{2}\right)^2-1$ at least two of them must yield the same remainder when divided by p. From what has been shown above for pairs of this kind, one number must be of the form x^2 and the other of the form $-y^2-1$. But if

$$x^2 = kp + r\,,$$
$$-y^2 - 1 = lp + r\,,$$

then

$$x^2 + y^2 = (k-l)p - 1 = mp - 1\,;$$

that is, $x^2 + y^2 + 1 = mp$ is divisible by p.

Remark: The problem could have required that neither of the integers x or y is to exceed $p/2$, that is, that the sum $x^2 + y^2 + 1$ be less than p^2 and the quotient m resulting from the division of $x^2 + y^2 + 1$ by p be less than p.

249. (a) The assertion of the problem follows from the following identity:

$$(x_1^2 + x_2^2 + x_3^2 + x_4^2)(y_1^2 + y_2^2 + y_3^2 + y_4^2)$$
$$= (x_1 y_1 + x_2 y_2 + x_3 y_3 + x_4 y_4)^2 + (x_1 y_2 - x_2 y_1 + x_3 y_4 - x_4 y_3)^2$$
$$+ (x_1 y_3 - x_3 y_1 + x_4 y_2 - x_2 y_4)^2 + (x_1 y_4 - x_4 y_1 + x_2 y_3 - x_3 y_2)^2\,,$$

the validity of which can readily be verified.

Remark: Since the identity just displayed is rather involved, let us note its relationship to the simpler identity in problem 247 (a). The identity in problem 247 (a) may be generalized in the following manner:

$$(aa' + bb')(a_1 a_1' + b_1 b_1') = (aa_1' + bb_1')(a_1 a' + b_1 b') + (ab_1 - ba_1)(a' b_1' - b' a_1')\,.$$

If in the last identity we now set

$$a = x_1 + ix_2\,, \qquad a_1 = y_1 + iy_2\,,$$
$$a' = x_1 - ix_2\,, \qquad a_1' = y_1 - iy_2\,,$$
$$b = x_3 + ix_4\,, \qquad b_1 = y_3 + iy_4\,,$$
$$b' = x_3 - ix_4\,, \qquad b_1' = y_3 - iy_4\,,$$

where $i = \sqrt{-1}$, then we arrive at the identity of the present problem.

[†] The quotient q and the (nonnegative) remainder r resulting upon the division of a positive or a negative integer a by p are determined by the formula $a = qp + r$, where $0 \leqq r < p$, and the quotient q is negative for negative a.

(b) Since each integer may be expressed as a product of prime numbers, the result of problem (a) reduces this problem to showing that every prime number p may be expressed as the sum of the squares of four integers.

The proof of this proposition is completely analogous to the solution of problem 247 (b). We know, from the result of problem 248, that there exists a number m such that mp may be expressed as the sum of the squares of at most four integers:

$$mp = x_1^2 + x_2^2 + x_3^2 + x_4^2$$

(we can consider $x_3 = 1$ and $x_4 = 0$, although we do not need that information). We can further consider $m < p$ (see the remarks on the solution of problem 248). We shall show that if $m > 1$, then m can be reduced; that is, we can always find some number $n < m$ such that np can also be expressed as the sum of at most four squares.

This proof is straightforward if m is even, since in that case

$$mp = x_1^2 + x_2^2 + x_3^2 + x_4^2$$

is even, and either all x_k $(k = 1, 2, 3, 4)$ are even, or two of them are odd and the other two even, or they are all odd. In every case the four numbers x_1, x_2, x_3, and x_4 can be grouped into two pairs (say, $x_1; x_2$ and x_3, x_4), each pair consisting of two even or two odd numbers. Then the numbers

$$\frac{x_1 + x_2}{2}, \qquad \frac{x_3 + x_4}{2},$$

$$\frac{x_1 - x_2}{2}, \qquad \frac{x_3 - x_4}{2}$$

will be integers, and we will have

$$\frac{m}{2} \cdot p = \left(\frac{x_1 + x_2}{2}\right)^2 + \left(\frac{x_1 - x_2}{2}\right)^2 + \left(\frac{x_3 + x_4}{2}\right)^2 + \left(\frac{x_3 - x_4}{2}\right)^2.$$

That is, the number $\frac{m}{2} \cdot p$ can also be expressed as the sum of the squares of at most four integers.

The case where m is odd is more involved. Let us substitute y_k $(k = 1, 2, 3, 4)$ for the remainder, smallest in absolute value, which can appear when x_k is divided by m $\Big($if when x_k is divided by m there is a positive remainder greater than $\frac{m}{2}$, then we increase the quotient by 1 and show a negative remainder, whose magnitude is then $< \frac{m}{2}$):

$$x_k = mq_k + y_k \qquad (k = 1, 2, 3, 4),$$

where y_k is a positive or negative integer and $|y_k| < \frac{m}{2}$ (none of the integers y_k can have magnitude $\frac{m}{2}$, since m is odd).

We now have

$$x_k^2 = m^2q_k^2 + 2mq_ky_k + y_k^2 = mQ_k + y_k^2 \qquad (k = 1, 2, 3, 4),$$

where $Q_k = mq_k^2 + 2q_ky_k$ is an integer. Therefore,

$$mp = x_1^2 + x_2^2 + x_3^2 + x_4^2 = mq + y_1^2 + y_2^2 + y_3^2 + y_4^2$$

(here $q = Q_1 + Q_2 + Q_3 + Q_4$) and

$$y_1^2 + y_2^2 + y_3^2 + y_4^2 = mn$$

(here $n = p - q$). In this connection $n < m$, since

$$mn = y_1^2 + y_2^2 + y_3^2 + y_4^2 < 4\left(\frac{m}{2}\right)^2 = m^2,$$

moreover, $n \neq 0$, since otherwise all x_k would be divisible by m and $x_1^2 + x_2^2 + x_3^2 + x_4^2 = mp$ would necessarily be divisible by m^2, which is impossible, since p is prime and m is different from 1 and less than p.

We now show that the number np also can be expressed as the sum of not more than four squares. We shall see that each of the numbers mp and mn can be expressed as the sum of not more than four squares. From the identity proven in problem (a), it follows that the product

$$mp \cdot mn = m^2np$$

may also be expressed as the sum of the squares of four numbers:

$$m^2np = (x_1y_1 + x_2y_2 + x_3y_3 + x_4y_4)^2$$
$$+ (x_1y_2 - x_2y_1 + x_3y_4 - x_4y_3)^2$$
$$+ (x_1y_3 - x_3y_1 + x_2y_4 - x_4y_2)^2$$
$$+ (x_1y_4 - x_4y_1 + x_2y_3 - x_3y_2)^2.$$

We shall show that both sides of the last equality may be divided by m^2. Let us substitute on the right side of the equality $mq_k + y_k$ for all x_k. We see that all the expressions in parentheses on the right side of the equality are divisible by m: the expression in the first set of parentheses is divisible by m, since $y_1^2 + y_2^2 + y_3^2 + y_4^2 = mn$ is divisible by m, and the expressions in the remaining three sets of parentheses are divisible by m because after the substitution $x_k = mq_k + y_k$ all the products of the form $y_1 y_2$, and so on, cancel. Now if we divide the last equality by m^2, we obtain

$$np = z_1^2 + z_2^2 + z_3^2 + z_4^2 \,,$$

as was to be shown.

Thus, if the number m in the equation

$$mp = x_1^2 + x_2^2 + x_3^2 + x_4^2$$

is not equal to 1, it can always be decreased; that is, there will always be a positive $n < m$ for which a similar equality exists. If $n \neq 1$, we can still decrease the number n. In this way we obtain a strictly decreasing sequence of positive integers $m > n > n_1 > \cdots$ until in at most a finite number of steps we obtain

$$p = X_1^2 + X_2^2 + X_3^2 + X_4^2 \,.$$

250. Let us assume that $4^n(8k-1) = X^2 + Y^2 + Z^2$, where X, Y, and Z are integers (one or even two of which may be zero). For $n > 0$, the numbers X, Y, and Z must all be even, for if precisely one is odd (and the other two even), then the sum $X^2 + Y^2 + Z^2$ will be odd, and if two are odd (for example, $X = 2k + 1$ and $Y = 2l + 1$) and the other (for example, $Z = 2m$) is even, then the sum

$$X^2 + Y^2 + Z^2 = (2k+1)^2 + (2l+1)^2 + (2m)^2$$
$$= 4(k^2 + k + l^2 + l + m^2) + 2$$

is not divisible by 4. Now if we set

$$\frac{X}{2} = X_1 \,, \qquad \frac{z}{2} = z \,,$$
$$\frac{Y}{2} = Y_1 \,,$$

we arrive at the equation

$$4^{n-1}(8k-1) = X_1^2 + Y_1^2 + Z_1^2 \,.$$

If $n > 1$ $(n - 1 > 0)$, it can be shown, exactly as before, that all three of the numbers X_1, Y_1 and Z_1 also must be even, from which we obtain the equation

$$4^{n-2}(8k-1) = X_2^2 + Y_2^2 + Z_2^2 \,,$$

where X_2, Y_2, Z_2 are integers. Continuing to reason in exactly the same way, we are finally led to the conclusion that the number $8k - 1$ also must be expressible as the sum of three integers:

$$8k - 1 = x^2 + y^2 + z^2 . \tag{1}$$

Either one, or else all three, of the numbers x, y, z must be odd; in any other case the sum $x^2 + y^2 + z^2$ would be even. But the square of an odd number $2n + 1$,

$$(2n + 1)^2 = 4n^2 + 4n + 1 = 4n(n + 1) + 1 ,$$

always has a remainder of 1 when divided by 8 [since one of the consecutive numbers n and $n + 1$ must be even, which means that $4n(n + 1)$ is divisible by 8]. The square of an even number has a remainder of 0 when divided by 8 (if the number itself is divisible by 4) or else a remainder of 4 (if the number itself is not divisible by 4). This implies that if all the numbers x, y, z are odd, then the sum $x^2 + y^2 + z^2$ has a remainder of 3 when divided by 8, and if two of them are even, and one is odd, then when $x^2 + y^2 + z^2$ is divided by 8 there must be a remainder of 1 or 5. Thus, the sum of the squares of three integers can never yield a remainder of 7 when it is divided by 8. This contradiction of (1) proves the theorem.

251. We employ the identity

$$(a + b)^4 + (a - b)^4 = 2a^4 + 12a^2b^2 + 2b^4 ,$$

which follows from the expansions

$$(a + b)^4 = a^4 + 4a^3b + 6a^2b^2 + 4ab^3 + b^4 ,$$
$$(a - b)^4 = a^4 - 4a^3b + 6a^2b^2 - 4ab^3 + b^4 .$$

It follows from this identity that

$$
\begin{aligned}
[(a + b)^4 &+ (a - b)^4] + [(a + c)^4 + (a - c)^4] \\
&+ [(a + d)^4 + (a - d)^4] + [(b + c)^4 + (b - c)^4] \\
&+ [(b + d)^4 + (b - d)^4] + [(c + d)^4 + (c - d)^4] \\
&= 6a^4 + 6b^4 + 6c^4 + 6d^4 + 12a^2b^2 + 12a^2c^2 + 12\,a^2d^2 \\
&\quad + 12b^2c^2 + 12b^2d^2 + 12c^2d^2 = 6(a^2 + b^2 + c^2 + d^2)^2 .
\end{aligned}
$$

Thus

$$
\begin{aligned}
6(a^2 + b^2 &+ c^2 + d^2)^2 \\
&= (a + b)^4 + (a - b)^4 + (a + c)^4 + (a - c)^4 + (a + d)^4 + (a - d)^4 \\
&\quad + (b + c)^4 + (b - c)^4 + (b + d)^4 + (b - d)^4 + (c + d)^4 + (c - d)^4
\end{aligned}
$$

or, expressed in words: *if a number can be expressed as the sum of four squares, then six times its square can be expressed as the sum of twelve integers, each raised to the fourth power.* But, from the result of problem 250, each integer can be expressed as the sum of four squared integers (some of which may be zero); this implies that six times the square of each integer can be expressed as the sum of twelve integers, each raised to the fourth power (some of these may be zeros).

An arbitrary integer N divided by 6 has a remainder of $0, 1, 2, 3, 4,$ or 5; that is,

$$N = 6n + r ,$$

where $r = 0, 1, 2, 3, 4,$ or 5. Further, from the theorem of problem 249 (b), the number n may be expressed as the sum of four squares of integers (some of which may be zeros):

$$n = x^2 + y^2 + z^2 + t^2 .$$

By what has been shown above, each of the numbers $6x^2, 6y^2, 6z^2,$ and $6t^2$ (which are expressed as six times the squares of integers) can be expressed as the sum of twelve integers each raised to the fourth power (some of which may be zeros). Thus the number

$$6n = 6x^2 + 6y^2 + 6z^2 + 6t^2$$

can be expressed as the sum of $4 \cdot 12 = 48$ integers, each raised to the fourth power. But since $r = 0, 1, 2, 3, 4,$ or 5, that is,

$$r = 0^4 + 0^4 + 0^4 + 0^4 + 0^4 ,$$
$$\text{or} \quad r = 1^4 + 0^4 + 0^4 + 0^4 + 0^4 ,$$
$$\text{or} \quad r = 1^4 + 1^4 + 0^4 + 0^4 + 0^4 ,$$
$$\text{or} \quad r = 1^4 + 1^4 + 1^4 + 0^4 + 0^4 ,$$
$$\text{or} \quad r = 1^4 + 1^4 + 1^4 + 1^4 + 0^4 ,$$
$$\text{or} \quad r = 1^4 + 1^4 + 1^4 + 1^4 + 1^4 ,$$

the integer $N = 6n + r$ can be expressed in the form of the sum of $48 + 5 = 53$ fourth powers of integers (zeros allowed).

252. Set

$$a = x^3 + y^3 + z^3 .$$

From the identity in the solution of problem 162 (b) we have

$$(x + y + z)^3 - a = (x + y + z)^3 - x^3 - y^3 - z^3$$
$$= 3(x + y)(x + z)(y + z)$$

or

$$a = (x + y + z)^3 - 3(x + y)(x + z)(y + z)$$

(see the hint for the present problem). This brings us to the new unknowns, $x + y + z = Z$, $x + y = Y$, and x. We have $y = Y - x$ and $z = Z - Y$, and, therefore,

$$a = (x + y + z)^3 - 3(x + y)(x + z)(y + z)$$
$$= Z^3 - 3Y(x + Z - Y)(Y - x + Z - Y)$$
$$= Z^3 - 3Y(Z + x - Y)(Z - x)$$
$$= Z^3 - 3Y(Z^2 - x^2) + 3Y^2(Z - x).$$

We can now simplify the equation considerably by supposing that the unknowns x, Y, Z are related by

$$Z^3 = 3Y(Z^2 - x^2),$$

or the equivalent

$$Z = 3Y\left[1 - \left(\frac{x}{Z}\right)^2\right]$$

(see the hint for this problem). In this case our equation will take the form

$$a = 3Y^2(Z - x) = 3Y^2 Z\left(1 - \frac{x}{Z}\right),$$

or, since

$$Z = 3Y\left[1 - \left(\frac{x}{Z}\right)^2\right],$$

the form

$$a = 9Y^3\left(1 - \frac{x}{Z}\right)^2\left(1 + \frac{x}{Z}\right).$$

Let us introduce, together with the unknown x, a new unknown, $X = \frac{x}{Z}$. We then obtain

$$a = 9Y^3(1 - X)^2(1 + X);$$
$$Z = 3Y(1 - X^2).$$

Finally, let us introduce, together with the unknown Y, a new unknown, $\bar{Y} = 3Y(1 - X)$. Then the relation concerning the unknowns will take the form

$$Z = \bar{Y}(1 + X),$$

and the equation will be in the form

$$a = \frac{1}{3} \bar{Y}^3 \frac{1 + X}{1 - X}.$$

We are now at the end of the solution. Indeed, from the last equation X is expressed rationally in a (and $\bar{Y}$):

$$X = \frac{3a - \bar{Y}^3}{3a + \bar{Y}^3}.$$

Thus if $X = \dfrac{3a - \bar{Y}^3}{3a + \bar{Y}^3}$ and $Z = \bar{Y}(1 + X)$, then

$$a = x^3 + y^3 + z^3,$$

where x, y, and z are as in the following formulas:

$$x = ZX,$$

$$y = Y - x = \frac{\bar{Y}}{3(1 - X)} - ZX,$$

$$z = Z - Y = Z - \frac{\bar{Y}}{3(1 - X)}.$$

From these formulas it follows, in particular, that x, y, and z are rational, if only the unknown $\bar{Y}$ is rational. We may choose any desired $\bar{Y}$ in the formulas. (This circumstance is analogous to that where we simplify the equation $x^3 + y^3 + z^3 = a$ using the relation $y = -z$; then the unknown z may be chosen arbitrarily. See the hints for this problem).

We have thus found the solution to the equation

$$x^3 + y^3 + z^3 = a$$

in rational numbers (and even as many solutions as are desired which are related to the chosen distinct rational values of $\bar{Y}$). We have only to show that $\bar{Y}$ may be so chosen that x, y, and z will be positive (here we may employ the positive number a, which we have not used as yet). Let us express $Z = x + y + z$, $Y = x + y$, and $Z - x = y + z$ in terms of X and $\bar{Y}$,

$$x + y + z = Z = \bar{Y}(1 + X),$$
$$x + y = Y = \frac{\bar{Y}}{3(1 - X)},$$
$$y + z = Z - x = (1 - X)Z = \bar{Y}(1 - X^2),$$

and in these formulas let us set

$$1 - X = 1 - \frac{3a - \bar{Y}^3}{3a + \bar{Y}^3} = \frac{2\bar{Y}^3}{3a + \bar{Y}^3},$$
$$1 + X = 1 + \frac{3a - \bar{Y}^3}{3a + \bar{Y}^3} = \frac{6a}{3a + \bar{Y}^3}.$$

We obtain

$$x + y + z = \frac{6a\bar{Y}}{3a + \bar{Y}^3},$$
$$x + y = \frac{3a + \bar{Y}^3}{6\bar{Y}^2},$$
$$y + z = \frac{12a\bar{Y}^4}{(3a + \bar{Y}^3)^2}.$$

In these formulas let us set $\bar{Y} = \sqrt[3]{3a}$, that is

$$3a = \bar{Y}^3$$

(this value of $\bar{Y}$ may, of course, be irrational). We obtain

$$x + y + z = \bar{Y},$$
$$x + y = \frac{1}{3}\bar{Y},$$
$$y + z = \bar{Y};$$

that is,

$$x = 0,$$
$$y = \frac{1}{3}\bar{Y} = \frac{1}{3}\sqrt[3]{3a},$$
$$z = \frac{2}{3}\bar{Y} = \frac{2}{3}\sqrt[3]{3a}.$$

Let us choose a $\bar{Y}$ such that it will be rational and sufficiently close to $\sqrt[3]{3a}$ (it is possible to find a rational $\bar{Y}$ as close as we wish to $\sqrt[3]{3a}$). In that case y and z remain close to $\frac{1}{3}\sqrt[3]{3a}$ and to $\frac{2}{3}\sqrt[3]{3a}$, respectively (they remain positive). Further, from the formulas we find

$$\frac{x+y+z}{y+z} = \frac{3a + \bar{Y}^3}{2\bar{Y}^3} .$$

Therefore, if it is still necessary that the chosen value of $\bar{Y}$ be less than $\sqrt[3]{3a}$ (so that $3a > \bar{Y}^3$ and $3a + \bar{Y} > 2\bar{Y}^3$), then we have

$$\frac{x+y+z}{y+z} = \frac{3a + \bar{Y}^3}{2\bar{Y}^3} > 1 ,$$

and therefore x will also be positive. This cancludes the proof of the theorem.

For example, consider the case $a = \frac{2}{3}$. By setting $\bar{Y} = 1$ in the formulas we can easily obtain

$$x = \frac{5}{9} ,$$

$$y = \frac{1}{18} ,$$

$$z = \frac{5}{6} ;$$

in fact,

$$\left(\frac{5}{9}\right)^3 + \left(\frac{1}{18}\right)^3 + \left(\frac{5}{6}\right)^3 = \frac{2}{3} .$$

253. It follows from the result of problem 159 that there must be an infinite number of prime numbers. (That problem implies that prime numbers occur in the sequence of integers sufficiently "often"; for example, they occur "more often than" squares. See the remark to that problem). Also, from problem 65 we see that there must be an infinite number of prime numbers: if there existed only n prime numbers, then there would be no more than n number-pairs relatively prime to each other.

A much simpler and more direct proof of the theorem of the infinitude of prime numbers, ascribed to Euclid, is the following one.

Let us suppose that in all there are n prime numbers $2, 3, 5, 7, 11,$ $\cdots, p_n$. Let us form the number $N = 2 \cdot 3 \cdot 5 \cdot 7 \cdot 11 \cdots p_n + 1$. The number N is greater than all the prime numbers $2, 3, 5, \cdots, p_n$ and must therefore be composite. But since $N - 1$ is divisible by $2, 3, 5, 7, \cdots, p_n$, it follows that N is relatively prime to all prime numbers. This contradiction proves the theorem.

254. (a) The proof given here will be quite similar to Euclid's proof of the existence of an infinite number of primes. The integers comprising the first sequence given in the problem are all those of form $4k - 1$. Assume that only a finite number of primes appear in the sequence, that is, $3, 7, 11, 19, 23, \cdots, p_n$. Consider the number

$$N = 4(3 \cdot 7 \cdot 11 \cdot 19 \cdot 23 \cdots p_n) - 1 .$$

This integer exceeds every prime which appears in the given progression, and so, being a number of form $4k - 1$ (hence belonging to the progression), it must be composite. Factor N into its prime factors. none of these factors can be of form $4k - 1$, since $N + 1 = 4(3 \cdot 7 \cdot 11 \cdot 15 \cdot 19 \cdots p_n)$ is divisible by all primes of form $4k - 1$, and, consequently, N is relatively prime to all these numbers. Since N is odd, it must then be representable as the product of primes of form $4k + 1$. This is impossible, since the product of numbers of form $4k + 1$ is again a number of this same form,

$$(4k_1 + 1)(4k_2 + 1) = 16k_1k_2 + 4k_1 + 4k_2 + 1$$
$$= 4(4k_1k_2 + k_1 + k_2) + 1 = 4k_3 + 1 ,$$

and N is of form $4k - 1$.

Thus the assumption that there is a finite number of integers of form $4k - 1$ produces a contradiction. Hence the number of primes in the given sequence must be infinite.

The proof for the second sequence, which contains all the integers of form $6k - 1$, is quite analogous.

(b) The proof of this problem is based on the same idea as that just presented in part (a). Assume that the sequence $5, 9, 13, 17, 21, 25, \cdots$ contains only a finite number of primes: $5, 13, 17, \cdots, p_n$. Consider the number

$$N = (5 \cdot 13 \cdot 17 \cdots p_n)^2 + 1 .$$

The number N is clearly not a perfect square (it is one more than a square). However, N is the sum of two squares; from problem 247 (b) it follows that N can only be of form $4k + 1$ (no number of form $4k - 1$ can be expressed as the sum of two squares). From this point the method of proof is analogous to that used in problem 253 or 254 (a).

(c) The proof is somewhat more complicated than the proofs of parts (a) and (b), although it is still based on the same idea.

Assume that in the sequence $11, 21, 31, 41, 51, 61, \cdots$ there exists only a finite number of primes:

$$11, 31, 41, 61, \cdots, p_n .$$

Consider the integer $N = (11 \cdot 31 \cdot 41 \cdot 61 \cdots p_n)^5 - 1$. This number is relatively prime to all the integers $11, 31, 41, \cdots, p_n$, since $N + 1$ is divisible by all these integers. We shall designate the product $11 \cdot 31 \cdot 41 \cdots p_n$ by a. Then

$$N = a^5 - 1 = (a - 1)(a^4 + a^3 + a^2 + a + 1) .$$

Let us investigate what factorization of N can produce a factor of $a^4 + a^3 + a^2 + a + 1$. Clearly, $a^4 + a^3 + a^2 + a + 1$ is not divisible by 2 (it is the sum of five odd numbers). Further, $a^4 + a^3 + a^2 + a + 1$ is divisible by 5, inasmuch as a itself terminates in the digit 1 (a is a product of numbers all ending with 1; a^2, a^3, and a^4 each end with the digit 1, and so the sum $a^4 + a^3 + a^2 + a + 1$ ends with the digit 5). Now let p be a prime divisor of $a^4 + a^3 + a^2 + a + 1$ differing from 5. Here, $a - 1$ cannot be divisible by p, since otherwise a would be of form $kp + 1$, and so a^2, a^3, and a^4 (which are equal, respectively, to $(kp + 1)^2$, $(kp + 1)^3$, and $(kp + 1)^4$) would be of the same form, and the number

$$a^4 + a^3 + a^2 + a + 1 = (kp + 1)^4 + (kp + 1)^3 + (kp + 1)^2 + (kp + 1) + 1$$

would yield a remainder of 5 upon division by p. It follows that $p - 1$ must be divisible by 5; in fact, suppose that $p - 1$ yields the remainder 4 when divided by 5:

$$p - 1 = 5k + 4 .$$

We note (Fermat's theorem, problem 240) that $a^{p-1} - 1$ is divisible by p. But in this case

$$a^{p-1} = a^{5k+4} - 1 = a^4(a^{5k} - 1) + (a^4 - 1) ,$$

and since $a^{5k} - 1 = (a^5)^k - 1^k$ is divisible by $a^5 - 1$, which means that it is divisible also by p, it follows that $a^4 - 1$ is divisible by p. However,

$$a^5 - 1 = a(a^4 - 1) + (a - 1) ;$$

consequently, if $a^5 - 1$ and $a^4 - 1$ are divisible by p, then $a - 1$ also must be divisible by p. This, as shown above, is impossible. It may be shown, in analogous fashion, that $p - 1$ cannot yield the remainders 1, 2, or 3 upon division by 5.

Thus, $p - 1$ is divisible by 5 and is an even number (p being odd). Consequently, $p - 1$ is divisible by 10, which means that p is of form $10k + 1$; that is, it belongs to the given progression. Therefore, it is established that the prime divisors of the number $a^4 + a^3 + a^2 + a + 1$ can be only 5 and prime numbers of form $10k + 1$.

However, the number $a^4 + a^3 + a^2 + a + 1$ is obviously larger than 5, and it is not divisible by $5^2 = 25$. In fact, the integer a ends with the digit 1 and is consequently of form $5k + 1$. Further, by the binomial theorem, we have

$$a^4 + a^3 + a^2 + a + 1$$
$$= (5k + 1)^4 + (5k + 1)^3 + (5k + 1)^2 + 5k + 1 + 1$$
$$= 625k^4 + 4 \cdot 125k^3 + 6 \cdot 25k^2 + 4 \cdot 5k + 1$$
$$\quad + 125k^3 + 3 \cdot 25k^2 + 3 \cdot 5k + 1$$
$$\quad + 25k^2 + 2 \cdot 5k + 1 + 5k + 1 + 1$$
$$= 625k^4 + 5 \cdot 125k^3 + 10 \cdot 25k^2 + 10 \cdot 5k + 5$$
$$= 5 \cdot [5(25k^4 + 25k^3 + 10k^2 + 2k) + 1] .$$

It follows that this number, and, consequently, $N = a^5 - 1$, must have at least one prime divisor of form $10^k + 1$. But, as noted above, N is relatively prime to all prime numbers of form $10k + 1$. This contradiction proves the theorem.

Remark: We note that this proof, almost as it stands, will allow us to prove that any infinite arithmetic progression consisting of integers of form $2pk + 1$, where p is an odd prime, contains an infinite number of primes.

255. (a) Let a and b be adjacent sides of a rectangle; then its perimeter is $P = 2(a + b)$ and its area is $S = ab$. From the so-called theorem of the geometric and arithmetic means [that is, $ab \leqq \left(\dfrac{a + b}{2}\right)^2$; see the discussion of Section 11 (Problems)], we have

$$S = ab \leqq \left(\frac{a + b}{2}\right)^2 = \left(\frac{P}{4}\right)^2 .$$

It is obvious that the area S will be maximal when $\leqq$ in this expression represents equality; this happens only for $a = b$. Therefore, of all rectangles having the given perimeter P, the rectangle of greatest area will be a square.

(b) The solution is analogous to that of part (a), except here S is fixed and P becomes least when $a = b$.

256. Let a and b be the lengths of the legs of our right triangle, and let c be its hypotenuse. Then

$$c^2 = a^2 + b^2 .$$

Let $d = a + b$ be the sum of the legs. From equation ($1'$) of the discussion of Section 11 (Problems), we have

$$\frac{a+b}{2} \leqq \frac{1}{\sqrt{2}}\sqrt{a^2+b^2} \; ;$$

that is,

$$\frac{d}{2} \leqq \frac{C}{\sqrt{2}} \; ,$$

or,

$$d \leqq \frac{2C}{\sqrt{2}} \; .$$

257. In view of equation (*1*) of the discussion of Section 11 (Problems), we may write

$$\tan \alpha + \cot \alpha \geqq 2\sqrt{\tan \alpha \cdot \cot \alpha} = 2 \; .$$

The equality holds only when $\tan \alpha = \cot \alpha$, that is, for $\alpha = 45°$.

258. We shall rewrite the inequality

$$\frac{x^2+y^2}{2} \geqq \left(\frac{x+y}{2}\right)^2 ,$$

using $x = a + \dfrac{1}{a}$ and $y = b + \dfrac{1}{b}$. Then

$$\frac{\left(a+\frac{1}{a}\right)^2 + \left(b+\frac{1}{b}\right)^2}{2} \geqq \frac{1}{4}\left(a+\frac{1}{a}+b+\frac{1}{b}\right)^2$$

$$= \frac{1}{4}\left(1+\frac{1}{a}+\frac{1}{b}\right)^2 = \frac{1}{4}\left(1+\frac{a+b}{ab}\right)^2 = \frac{1}{4}\left(1+\frac{1}{ab}\right)^2 .$$

The fraction $\dfrac{1}{ab}$ is least when ab is greatest; hence by inequality (*1*) of the discussion in Section 11 (Problems), we have

$$ab \leqq \left(\frac{a+b}{2}\right)^2 = \frac{1}{4} \; .$$

Thus, $\dfrac{1}{ab} \geqq 4$ and $1+\dfrac{1}{ab} \geqq 5$. It follows that

$$\left(a+\frac{1}{a}\right)^2 + \left(b+\frac{1}{b}\right)^2 \geqq \frac{1}{2}\left(1+\frac{1}{ab}\right)^2 \geqq \frac{1}{2}5^2 = \frac{25}{2} ,$$

which was to be shown. The equality is attained only for $a=b=\dfrac{1}{2}$.

259. In view of inequality (*1*) of the discussion of Section 11 (Problems), we have

$$\frac{a+b}{2} \geqq \sqrt{ab} .$$

$$\frac{b+c}{2} \geqq \sqrt{bc} ,$$

$$\frac{c+a}{2} \geqq \sqrt{ca} .$$

If these three inequalities are multiplied together, we obtain

$$\frac{(a+b)(b+c)(c+a)}{8} \geqq \sqrt{a^2b^2c^2} = abc .$$

Equality takes place only if all three inequalities are equalities, that is, for $a = b = c$.

260. By inequality *(1)* of the discussion of Section 11 (Problems), we have

$$\frac{a+bx^4}{x^2} = \frac{a}{x^2} + bx^2 \geqq 2\sqrt{\frac{a}{x^2} \cdot bx^2} = 2\sqrt{ab} .$$

Equality holds only if

$$\frac{a}{x^2} = bx^2 ,$$

$$x^2 = \sqrt{\frac{a}{b}} .$$

261. Let the lengths of the two beams of the scale be a and b, respectively. Then in order to balance a weight of 1 pound placed in one of the pans—say on the pan hung from the beam of length a—the butcher must place in the other pan an amount of meat weighing actually $x = \frac{a}{b}$ pounds (since the moments $a \cdot 1$ and $b \cdot x$ must be equal in order to produce equilibrium). Similarly, if a one-pound weight is placed on the other pan, it will be balanced by $\frac{b}{a}$ pounds of meat. Therefore, the butcher gives out $\frac{a}{b} + \frac{b}{a}$ pounds of meat, and this is weighed out as 2 pounds. However, $\frac{a}{b} + \frac{b}{a} > 2$ $\left[\text{since } \frac{a}{b} + \frac{b}{a} - 2 = \frac{a^2 - 2ab + b^2}{ab} = \frac{(a-b)^2}{ab} \geqq 0, \text{ and equality can} \right.$ hold only if $a = b \Big]$. This means that the butcher gives out more meat than he charges for.

On the other hand, suppose the butcher sells his meat in the following way: A given piece is divided into two equal parts, and each part is weighed on a different pan. Let us assume (since other cases can be easily handled by similar reasoning) that the true weight of the whole is exactly 2 pounds; thus, each piece has true weight, by assumption, of 1 pound. When one piece of meat is placed on one of the pans, the total of the weights needed to balance it comes to $\frac{a}{b}$ pounds, and when the other piece is placed on the other pan, the total of the weights needed to balance that piece comes to $\frac{b}{a}$ pounds. Thus in this case, the sum of the markings on the weights *exceeds* the total weight of the meat—that is the butcher is short-weighing his customers.

Thus, whether the customer gains or the butcher gains depends upon the procedure used. The reader is invited to answer the equestion: Is a paradox involved here? [*Editor.*]

262. (a)   The harmonic mean H of two numbers a and b is defined by

$$\frac{1}{H} = \frac{\frac{1}{a} + \frac{1}{b}}{2} = \frac{a+b}{2ab} \ ;$$

thus $H = \frac{2ab}{a+b}$. We see that

$$\sqrt{\frac{a+b}{2} \cdot \frac{2ab}{a+b}} = \sqrt{ab} \ .$$

(b)   The result called for follows from part (a) and from inequality (*1*) of the discussion in Section 11 (Problems).

263.   It is to be shown that

$$a + b + c - 3\sqrt[3]{abc} \geqq 0 \ .$$

If we refer to problem 162 (a), we find

$$a + b + c - 3\sqrt[3]{abc} = (\sqrt[3]{a} + \sqrt[3]{b} + \sqrt[3]{c})$$
$$\times (\sqrt[3]{a^2} + \sqrt[3]{b^2} + \sqrt[3]{c^2} - \sqrt[3]{ab} - \sqrt[3]{bc} - \sqrt[3]{ac})$$
$$= \frac{1}{2}(\sqrt[3]{a} + \sqrt[3]{b} + \sqrt[3]{c})(\sqrt[3]{a^2} - 2\sqrt[3]{ab} + \sqrt[3]{b^2}$$
$$+ \sqrt[3]{b^2} - 2\sqrt[3]{bc} + \sqrt[3]{c^2} + \sqrt[3]{a^2} - 2\sqrt[3]{ac} + \sqrt[3]{c^2})$$
$$= \frac{1}{2}(\sqrt[3]{a} + \sqrt[3]{b} + \sqrt[3]{c})$$
$$\times [(\sqrt[3]{a} - \sqrt[3]{b})^2 + (\sqrt[3]{b} - \sqrt[3]{c})^2 + (\sqrt[3]{c} - \sqrt[3]{a})^2] \geqq 0 \ .$$

The equality sign is obtained only if all three differences

$$\sqrt[3]{a} - \sqrt[3]{b} ,$$
$$\sqrt[3]{b} - \sqrt[3]{c} ,$$
$$\sqrt[3]{c} - \sqrt[3]{a}$$

are zero, which happens only when $a = b = c$.

264. By Heron's formula, the area of a triangle having sides a, b, c, is equal to

$$S = \sqrt{p(p - a)(p - b)(p - c)}$$
$$= \sqrt{p} \cdot \sqrt{(p - a)(p - b)(p - c)} ,$$

where $p = \dfrac{a + b + c}{2}$ is the semi-perimeter. In view of problem 263,

$$(p - a)(p - b)(p - c) \leqq \left(\frac{p - a + p - b + p - c}{3} \right)^3$$
$$= \left(\frac{3p - 2p}{3} \right)^3 = \left(\frac{1}{3} p \right)^3 .$$

Since the sides of an equilateral triangle having perimeter $2p$ are each equal to $\dfrac{2}{3}p$, we have

$$(p - a)(p - b)(p - c) = \left(p - \frac{2}{3} p \right)^3 = \left(\frac{1}{3} p \right)^3 ,$$

which yields the greatest possible area for the triangle.

265. The volume of the pyramid is

$$V = \frac{xyz}{6} .$$

However, from problem 263 we see that

$$xyz \leqq \left(\frac{x + y + z}{3} \right)^3 = \frac{a^3}{27} .$$

Since the number $\dfrac{a^3}{27}$ is independent of $x, y,$ and z, the condition for maximal volume is

$$x = y = z = \frac{a}{3} .$$

266. It suffices to prove that

$$(a_1 + b_1)((a_2 + b_2)(a_3 + b_3) \geqq (\sqrt[3]{a_1 a_2 a_3} + \sqrt[3]{b_1 b_2 b_3})^3$$

We have

$$(a_1 + b_1)(a_2 + b_2)(a_3 + b_3)$$
$$= a_1 a_2 a_3 + b_1 b_2 b_3 + (a_1 a_2 b_3 + a_1 a_3 b_2 + a_2 a_3 b_1)$$
$$+ (a_1 b_2 b_3 + a_2 b_1 b_3 + a_3 b_1 b_2) .$$

However,

$$(\sqrt[3]{a_1 a_2 a_3} + \sqrt[3]{b_1 b_2 b_3})^3$$
$$= a_1 a_2 a_3 + b_1 b_2 b_3 + 3 \sqrt[3]{a_1^2 a_2^2 a_3^2 b_1 b_2 b_3} + 3 \sqrt[3]{a_1 a_2 a_3 b_1^2 b_2^2 b_3^2} .$$

In view of the inequality of problem 263, we obtain

$$\frac{a_1 a_2 b_3 + a_1 a_3 b_2 + a_2 a_3 b_1}{3} \geq \sqrt[3]{a_1^2 a_2^2 a_3^2 b_1 b_2 b_3} ,$$

$$\frac{a_1 b_2 b_3 + a_2 b_3 b_1 + a_3 b_1 b_2}{3} \geq \sqrt[3]{a_1 a_2 a_3 b_1^2 b_2^2 b_3^2} .$$

A comparison of the last three formulas yields the result sought.

267. The inequality of the problem may be written in the form

$$a_1 a_2 \cdots a_{2^m} \leq \left(\frac{a_1 + a_2 + \cdots + a_{2^m}}{2^m} \right)^{2^m} .$$

From inequality (1) of the discussion on of Section 11 (Problems), it follows that

$$a_1 a_2 \leq \left(\frac{a_1 + a_2}{2} \right)^2 , \tag{1}$$

and we also have

$$a_3 a_4 \leq \left(\frac{a_3 + a_4}{2} \right)^2 ,$$

from which we obtain

$$a_1 a_2 a_3 a_4 \leq \left[\frac{a_1 + a_2}{2} \cdot \frac{a_3 + a_4}{2} \right]^2 \leq \left[\left(\frac{\frac{a_1 + a_2}{2} + \frac{a_3 + a_4}{2}}{2} \right)^2 \right]^2$$
$$= \left(\frac{a_1 + a_2 + a_3 + a_4}{4} \right)^4 .$$

We find, in an analogous manner, that

$$a_5 a_6 a_7 a_8 \leq \left(\frac{a_5 + a_6 + a_7 + a_8}{4} \right)^4 ,$$

and from this we obtain

$$a_1 a_2 \cdots a_8 \leq \left[\frac{a_1 + \cdots + a_4}{4} \cdot \frac{a_5 + \cdots + a_8}{4} \right]^4$$

$$\leq \left[\left(\frac{\dfrac{a_1 + \cdots + a_4}{4} \cdot \dfrac{a_5 + \cdots + a_8}{4}}{2} \right)^2 \right]^4$$

$$= \left(\frac{a_1 + \cdots + a_8}{8} \right)^8 .$$

Repetition of this process finally yields

$$a_1 a_2 \cdots a_{2^m} \leq \left(\frac{a_1 + a_2 + \cdots + a_{2^m}}{2} \right)^{2^m} , \tag{2}$$

which is what we wished to show. We note that in *(1)* the equality is obtained for $a_1 = a_2$. If this fact is employed repeatedly, we find that the equality holds for (2) only for

$$a_1 = a_2 = \cdots = a_{2^m} .$$

268. There exist many proofs of this theorem, some of them not elementary. In view of the importance of this theorem, three different elementary proofs will be given. The first will be based on the result of problem 267, and the second and third will use mathematical induction.

First Proof. We show first that if the statement of the theorem concerning arithmetic and geometric means holds for $n + 1$ positive numbers, then it holds for n positive numbers. Assume that for any set of $n + 1$ positive numbers $a_1, a_2, \cdots, a_n, a_{n+1}$

$$\frac{a_1 + a_2 + \cdots + a_n + a_{n+1}}{n + 1} \geq \sqrt[n+1]{a_1 a_2 \cdots a_n a_{n+1}} ,$$

or,

$$\left(\frac{a_1 + a_2 + \cdots + a_n + a_{n+1}}{n + 1} \right)^{n+1} \geq a_1 a_2 \cdots a_n a_{n+1} .$$

If we substitute into this inequality

$$a_{n+1} = \frac{a_1 + a_2 + \cdots + a_n}{n} ,$$

then we obtain

$$\frac{a_1 + a_2 + \cdots + a_n + a_{n+1}}{n + 1} = \frac{a_1 + a_2 + \cdots + a_n + \dfrac{a_1 + a_2 + \cdots + a_n}{n}}{n + 1}$$

$$= \frac{\dfrac{n + 1}{n} (a_1 + a_2 + \cdots a_n)}{n + 1} = \frac{a_1 + a_2 + \cdots + a_n}{n} .$$

Consequently, we have

$$\left(\frac{a_1 + a_2 + \cdots + a_n}{n}\right)^{n+1} \geqq a_1 a_2 \cdots a_n \left(\frac{a_1 + a_2 + \cdots + a_n}{n}\right).$$

If we cancel $\frac{a_1 + a_2 + \cdots + a_n}{n}$ on both sides and take the nth root, we arrive at the desired result,

$$\frac{a_1 + a_2 + \cdots + a_n}{n} \geqq \sqrt[n]{a_1 a_2 \cdots a_n}.$$

But in view of the result of the preceding problem we know that the theorem of arithmetic and geometric means holds for arbitrarily large integers, in particular, for integers of form 2^m. Therefore, the theorem holds for arbitrary n, since given any n it is dominated by some number 2^m, and, by what has just been shown, the theorem holds for $2^m - 1, 2^m - 2$, and so on, and thus for n.

Also, if for $n + 1$ numbers equality holds only if all the numbers are equal, then the same condition must clearly hold for n numbers. Since this condition for equality was shown necessary in problem 267 for $n = 2^m$, it also applies in this problem.

Remark: This method of proof, which somewhat resembles that of mathematical induction, is sometimes called transfinite induction. It may be stated as follows: if the truth of the theorem for $n = k + 1$ implies that it holds for $n = k$, and if, no matter what positive integer n is chosen, the theorem can always be demonstrated for some integer exceeding n, then the theorem holds for all natural numbers n.

Second Proof (*mathematical induction*). For $n = 2$, we have

$$\frac{a_1 + a_2}{2} \geqq \sqrt{a_1 a_2}$$

Assume that the inequality of the problem has been proved for some number n of positive numbers; we shall show that it must hold for $n + 1$ positive numbers. Let $a_1, a_2, \cdots, a_n, a_{n+1}$ be $n + 1$ positive numbers, arranged such that a_{n+1} is the largest. Since

$$a_{n+1} \geqq a_1,$$
$$a_{n+1} \geqq a_2, \cdots, a_{n+1} \geqq a_n,$$

it follows that

$$a_{n+1} \geqq \frac{a_1 + a_2 + \cdots + a_n}{n}.$$

We shall write

$$\frac{a_1 + a_2 + \cdots + a_n}{n} = A_n,$$

$$\frac{a_1 + a_2 + \cdots + a_n + a_{n+1}}{n+1} = A_{n+1}.$$

Then

$$A_{n+1} = \frac{nA_n + a_{n+1}}{n+1}.$$

But since $a_{n+1} \geqq A_n$, we can write $a_{n+1} = A_n + b$, where $b \geqq 0$, then

$$A_{n+1} = \frac{nA_n + A_n + b}{n+1} = A_n + \frac{b}{n+1}.$$

If both members of the last equation are raised to the $(n+1)$st power we have

$$(A_{n+1})^{n+1} = \left(A_n + \frac{b}{n+1} \right)^{n+1}$$

$$= (A_n)^{n+1} + C_{n+1}^1 (A_n)^n \frac{b}{n+1} + \cdots$$

$$\geqq (A_n)^{n+1} + (A_n)^n b = (A_n)^n(A_n + b) = (A_n)^n a_{n+1}.$$

Since, by the induction assumption, the inequality holds for n numbers, that is,

$$(A_n)^n \geqq a_1 a_2 \cdots a_n,$$

it follows that

$$(A_{n+1})^{n+1} \geqq (A_n)^n a_{n+1} \geqq a_1 a_2 \cdots a_n a_{n+1},$$

and, consequently,

$$A_{n+1} \geqq \sqrt[n+1]{a_1 a_2 \cdots a_{n+1}}..$$

Now, if not all the numbers are equal, then $b > 0$, and the strict inequality holds.

Third Solution (mathematical induction). Designate a_1 by b_1^n, and so on; then the inequality whose proof we seek takes on the form

$$b_1^n + b_2^n + \cdots + b_n^n \geqq n b_1 b_2 \cdots b_n..$$

Suppose that this inequality is assumed to hold for any n positive numbers; we shall show that it holds for $n+1$ numbers:

$$b_1^{n+1} + b_2^{n+1} + \cdots + b_n^{n+1} + b_{n+1}^{n+1} \geqq (n+1) b_1 b_2 \cdots b_{n+1}.$$

Dividing both members of this inequality by b_{n+1}^{n+1}, and designating $\dfrac{b_1}{b_{n+1}}$ by c_1, and so on, yields

$$c_1^{n+1} + c_2^{n+1} + \cdots + c_n^{n+1} + 1 \geqq (n+1)c_1 c_2 \cdots c_n\,,$$

or,

$$c_1^{n+1} + c_2^{n+1} + \cdots + c_n^{n+1} \geqq (n+1)c_1 c_2 \cdots c_n - 1\,.$$

But by the induction hypothesis,

$$c_1^{n+1} + c_2^{n+1} + \cdots + c_n^{n+1} \geqq n(c_1 c_2 \cdots c_n)^{(n+1)/n}\,.$$

Hence it suffices to show that

$$(n+1)c_1 c_2 \cdots c_n - 1 \leqq n(c_1 c_2 \cdots c_n)^{(n+1)/n}\,,$$

or, designating $\sqrt[n]{c_1 c_2 \cdots c_n}$ by k, that

$$(n+1)k^n - 1 \leqq nk^{n+1}\,.$$

The last inequality follows from the following computation:

$$
\begin{aligned}
(n+1)k^n - 1 - nk^{n+1} &= -nk^n(k-1) + (k^n - 1) \\
&= (k-1)(-nk^n + k^{n-1} + k^{n-2} + \cdots + k + 1) \\
&= -(k-1)[(k^n - k^{n-1}) + (k^n - k^{n-2}) + \cdots + (k^n - 1)] \\
&= -(k-1)^2[k^{n-1} + k^{n-2}(k+1) + \cdots \\
&\quad + k(k^{n-2} + k^{n-3} + \cdots + k + 1) + (k^{n-1} + k^{n-2} + \cdots + k + 1)] \leqq 0
\end{aligned}
$$

(for $k > 0$, the expression in brackets is clearly positive).

The final inequality shown reduces to equality for $k = 1$. Mathematical induction may be employed to show that equality will hold only if

$$c_1 = c_2 = \cdots = c_n = 1\,;$$

that is, if

$$b_1 = b_2 = \cdots = b_n = b_{n+1}\,.$$

Remark: Other proofs may be found in the volumes referred to at the beginning of Section 11 (Problems); see, for examples, the volume by Hardy, Littlewood, and Polya. Moreover, the inequality of this problem is easily established from the results of problem 269 (a) and (b) and problem 270, provided problem 268 is not used to establish it. (See, for example, the second proof of problem 269 (a) and (b).

269. (a) *First Solution*. Designate the n positive numbers by $x_1, x_2, \cdots, x_n$. By the imposed conditions, the sum $x_1 + x_2 + \cdots + x_n$ is given, and hence so is the arithmetic mean

$$\frac{x_1 + x_2 + \cdots + x_n}{n} = A .$$

By the theorem of arithmetic and geometric means

$$x_1 x_2 \cdots x_n \leqq A^n ,$$

for which the equality is obtained only for

$$x_1 = x_2 = \cdots = x_n = A .$$

Therefore, this product assumes its maximal value A^n for

$$x_1 = x_2 = \cdots = x_n$$

Second Solution. Arrange the n positive numbers in increasing (non-decreasing) order:

$$x_1 \leqq x_2 \leqq x_3 \leqq \cdots \leqq x_n .$$

If all the numbers are equal, we have

$$x_1 x_2 \cdots x_n = \left(\frac{x_1 + x_2 + \cdots + x_n}{n}\right)^n .$$

Assume that not all these numbers are equal. We shall show, under this assumption, that there exists another set of n numbers having the same sum but whose product exceeds $x_1 x_2 \cdots x_n$.

Let A be the arithmetic mean of the numbers $x_1, x_2, \cdots, x_n$. We have, from our assumptions,

$$x_1 < x_n ;$$
$$x_1 < A ;$$
$$x_n > A .$$

Replace the numbers x_1 and x_n, respectively, by new numbers x_1' and x_n' which preserve the arithmetic mean:

$$\frac{x_1' + x_2 + x_3 + \cdots + x_{n-1} + x_n'}{n} = A .$$

To do this, we can set

$$x_n' = A ,$$
$$x_1' = x_1 + (x_n - x_n') .$$

In the product $x_1 x_2 \cdots x_n$ all the factors except the first and last are left unchanged; we shall show that the product $x_1' x_n'$ exceeds $x_1 x_n$:

$$x_1' x_n' > x_1 x_n .$$

We designate $x_n - x'_n$ by t; then $x'_n = x_n - t$ and $x'_1 = x_1 + t$. We have

$$x'_n x'_1 = (x_n - t)(x_1 + t) = x_n x_1 + (x_n - x_1) t - t^2 .$$

But since $x'_n = A > x_1$, we have

$$x_n - x_1 > x_n - x'_n = t ,$$

from which we obtain

$$(x_n - x_1) - t > 0 ,$$
$$(x_n - x_1)t - t^2 > 0 ,$$

and, consequently,

$$x'_n x'_1 = x_n x_1 + (x_n - x_1)t - t^2 > x_n x_1 .$$

If the new set $x'_1, x_2, x_3, \cdots, x_{n-1}, A$ is now not yet composed of equal numbers, then, arranging them again in increasing order, we may repeat the previous procedure to find a new set whose sum is the same but whose product is greater: In this new sequence at least two of the numbers will be equal to A. Repetition of this reasoning must finally produce a set all of whose elements are the number A.

(b) *First Solution.* Set

$$\frac{a_1}{a_2} = x_1 , \qquad \cdots\cdots\cdots\cdots ,$$

$$\frac{a_2}{a_3} = x_2 , \qquad \frac{a_{n-1}}{a_n} = x_{n-1}$$

$$\frac{a_3}{a_4} = x_3 , \qquad \frac{a_n}{a_1} = x_n$$

The geometric mean of the n numbers $x_1, x_2, x_2 \cdots, x_n$ is equal to 1. Hence, by problem 268, we have

$$\frac{x_1 + x_2 + \cdots + x_n}{n} \geq 1 ,$$

that is,

$$x_1 + x_2 + \cdots + x_n \geq n .$$

Second Solution. The given inequality is readily proved by mathematical induction, without recourse to the theorem of arithmetic and geometric means. Assume the inequality valid for $n - 1$ positive numbers, that is,

$$\frac{a_1}{a_2} + \frac{a_2}{a_3} + \cdots + \frac{a_{n-2}}{a_{n-1}} + \frac{a_{n-1}}{a_1} \geq n - 1 \, . \tag{1}$$

We shall show that the inequality is then implied for n positive numbers.

Let a_n be the least of the n numbers $a_1, a_2, \cdots, a_n$. Then

$$a_1 - a_n \geq 0 \, ,$$

$$a_{n-1} \geq a_n \, .$$

Hence, $a_{n-1}(a_1 - a_n) \geq a_n(a_1 - a_n)$, and then

$$a_1 a_{n-1} + a_n^2 - a_n a_{n-1} \geq a_n a_1 \, .$$

Division by $a_n a_1$ yields

$$\frac{a_{n-1}}{a_n} + \frac{a_n}{a_1} - \frac{a_{n-1}}{a_1} \geq 1 \, . \tag{2}$$

If inequalities (1) and (2) are added, we obtain

$$\frac{a_1}{a_2} + \frac{a_2}{a_3} + \cdots + \frac{a_{n-2}}{a_{n-1}} + \frac{a_{n-1}}{a_1} + \frac{a_{n-1}}{a_n} + \frac{a_n}{a_1} - \frac{a_{n-1}}{a_1}$$

$$= \frac{a_1}{a_2} + \frac{a_2}{a_3} + \cdots + \frac{a_{n-1}}{a_n} + \frac{a_n}{a_1} \geq (n-1) + 1 = n \, ;$$

the inequality holds for the n numbers $a_1, a_2, \cdots, a_n$. Therefore, the inequality holds for all natural numbers n.

270. By the theorem of the arithmetic and geometric means (problem 268), we have

$$\frac{\frac{1}{a_1} + \frac{1}{a_2} + \cdots + \frac{1}{a_n}}{n} \geq \sqrt[n]{\frac{1}{a_1} \frac{1}{a_2} \cdots \frac{1}{a_n}} \, ,$$

from which we obtain

$$n : \left(\frac{1}{a_1} + \frac{1}{a_2} + \cdots + \frac{1}{a_n} \right) \leq \sqrt[n]{a_1 a_2 \cdots a_n} \, .$$

The equality holds only if $a_1 = a_2 = \cdots = a_n$.

Remark: The inequality here can be proven in many other ways, independently of problem 268 (it is recommended that the reader try to find such a proof). Then this result may be used to find a new proof for problem 268.

271. By problem 268, we can write

$$^{n+1}\sqrt{ab^n} = {}^{n+1}\sqrt{a\underbrace{bb\cdots b}}$$

$$\leqq \frac{a+b+b+\cdots+b}{n+1} = \frac{a+nb}{n+1}\;.$$

The equals sign holds only if $a = b$.

272. Since the arithmetic mean of n positive numbers exceeds the geometric mean, and the harmonic mean is less than the geometric mean, we have

$$A(a) \geqq H(a)\;,$$

or

$$\frac{a_1 + a_2 + \cdots + a_n}{n} \geqq \frac{n}{\dfrac{1}{a_1} + \dfrac{1}{a_2} + \cdots + \dfrac{1}{a_n}}\;.$$

which yields the required result.

Equality is attained only if $a_1 = a_2 = \cdots = a_n$.

273. Use the theorem of the arithmetic and geometric means, making the following substitutions:

$$a_1 = 1\;,$$
$$a_2 = 2\;,$$
$$\cdots\cdots\;,$$
$$a_n = n\;.$$

We obtain

$$^n\sqrt{n!} = {}^n\sqrt{1 \cdot 2 \cdots n} \leqq \frac{1 + 2 + \cdots + n}{n}$$

$$= \frac{n(n+1)}{2n} = \frac{n+1}{2}\;.$$

If the first and last members shown are raised to the power n, the statement of the problem follows.

274. By the theorem of the arithmetic and geometric means, we have

$$a_1 a_2^2 a_3^3 a_4^4 = a_1 a_2 a_2 a_3 a_3 a_3 a_4 a_4 a_4 a_4$$

$$\leqq \left(\frac{a_1 + a_2 + a_2 + a_3 + a_3 + a_3 + a_4 + a_4 + a_4 + a_4}{10}\right)^{10}$$

$$= \left(\frac{a_1 + 2a_2 + 3a_3 + 4a_4}{10}\right)^{10}\;.$$

Equality holds only for $a_1 = a_2 = a_3 = a_4$.

275. (a) We first note that the left side of the given inequality contains two factors $\frac{1}{2}$, three factors $\frac{1}{3}$ and so on, and finally n factors $\frac{1}{n}$; in all, there are $1 + 2 + 3 + \cdots + n = \frac{n(n+1)}{2}$ such factors. The geometric mean of these factors is equal to the $\frac{n(n+1)}{2}$ root of this product; the arithmetic mean is

$$\frac{1 \cdot 1 + 2 \cdot \frac{1}{2} + 3 \cdot \frac{1}{3} + \cdots + n \cdot \frac{1}{n}}{\frac{n(n+1)}{2}} = \frac{n}{\frac{n(n+1)}{2}} = \frac{2}{n+1}.$$

The validity of the given inequality follows immediately from the theorem of the arithmetic and geometric means (problem 268).

(b) The solution here is analogous to that of part (a). The arithmetic mean of the factors of the left member is equal to

$$\frac{1 \cdot 1 + 2 \cdot 2 + 3 \cdot 3 + \cdots + n \cdot n_1}{\frac{n(n+1)}{2}} = \frac{1^2 + 2^2 + 3^2 + \cdots + n^2}{\frac{n(n+1)}{2}}$$

$$= \frac{n(n+1)(2n+1)}{6} : \frac{n(n+1)}{2} = \frac{2n+1}{3}$$

[see the solution of problem 134 (a)].

276. By the theorem of the arithmetic and geometric means we have

$$(1 + a_1)(1 + a_2) \cdots (1 + a_n) \leq \left(\frac{n + a_1 + a_2 + \cdots + a_n}{n}\right)^n$$

$$= \left(1 + \frac{s}{n}\right)^n = 1 + n\left(\frac{s}{n}\right) + \frac{n(n-1)}{2}\left(\frac{s}{n}\right)^2 + \cdots + \left(\frac{s}{n}\right)^n,$$

where the last equality is an application of the binomial theorem. We note that the coefficient of s^m will be

$$\frac{n!}{m!(n-m)!} \frac{1}{n^m}.$$

However, $(n - m)! \, n^m \geq n!$, whence

$$\frac{n!}{m!(n-m)!} \frac{1}{n^m} \leq \frac{n!}{m! \, n!} = \frac{1}{m!},$$

from which the assertion of the problem follows directly.

The inequality reduces to equality only for $n = 1$.

277. Rewrite the left member of the inequality in the form

$$1^{\alpha/2^n}2^{1/2}(2^2)^{1/2^2}\cdots(2^n)^{1/2^n},$$

where α is an arbitrary integer.

Now the solution of this problem is analogous to that of problem 275 (a) and (b). It follows from the theorem of the arithmetic and geometric means that

$$1^{\alpha/2^n}2^{1/2}(2^2)^{1/2^2}\cdots(2^n)^{1/2^n}[1^{\alpha}2^{2^{n-1}}(2^2)^{2^{n-2}}\cdots(2^n)]^{(1/2)n}$$

$$\leqq\left[\left(\frac{\alpha + 2\cdot2^{n-1} + 2^2\cdot2^{n-2} + \cdots + 2^n\cdot1}{\alpha + 2^{n-1} + 2^{n-2} + \cdots + 1}\right)^{\alpha+2^{n-1}+\ldots+1}\right]\frac{1}{2}n$$

$$=\left(\frac{\alpha + n\cdot2^n}{\alpha + 2^n - 1}\right)^{\frac{\alpha+2^n-1}{2n}}.$$

If $\alpha = 1$, the last exponent shown above is 1, and the expression on the right becomes

$$\frac{1}{2^n} + n.$$

The statement of the problem immediately follows.

278. The expression $(1 - x^5(1 + x)(1 + 2x))^2$ is negative for $|x|>1$, and is positive for $|x| < 1$. In fact,

$$(1 - x)^5(1 + x)(1 + 2x)^2 = (1 - x)^4(1 - x^2)(1 + 2x)^2.$$

The first and third factors on the right are always positive for $x \neq 1$, and the second factor is positive or negative, respectively, depending upon whether $|x| < 1$ or $|x| > 1$. Accordingly, we shall consider only values of x such that $|x| < 1$. We now use the theorem of arithmetic and geometric means on the five factors $1 - x$, the single factor $1 + x$, and the two factors $1 + 2x$:

$$(1 - x)^5(1 + x)(1 + 2x)^2$$

$$\leqq \frac{5(1 - x) + (1 + x) + 2(1 + 2x)}{5 + 1 + 2}\bigg)^{5+1+2}$$

$$=\left(\frac{5 + 1 + 2}{5 + 1 + 2}\right)^{5+1+2} = 1.$$

The right side of this inequality is independent of the value assigned to x. Hence, the left side will be maximal for that value of x which makes all the factors equal. The only value of x which will accomplish this is $x = 0$. Then for this value of x the given product reaches its maximal value 1.

279. Let r be the radius of the circle; designate the known length OM by a, and the unknown length ON by x (Figure 27). Then we can write:

$$MN = x - a ,$$
$$NQ = \sqrt{r^2 - x^2} ,$$

and, consequently, the squared area of the rectangle is equal to

$$4(x - a)^2 (r^2 - x^2) .$$

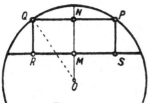

Figure 27

We must determine for which value of x this expression is maximal. We rewrite our product in the form

$$\frac{4}{\alpha\beta}[(x - a) \cdot (x - a) \cdot \alpha(r - x) \cdot \beta(r + x)] ,$$

where α and β are chosen such that the sum of the factors in the brackets, that is,

$$(x - a) + (x - a) + \alpha(r - x) + \beta(r + x)$$
$$= (2 - \alpha + \beta)x + (\alpha + \beta)r - 2a ,$$

will be independent of x (such that $\alpha - \beta = 2$).

The product (1) attains its maximal value if

$$\alpha(r - x) = \beta(r + x) = x - a$$

[see the solution of problem 269 (a)]. But the equation $\alpha(r - x) = \beta(r + x)$ implies that

$$\alpha + \beta = \frac{(\alpha - \beta)r}{x} = \frac{2r}{x} .$$

From this, and from the condition $\alpha - \beta = 2$, it is readily found that

$$\alpha = \frac{r}{x} + 1 = \frac{r+x}{x} ,$$

$$\beta = \frac{r-x}{x} .$$

Substituting this value of α in the equation $\alpha(r-x) = x - \alpha$, we obtain

$$\frac{r^2 - x^2}{x} = x - a;$$

$$2x^2 - ax - r^2 = 0 ,$$

from which we have

$$x = \frac{a + \sqrt{a^2 + 8r^2}}{4}$$

(the positive sign is taken for the root, since $x > 0$).

The segment of length x, and hence the rectangle sought, can be constructed using ruler and compass.

280. The volume of the box is equal to

$$(2a - 2b)^2 \cdot b = 4b(a - b)^2 .$$

We can write the right member of this equality in the form

$$\frac{4}{\alpha^2} [b \cdot \alpha(a - b) \cdot \alpha(a - b)] ,$$

and we shall select α such that the sum of the factors within the brackets, that is,

$$b + 2\alpha(a - b) = 2\alpha a + (1 - 2\alpha)b ,$$

does not depend upon $b \left(\text{we take } \alpha = \frac{1}{2} \right)$.

The maximal value of the product (1) [see the solution of problem 269 (a)] is obtained when

$$b = \alpha(a - b) .$$

Thus, we find

$$b = \frac{\alpha a}{1 + \alpha} = \frac{\frac{1}{2}a}{\frac{3}{2}} = \frac{a}{3} .$$

281. (a) The inequality of this problem has the same relationship to inequality $(1')$ in the discussion of § 11 (problems) as does the theorem of arithmetic and geometric means (problem 268) to inequality (1) of that discussion. A proof of this is possible in any one of several ways, all of them analogous to the proofs given for problem 268.

The solution here will be analogous to the first solution of problem 268. Inasmuch as

$$\left(\frac{a_1 + a_2}{2}\right)^2 \leqq \frac{a_1^2 + a_2^2}{2} ,$$

and

$$\left(\frac{a_3 + a_4}{2}\right)^2 \leqq \frac{a_3^2 + a_4^2}{2} ,$$

we have

$$\left(\frac{a_1 + a_2 + a_3 + a_4}{4}\right)^2 = \left(\frac{\dfrac{a_1 + a_2}{2} + \dfrac{a_3 + a_4}{2}}{2}\right)^2$$

$$\leqq \frac{\left(\dfrac{a_1 + a_2}{2}\right)^2 + \left(\dfrac{a_3 + a_4}{2}\right)^2}{2} \leqq \frac{\dfrac{a_1^2 + a_2^2}{2} + \dfrac{a_3^2 + a_4^2}{2}}{2}$$

$$= \frac{a_1^2 + a_2^2 + a_3^2 + a_4^2}{4} .$$

Also, since

$$\left(\frac{a_1 + a_2 + a_3 + a_4}{4}\right)^2 \leqq \frac{a_1^2 + a_2^2 + a_3^2 + a_4^2}{4}$$

and

$$\left(\frac{a_5 + a_6 + a_7 + a_8}{4}\right)^2 \leqq \frac{a_5^2 + a_6^2 + a_7^2 + a_8^2}{4} ,$$

we can conclude that

$$\left(\frac{a_1 + a_2 + \cdots + a_8}{8}\right)^2 \leqq \frac{a_1^2 + a_2^2 + \cdots + a_8^2}{8} .$$

Continuation of this process enables us to prove the theorem for 2^m numbers, where m is any arbitrary natural number.

We now shall show that if the theorem is valid for $n + 1$ positive numbers, that is, if

$$\left(\frac{a_1 + a_2 + \cdots + a_n + a_{n+1}}{n + 1}\right)^2 \leqq \frac{a_1^2 + a_2^2 + \cdots + a_n^2 + a_{n+1}^2}{n + 1} , \quad (1)$$

then it holds for n numbers. Substituting in (1)

$$a_{n+1} = \frac{a_1 + a_2 + \cdots + a_n}{n},$$

we obtain

$$\left(\frac{a_1 + a_2 + \cdots + a_n}{n}\right)^2 \leq \frac{a_1^2 + a_2^2 + \cdots + a_n^2 + \left(\dfrac{a_1 + a_2 + \cdots + a_n}{n}\right)^2}{n+1}$$

(compare with the first solution of problem 268), from which we conclude that

$$\left(\frac{a_1 + a_2 + \cdots + a_n}{n}\right)^2 \leq \frac{a_1^2 + a_2^2 + \cdots + a_n^2}{n}.$$

The remainder of the proof does not differ significantly from that of the first solution of problem 268.

It is easily established that equality holds only if $a_1 = a_2 = \cdots = a_n$.

(b) The validity of the inequality will first be established for two numbers; that is, we shall establish that

$$\left(\frac{a_1 + a_2}{2}\right)^k \leq \frac{a_1^k + a_2^k}{2}. \tag{1}$$

For $k = 2$ this inequality has already been proved [see $(1')$ of the discussion of Section 11 (Problems)]. Assume now that the inequality holds for some given k. We have

$$\left(\frac{a_1 + a_2}{2}\right)^{k+1} = \left(\frac{a_1 + a_2}{2}\right)^k \frac{a_1 + a_2}{2} < \frac{a_1^k + a_2^k}{2} \cdot \frac{a_1 + a_2}{2}$$

$$= \frac{a_1^{k+1} + a_2^{k+1}}{2} - \frac{a_1^{k+1} + a_2^{k+1} - a_1^k a_2 - a_1 a_2^k}{4}$$

$$= \frac{a_1^{k+1} + a_2^{k+1}}{2} - \frac{(a_1^k - a_2^k)(a_1 - a_2)}{4} \leq \frac{a_1^{k+1} + a_2^{k+1}}{2},$$

from which it follows that the inequality holds for $k + 1$. By the principle of mathematical induction, the proposition holds for all natural numbers. The remainder of the proof follows as in (a).

282. Let $\alpha > 0$. Using the theorem of arithmetic and geometric means, we have

$$\sqrt[n]{a_1^\alpha a_2^\alpha \cdots a_n^\alpha} \leq \frac{a_1^\alpha + a_2^\alpha + \cdots + a_n^\alpha}{n}.$$

Taking the $\dfrac{1}{\alpha} > 0$ power of both sides produces the desired result.

The case in which $\alpha < 0$ is proved in a similar manner (see the solution to problem 270).

283. If α and β are of different sign, the theorem of power means follows from the result of problem 282. Hence we shall investigate only the case in which α and β have the same sign. Assume now that $0 < \alpha < \beta$. Designate $S_\beta(a)$ by K, and divide both sides of the investigated inequality by K; designate $\left(\dfrac{a_1}{K}\right)^\beta$ by b_1; and so on. The inequality assumes the form

$$\left(\frac{b_1^{\alpha/\beta} + b_2^{\alpha/\beta} + \cdots + b_n^{\alpha/\beta}}{n}\right)^{1/\alpha} \leqq 1 .$$

Here we have

$$\frac{b_1 + b_2 + \cdots + b_n}{n} = \frac{1}{K^\beta}\frac{a_1^\beta + a_2^\beta + \cdots + a_n^\beta}{n} = \frac{1}{K^\beta}K^\beta = 1 ;$$

$$b_1 + b_2 + \cdots + b_n = n .$$

Assume now that $b_1 = 1 + x_1, b_2 = 1 + x_2, \cdots, b_n = 1 + x_n$; then the equality $b_1 + b_2 + \cdots + b_n = n$ becomes

$$x_1 + x_2 + \cdots + x_n = 0 .$$

Suppose that $\dfrac{\alpha}{\beta} = \dfrac{k}{l}$ (a rational number). Then we have

$$b^{\alpha/\beta} = {}^l\!\sqrt{(1 + x_1)^k} = {}^l\!\sqrt{\underbrace{(1 + x_1)(1 + x_1)\cdots(1 + x_1)}_{k \text{ times}}\underbrace{1\cdot 1\cdots 1}_{(l-k)\text{ times}}}$$

$$\leqq \frac{k(1 + x_1) + (l - k)\cdot 1}{l} = 1 + \frac{k}{l}x_1 = 1 + \frac{\alpha}{\beta}\,x_1 ,$$

for which the equality holds only if $1 + x_1 = 1$ and $b = 1$. Using a limiting process we find that for an irrational ratio $\dfrac{\alpha}{\beta}$

$$b_1^{\alpha/\beta} \leqq 1 + \frac{\alpha}{\beta}x_1 .$$

Equality holds only if $x_1 = 0$ and $b_1 = 1$.[†]

[†] Let $x_1 \neq 0$, and let r be rational, so that $\dfrac{\alpha}{\beta} < r < 1$.

$$b_1^{\alpha/\beta} = (1 + x_1)^{\alpha/\beta} = [(1 + x_1)^{(\alpha/\beta)/r}]^r \leqq \left[1 + \frac{\alpha/\beta}{r}x_1\right]^r \leqq 1 + r\cdot\frac{\alpha/\beta}{r}x_1$$

$$= 1 + \frac{\alpha}{\beta}x_1 .$$

Similarly, we have

$$b_2^{\alpha/\beta} \leqq 1 + \frac{\alpha}{\beta} x_2, \cdots, b_n^{\alpha/\beta} \leqq 1 + \frac{\alpha}{\beta} x_n .$$

As a result, we find

$$\frac{(b_1^{\alpha/\beta} + b_2^{\alpha/\beta} + \cdots + b_n^{\alpha/\beta})^{1/\alpha}}{n}$$

$$\leqq \left[\frac{\left(1 + \frac{\alpha}{\beta} x_1\right) + \left(1 + \frac{\alpha}{\beta} x_2\right) + \cdots + \left(1 + \frac{\alpha}{\beta} x_n\right)}{n} \right]^{1/\alpha}$$

$$= \left(1 + \frac{\alpha}{\beta} \frac{x_1 + x_2 + \cdots + x_n}{n}\right)^{1/\alpha} = 1 ,$$

which concludes the proof.

Equality holds only if $b_1 = b_2 = \cdots = b_n = 1$, that is, if $a_1 = a_2 = \cdots = a_n$. Proof of the case in which $\alpha < \beta < 0$ is quite analogous.

284. (a)   Since $S_1 = 2$, we have

$$\frac{a_1^2 + a_2^2 + a_3^2}{3} = (S_2)^2 \geqq (S_1)^2 = 4 ;$$

$$a_1^2 + a_2^2 + a_3^2 \geqq 12 .$$

Equality holds only for $a_1 = a_2 = a_3 = 2$.

Similarly,

$$\frac{a_1^3 + a_2^3 + a_3^3}{3} = (S_3)^3 \geqq (S_1)^3 = 8 ;$$

$$a_1^3 + a_2^3 + a_3^3 \geqq 24 .$$

(b)   Proceeding as in part (a), we obtain $S_2 = \sqrt{\frac{18}{3}} = \sqrt{6}$, from which we obtain

$$\frac{a_1^3 + a_2^3 + a_3^3}{3} = (S_3)^3 \geqq (S_2)^3 = 6\sqrt{6};$$

$$a_1^3 + a_2^3 + a_3^3 \geqq 18\sqrt{6}$$

and

$$\frac{a_1 + a_2 + a_3}{3} = S_1 \leqq S_2 = \sqrt{6};$$

$$a_1 + a_2 + a_3 \leqq 3\sqrt{6} .$$

Equality in both cases is obtained only $a_1 = a_2 = a_3 = \sqrt{6}$.

285. A proof by induction will be given. It will be convenient to introduce a special designation for $(\Sigma_k)^k$ (the arithmetic mean of all possible products of the n numbers $a_1, a_2, \cdots, a_n$ taken k at a time); this summation will be designated as P_k. The inequality we wish to prove may then be written

$$P_k^2 \geqq P_{k+1} \cdot P_{k-1} \,.$$

For convenience we also introduce the notation $P_0(a)$, which will be taken equal to 1.

The inequality is obviously true if there are only two numbers, a_1 and a_2. In fact, in this case there are only three expressions $P_k(a)$:

$$P_0(a) = 1 \,,$$

$$P_1(a) = \frac{a_1 + a_2}{2} \,,$$

$$P_2(a) = a_1 a_2 \,,$$

and the inequality assumes the form

$$P_1(a)^2 > P_2(a) P_0(a) = P_2(a) \,,$$

or

$$\left(\frac{a_1 + a_2}{2} \right)^2 > a_1 a_2$$

(which we have already encountered).

Assume now that the inequality has been established for $n-1$ positive numbers $a_1, a_2, \cdots, a_{n-1}$; we shall show that the inequality must also be valid for n positive numbers $a_1, a_2, \cdots, a_n$. Designate the sum of all the possible products of the n numbers, taken k at a time, by $S_k(a)$, and the sum of the products of the $n-1$ numbers $a_1, a_2 \cdots, a_{n-1}$, taken k at a time, by $\overline{S_k(a)}$. If we factor a_n out of each term of $S_k(a)$ which contains this number, we derive the identity

$$S_k(a) = \overline{S_k(a)} + a_n \overline{S_{k-1}(a)} \,.$$

Further, let us write $\overline{P_k}$ to designate the P_k-expressions for the $n-1$ numbers $a_1, a_2, \cdots, a_{n-1}$. We have

$$P_k = \frac{S_k(a)}{C_n^k} = \frac{\overline{S_k(a)}}{C_n^k} + a_n \frac{\overline{S_{k-1}(a)}}{C_n^k}$$

$$= \frac{\overline{S_k(a)}}{C_{n-1}^k \dfrac{n}{n-k}} + a_n \frac{\overline{S_{k-1}(a)}}{C_{n-1}^{k-1} \dfrac{n}{k}} = \frac{n-k}{n} \overline{P_k} + \frac{k}{n} a_n \overline{P_{k-1}} \,.$$

We now consider the difference

$$P_k^2 - P_{k+1}P_{k-1}$$

$$= \left[\frac{n-k}{n}\,\overline{P_k} + \frac{k}{n}\,a_n\overline{P_{k-1}}\right]^2 - \left[\frac{n-k-1}{n}\,\overline{P_{k+1}} + \frac{k+1}{n}\,a_n\overline{P_k}\right]$$

$$\times \left[\frac{n-k+1}{n}\,\overline{P_{k-1}} + \frac{k-1}{n}\,a_n\overline{P_{k-2}}\right]$$

$$= \frac{1}{n^2}\{[(n-k)^2\overline{P}_k^2 - (n-k-1)(n-k+1)\overline{P_{k+1}P_{k-1}}]$$

$$+ a_n[2k(n-k)\overline{P_kP_{k-1}} - (k-1)(n-k-1)\overline{P_{k+1}P_{k-2}}$$

$$- (k+1)(n-k+1)\overline{P_kP_{k-1}}] + a_n^2[k^2\overline{P_{k-1}}^2$$

$$- (k+1)(k-1)\overline{P_kP_{k-2}}]\} = \frac{1}{n^2}(A + a_nB + a_n^2C)\,,$$

where A, B, and C designate, respectively, the expressions enclosed by brackets.

The induction hypothesis implies that

$$\overline{P}_k^2 > \overline{P_{k+1}P_{k-1}}, \ \overline{P_{k-1}}^2 > \overline{P_kP_{k-2}}\,,$$

from which, by multiplying these two inequalities, we obtain

$$\overline{P_kP_{k-1}} > \overline{P_{k+1}P_{k-2}}\,.$$

Therefore, in the expression for $P_k^2 - P_{k+1}P_{k-1}$ we have

$$A = (n-k)^2\overline{P}_k^2 - [(n-k)^2 - 1]\overline{P_{k+1}P_{k-1}}$$

$$= \overline{P}_k^2 + [(n-k)^2 - 1][\overline{P}_k^2 - \overline{P_{k+1}P_{k-1}}] > \overline{P}_k^2,$$

$$B = 2k(n-k)\overline{P_kP_{k-1}} - (k-1)(n-k-1)\overline{P_{k+1}P_{k-2}}$$

$$- (k+1)(n-k+1)\overline{P_kP_{k-1}} = -2\overline{P_kP_{k-1}}$$

$$+ (k-1)(n-k-1)[\overline{P_kP_{k-1}} - \overline{P_{k+1}P_{k-2}}] > 2\overline{P_kP_{k-1}}\,,$$

$$C = k^2\overline{P_{k-1}}^2 - (k^2 - 1)\overline{P_kP_{k-2}}$$

$$= \overline{P_{k-1}}^2 + (k^2 - 1)[\overline{P_{k-1}}^2 - \overline{P_kP_{k-2}}] > \overline{P_{k-1}}^2\,.$$

As a result,

$$P_k^2 - P_{k+1}P_{k-1} > \overline{P}_k^2 - 2a_n\overline{P_kP_{k-1}} + a_n^2\overline{P_{k-1}}^2$$

$$= (\overline{P}_k - a_n\overline{P_{k-1}})^2 \geqq 0\,,$$

which proves the theorem.

286. Excluding as obvious the case in which $a_1 = a_2 = \cdots = a_n$, we shall show that $\Sigma_k < \Sigma_l$. First, since $P_0(a) = 1$, it follows from problem 285 that

$$P_1^2 > P_2 P_0 = P_2 \ ;$$
$$\Sigma_1^2 > \Sigma_2^2 \quad \text{or} \quad \Sigma_1 > \Sigma_2 \ .$$

Further, by multiplying the inequalities $\Sigma_1 > \Sigma_2$, $\Sigma_2^4 > \Sigma_3^3 \cdot \Sigma_1$, we obtain $\Sigma_2^3 > \Sigma_3^3$, $\Sigma_2 > \Sigma_3$. Similarly, taking the products of the inequalities $\Sigma_1 > \Sigma_2$, $\Sigma_2^4 > \Sigma_3^3 \cdot \Sigma_1$, $\Sigma_3^9 > \Sigma_4^6 \cdot \Sigma_2^3$, we arrive at $\Sigma_3^6 > \Sigma_4^6$, $\Sigma_3 > \Sigma_4$. Similar procedures show, finally, that

$$\Sigma_1 > \Sigma_2 > \Sigma_3 > \cdots > \Sigma_n \ .$$

287. We have, $\Sigma_2 = \sqrt{\dfrac{24}{6}} = 2$ Further, by the theorem of the symmetric mean, we have

$$\Sigma_1(a) \geqq \Sigma_2(a) = 2 \ ,$$
$$a_1 + a_2 + a_3 + a_4 \geqq 4 \cdot 2 = 8$$

and

$$\Sigma_4(a) \leqq \Sigma_2(a) = 2 \ ,$$
$$a_1 a_2 a_3 a_4 \leqq 2^4 = 16 \ .$$

In both cases, equality holds only if $a_1 = a_2 = a_3 = a_4 = 2$.

288. Since the sum of the angles $\alpha + \beta + \gamma = \pi$, $\tan \dfrac{\alpha + \beta}{2} = \cot \dfrac{\gamma}{2}$, or

$$\frac{\tan \dfrac{\alpha}{2} + \tan \dfrac{\beta}{2}}{1 - \tan \dfrac{\alpha}{2} \tan \dfrac{\beta}{2}} = \frac{1}{\tan \dfrac{\gamma}{2}} \ ,$$

from which we obtain, after simplification,

$$\tan \frac{\alpha}{2} \tan \frac{\beta}{2} + \frac{\alpha}{2} \tan \frac{\gamma}{2} + \tan \frac{\beta}{2} \tan \frac{\gamma}{2} = 1 \ .$$

We write

$$\tan \frac{\alpha}{2} = a_1 \ ,$$

$$\tan \frac{\beta}{2} = a_2 \ ,$$

$$\tan \frac{\gamma}{2} = a_3 \ .$$

Then we see that the symmetric mean Σ_2 of the numbers a_1, a_2, and a_3 is equal to $\sqrt{\frac{1}{3}} = \frac{\sqrt{3}}{3}$. It follows, by reasoning analogous to that used for problem 287, that

(a) $\tan\frac{\alpha}{2} + \tan\frac{\beta}{2} + \tan\frac{\gamma}{2} \geqq 3\cdot\frac{\sqrt{3}}{3} = \sqrt{3}$;

(b) $\tan\frac{\alpha}{2}\cdot\tan\frac{\beta}{2}\cdot\tan\frac{\gamma}{2} \leqq \left(\frac{\sqrt{3}}{3}\right)^2 = \frac{\sqrt{3}}{9}$.

In both cases, equality holds only if $\alpha = \beta = \gamma = 60°$.

289.   The Cauchy-Buniakowski inequality is of sufficient importance to justify the giving of four proofs of its validity.

First Proof.   We can write

$$(xa_1 + b_1)^2 + (xa_2 + b_2)^2 + \cdots + (xa_n + b_n)^2$$
$$= (x^2a_1^2 + 2xa_1b_1 + b_1^2)$$
$$+ (x^2a_2^2 + 2xa_2b_2 + b_2^2) + \cdots + (x^2a_n^2 + 2xa_nb_n + b_n^2)$$
$$= Ax^2 + 2Bx + C ,$$

where

$$A = a_1^2 + a_2^2 + \cdots + a_n^2 ,$$
$$B = a_1b_1 + a_2b_2 + \cdots + a_nb_n ,$$
$$C = b_1^2 + b_2^2 + \cdots + b_n^2 .$$

The left member of this equation is, as a sum of squares, non-negative for all x; in particular, it is non-negative for $x = -\frac{B}{A}$. Substitution of this value for x in the equation yields

$$A\frac{B^2}{A^2} - 2B\frac{B}{A} + C = \frac{AC - B^2}{A} \geqq 0 .$$

Since $A > 0$, $AC - B^2 \geqq 0$, or $B^2 \leqq AC$.

We obtain the inequality sought by substituting for A, B, and C their expressions in terms of a_i and b_i.

The equality sign is possible only if

$$xa_1 + b_1 = xa_2 + b_2 = xa_3 + b_3 = \cdots = xa_n + b_n = 0 ,$$

from which we find

$$\frac{b_1}{a_1} = \frac{b_2}{a_2} = \frac{b_3}{a_3} = \cdots = \frac{b_n}{a_n} (= -x) .$$

Second Proof.   Given two numbers a and b, we have $(a - b)^2 \geqq 0$, or $a^2 + b^2 \geqq 2ab$, and so

$$ab \leq \frac{1}{2}a^2 + \frac{1}{2}b^2 .$$

Now let

$$A = \sqrt{a_1^2 + \cdots + a_n^2} ,$$
$$B = \sqrt{b_1^2 + \cdots + b_n^2} ,$$
$$\bar{a}_i = \frac{a_i}{A} ,$$
$$\bar{b}_i = \frac{b_i}{B} \qquad (i = 1, 2, \cdots, n);$$

then,

$$\bar{a}_1^2 + \cdots + \bar{a}_n^2 = \frac{a_1^2 + \cdots + a_n^2}{A^2} = 1 ,$$
$$\bar{b}_1^2 + \cdots + \bar{b}_n^2 = \frac{b_1^2 + \cdots + b_n^2}{B^2} = 1 .$$

We can write the n inequalities

$$\bar{a}_1\bar{b}_1 \leq \frac{1}{2}\bar{a}_1^2 + \frac{1}{2}\bar{b}_1^2, \cdots, \bar{a}_n\bar{b}_n \leq \frac{1}{2}\bar{a}_n^2 + \frac{1}{2}b_n^2 .$$

If we add these together, we arrive at

$$\bar{a}_1\bar{b}_1 + \cdots + \bar{a}_n\bar{b}_n \leq \frac{1}{2} + \frac{1}{2} = 1 .$$

Substitution of the appropriate quantities for $\bar{a}_i$ and b_i yields

$$\frac{a_1 b_1}{AB} + \cdots + \frac{a_n b_n}{AB} \leq 1 ,$$
$$a_1 b_1 + \cdots + a_n b_n \leq AB ,$$
$$(a_1 b_1 + \cdots + a_n b_n)^2 \leq (a_1^2 + \cdots + a_n^2)(b_1^2 + \cdots + b_n^2) ,$$

which is what we wished to prove.

Equality holds only if

$$\bar{a}_1 - \bar{b}_1 = \bar{a}_2 - \bar{b}_2 = \cdots = \bar{a}_n - \bar{b}_n = 0 ,$$

which implies

$$\frac{b_1}{a_1} = \frac{b_2}{a_2} = \cdots = \frac{b_n}{a_n}\left(= \frac{B}{A}\right) .$$

Third Proof. The inequality holds, trivially, for $n = 1$; that is,

$$(a_1 b_1)^2 \leq a_1^2 b_1^2 .$$

We shall show that if the inequality is assumed to hold for n pairs of numbers, that is,

$$C^2 \leqq AB,$$

where

$$A = a_1^2 + a_1^2 + \cdots + a_n^2,$$
$$B = b_1^2 + b_2^2 + \cdots + b_n^2,$$
$$C = a_1b_1 + a_2b_2 + \cdots + a_nb_n,$$

then it must hold also for $n+1$ pairs of numbers:

$$(C + a_{n+1}b_{n+1})^2 \leqq (A + a_{n+1}^2)(B + b_{n+1}^2).$$

In fact, we can write

$$
\begin{aligned}
&(A + a_{n+1}^2)(B + b_{n+1}^2) - (C + a_{n+1}b_{n+1})^2 \\
&= AB + Ab_{n+1}^2 + Ba_{n+1}^2 + a_{n+1}^2b_{n+1}^2 - C^2 - 2Ca_{n+1}b_{n+1} \\
&\quad - (a_{n+1}b_{n+1})^2 \\
&= (AB - C) + (Ab_{n+1}^2 + Ba_{n+1}^2 - 2Ca_{n+1}b_{n+1}) \\
&= (AB - C) + (\sqrt{A}\,b_{n+1} - \sqrt{B}\,a_{n+1})^2 \\
&\quad + 2(\sqrt{AB} - \sqrt{C^2})a_{n+1}b_{n+1} \geqq 0
\end{aligned}
$$

(since each of the three terms is equal to or greater than zero).

Therefore, by the principle of mathematical induction, the inequality is valid for all natural numbers n.

The equality sign holds only if

$$\frac{a_1}{b_1} = \frac{a_2}{b_2} = \cdots = \frac{a_n}{b_n}.$$

Fourth Proof. It can be proved by mathematical induction that

$$\left(\sum_{k=1}^{n} a_k^2\right)\left(\sum_{k=1}^{n} b_k^2\right) - \left(\sum_{k=1}^{n} a_kb_k\right)^2 = \frac{1}{2}\sum_{k=1}^{n}\sum_{l=1}^{n}(a_kb_l - a_lb_k)^2.$$

The right member of this inequality is nonnegative and vanishes only in the event of equality of the ratios $a_k : b_k$. If this expression is not zero, we have

$$\left(\sum_{k=1}^{n} a_k^2\right)\left(\sum_{k=1}^{n} b_k^2\right) - \left(\sum_{k=1}^{n} a_kb_k\right)^2 > 0,$$

which proves the theorem.

290. By the Cauchy-Buniakowski inequality we have

$$\left(a_1 + a_2 + \cdots + a_n\right)\left(\frac{1}{a_1} + \frac{1}{a_2} + \cdots + \frac{1}{a_n}\right)$$

$$= \left[(\sqrt{a_1})^2 + (\sqrt{a_2})^2 + \cdots + (\sqrt{a_n})^2\right]$$

$$\times \left[(\sqrt{1/a_1})^2 + (\sqrt{1/a_2})^2 + \cdots + (\sqrt{1/a_n})^2\right]$$

$$\geq (\sqrt{a_1}\ \sqrt{1/a_1} + \sqrt{a_2}\ \sqrt{1/a_2} + \cdots + \sqrt{a_n}\ \sqrt{1/a_n})^2 = n^2 ,$$

which yields the result sought. The equality sign holds only if $a_1 = a_2 = \cdots a_n$.

291. By the Cauchy-Buniakowski inequality we have

$$(a_1 \cdot 1 + a_2 \cdot 1 + \cdots + a_n \cdot 1)^2$$

$$\leq (a_1^2 + a_2^2 + \cdots + a_n^2)(1 + 1 + \cdots + 1) ,$$

from which we obtain

$$\left(\frac{a_1 + a_2 + \cdots + a_n}{n}\right)^2 \leq \frac{a_1^2 + a_2^2 + \cdots + a_n^2}{n} .$$

The equality holds only for $a_1 = a_2 = \cdots = a_n$.

292. In the solution of problem 288 it was shown that

$$\tan\left(\frac{\alpha}{2}\right)\tan\left(\frac{\beta}{2}\right) + \tan\left(\frac{\beta}{2}\right)\tan\left(\frac{\gamma}{2}\right) + \tan\left(\frac{\gamma}{2}\right)\tan\left(\frac{\alpha}{2}\right) = 1 .$$

By the Cauchy-Buniakowski inequality, we can write

$$\left[\tan^2\left(\frac{\alpha}{2}\right) + \tan^2\left(\frac{\beta}{2}\right) + \tan^2\left(\frac{\gamma}{2}\right)\right]\left[\tan^2\left(\frac{\beta}{2}\right) + \tan^2\left(\frac{\gamma}{2}\right) + \tan^2\left(\frac{\gamma}{2}\right)\right]$$

$$\geq \left[\tan\left(\frac{\alpha}{2}\right)\tan\left(\frac{\beta}{2}\right) + \tan\left(\frac{\beta}{2}\right)\tan\left(\frac{\gamma}{2}\right) + \tan\left(\frac{\gamma}{2}\right)\tan\left(\frac{\alpha}{2}\right)\right]^2 = 1 ,$$

from which the desired result follows.
Equality can hold only for $\alpha = \beta = \gamma = 60°$.

293. We can write the following expansion:

$$(x_1 + y_1)^2 + \cdots + (x_n + y_n)^2 = (x_1 + y_1)x_1$$

$$+ \cdots + (x_n + y_n)x_n + (x_1 + y_1)y_1 + \cdots + (x_n + y_n)y_n .$$

Let us make the following substitutions in the Cauchy-Buniakowski inequality:

$$x_1 + y_1 = a_1, \cdots, x_n + y_n = a_n ,$$

$$x_1 = b_1, \cdots, x_n = b_n .$$

The following inequality is then valid:

$$(x_1 + y_1)x_1 + \cdots + (x_n + y_n)x_n$$
$$\leqq [(x_1 + y_1)^2 + \cdots + (x_n + y_n)^2]^{1/2}[x_1^2 + \cdots + x_n^2]^{1/2}.$$

Analogously, we have

$$(x_1 + y_1)y_1 + \cdots + (x_n + y_n)y_n$$
$$\leqq [(x_1 + y_1)^2 + \cdots + (x_n + y_n)^2]^{1/2}[y_1^2 + \cdots + y_n^2]^{1/2}.$$

If we combine these two inequalities, we obtain

$$(x_1 + y_1)^2 + \cdots + (x_n + y_n)^2$$
$$\leqq [(x_1 + y_1)^2 + \cdots + (x_n + y_n)^2]^{1/2}$$
$$\times [(x_1^2 + \cdots + x_n^2)^{1/2} + (y_1^2 + \cdots + y_n^2)^{1/2}].$$

If both members of this last inequality are divided by

$$[(x_1 + y_1)^2 + \cdots + (x_n + y_n)^2]^{1/2},$$

we obtain

$$\sqrt{(x_1 + y_1)^2 + \cdots + (x_n + y_n)^2}$$
$$\leqq \sqrt{x_1^2 + \cdots + x_n^2} + \sqrt{y_1^2 + \cdots + y_n^2},$$

which is what we set out to prove.

The equality can hold only if $\frac{x_1}{y_1} = \frac{x_2}{y^2} = \cdots = \frac{x_n}{y_n}$.

294. The Cauchy-Buniakowski inequality yields

$$4Q^2 = (a_1a_2 + a_2a_1 + a_1a_3 + a_3a_1 + \cdots + a_{n-1}a_n + a_na_{n-1})^2$$
$$\leqq (a_1^2 + a_2^2 + a_1^2 + \cdots + a_n^2)(a_2^2 + a_1^2 + a_3^2 + \cdots + a_{n-1}^2)$$
$$= (n-1)P \cdot (n-1)P,$$

from which the result immediately follows.

We have equality only if $a_1 = a_2 = \cdots a_n$.

295. From the Cauchy-Buniakowski inequality, we can write

$$(\sqrt{p_1} \cdot \sqrt{p_1}x_1 + \sqrt{p_2} \cdot \sqrt{p_2}x_2 + \cdots + \sqrt{p_n} \cdot \sqrt{p_n}x_n)^2$$
$$\leqq (\sqrt{p_1^2} + \sqrt{p_2^2} + \cdots + \sqrt{p_n^2})(\sqrt{p_1^2}\,x_1^2$$
$$+ \sqrt{p_2^2}\,x_2^2 + \cdots + \sqrt{p_n^2}\,x_n^2).$$

The equality sign can be used only if $x_1 = x_2 = \cdots = x_n$.

296. We make the following substitutions in the inequality of problem 295:

$$p_1 = \frac{1}{2},$$

$$p_2 = \frac{1}{3},$$

$$p_3 = \frac{1}{6}.$$

We then have

$$\left(\frac{1}{2}x_1 + \frac{1}{3}x_2 + \frac{1}{6}x_3\right)^2 \leqq \left(\frac{1}{2} + \frac{1}{3} + \frac{1}{6}\right)\left(\frac{1}{2}x_1^2 + \frac{1}{3}x_2^2 + \frac{1}{6}x_3^2\right).$$

The equality can hold only for equality of the ratios

$$x_1 \colon x_2 \colon x_3 = \frac{1}{2} \colon \frac{1}{3} \colon \frac{1}{6}.$$

297. This inequality is merely a restatement, in another notation, of the Cauchy-Buniakowski inequality. In fact, if we substitute in the latter $a_k^2 = x_k$ and $b_k^2 = y_k$, and take the square root of both sides, we obtain the inequality of the problem.

298. By two applications of the Cauchy-Buniakowski inequality, first to the pairs $a_1b_1, a_2b_2, \cdots, a_nb_n$ and $c_1d_1, c_2d_2, \cdots, c_nd_n$, and then to the squares of the numbers $a_1, a_2, \cdots, a_n, b_1, b_2, \cdots, b_n; c_1, c_2, \cdots, c_n; d_1, d_2, \cdots, d_n$, we obtain

$$(a_1b_1c_1d_1 + a_2b_2c_2d_2 + \cdots + a_nb_nc_nd_n)^4$$
$$\leqq (a_1^2b_1^2 + a_2^2b_2^2 + \cdots + a_n^2b_n^2)^2(c_1^2d_1^2 + c_2^2d_2^2 + \cdots + c_n^2d_n^2)^2$$
$$\leqq (a_1^4 + a_2^4 + \cdots + a_n^4)(b_1^4 + b_2^4 + \cdots + b_n^4)$$
$$\times (c_1^4 + c_2^4 + \cdots + c_n^4)(d_1^4 + d_2^4 + \cdots + d_n^4).$$

Equality holds only if

$$a_1 \colon b_1 \colon c_1 \colon d_1 = a_2 \colon b_2 \colon c_2 \colon d_2 = \cdots = a_n \colon b_n \colon c_n \colon d_n.$$

299. We shall rearrange the a_i in nondecreasing order, assuming (renumbering if necessary) that

$$a_1 \leqq a_2 \leqq a_3 \leqq \cdots \leqq a_n.$$

Now we may consider the b_i to be in nonincreasing order; that is, we assume that

$$b_1 \geqq b_2 \geqq b_3 \geqq \cdots \geqq b_n.$$

[Once we have rearranged the a_i and written the given fraction

$$\frac{(a_1^2 + a_2^2 + \cdots + a_n^2)(b_1^2 + b_2^2 + \cdots + b_n^2)}{(a_1 b_1 + a_2 b_2 + \cdots + a_n b_n)^2},$$

if $b_i < b_j$, for $i < j$, for any i, an interchange of these two b within the fraction can only increase its value, inasmuch as the value of the numerator will remain unaltered, but the denominator will become less (we will have supplied a greater a_i with a lesser b_i factor; consider the difference

$$(a_i b_j + a_j b_i) - (a_i b_i + a_j b_j) = (a_i - a_j)(b_j - b_i) < 0)].$$

Now, in the event that all a_i are equal and all b_i are equal, the fraction has a value of 1. Hence we may assume that either not all the a_i are equal or else not all the b_i are the same (or both). Let us write the system:

$$a_i^2 = \alpha_i a_1^2 + \beta_i a_n^2,$$
$$b_i^2 = \alpha_i b_1^2 + \beta_i b_n^2 \qquad (i = 2, 3, \cdots, n - 1).$$

This system may be solved for α_i and β_i:

$$\alpha_i = \frac{a_n^2 b_i^2 - a_i^2 b_n^2}{a_n^2 b_1^2 - a_1^2 b_n^2},$$

$$\beta_i = \frac{a_i^2 b_1^2 - a_1^2 b_i^2}{a_n^2 b_1^2 - a_1^2 b_n^2}.$$

The denominators of these fractions are positive,

$$a_n^2 b_1^2 - a_1^2 b_n^2 > 0,$$

or

$$\frac{a_n^2}{a_1^2} > \frac{b_n^2}{b_1^2},$$

and for the numerators we have

$$a_n^2 b_i^2 - a_i^2 b_n^2 \geqq 0,$$

or

$$\frac{a_n^2}{a_i^2} \geqq \frac{b_n^2}{b_i^2};$$

and also

$$a_i^2 b_1^2 - a_1^2 b_i^2 \geqq 0.$$

or

$$\frac{a_i^2}{a_1^2} \geqq \frac{b_i^2}{b_1^2}.$$

Accordingly, $\alpha_i \geqq 0$, and $\beta_i \geqq 0$. Now, $\alpha_i = 0$ only if

$$a_i = a_n,$$
$$b_i = b_n,$$
$$\beta_i = 1,$$

and, analogously,

$$\beta_i = 0$$

if $a_i = a_1, b_i = b_1, \alpha_i = 1$.

We introduce the terminology

$$1 + \alpha_2 + \alpha_3 + \cdots + \alpha_{n-1} = A,$$
$$\beta_2 + \beta_3 + \cdots + \beta_{n-1} + 1 = B.$$

Then the numerator of the fraction of the problem can be rewritten in the form

$$(Aa_1^2 + Ba_n^2)(Ab_1^2 + Bb_n^2).$$

As for the denominator of the given fraction, we shall use the Cauchy-Buniakowski inequality $(a_1b_1 + a_2b_2)^2 \leqq (a_1^2 + a_2^2)(b_1^2 + b_2^2)$ (see the discussion immediately following problem 288) to obtain

$$a_ib_i = \sqrt{(\sqrt{\alpha_i}\,a_1)^2 + (\sqrt{\beta_i}\,a_n)^2} \cdot \sqrt{(\sqrt{\alpha_i}\,b_1)^2 + (\sqrt{\beta_i}\,b_n)^2}$$
$$\geqq \alpha_ia_1b_1 + \beta_ia_nb_n. \tag{1}$$

If the inequalities of (1) are added, for $i = 1, 2, 3, \cdots, n-1, n$ (here we assume $\alpha_1 = 1, \beta_1 = 0$ and $\alpha_n = 0, \beta_n = 1$), we obtain

$$a_1b_1 + a_2b_2 + \cdots + a_nb_n \geqq Aa_1b_1 + Ba_nb_n.$$

Thus, the given fraction does not exceed

$$\frac{(Aa_1^2 + Ba_n^2)(Ab_1^2 + Bb_n^2)}{(Aa_1b_1 + Ba_nb_n)^2}.$$

However,

$$\frac{(Aa_1^2 + Ba_n^2)(Ab_1^2 + Bb_n^2)}{(Aa_1b_1 + Ba_nb_n)^2} = 1 + AB\frac{(a_nb_1 - a_1b_n)^2}{(Aa_1b_1 + Ba_nb_n)^2}.$$

In view of the theorem of arithmetic and geometric means, we have

$$Aa_1b_1 + Ba_nb_n \geqq 2\sqrt{Aa_1b_1 \cdot Ba_nb_n}. \tag{2}$$

Finally, we obtain

$$\frac{(a_1^2 + a_2^2 + \cdots + a_n^2)(b_1^2 + b_2^2 + \cdots + b_n^2)}{(a_1b_1 + a_2b_2 + \cdots + a_nb_n)^2}$$

$$\leq 1 + AB\frac{(a_nb_1 - a_1b_n)^2}{(Aa_1b_1 + Ba_nb_n)^2} \leq 1 + AB\frac{(a_nb_1 - a_1b_n)^2}{(2\sqrt{ABa_1b_1a_nb_n})^2}$$

$$= 1 + \left(\frac{\sqrt{a_nb_1/a_1b_n} - \sqrt{a_1b_n/a_nb_1}}{2}\right)^2 ,$$

which is what we wished to show.

Inequality (*1*) cannot become an equality for positive numbers, since $\frac{a_1}{a_n} \neq \frac{b_1}{b_n}$. Hence this possibility exists only if all α_i are zero or 1 (in the latter case, $\beta_i = 0$), that is, if

$$a_1 = a_2 = \cdots = a_k < a_{k+1} = a_{k+2} = \cdots = a_n ,$$
$$b_1 = b_2 = \cdots = b_k > b_{k+1} = b_{k+1} = \cdots = b_n .$$

Then for (*2*) to become an equality it is necessary that

$$Aa_1b_1 = Ba_nb_n ,$$

that is,

$$ka_1b_1 = (n - k)a_nb_n ,$$

or,

$$\frac{a_1}{a_n} : \frac{b_n}{b_1} = \frac{n - k}{k} .$$

This determines a condition for the equality

$$\frac{(a_1^2 + a_2^2 + \cdots + a_n^2)(b_1^2 + b_2^2 + \cdots + b_n^2)}{(a_1b_1 + a_2b_2 + \cdots + a_nb_n)^2}$$

$$= 1 + \left(\frac{\sqrt{M_1M_2/m_1m_2} - \sqrt{m_1m_2/M_1M_2}}{2}\right)^2$$

300. We can write

$$n(a_1b_1 + a_2b_2 + \cdots + a_nb_n)$$
$$\quad - (a_1 + a_2 + \cdots + a_n)(b_1 + b_2 + \cdots + b_n)$$
$$= \frac{1}{2}[(a_1 - a_2)(b_1 - b_2) + (a_1 - a_3)(b_1 - b_2)$$
$$\quad + \cdots + (a_{n-1} - a_n)(b_{n-1} - b_n)] \geq 0 ,$$

from which the required inequality immediately follows.

301. The condition $\frac{1}{p} + \frac{1}{q} = 1$ implies that $p + q = pq$, and so

$$p = \frac{p+q}{q} = \frac{p_1 + q_1}{q_1},$$

$$q = \frac{p+q}{p} = \frac{p_1 + q_1}{p_1},$$

where p_1 and q_1 may be any positive integers proportional to p and q $\left(\text{if } p = \frac{\alpha}{\beta} \text{ and } q = \frac{\gamma}{\delta}, \text{ where } \alpha, \beta, \gamma \text{ and } \delta \text{ are integers, then we}\right.$ may use, for example, $p_1 = \alpha\delta$ and $q_1 = \gamma\beta$).

Accordingly, the inequality we seek to prove may be rewritten in the form

$$xy \leqq \frac{qx^{(p_1+q_1)/q_1} + py^{(p_1+q_1)/p_1}}{p+q}.$$

We now set

$$x_1 = x^{(p_1+q_1)/q_1}$$
$$y_1 = y^{(p_1+q_1)/p_1}.$$

Then, using the theorem of arithmetic and geometric means (problem 268), we obtain

$$xy = x_1^{q_1/(p_1+q_1)} y_1^{p_1/(p_1+q_1)} = (\underbrace{x_1 x_1 \cdots x_1}_{q_1 \text{ times}} \underbrace{y_1 y_1 \cdots y_1}_{p_1 \text{ times}})^{1/(p_1+q_1)}$$

$$\leqq \left[\left(\frac{q_1 x_1 + p_1 y_1}{q_1 + p_1}\right)^{q_1+p_1}\right]^{1/(p_1+q_1)} = \frac{qx_1 + py_1}{q+p}$$

$$= \frac{p+q}{q} x^{(p_1+q_1)/q_1} + \frac{p}{p+q} y^{(p_1+q_1)/p_1} = \frac{1}{p} x^p + \frac{1}{q} y^q.$$

We have equality only if $x^p = y^q$.

302. We shall use the designations

$$a_1 + a_2 + \cdots + a_n = a,$$
$$b_1 + b_2 + \cdots + b_n = b$$

and divide the given inequality by $a^\alpha b^\beta$. This yields the following inequality, which is equivalent to that of the problem:

$$\left(\frac{a_1}{a}\right)^\alpha \left(\frac{b_1}{b}\right)^\beta + \left(\frac{a_2}{a}\right)^\alpha \left(\frac{b_2}{b}\right)^\beta + \cdots + \left(\frac{a_n}{a}\right)^\alpha \left(\frac{b_n}{b}\right)^\beta \leqq 1.$$

From the theorem of arithmetic and geometric means (problem 268) it is possible to derive the result that, for every $k = 1, 2, \cdots, n$,

$$\left(\frac{a_k}{a}\right)^\alpha \left(\frac{b_k}{b}\right)^\beta \leqq \left(\frac{\alpha\dfrac{a_k}{a} + \beta\dfrac{b_k}{b}}{\alpha + \beta}\right)^{\alpha+\beta} = \alpha\frac{a_k}{a} + \beta\frac{b_k}{b}$$

(compare with the solution of problem 301).

If these inequalities are added for $k = 1, 2, \cdots, n$, we have

$$\left(\frac{a_1}{a}\right)^\alpha\left(\frac{b_1}{b}\right)^\beta + \left(\frac{a_2}{a}\right)^\alpha\left(\frac{b_2}{b}\right)^\beta + \cdots + \left(\frac{a_n}{a}\right)^\alpha\left(\frac{b_n}{b}\right)^\beta$$

$$\leqq \frac{\alpha a_1}{a} + \frac{\beta b_1}{b} + \frac{\alpha a_2}{a} + \frac{\beta b_2}{b} + \cdots + \frac{\alpha a_n}{a} + \frac{\beta b_n}{b}$$

$$= \frac{\alpha}{a}(a_1 + a_2 + \cdots + a_n) + \frac{\beta}{b}(b_1 + b_2 + \cdots + b_n)$$

$$= \frac{\alpha a}{a} + \frac{\beta b}{b} = \alpha + \beta = 1 .$$

which is what we set out to prove.

Equality holds only if $a_1 : b_1 = a_2 : b_2 = \cdots = a_n : b_n$.

Remark: This result can be derived from the inequality of Cauchy-Buniakowski, but the proof is more complicated.

303. Two proofs will be given for this inequality, which has frequent application in analysis.

First Proof. If we substitute into the inequality of problem 302

$$\alpha = \frac{1}{p} ,$$

$$\beta = \frac{1}{q}$$

and, further,

$$a_1 = x_1^p, a_2 = x_2^p, \cdots, a_n = x_n^p ;$$

$$b_1 = y_1^q, b_2 = y_2^q, \cdots b_n = y_n^q ,$$

we obtain Hölder's inequality.

We arrive at equality only if $x_1 : y_1 = x_2 : y_2 = \cdots = x_n : y_n$.

Second Proof. We write

$$(x_1^p + x_2^p + \cdots + x_n^p)^{1/p} = X ,$$

$$(y_1^q + y_2^q + \cdots + y_n^q)^{1/q} = Y ,$$

and divide both members of the inequality by XY. If we use the terminology,

$$\frac{x_k}{X} = z_k ,$$

$$\frac{y_k}{Y} = t_k \quad (k = 1, 2, \cdots, n) ,$$

then the inequality we are considering becomes equivalent to

$$z_1 t_1 + z_2 t_2 + \cdots + z_n t_n \leqq 1 . \tag{1}$$

where the following conditions are to hold:

$$z_1^p + z_2^p + \cdots + z_n^p = 1 ,$$
$$t_1^q + t_2^q + \cdots + t_n^q = 1 . \tag{2}$$

By using the inequality of problem 301, we can write the n inequalities:

$$z_1 t_1 \leqq \frac{1}{p} z_1^p + \frac{1}{q} t_1^q ,$$

$$z_2 t_2 \leqq \frac{1}{p} z_2^p + \frac{1}{q} t_2^q ,$$

$$\cdots\cdots\cdots\cdots\cdots ,$$

$$z_n t_n \leqq \frac{1}{p} z_n^p + \frac{1}{q} t_n^q$$

The sum of all these inequalities yields

$$z_1 t_1 + z_2 t_2 + \cdots + z_n t_n$$
$$\leqq \frac{1}{p}(z_1^p + z_2^p + \cdots + z_n^p) + \frac{1}{q}(t_1^q + t_2^q + \cdots + t_n^q) .$$

Now, using condition (2), and the fact that $\frac{1}{p} + \frac{1}{q} = 1$, we obtain inequality (1), which is equivalent to that of the problem.

The equality holds only if

$$x_1 : y_1 = x_2 : y_2 = \cdots = x_n : y_n .$$

304. The proof is analogous to that of problem 302. We write

$$a_1 + a_2 + \cdots + a_n = a ,$$
$$b_1 + b_2 + \cdots + b_n = b ,$$
$$\cdots\cdots\cdots\cdots\cdots\cdots , $$
$$l_1 + l_2 + \cdots + l_n = l$$

If the inequality of the problem is divided by $a^\alpha b^\beta \cdots l^\lambda$, an equivalent inequality is found:

$$\left(\frac{a_1}{a}\right)^{\alpha}\left(\frac{b_1}{b}\right)^{\beta}\cdots\left(\frac{l_1}{l}\right)^{\lambda}+\cdots+\left(\frac{a_n}{a}\right)^{\alpha}\left(\frac{b_n}{b}\right)^{\beta}\cdots\left(\frac{l_n}{l}\right)^{\lambda}\leqq 1\,.$$

From the theorem of arithmetic and geometric means we obtain (compare with the solutions of problems 301 and 302)

$$\left(\frac{a_k}{a}\right)^{\alpha}\left(\frac{b_k}{b}\right)^{\beta}\cdots\left(\frac{l_k}{l}\right)^{\lambda}\leqq\left[\frac{\alpha\frac{a_k}{a}+\beta\frac{b_k}{b}+\cdots+\lambda\frac{l_k}{l}}{\alpha+\beta+\cdots+\lambda}\right]^{\alpha+\beta+\ldots+\lambda}$$

$$=\alpha\frac{a_k}{a}+\beta\frac{b_k}{b}+\cdots+\lambda\frac{l_k}{l}\,.$$

The sum of these inequalitites for $k=1,2,\cdots,n$ yields

$$\left(\frac{a_1}{a}\right)^{\alpha}\left(\frac{b_1}{b}\right)^{\beta}\cdots\left(\frac{l_1}{l}\right)^{\lambda}+\cdots+\left(\frac{a_n}{a}\right)^{\alpha}\left(\frac{b_n}{b}\right)^{\beta}\cdots\left(\frac{l_n}{l}\right)^{\lambda}$$

$$\leqq\left(\frac{\alpha a_1}{a}+\frac{\beta b_1}{b}+\cdots+\frac{\lambda l_1}{l}\right)+\cdots+\left(\frac{\alpha a_n}{a}+\frac{\beta b_n}{b}+\cdots+\frac{\lambda l_n}{l}\right)$$

$$=\frac{\alpha}{a}(a_1+a_2+\cdots+a_n)+\frac{\beta}{b}(b_1+b_2+\cdots+b_n)$$

$$+\cdots+\frac{\lambda}{l}(l_1+l_2+\cdots+l_n)=\frac{\alpha a}{a}+\frac{\beta b}{b}+\cdots+\frac{\lambda l}{l}=1\,,$$

which is what we wished to show.

Equality results only if

$$a_1:b_1:\cdots:l_1=a_2:b_2::\cdots:l_2=\cdots=a_n:b_n:\cdots:l_n.$$

Remark: If we write

$$\alpha=\frac{1}{p}\,,\ \beta=\frac{1}{q}\,,\cdots,\lambda=\frac{1}{t}\,,$$

and

$$a_i=x_i^p,b_i=y_i^q,\cdots,l_i=u_i^t\quad(i=1,2,\cdots,n)\,,$$

we derive a new form of Hölder's inequality: If $\frac{1}{p}+\frac{1}{q}+\cdots+\frac{1}{u}=1$, then for arbitrary sets of positive numbers $x_1,\cdots,x_n;y_1,\cdots,y_n;\cdots;u_1,\cdots,u_n$; we have

$$x_1y_1\cdots u_1+x_2y_2\cdots u_2+x_ny_n\cdots u_n$$

$$\leqq(x_1^p+x_2^p+\cdots+x_n^p)^{1/p}(y_1^q+y_2^q+\cdots+y_n^q)^{1/q}$$

$$\times\cdots\times(u_1^t+u_2^t+\cdots+u_n^t)^{1/t}\,.$$

305. Use, in the inequality of problem 304, the n (two-number) sequences

$$1, a_1; 1, a_2; \cdots; 1, a_n;$$

and in that problem let $\alpha = \beta = \cdots = \dfrac{1}{n}$. We then have

$$(1 + a_1)^{1/n}(1 + a_2)^{1/n} \cdots (1 + a_n)^{1/n} \geqq 1 + a_1^{1/n} a_2^{1/n} \cdots a_n^{1/n} \,,$$

which proves the inequality of this problem.

We have equality only for $a_1 = a_2 = \cdots a_n$.

306. The inequality here is a special case of problem 304, with $\beta = \alpha = \cdots = \lambda = \dfrac{1}{n}$.

Equality holds only if

$$a_1 : b_1 : \cdots : l_1 = \cdots = a_n : b_n : \cdots : l_n \,.$$

307. In the inequality of problem 304 set

$$\alpha = \beta = \gamma = \frac{1}{3} \,,$$

$$a_1 = b_1 = c_1 = 1 \,,$$

$$a_2 = \frac{1}{x} \,,$$

$$b_2 = \frac{1}{y} \,,$$

$$c_2 = \frac{1}{z} \,.$$

Then we have

$$(a_1 + a_2)^{1/3}(b_1 + b_2)^{1/3}(c_1 + c_2)^{1/3} \geqq a_1^{1/3} b_1^{1/3} c_1^{1/3} + a_2^{1/3} b_2^{1/3} c_2^{1/3} \,,$$

or

$$\sqrt[3]{\left(1 + \frac{1}{x}\right)\left(1 + \frac{1}{y}\right)\left(1 + \frac{1}{z}\right)} \geqq 1 + \frac{1}{\sqrt[3]{xyz}} \,.$$

By the theorem of arithmetic and geometric means (see problem 263), we have

$$\sqrt[3]{xyz} \leqq \frac{x + y + z}{3} = \frac{1}{3} \,.$$

from which we obtain

$$\frac{1}{\sqrt[3]{xyz}} \geqq 3 \,.$$

Therefore,

$$\sqrt[3]{\left(1 + \frac{1}{x}\right)\left(1 + \frac{1}{y}\right)\left(1 + \frac{1}{z}\right)} \geqq 1 + 3 = 4 \,.$$

If both sides of this inequality are cubed, the required inequality ensues.

308. Designate by S the right member of the inequality. We then have

$$\begin{aligned}
S^2 &= (a_1 + b_1 + \cdots + l_1)^2 + (a_2 + b_2 + \cdots + l_2)^2 \\
&\quad + \cdots + (a_n + b_n + \cdots + l_n)^2 = [a_1(a_1 + b_1 + \cdots + l_1) \\
&\quad + a_2(a_2 + b_2 + \cdots + l_2) + \cdots + a_n(a_n + b_n + \cdots + l_n)] \\
&\quad + [b_1(a_1 + b_1 + \cdots + l_1) + b_2(a_2 + b_2 + \cdots + l_2) \\
&\quad + \cdots + b_n(a_n + b_n + \cdots + l_n)] + \cdots + [l_1(a_1 + b_1 + \cdots + l_1) \\
&\quad + l_2(a_2 + b_2 + \cdots + l_2) + \cdots + l_n(a_n + b_n + \cdots + l_n)] \,.
\end{aligned}$$

Application of Hölder's inequality (problem 303) for $p = q = 2$ to each of the expressions in brackets produces the inequality

$$\begin{aligned}
S^2 &\leqq \sqrt{a_1^2 + a_2^2 + \cdots + a_n^2}\, S + \sqrt{b_1^2 + b_2^2 + \cdots + b_n^2}\, S \\
&\quad + \cdots + \sqrt{l_1^2 + l_2^2 + \cdots + l_u^2}\, S, \cdots
\end{aligned}$$

The required result follows immediately.

We have equality only if $a_1 : b_1 : \cdots : l_1 = a_2 : b_2 : \cdots : l_2 = \cdots = a_n : b_n : \cdots : l_u$.

309. (a) We have $u_{n+1} - u_n = a_0[(n+1)^k - n^k] + a_1[(n+1)^{k-1} - n^{k-1}] + \cdots + a_{k-1}[(n+1) - n]$. However, for arbitrary a and b the following identity holds:

$$a^k - b^k = (a - b)(a^{k-1} + a^{k-2}b + \cdots + ab^{k-2} + b^{k-1}) \,.$$

If this identity is applied to the difference $(n+1)^k - n^k$, we have

$$\begin{aligned}
(n+1)^k - n^k &= (n+1)^{k-1} + (n+1)^{k-2}n \\
&\quad + \cdots + (n+1)n^{k-2} + n^{k-1} = k \cdot n^{k-1} + \cdots,
\end{aligned}$$

where the dots designate only terms of less than $(k-1)$st degree in n. These terms can not increase the order of the sequence; it follows that the order of the sequence $u_n^{(1)} = u_{n+1} - u_n$ is equal to $k-1$:

$$u_n^{(1)} = a_0 k \cdot n^{k-1} + \cdots \,.$$

(b) Upon transition from the sequence $u_n^{(s)}$ to the sequence $u_n^{(s+1)}$, the order of the sequence is decreased by 1. Therefore $u_n^{(k)}$ is a sequence of zero order; that is, all the elements of the sequence $u_n^{(k)}$ are the same number. This means that $u_n^{(k+1)} = 0$ for all n, which is what we wished to show.

Remark: It also follows from the solution of this problem that a sequence of order k has sequences of first-order, second-order, and so on, differences, up to the kth order, which do not include a sequence all of whose terms vanish [that is, $u_n^{(k+1)}$ is the first difference sequence comprising all zeros]. In fact, a sequence of differences of s-order of the sequence, where $s < k$, is a sequence of order $k - s$ whose terms cannot all reduce to 0 (a polynomial of degree $k-s$ can have value 0 for not more than $k - s$ values of its variable). See problem 310 for the sequence of differences of order k.

310. *First Solution.* In the solution of problem 309 (a) we saw that if $u_n = a_0 n^k + a_1 n^{k-1} + \cdots + a_k$ is a sequence of order k, then $u_n^{(1)}$ is a sequence of order $k - 1$ of form $u_n^{(1)} = a_0 k n^{k-1} + \cdots$. It follows that the sequence $u_n^{(2)}$ has the form

$$u_n^{(2)} = a_0 k(k - 1)n^{k-2} + \cdots,$$

the sequence $u_n^{(3)}$ has the form

$$u_n^{(3)} = a_0 k(k - 1)(k - 2)n^{k-3} + \cdots;$$

and so on. Finally, the sequence $u_n^{(k-1)}$ has the form

$$a_0 k(k - 1)\cdots 2 \cdot n + \cdots,$$

where the final dots indicate only a constant. The assertion of the problem follows at once.

Second Solution. By problem 309 (a), the sequence of differences of kth order of the sequence of degree k is of zero order, that is, all terms are the same, independent of n. Hence it suffices to determine $u_0^{(k)}$.

For u_n we have the formula

$$u_n = u_0 + C_n^1 u_0^{(1)} + \cdots + C_n^k u_0^{(k)}$$

(see problem 313). Equating this with the given polynomial, we have

$$a_0 n^k + a_1 n^{k-1} + \cdots + a_k = u_0 + C_n^1 u_0^{(1)} + \cdots + C_n^k u_0^{(k)} .$$

The two members of this equation represent the same polynomial in n; but it is necessary to expand the C's on the right to enable comparison of the coefficients of like-degree terms. The term $a_0 n^k$, however, can be obtained from the term $C_n^k u_0^{(k)}$ alone:

$$C_n^k u_0^{(k)} = \frac{n(n-1)(n-2)\cdots(n-k+1)}{k!} u_0^{(k)} \, .$$

The coefficient of n^k is equal to $u_0^{(k)}/k!$. Equating the coefficients for n_k on both sides, we obtain

$$a_0 = \frac{u_0^{(k)}}{k!} \, ,$$

from which we obtain

$$u_0^{(k)} = a_0 \cdot k! \, .$$

311. (a) We shall write the sum-sequence in the triangular form explained in the introduction preceding problem 309; each $\bar{u}$ will be the sum of the two elements flanking it in the previous line:

$$
\begin{array}{cccccccc}
u_0 & u_1 & u_2 & u_3 & \cdots u_n & u_{n+1} & \cdots \\
\bar{u}_0^{(1)} & \bar{u}_1^{(1)} & \bar{u}_2^{(1)} & \bar{u}_3^{(1)} & \cdots \bar{u}_n^{(1)} & \bar{u}_{n+1}^{(1)} & \cdots \\
\bar{u}_0^{(2)} & \bar{u}_1^{(2)} & \bar{u}_2^{(2)} & \bar{u}_3^{(2)} \cdots & \bar{u}_n^{(2)} & \bar{u}_{n+1}^{(2)} \cdots \\
& & \cdot & \cdot & \cdot & \cdot & \\
\bar{u}_0^{(k)} & \bar{u}_1^{(k)} & \bar{u}_2^{(k)} & \bar{u}_3^{(k)} \cdots\cdots\cdots \bar{u}_n^{(k)} & & \bar{u}_{n+1}^{(k)}
\end{array}
$$

Note that $\bar{u}_n^{(k)}$ depends only upon $u_n, u_{n+1}, \cdots, u_{n+k}$ and not upon any other numbers of the initial sequence comprising the first row. We shall now show by mathematical induction that

$$\bar{u}_n^{(k)} = C_k^0 u_n + C_k^1 u_{n+1} + C_k^2 u_{n+2} + \cdots + C_k^k u_{n+k} \, .$$

The formula is valid for $k = 1$, since here it assumes the form

$$\bar{u}_n^{(1)} = C_1^0 u_n + C_1^1 u_{n+1} = u_n + u_{n+1} \, ,$$

which, by definition, is obvious. Now, assume the formula is valid for $k - 1$; we shall show that validity for k is implied:

$$
\begin{aligned}
\bar{u}_n^{(k)} &= \bar{u}_n^{(k-1)} + \bar{u}_{n+1}^{(k-1)} \\
&= (C_{k-1}^0 u_n + C_{k-1}^1 u_{n+1} + C_{k-1}^2 u_{n+2} + \cdots + C_{k-1}^{k-1} u_{n+k-1}) \\
&\quad + (C_{k-1}^0 u_{n+1} + C_{k-1}^1 u_{n+2} + C_{k-1}^2 u_{n+3} + \cdots + C_{k-1}^{k-1} u_{n+k}) \\
&= C_{k-1}^0 u_n + (C_{k-1}^0 + C_{k-1}^1) u_{n+1} + (C_{k-1}^1 + C_{k-1}^2) u_{n+2} \\
&\quad + \cdots + (C_{k-1}^{k-2} + C_{k-1}^{k-1}) u_{n+k-1} + C_{k-1}^{k-1} u_{n+k} \, .
\end{aligned}
$$

From the definition of the symbol C_n^m, it follows that

$$
\begin{aligned}
C_{k-1}^0 &= C_k^0 \, , \\
C_{k-1}^0 + C_{k-1}^1 &= C_k^1 \, , \\
C_{k-1}^1 + C_{k-1}^2 &= C_k^2 \, ,
\end{aligned}
$$

and so on. Therefore,

$$\bar{u}_n^{(k)} = C_k^0 u_n + C_k^1 u_{n+1} + \cdots + C_k^{k-1} u_{n+k-1} + C_k^k u_{n+k},$$

which is what we set out to prove.

(b) The proof here is entirely analogous to that of problem (a) and is left for the reader.

312. The proof will be given by mathematical induction. (Refer to the terminology of the Pascal Triangle in the introduction to this series of problems.). The proposition holds for $k = 1$; that is, $C_1^1 = \frac{1}{1!}$ in the first row of the triangular array.

Assume validity for the $(n-1)$st row; we must show that this holds for the nth row. From the definition of the Pascal triangle, we have

$$C_n^k = C_{n-1}^{k-1} + C_{n-1}^k.$$

Therefore, for $k > 1$, we have

$$C_n^k = \frac{(n-1)(n-2)\cdots(n-k+1)}{(k-1)!} + \frac{(n-1)(n-2)\cdots(n-k)}{k!}.$$

We can write this in the form

$$C_n^k = \frac{(n-1)(n-2)\cdots(n-k+1)}{(k-1)!}\left(1 + \frac{n-k}{k}\right)$$
$$= \frac{n(n-1)(n-2)\cdots(n-k+1)}{k!},$$

which is the desired result.

If $k = 1$, then we have

$$C_n^1 = C_{n-1}^0 + C_{n-1}^1 = 1 + (n-1) = n.$$

Remark: It follows from this formula that the numbers standing in the $(k+1)$st row of Pascal's triangle, are of degree k in n. In fact, if we take k fixed and let n vary, then

$$C_n^k = \frac{n(n-1)\cdots(n-k+1)}{k!}$$

is a polynomial in n of degree k.

313. It suffices here to note that the number triangle

$$u_0^{(n)} \quad u_0^{(n-1)} \quad u_0^{(n-2)} \;\cdot\;\cdot\; u_0^{(1)} \quad u_0$$

$$u_1^{(n-1)} \quad u_1^{(n-2)} \quad u_1^{(n-3)} \;\cdot\;\cdot\; u_1$$

$$u_2^{(n-2)} \quad u_2^{(n-3)} \;\cdot\;\cdot\; u_2$$

$$\cdot \quad \cdot \quad \cdot \quad \cdot \quad \cdot \quad \cdot$$

$$\cdot \quad \cdot \quad \cdot \quad \cdot$$

$$u_n$$

is an arithmetic triangle. That is, it can be written in the form

$$v_0 \quad v_1 \quad v_2 \;\cdot\;\cdot\; v_{n-1} \quad v_n$$

$$\bar{v}_0^{(1)} \quad \bar{v}_1^{(1)} \quad \bar{v}_2^{(1)} \;\cdot\;\cdot\; \bar{v}_{n-1}^{(1)}$$

$$\bar{v}_0^{(2)} \quad \bar{v}_1^{(2)} \;\cdot\;\cdot\; \bar{v}_{n-2}^{(2)}$$

$$\cdot \quad \cdot \quad \cdot \quad \cdot \quad \cdot \quad \cdot$$

$$\cdot \quad \cdot \quad \cdot \quad \cdot$$

$$\bar{v}_0^{(n)}$$

on which we can now use the result of problem 311 (a).

314. If all the numbers of the kth (and higher) sequence of differences are zeros, then the members of the $(k-1)$st row of successive differences are all the same number (which is not zero, according to the conditions of the problem). Further, since

$$u_0^{(k+1)} = u_0^{(k+2)} = \cdots = 0\,,$$

$$u_0^{(k)} \neq 0\,,$$

we have

$$u_n = u_0 + C_n^1 u_0^{(1)} + C_n^2 u_0^{(2)} + \cdots + C_n^k u_0^{(k)} = u_0 + n \cdot u_0^{(1)}$$

$$+ \frac{n(n-1)}{2} \cdot u_0^{(2)} + \cdots + \frac{n(n-1)\cdots(n-k+1)}{k!} u_0^{(k)}$$

(see problem 313). But this formula expresses u_n as a polynomial in n of degree k, and this proves the assertion of the problem.

315. We consider the series

$$u_n = 1^4 + 2^4 + 3^4 + 4^4 + 5^4 + \cdots + n^4$$

and compute the first five difference sequences:

$$
\begin{array}{cccccc}
0 & u_1 & u_2 & u_3 & u_4 & u_5 \cdots \\
1 & 16 & 81 & 256 & 625 \cdots \\
& 15 & 65 & 175 & 369 \cdots \\
& & 50 & 110 & 194 \cdots \\
& & & 60 & 84 \cdots \\
& & & & 24 \cdots
\end{array}
$$

We note that $u_n^{(1)} = (n + 1)^4$ is a fourth-degree polynomial; therefore, $u_0^{(k)} = 0$, if only $k > 5$ [see problem 309 (b)]:

$$
\begin{aligned}
u_0^{(1)} &= 1; & u_0^{(4)} &= 60 ; \\
u_0^{(2)} &= 15; & u_0^{(5)} &= 24 ; \\
u_0^{(3)} &= 50 .
\end{aligned}
$$

Therefore, in view of problem 313, we have

$$
u_n = 0 + n + 15\frac{n(n - 1)}{1} + 50\frac{n(n - 1)(n - 2)}{6}
$$

$$
+ 60\frac{n(n - 1)(n - 2)(n - 3)}{24} + 24\frac{n(n - 1)(n - 2)(n - 3)(n - 4)}{120} .
$$

This is the formula we seek. Simplification yields

$$
1^4 + 2^4 + 3^4 + 4^4 + \cdots + n^4 = \frac{n(n + 1)(2n + 1)(3n^2 + 3n - 1)}{30} .
$$

Remark: Analogous reasoning enables computation of the sum

$$
1^k + 2^k + 3^k + \cdots + n^k ,
$$

where k is any positive integer.

316. (a) The first row of differences of the sequence

$$
U_n = 1^k + 2^k + 3^k + \cdots + n^k ,
$$

such that all members of the $(k + 1)$st row vanish, is the sequence of kth degree polynomials $(n + 1)^k$. This means that $u_n^{(k+2)} = 0$. If the initial sequence provides $k + 1$ sequences of differences before the zero row is reached, then it must have been of $(k + 1)$st order, that is, of $(k + 1)$st degree in n (see problem 314).

(b) Set

$$
u_n = 1^k + 2^k + 3^k + \cdots + n^k = A_0 n^{k+1} + A_1 n^k + \cdots .
$$

Then $u_n^{(k+1)} = A_0(k + 1)!$. (See problem 310.) But $u_n^{(k+1)}$ is the kth row of differences of the sequence $u_n^{(1)}$ of kth degree: $u_n^{(1)} = (n+1)^k = n^k + \cdots$. It follows from problem 310 that $u_n^{(k+1)} = k!$. Hence, we have $k! = A_0(k+1)!$; $A_0 = \dfrac{1}{k+1}$ is the coefficient of n^{k+1} in the summation of

$$u_k = 1^k + 2^k + 3^k + \cdots + n^k .$$

Determining the coefficient of n^k is more involved. We refer to the formula derived in problem 313:

$$u_n = a_0 n^k + a_1 n^{k-1} + \cdots + a_k$$
$$= u_0 + C_n^1 u_0^{(1)} + \cdots + C_n^{k-1} u_0^{(k-1)} + C_n^k u_0^{(k)} .$$

Equating the coefficients of n^{k-1} from both sides, we obtain

$$a_1 = -\frac{1 + 2 + \cdots + (k-1)}{k!} u_0^{(k)} + \frac{1}{(k-1)!} u_0^{(k-1)}$$
$$= -\frac{(k-1)k}{2k!} u_0^{(k)} + \frac{1}{(k-1)!} u_0^{(k-1)}$$
$$= -\frac{1}{2(k-2)!} u_0^{(k)} + \frac{1}{(k-1)!} u_0^{(k-1)}$$

(compare with the second solution of problem 310). Now, applying this formula to $(n+1)^k = n^k + C_k^1 n^{k-1} + \cdots + 1$, we obtain

$$C_k^1 = -\frac{1}{2(k-2)!} \cdot k! + \frac{1}{(k-1)!} u_0^{(k-1)} ;$$
$$u_0^{(k-1)} = (k-1)! \left[k + \frac{k(k-1)}{2} \right] = (k-1)! \, \frac{k(k+1)}{2} = \frac{(k+1)!}{2} .$$

Going over to the series

$$u_n = 1^k + 2^k + 3^k + \cdots + n^k = A_0 u^{k+1} + A_1 u^k + \cdots A_{n+1} ,$$

of which the sequence of $(n+1)^k$ is the first difference sequence, we obtain

$$A_1 = -\frac{1}{2(k-1)!} u_0^{(k+1)} + \frac{1}{k!} u_0^{(k)}$$
$$= -\frac{1}{2(k-1)!} k! + \frac{1}{k!} \cdot \frac{(k+1)!}{2} = -\frac{k}{2} + \frac{k+1}{2} = \frac{1}{2} .$$

[Here, $u_0^{(k+1)}$ and $u_0^{(k)}$ designate the quantities described above by $u_0^{(k)}$ and $u_0^{(k-1)}$, respectively.]

Therefore, the coefficient A_1 of n^k is $\dfrac{1}{2}$.

Remark: In the solution of problems 134 (a, b), we now have, in agreement with the above problem,

$$1^2 + 2^2 + \cdots + n^2 = \frac{n(n+1)(2n+1)}{6} = \frac{1}{3}n^3 + \frac{1}{2}n^2 + \frac{1}{6}n$$

$$1^3 + 2^3 + \cdots + n^3 = \frac{n^2(n+1)^2}{4} = \frac{1}{4}n^4 + \frac{1}{2}n^3 + \frac{1}{4}n^2 ,$$

$$1^4 + 2^4 + \cdots + n^4 = \frac{n(n+1)(2n+1)(3n^2 + 3n - 1)}{30}$$

$$= \frac{1}{5}n^5 + \frac{1}{2}n^4 + \frac{1}{3}n^3 - \frac{1}{30}n .$$

For $k = 1$, we obtain

$$1 + 2 + \cdots + n = \frac{n(n+1)}{2} = \frac{1}{2}n^2 + \frac{1}{2}n .$$

In the same manner we obtain, for example,

$$1^{100} + 2^{100} + \cdots + n^{100} = \frac{1}{100}n^{101} + \frac{1}{2}n^{100} + \cdots ,$$

$$1^{1000} + 2^{1000} + \cdots + n^{1000} = \frac{1}{1001}n^{1001} + \frac{1}{2}n^{1000} + \cdots .$$

317. If $a_0 = 1$, the problem has a simple solution. It was shown in problem 310 that in this case the integers of the kth row of differences were equal to $k!$. If all the numbers of a given row are divisible by d, then the numbers of all successive rows are divisible by d. Therefore, $k!$ is divisible by d.

Consider now the more general case, $a_0 \neq 1$. We have, from problem 313,

$$u_n = u_0 + nu_0^{(1)} + \cdots + \frac{n(n+1)\cdots(n-k+1)}{k!}u_0^{(k)} .$$

Since the integers u_n are all divisible by d, all the numbers $u_0^{(1)}, u_0^{(2)}, \cdots, u_0^{(k)}$ are also divisible by d. If d is factored out and placed as a multiplier, we have

$$u_n = d\left[v_0 + nv_0^{(1)} + \cdots + \frac{n(n-1)\cdots(n-k+1)}{k!}v_0^{(k)}\right] ,$$

where $v_0, v_0^{(1)}, \cdots, v_0^{(k)}$ designate the quotients obtained upon dividing $u_0, u_0^{(1)}, \cdots, u_0^{(k)}$ respectively, by d. These v's are integers.

We have above, in the brackets, only Pascal numbers: $\frac{n(n-1)}{2}, \cdots,$ $\frac{n(n-1)\cdots(n-k+1)}{k!}$. When we give these a common denominator and add, we arrive at the following form:

$$u_n = d\left(\frac{b_0 n^k + b_1 n^{k-1} + \cdots + b_k}{k!}\right).$$

The numbers $b_0, b_1, \cdots, b_k$ are integers, and we have

$$\frac{db_0}{k!} = a_0, \qquad \frac{db_1}{k!} = a_1, \qquad \cdots, \qquad \frac{db_k}{k!} = a_k.$$

We divide d and $k!$ by their greatest common divisor and write $\frac{d}{k!} = \frac{d_1}{m}$, where d_1 and m have no common factor:

$$\frac{d_1 b_0}{m} = a_0, \qquad \frac{d_1 b_1}{m} = a_1, \qquad \cdots, \qquad \frac{d_1 b_k}{m} = a_k,$$

or,

$$b_0 = \frac{a_0 m}{d_1}, \qquad b_1 = \frac{a_1 m}{d_1}, \cdots, b_k = \frac{a_k m}{d_1}.$$

It follows that the integers $a_0, a_1, \cdots, a_k$ are all divisible by d_1. Now, assume that the integers $a_0, a_1, \cdots, a_k$ do not have a nontrivial common divisor. This means that $d_1 = 1, \frac{d}{k!} = \frac{1}{m}$, or $\frac{k!}{d} = m$; that is, $k!$ is divisible by d. This completes the proof.

318. We construct the sum triangle for the sequence $C_n^0, C_u, C_n^2, \cdots, C_n^n$:

$$
\begin{array}{ccccc}
C_n^0 & C_n^1 & C_n^2 & \cdots & C_n^n \\
C_{n+1}^1 & C_{n+1}^2 & C_{n+1}^3 & \cdots & C_{n+1}^n \\
C_{n+2}^2 & C_{n+2}^3 & & \cdots \ C_{n+2}^n \\
\cdots & \cdots & & \cdots \\
& C_{2n-1}^{n-1} & & C_{2n-1}^n \\
& C_{2n}^n & &
\end{array}
$$

We have at the apex C_{2n}^n, inasmuch as we are dealing with Pascal triangles as elements.

On the other hand, using the general addition formula for these elements [see problem 311 (a)], we arrive at the following expression for the apex element:

$$(C_n^0)^2 + (C_n^1)^2 + (C_n^2)^2 + \cdots + (C_n^n)^2.$$

Thus, we have

$$(C_n^0)^2 + (C_n^1)^2 + (C_n^2)^2 + \cdots + (C_n^n)^2 = C_{2n}^n.$$

319. Consider the sequence $1, \dfrac{a}{b}, \dfrac{a^2}{b^2}, \dfrac{a^3}{b^3}, \cdots$, for which we construct the sum triangle

$$
\begin{array}{cccc}
1 & \dfrac{a}{b} & \dfrac{a^2}{b^2} & \dfrac{a^3}{b^3} \cdots \\[2mm]
1+\dfrac{a}{b} & \dfrac{a}{b}+\dfrac{a^2}{b^2} & \dfrac{a^2}{b^2}+\dfrac{a^3}{b^3} \cdots \\[2mm]
1+2\dfrac{a}{b}+\dfrac{a^2}{b^2} & \dfrac{a}{b}+2\dfrac{a^2}{b^2}+\dfrac{a^3}{b^3} \cdots \\[2mm]
1+3\dfrac{a}{b}+3\dfrac{a^2}{b^2}+\dfrac{a^3}{b^3} \cdots
\end{array}
$$

Let u_n designate the nth terms of the given sequence: $u_n = \dfrac{a^n}{b^n}$. We then have

$$\bar{u}_n^{(1)} = \left(1 + \frac{a}{b}\right) u_n .$$

By induction, we can conclude that

$$\bar{u}_n^{(k)} = \left(1 + \frac{a}{b}\right)^k u_n .$$

This means that $\bar{u}_0^{(k)} = \left(1 + \dfrac{a}{b}\right)^k$. However,

$$
\bar{u}_0^{(k)} = C_k^0 u_0 + C_k^1 u_1 + C_k^2 u_2 + \cdots + C_k^k u_k
$$
$$
= 1 + k\frac{a}{b} + \frac{k(k-1)}{2}\frac{a^2}{b^2} + \cdots + \frac{k(k-1)\cdots 2\cdot 1}{k!}\frac{a^k}{b^k} .
$$

Hence, we have

$$\left(1 + \frac{a}{b}\right)^k = 1 + k\frac{a}{b} + \frac{k(k-1)}{2}\frac{a^2}{b^2} + \cdots + \frac{k(k-1)\cdots 2\cdot 1}{k!}\frac{a^k}{b^k} .$$

If both members of the above equality are multiplied by b^k, we obtain

$(a+b)^k$
$$
= b^k + kab^{k-1} + \frac{k(k-1)}{2}a^2 b^{k-2} + \cdots + \frac{k(k-1)\cdots 2\cdot 1}{k!}a^k
$$
$$
= a^k + ka^{k-1}b + \frac{k(k-1)}{2}a^{k-2}b^2 + \cdots + \frac{k(k-1)\cdots 2\cdot 1}{k!}b^k ,
$$

which proves the formula.

320. Consider the following triangle:

$$\frac{1}{C_0^0} \qquad \frac{1}{2C_1^1} \qquad \frac{1}{3C_2^2} \qquad \frac{1}{4C_3^3} \cdots$$

$$-\frac{1}{2C_1^0} \qquad -\frac{1}{3C_2^1} \qquad -\frac{1}{4C_3^2} \cdots$$

$$\frac{1}{3C_2^0} \qquad \frac{1}{4C_3^1} \cdots$$

$$-\frac{1}{4C_3^0} \cdots$$

(The minus sign appears in the second, fourth, sixth, $\cdots$ rows.)

We shall prove that this is the difference triangle for the sequence $1, \frac{1}{2} \cdots, \frac{1}{n}, \cdots$. It suffices to show that for any $k \leqq n - 1$ the following equation holds:

$$\frac{1}{nC_{n-1}^k} - \frac{1}{(n+1)C_n^{k+1}} = \frac{1}{(n+1)C_n^k}.$$

We have

$$\frac{1}{nC_{n-1}^k} - \frac{1}{(n+1)C_n^{k+1}}$$

$$= \frac{k!}{n(n-1)(n-2)\cdots(n-k)} - \frac{(k+1)!}{(n+1)n(n-1)\cdots(n-k)}$$

$$= \frac{k!}{n(n-1)(n-2)\cdots(n-k)}\left(1 - \frac{k+1}{n+1}\right)$$

$$= \frac{k!}{n(n-1)(n-2)\cdots(n-k)} \cdot \frac{n-k}{n+1} = \frac{1}{(n+1)C_n^k}.$$

Now, having shown that the triangle is the difference triangle of the sequence $1, \frac{1}{2}, \cdots, \frac{1}{n}, \cdots$, we may carry out on it all the operations indicated by the conditions of the problem. The result is a Pascal triangle.

ANSWERS AND HINTS

---

1. First show that the total of all handshakes made up to any time is an even number.

2. Calculate the number of moves which the knight must make in order to arrive at every square exactly once (this number is odd.)

3. (a) Use mathematical induction.
Answer: $k = 2^n - 1$.

    (b) Designate by $k(n)$ the number of moves needed to remove n rings from the loop, and express $k(n)$ in terms of $k(n-2)$.

$$\textit{Answer:}\quad k(n) = \begin{cases} \dfrac{1}{3}(2^{n+1} - 2), & \text{if } n \text{ is even};\\[2mm] \dfrac{1}{3}(2^{n+1} - 1), & \text{if } n \text{ is odd}. \end{cases}$$

4. (a) For a first weighing, put 27 coins on each pan.
    (b) The number k, which we seek, can be determined from the inequality $3^{k-1} < n \leq 3^k$.

5. First, place one block on each pan of the balance. Then put both of these blocks on one pan, and put pairs of remaining blocks, successively, on the other pan.

423

6. (a) For a first weighing, put four coins on each pan.

(b) $k = 7$. Prove that if the number n of coins does not exceed $\frac{3^n - 3}{2}$, then it is always possible to detect the counterfeit by n weighings and simultaneously to determine whether it is light or heavy (of course, $n > 2$ is understood.) If $n > \frac{3^n - 3}{2}$, then n weight trials may not suffice. Then proceed by mathematical induction, in several stages. Specifically, prove the following propositions:

(A) Divide N coins into two sets, X and Y; it is understood that the counterfeit coin is in one of these sets. If this coin is in set X, it is light; and if it is in set Y, it is heavy. Then if $N \leqq 3^n$, the counterfeit can be detected by N weight trials; and if $N > 3^n$, there is no procedure which can guarantee success in N trials.

(B) Given N coins, among which there is one counterfeit which differs in weight from the others (although it is not known whether it is lighter or heavier than a genuine coin), then we know we have at least one genuine coin. Now, if $N \leqq \frac{3^n - 1}{2}$, then by N weight trials we can detect the counterfeit and know whether it is light or heavy. But if $N > \frac{3^n - 1}{2}$, then this number of weighings may not suffice.

(C) See the suggestion given at the beginning of this hint.

7. (a) One link. (b) Seven links.

8. The second.

9. Prove first that all the weights are even or odd.

10. The evenness or oddness of the first four numbers of each row of the number triangle depends only upon the evenness or the oddness of the first four numbers of the preceding row.

11. Change the order of the squares themselves in such a way that the problem becomes one of moving chips from one square to an adjoining one.

12. 15,621.

13. 2 roubles.

14. (a) The rule for the alternation of the number of days in the year is as follows: Each year whose total number of days is divisible by 4 is a leap year (has an extra day), with the exception of those years divisible by 100 but not by 400. It is easy to see that

400 years contain an integral number of weeks. Consequently, it remains to prove whether the new year begins most often with a Saturday or a Sunday for any 400-year period.
Answer: *With Sunday.*
(b) On Saturday.

15. All numbers ending in 0, and also the two-digit numbers 11, 22, 33, 44, 55, 66, 77, 88, 99, 12, 24, 36, 48, 13, 26, 39, 14, 28, 15, 16, 17, 18, 19.

16. (a) $\underbrace{6250 \cdots 0}_{n \text{ times}}, n = 0, 1, 2, \cdots$.

(b) Try to solve the following: Find an integer beginning with a given digit and which is reduced to $\frac{1}{35}$ its original value if the initial digit is deleted.

17. (a) Prove the number can be reduced to $\frac{1}{9}$ its value only on deleting the digit zero, which must stand in the second position.

(b) 10,125, 2025, 30,375, 405, 50,625, 6075, 70,875 (to each of these numbers may be attached, at the end, an arbitrary number of zeros).

18. (a) Numbers in which all but the first two digits are zeros.

(b) Investigate separately the cases in which the first digit of the number sought is: $1, 2, 3, \cdots, 9$. There is a total of 104 numbers satisfying the conditions of the problem. To each may be attached (at the end) an arbitrary number of zeros.

19. (a) The smallest possible number is 142,857.

(b) The digits 1 or 2. The smallest number beginning with 2 is 285,714.

20. Recall that numbers which are divisible by 5 must end in 0 or 5, numbers divisible by 6 or 8 in an even number.

21. Try to solve the following problem: Find a number with a given initial digit which is doubled upon transferring the initial digit to the end.

22. This solution is analogous to that of problem 21.

23. The smallest number satisfying the conditions of the problem is 7, 241, 379, 310, 344, 827, 586, 206, 896, 551.

24. (a) A number whose inversion exceeds it by a factor of 5, 6, 7, or 8 must necessarily have 1 as its first digit; a number whose inversion exceeds it by a factor of 2 or 3 can begin only with digits 1, 2, 3, 4, or, respectively, 1, 2, or 3.

(b) Some numbers which are $\frac{1}{4}$ of their inversions are: 0, 2178; 21,978; 2,199,978, $\cdots$ (∗). All other such numbers are of form

$$\overline{P_1 P_2 \cdots P_{n-1} P_n P_{n-1} \cdots P_2 P_1} \, ,$$

where P_1, P_2, $\cdots$, P_n represent numbers given by the (∗)-display.

25. (a) 142,857.

(b) Set up the equations to be satisfied by an eight-digit number which is multiplied by 6 if we transfer the final four digits to the front (maintaining the same order for these).

26. 142,857.

27. Factor the given polynomials. See what remainders are possible upon dividing n by 3 (and, respectively, by 4, 5, 7, and so on).

28. (a) Use the fact that the difference $a^{2n} - b^{2n}$ is divisible by $a + b$.

(b) See the hint to problem 27.

(c) $56,786,730 = 2 \cdot 3 \cdot 5 \cdot 7 \cdot 11 \cdot 13 \cdot 31 \cdot 61$. Further, use the propositions of problem 27 (a–d) and also the results of Fermat's theorem (problem 240).

29. Note that $n^2 + 3n + 5 = (n + 7)(n - 4) + 33$.

30. Factor the given expression and compare the number of factors with the number of factors of 33.

31. Use the fact that every integer not divisible by 5 can be represented in the form $5k \pm 1$ or else $5k \pm 2$.
Answer: 0 or 1.

32. Use the result of the problem 31.

33. 625 and 376.

34. Find the last two digits of n^{20}; and the three final digits of n^{200}.
Answer: 7; 3.

35. $1 + 2 + 3 + \cdots + n = \dfrac{n(n + 1)}{2}$. Grouping separate terms of the sum $1^k + 2^k + 3^k + \cdots + n^k$, show that this sum is divisible by $\dfrac{n}{2}$ and by $(n + 1)$, or by n and $\dfrac{n + 1}{2}$.

36. The sum of alternate digits of the second, fourth, and so on, places minus the sum of the alternate digits of the first, third, fifth, and so on, places must be a number divisible by 11.

37. The number is divisible by 7.

38. It is always possible to find an integer with first two digits 1, 0 and divisible by k. If the difference of this number and its inversion is again divisible by k, then it can be shown that 9 is divisible by k.

39. $26,460 = 2^2 \cdot 3^3 \cdot 5 \cdot 7^2$. Show that the given expression is divisible by $5 \cdot 7^2$ and also by $2^2 \cdot 3^3$.

40. Use the identity
$$11^{10} - 1 = (11 - 1)(11^9 + 11^8 + 11^7 + \cdots + 11 + 1).$$

41. Write the given number in the form
$$(2222^{5555} + 4^{5555}) + (5555^{2222} - 4^{2222}) - (4^{5555} - 4^{2222}).$$

42. Use mathematical induction.

43. Use the facts that $10^6 - 1 = 999,999$ is divisible by 7, and that any power of 10 yields a remainder of 4 when divided by 6.
Answer: 5.

44. (a) 9; 2. (b) 88; 67. (c) Find the final two digits of $7^{(14^{14})}$ and $2^{(14^{14})}$.
Answer: 36.

45. (a) 7; 07. (b) 3; 43.

46. Investigate the numbers
$$Z_1 = 9, \ Z_2 = 9^{Z_1}, \ Z_3 = 9^{Z_2}, \ \cdots, \ Z_{1001} = 9^{Z_{1000}} = N$$
and determine the final digit of the number Z_1, the final two digits of Z_2, the final three digits of Z_3, the final four digits of Z_4, the final five digits of Z_5, and the final five digits of $Z_6, Z_7, \cdots, Z_{1001} = N$.
Answer: 45,289.

47. The 1000 digits sought will be $p\overline{\underbrace{PP \cdots P}_{23 \text{ times}}}$, where
$$P = 020408163265306122448979591836734693877551$$
is the periodic part of the fraction 1/49, and p is the number comprising the final 34 digits of the periodic part of P. For the proof, use the fact that

$$N = \frac{50^{1000} - 1}{50 - 1} = \frac{50^{1000} - 1}{49} \, .$$

48.   24.

49.   (a)   Compare the powers of any prime p in the product $a!$ and in the product $(t + 1)(t + 2)\cdots(t + a)$.

    (b, c)   Use the results of part (a).

    (d)   First prove the existence of a number k such that kd, where d is the common difference of a progression, yields a remainder of 1 upon division by $n!$.

50.   Not divisible.

51.   (a) $(n - 1)!$ is not divisible by n if n is a prime number or if $n = 4$.

    (b) $(n - 1)!$ is not divisible by n^2 if n is a prime or twice a prime, or if $n = 8$ or $n = 9$.

52.   Prove that all such numbers are less than $7^2 = 49$.
Answer: $24, 12, 8, 6, 4,$ and 2.

53.   (a)   Show that the sum of the squares of five consecutive integers is divisible by 5 but not by 25.

    (b)   Determine the remainder obtained when the sum of the even powers of the three consecutive integers is divided by 3.

    (c)   Find the remainder obtained when the sum of the same even power of nine consecutive integers is divided by 9.

54.   (a)   Find the remainders upon division of each of the numbers A and B by 9.

    (b)   192, 384, 576; or 273, 546, 819; or 327, 654, 981; or 219, 438, 657.

55.   The square must end with four zeros.

56.   Use the Pythagorean theorem. Prove, separately, that the area of the rectangle is divisible by 3 and 4.

57.   Investigate what remainders are possible if $b^2 - 4ac$ is divided by 8.

58.   Note that after the addition and reduction to lowest terms the denominator is divisible by 3 and by 2.

59.   To show that M and N cannot be integers, it would suffice to show that after the indicated addition is made the denominator is divisible by a higher power of 2 than is the numerator. For the fraction K, use 3 instead of 2.

60. (a) The denominator of the sum of the fractions is $(p-1)!$. The numerator is the sum of all possible products of the first $(p-1)$ integers taken $(p-2)$ at a time. Designate the sum of all possible products of n numbers $1, 2, \cdots, n$, taken k at a time, by π_n^k. It is to be shown that π_{p-1}^{p-2}, where p is a natural number, is divisible by p^2.

Using the formula (relation between roots and coefficients)

$$(x-1)(x-2)(x-3)\cdots(x-p+1)$$
$$= x^{p-1} - \pi_{p-1}^1 x^{p-2} + \cdots + \pi_{p-1}^{p-2},$$

and

$$[(x-1)(x-2)(x-3)\cdots(x-p+1)](x-p)$$
$$= (x-1)[(x-2)(x-3)\cdots(x-p+1)(x-p)],$$

prove that the two polynomials, whose coefficients depend upon $\pi_{p-1}^1, \pi_{p-1}^2, \cdots, \pi_{p-1}^{p-1}$ are equal. Equating coefficients of like powers of the two polynomials produces a set of relationships between the values π_{p-1}^k, from which the result asked for in the problem emerges.

(b) When the fractions are added we have the denominator $[(p-1)!]^2$ and the numerator $(\pi_{p-1}^{p-2})^2 - 2(p-1)! \, \pi_{p-1}^{p-3}$. Refer to the hint for part (a).

61. Use the fact that the fraction p/q is reducible to lower terms if and only if q/p is.

62. Prove that for any integer b the number of differences $a_k - a_l$ which are divisible by b is not less than the number of differences $k - l$ which are divisible by b. Investigate first the case in which n is a multiple of b.

63. Prove, first, the following equality:

$$(1 + 10^4 + 10^8 + \cdots + 10^{4k}) \cdot 101$$
$$= (1 + 10^2 + 10^4 + \cdots + 10^{2k})(10^{2k+2} + 1).$$

64. (a) $a - b$. (b) $a - b$.

65. Show that $(2^{2^n} + 1) - 2$ is divisible by all the preceding numbers of the given sequence. It then follows that $2^{2^n} + 1$ and those preceding numbers cannot have a common factor differing from 2.

66. Investigate what remainders are possible upon division of $2^n - 1$ and $2^n + 1$ by 3.

67. (a) Investigate what remainders are possible if p, $8p - 1$, and $8p + 1$ are divided by 3.

(b) Consider the possible remainders if p, $8p^2 + 1$, and $8p^2 - 1$ are divided by 3.

68. Show that $p^2 - 1$ is divisible by 12.

69. Consider the possible remainders upon division by 6.

70. (a) Prove that the common difference of the progression must be divisible by $2 \cdot 3 \cdot 5 \cdot 7 = 210$.
Answer: 199, 409, 619, $\cdots$, 2089.

(b) Prove that if the first number of the progression is not 11, then the common difference must be divisible by $2 \cdot 3 \cdot 5 \cdot 7 \cdot 11 = 2310$; if the first number is 11, then the common difference is divisible by 210.
For problems like problem 70 (a, b) it is convenient to refer to a table of prime numbers.

71. (a) This will be an odd number not divisible by 3.

(b) It suffices to find a number not having a common factor of 2, 3, 5, 7, 11, or 13 with respect to the other fifteen numbers of the sequence.

72. The product is equal to $22 \underbrace{\cdots}_{\text{665 times}} 2177 \underbrace{\cdots}_{\text{665 times}} 78$.

73. The quotient is 777 000 777 000 $\cdots$ 777 000 77 (there are 166 repetitions of 777 000); the remainder is 700.

74. $222{,}222{,}674{,}025 = 471{,}405^2$.

75. Prove that if $\sqrt{\alpha} < 1 - \left(\dfrac{1}{10}\right)^{100}$, then $\alpha < 1 - \left(\dfrac{1}{10}\right)^{100}$.

76. 523,152 and 523,656.

77. 1946.

78. (a) Consider the sum of the arithmetic progression whose common difference is 1 and whose first term is 10^{n-1} and whose last term is $10^n - 1$.

(b) 1,769,580.

79. Consider, first, all the integers from 0 to 99,999,999, putting enough zeros before all those integers having fewer than eight digits to make them eight-digit "numbers".

80. 7.

81. No.

82. Use the fact that nine weights of magnitude $n^2, (n + 1)^2, \cdots,$ $(n + 8)^2$ can be divided into three groups, the first two of which total the same weight and the third set is lighter by 18 units.

83. 240.

84. (a) $147, 258, 369$. (b) $941, 852, 763$.

85-86. Use the formula for the sum of an arithmetic progression.

87. Put the expression $n(n + 1)(n + 2)(n + 3) + 1$ into the form of a quadratic polynomial.

88. Show that among the considered numbers there cannot be more than four which are pairwise distinct.

89. Prove, first, that no matter what numbers we begin with, after at most four steps we arrive at four even numbers.

90. (b) Having made some arrangement of the numbers from 1 to 101, choose the longest increasing sequence starting with the first term of this arrangement. If fewer than eleven numbers can be found to make up such an increasing sequence, then cross out these numbers from the arrangement, and start over, beginning with the first remaining number. If again there are fewer than eleven numbers, then cross out also these numbers, and start over as before. Every sequence which has been deleted consists of fewer than eleven numbers, and, unless the problem is solved for this arrangement, we will obtain no less than eleven sequences. This circumstance will allow the construction of the desired sequence.

91. Factor out of each of the selected numbers the greatest power of 2 contained as a factor.

92. (a) Consider that number which when divided by 100 yields the remainder of least absolute value.

(b) let $a_1, a_2, \cdots, a_{100}$ be the given numbers. Investigate the remainders, upon division by 100, of the numbers $a_1, a_1 + a_2, a_1 + a_2 + a_3, \cdots$.

93. Assume that the player, on some day, has played a_1 matches, and at the end of the following day he has played, along with the previous day's games, a_2 matches, and for three consecutive days he has played a total of a_3 matches, and so on. Consider the 154 numbers $a_1, a_2, \cdots, a_{77}, a_1 + 20, a_2 + 20, \cdots, a_{77} + 20$.

94. Investigate the remainders, upon division by N, of the numbers $1, 11, 111, \cdots$.

95. The final digit of the difference of two successive numbers of the sequence yields the final digit of the number which precedes those two. Show that there exist natural numbers n and k such that the final four digits of the $(n + k)$th and $(n + k + 1)$st terms of the Fibonacci sequence are the same, respectively, as the final four digits of the kth and $(k + 1)$st terms. Then the last four digits of the $(n + k - 1)$st term will be the same as those last four digits of the $(k - 1)$st term of the sequence, and so on. In this way it may be shown that it is possible to find a term of the Fibonacci sequence whose last four digits coincide with the first term, that is, zero.

96. Investigate the fractional parts of the numbers $0, \alpha, 2\alpha, \cdots$, $1000 \, \alpha$.

97. Prove that an interval $(0, A)$ of the number axis (where A is an arbitrary natural number less than $m + n$) contains exactly $A - 1$ of these fractions.

98. Designate by $k_i (i = 1, 2, 3, \cdots)$ those of the numbers which are included between $\dfrac{1000}{i}$ and $\dfrac{1000}{i + 1}$, and select those numbers (less than 1000) which are multiples of at least one of the numbers $a_1 a_2$, $\cdots, a_n$.

99. The length k of the period of $\dfrac{p}{q}$ can be determined as the least power k for which $10^k - 1$ is divisible by q. If $k = 2l$, then it follows that $10^l + 1$ is divisible by q; that is, $\dfrac{10^l + 1}{q}$ is an integer. From this, it is possible to show that in the period $\overline{a_1 a_2 \cdots, a_l, a_{l+1}, a_{l+2}, \cdots a_k}$ of the fraction $\dfrac{p}{q}$ we have

$$a_1 + a_{l+1} = a_2 + a_{l+2} = \cdots = a_l + a_{2l} = 9 \, .$$

100. Use the fact that the number of digits in the period of each of the fractions $\dfrac{a_n}{p^n}$ and $\dfrac{a_{n+1}}{p^{n+1}}$ is equal to the smallest of the natural numbers k and l such that $10^k - 1$ is divisible by p^n and, respectively,

$10^l - 1$ is divisible by p^{n+1}.

101. Every number x can be represented in the form $[x] + \alpha$, where $0 \leq \alpha < 1$.

102. Investigate all points in the plane, with integer coordinates, located inside the rectangle $0 < x < q, 0 < y < p$, (where x, y are coordinates of points in the plane) and under the diagonal, excepting the point $(0, 0)$.

103. Use mathematical induction. (The problem can also be solved geometrically. To do this, investigate all points of the first quadrant with integer coordinates under the hyperbola $xy = n$.)

104–106. In solving problem 104, we must use the fact that
$$[(2 + \sqrt{2})^n] = (2 + \sqrt{2})^n + (2 - \sqrt{2})^n - 1.$$
Similar comment holds for problems 105 (a, b) and 106.

107. In the set of p consecutive integers, $n, n - 1, n - 2, \cdots$, $n - p + 1$, one (and only one) is divisible by p. If this integer is N, then $\left[\dfrac{n}{p}\right] = \dfrac{N}{p}$. Then the difference $C_n^p - \left[\dfrac{n}{p}\right]$ can be written
$$\frac{n(n - 1)\cdots(N + 1)N(N - 1)\cdots(n - p + 1)}{p!} - \frac{N}{p}.$$

108. Prove, first, that if the number n of elements in the sequence $[\alpha], [2\alpha], [3\alpha], \cdots$ does not exceed N, then $n\alpha = N + l$, where $1 - \alpha \leq l < 1$.

109. Prove that $\left(\dfrac{N}{2^k}\right)$ is equal to the number of integers, not exceeding N, which are divisible by 2^{k-1} but not by 2^k.

110. (a) 7744. (b) 29, 38, 47, 56, 65, 74, 83, 92.

111. If a is the number made up of the first two digits of the number sought, and if b is the number made up of the last two digits, then $99a = (a + b)^2 - (a + b) = (a + b)(a + b - 1)$.
Answer: 9801, 3025, 2025.

112. (a) 4624, 6084, 6400, 8464. (b) There are no such numbers.

113. (a) 145. (b) Only the number 1.

114. (a) 1, 81. (b) 1, 8, 17, 18, 26, 27.

115. (a) x cannot be greater than 4.
Answer: $x = 1, y = \pm 1; x = 3, y = \pm 3$.
    (b) $x = 1, y = \pm 1, z$ is any even number; $x = 3, y = \pm 3, z = 2$;

$x = 1, y = 1, z$ is any odd number; x is any positive integer,

$$y = 1! + 2! + \cdots + x!, z = 1 .$$

116. Find out by what powers of 2 it is possible to divide the four numbers sought.

Answer: For odd n, the representation is not possible. For even n, there exists just one representation:

$$2^n = [2^{(n/2)-1}]^2 + [2^{(n/2)-1}]^2 + [2^{(n/2)-1}]^2 + [2^{(n/2)-1}]^2 .$$

117. See the hint to problem 116. In part (b), the solution of which is analogous to that of part (a), there is only the answer $x = y = z = v = 0$.

118. (a) It is possible to show that if $x, y,$ and z satisfy the given equality, and if $z > \dfrac{kxy}{2}$, then these numbers can be reduced in such a way as to continue to satisfy the equality. If $x \leqq \dfrac{kyz}{2}$, $y \leqq \dfrac{kxz}{2}$, $z \leqq \dfrac{kxy}{2}$, and $x \leqq y \leqq z$, then it must follow that $2 \leqq kx \leqq 3$.

Answer: $k = 1$ and $k = 3$.

(b) All such triples of integers can be obtained by using substitutions of the form $x_1 = x, y_1 = y, z_1 = kxy - z$ in one of the triples $1, 1, 1$ and $3, 3, 3$. In all, there are (considering integers less than 1000) 23 number triples satisfying the conditions of the problem.

119. Let x and y be a pair of numbers satisfying the conditions of the problem. Then $\dfrac{x^2 + 125}{y} = u$ and $\dfrac{y^2 + 125}{x} = v$, where u and v are integers. Show that the number pairs (x, u) and (y, v) also satisfy the condition that the square of one number when increased by 125 is divisible by the other. It follows that if one pair of prime numbers is found to satisfy the condition, it is possible to construct an infinite chain of numbers in which adjacent numbers have this property. To find all such chains, prove that every chain contains a number not exceeding $\sqrt{125} < 12$.

There are 31 number pairs which satisfy the conditions of the problem.

120. Investigate the cases in which all four numbers are different, two numbers are the same but the other two are different, two pairs of equal numbers appear, and so on.

Answers: 96, 96, 57, 40; 11, 11, 6, 6; $k(3k \pm 2), k(3k \pm 2), k(3k \pm 2), 1,$

[k is an arbitrary integer, but such that $k(3k \pm 2)$ is positive]; 1, 1, 1, 1.

121. 2, 2 and 0, 0.

122. $1 = \dfrac{1}{2} + \dfrac{1}{4} + \dfrac{1}{4} = \dfrac{1}{2} + \dfrac{1}{3} + \dfrac{1}{6} = \dfrac{1}{3} + \dfrac{1}{3} + \dfrac{1}{3}$.

123. (a) Reduce the problem to the following: To find two numbers whose product is a multiple of their sum.

(b) $x = m(m + n)t, y = n(m + n)t, z = mnt$, where m, n, and y are arbitrary whole numbers.

124. (a) Let $y > x$; prove that then y is divisible by x.
Answer: $x = 2, y = 4$.

(b) $x = \left(\dfrac{p + 1}{p}\right)^p$, $y = \left(\dfrac{p + 1}{p}\right)^{p+1}$, where p is an arbitrary integer differing from 0 and -1.

125. 7 or 14.

126. By comparing the number of points won by the ninth graders with the number of games played by them, it is possible to conclude that all the ninth grade students won every match played by them. It follows from this that only one ninth-grade student could have participated in the tournament.

127. Use Heron's formula for the area of a triangle. Let $p - a = x$, $p - b = y$, and $p - c = z$, where a, b, and c are the sides of the triangle and p is its semi-perimeter. Then the problem reduces to the solution in integers of $xyz = 4(x + y + z)$, or $x = \dfrac{4y + 4z}{yz - 4}$. Assuming $x \geqq y$, we can investigate the squared inequality in y (with coefficients in terms of z); this enables us to found bounds for z and y.
Answer: The sides of the triangle can be equal to 6, 25, 29; 7, 15, 20; 9, 10, 17; 5, 12, 13; or 6, 8, 10 (all five solutions).

128. (a) The problems lead to the solution of the equation $x^2 + y^2 = z^2$. If x, y, and z are not relatively prime, then the equation can be divided through by the greatest common factor; hence we may consider the case for which x, y, and z are pairwise relatively prime. Then z is odd, and one of x or y is even and the other odd. Write the equation in the form $\left(\dfrac{x}{2}\right)^2 = \dfrac{z + y}{2} \cdot \dfrac{z - y}{2}$ and use the fact that $\dfrac{z + y}{2}$ and $\dfrac{z - y}{2}$ are relatively prime.
Answer: $x = 2tab, y = t(a^2 - b^2), z = t(a^2 + b^2)$, where a and b are

arbitrary relatively prime numbers $(a > b)$, and t is any natural number.

(b) The problem leads to the solution in integers of the equation $z^2 = x^2 + y^2 - xy$, or $[4z + (x + y)]^2 = [2x + 2(x + y)]^2 + [3(x-y)]^2$. Now use the result of problem (a).

Answer: $x = \frac{1}{3} tb(2a - b)$, $y = \frac{1}{3} ta(2b - a)$, $z = \frac{1}{3} t(a^2 + b^2 - ab)$,

where a and b are relatively prime $\left(\frac{a}{2} < b < 2a\right)$, and at least one of the numbers t or $a + b$ is divisible by 3; otherwise, t is arbitrary.

(c) The solution is analogous to that of problem (b).

Answer: $x = ta(a - 2b)$, $y = tb(2a - b)$, $z = t(a^2 + b^2 - ab)$, where a and b are relatively prime $(a > 2b)$, and at least one of the numbers t or $a + b$ is divisible by 3.

129. Begin with the formulas $\frac{b}{a} = \frac{\sin nA}{\sin A}$, $\frac{c}{a} = \frac{\sin (n + 1)A}{\sin A}$ (angle $B = nA$), where $n = 2, 5$, or 6. The right members of these equations can be expressed in terms of $2 \cos A$; use the fact that $2 \cos A$ is rational.

Answers: (a) $a = 4$, $b = 6$, $c = 5$. (b) $a = 1024$, $b = 1220$, $c = 231$. (c) $a = 46,656$, $b = 72,930$, $c = 30,421$.

130. Use the results of problem 128 (a) to show that if $x^4 + y^4 = z^2$, where x, y, and z are natural numbers, then there must exist positive integers x_1, y_1, z_1, where $z_1 < z$, such that $x_1^4 + y_1^4 = z_1^2$. But this allows us to prove the impossible equation $\bar{x}^4 + \bar{y}^4 = 1$, where $\bar{x}$ and $\bar{y}$ are positive integers.

131. Multiply both sides of the equation by $n!$ (factorial n; see note to problem 28).

132. (a) $1 - \frac{1}{n}$.   (b) $\frac{1}{2}\left[\frac{1}{2} - \frac{1}{n(n - 1)}\right]$.

(c) $\frac{1}{3}\left[\frac{1}{6} - \frac{1}{n(n - 1)(n - 2)}\right]$.

133. Use mathematical induction.

134. (a) $\dfrac{n(n + 1)(2n + 1)}{6}$.   (b) $\dfrac{n^2(n + 1)^2}{4}$.

(c) $\dfrac{n(n + 1)(2n + 1)(3n^2 + 3n - 1)}{30}$.   (d) $n^2(2n^2 - 1)$.

135. Transfer the 1 to the left side.

136. (a) $(n + 1)! - 1$. (b) $C_{n+k+1}^k - 1$.

137. Use the fact that $\dfrac{1}{\log_b a} = \log_a b$.

138. $\dfrac{1}{a_1 a_2 \cdots a_n}$.

139. (a) $\dfrac{3}{2}\left(1 - \dfrac{1}{3^{2^{n+1}}}\right)$. (b) $\dfrac{\sin 2^{n+1}\alpha}{2^{n+1}\sin\alpha}$.

140. 31.

141. (a) Compare the given product with $\dfrac{2}{3}\cdot\dfrac{4}{5}\cdot\dfrac{6}{7}\cdot\ldots\cdot\dfrac{98}{99}$, or square the given inequalities.

(b) Prove by mathematical induction that

$$\frac{1}{2}\cdot\frac{3}{4}\cdot\frac{5}{6}\cdots\frac{2n-1}{2n} < \frac{1}{\sqrt{3n+1}} .$$

142. Use the inequalities of problem 141 (a).

143. $99^n + 100^n$ exceeds 101^n if $n \leqq 48$ and is less than 101^n if $n > 48$.

144. 300!

145. Prove, first, that for any positive integer $k \leqq n$

$$1 + \frac{k}{n} \leqq \left(1 + \frac{1}{n}\right)^k \leqq 1 + \frac{k}{n} + \frac{k^2}{n^2} .$$

146–147. Use the results of problem 145.

148. Use mathematical induction.

149. Use the binomial theorem.

150. Use mathematical induction.

151. Use the inequalities

$$(k + 1)x^k(x - 1) > x^{k+1} - 1 > (k + 1)(x - 1) ,$$

from which it is possible to derive the following result for all positive integers p:

$$(p + 1)^{k+1} - p^{k+1} > (k + 1)p^k > p^{k+1} - (p - 1)^{k+1} .$$

152. These inequalities may be obtained by using "majorizing series"—that is, by replacing terms by larger (or, respectively, smaller) terms, and possibly using convenient new groupings, and comparing the sums of the new series with the bounding values.

153. Prove first that

$$2\sqrt{n+1} - 2\sqrt{n} < \frac{1}{\sqrt{n}} < 2\sqrt{n} - 2\sqrt{n-1} \ .$$

Answers: (a) 1998. (b) 1800.

154. First show that

$$\frac{3}{2} [\sqrt[3]{(n+1)^2} - \sqrt[3]{n^2}] < \frac{1}{\sqrt[3]{n}} < \frac{3}{2} [\sqrt[3]{n^2} - \sqrt[3]{(n-1)^2}] \ .$$

Answer: 14,996.

155. (a) 0.105.
(b) Use the fact that

$$\frac{1}{10!} + \frac{1}{11!} + \frac{1}{12!} + \cdots + \frac{1}{1000!} < \frac{1}{9} \left(\frac{9}{10!} + \frac{10}{11!} + \frac{11}{12!} + \cdots + \frac{999}{1000!} \right) .$$

Answer: 0.00000029.

156. Use the result of problem 152 (a).

157. First determine the sum of the undeleted terms between $\frac{1}{10^k}$ and $\frac{1}{10^{k+1}}$.

158. (a) Solve in a manner analogous to that used for problem 156.
(b) Use the result of problem 132 a).

159. Prove that for all integers k and primes $p \geq 2$

$$\log \left(1 + \frac{1}{p} + \frac{1}{p^2} + \frac{1}{p^3} + \cdots + \frac{1}{p^k} \right) < \frac{2 \log 3}{p} \ .$$

Using this result, deduce that

$$\log \left(1 + \frac{1}{2} + \frac{1}{3} + \frac{1}{4} + \cdots + \frac{1}{n-1} + \frac{1}{n} \right)$$
$$\leq 2 \log 3 \left(\frac{1}{2} + \frac{1}{3} + \frac{1}{5} + \cdots + \frac{1}{p_l} \right) ,$$

where p_l is the greatest prime between 1 and n.

160. 9.

161. If $a + b + c = 0$, then $0 = (a + b + c)^3 = a^3 + b^3 + c^3 - 3abc$.

162. (a) $a^3 + b^3 + c^3 - 3abc = (a + b + c)(a^2 + b^2 + c^2 - ab - ac - bc)$.
(b) $(a + b + c)^3 - a^3 - b^3 - c^3 = 3(a + b)(a + c)(b + c)$.

163. Use the result of problem 162 (a).

164. Use the result of problem 162 (b).

165. Use the facts that $a^{10} + a^5 + 1 = \dfrac{(a^5)^3 - 1}{a^5 - 1}$, and that $a^{15} - 1 = (a^3)^5 - 1$.

Answer: $a^{10} + a^5 + 1 = (a^2 + a + 1)(a^8 - a^7 + a^5 - a^4 + a^3 - a + 1)$.

166. Prove that the difference $(x^{9999} + x^{8888} + \cdots + x^{1111} + 1) - (x^9 + x^8 + \cdots + x + 1)$ is divisible by $x^9 + x^8 + \cdots + x + 1$.

167. $x_1 = -a - b,\ x_2 = \dfrac{a+b}{2} + \dfrac{a - b\sqrt{3}}{2}\,i,\ x_3 = \dfrac{a+b}{2} - \dfrac{a - b\sqrt{3}}{2}\,i$,

where $a = \sqrt[3]{\dfrac{q}{2} + \sqrt{\dfrac{q^2}{4} + \dfrac{p^3}{27}}},\qquad b = \sqrt[3]{\dfrac{q}{2} - \sqrt{\dfrac{q^2}{4} + \dfrac{p^3}{27}}}$.

168. Square to eliminate radicals and solve the resulting equation for x in terms of a.

Answer:

$$x_{1,2} = \frac{1}{2} \pm \sqrt{a + \frac{1}{4}}\,.$$

$$x_{3,4} = -\frac{1}{2} \pm \sqrt{a - \frac{3}{4}}\,.$$

169. Use the fact that if $x^2 + 2ax + \dfrac{1}{16} = y$, then

$$x = -a + \sqrt{a^2 + y - \frac{1}{16}}\,,$$

and investigate the graphs of the functions $y = x^2 + 2ax + \dfrac{1}{16}$ and $y_1 = -a + \sqrt{a^2 + x - \dfrac{1}{16}}$.

Answer: $x_{1,2} = \dfrac{1 - 2a}{2} \pm \sqrt{\left(\dfrac{1 - 2a}{2}\right)^2 - \dfrac{1}{16}}$.

170. (a) Prove that, necessarily, $3x = x^2$

Answer: $x_1 = 3,\ x_2 = 0$.

(b) Prove that, necessarily, $\dfrac{1}{1 + x} = x$.

Answer: $x_1 = \dfrac{1 + \sqrt{5}}{2},\ x_2 = \dfrac{1 - \sqrt{5}}{2}$.

171. The roots of the equation are all the numbers between 5 and 10 ($5 \leqq x \leqq 10$).

172. The roots are -2 and all numbers not less than 2 ($x = -2$ and $x \geqq 2$).

173. For $a = \pm 1$, the system has three solutions; if $a = \pm \sqrt{2}$, the system has two solutions.

174. (a) If $a = -1$, the system has no solutions; if $a = 1$, the system has an infinite number of solutions.

(b) If $a = \pm 1$, the system has an infinite number of solutions.

(c) If $a = 1$, the system has infinitely many solutions; if $a = -2$, there are no solutions.

175. For the system to have solutions it is necessary that three of the four numbers $\alpha_1, \alpha_2, \alpha_3, \alpha_4$ be equal to each other. If $\alpha_1 = \alpha_2 = \alpha_3 = \alpha$, and $\alpha_4 = \beta$, then $x_1 = x_2 = x_3 = \dfrac{\alpha^2}{2}$ and $x_4 = \alpha\left(\beta - \dfrac{\alpha}{2}\right)$.

176. The only real solution is $x = 1, y = 1, z = 0$.

177. (a) The number of real roots of the equation is the number of points of intersection of the straight line $y = \dfrac{x}{100}$ and the curve $y = \sin x$.

Answer: 63 roots.

(b) The number of points at which $y = \sin x$ and $y = \log x$ will meet is 3.

178. That $x_1^n + x_2^n$ is a whole number can be shown by mathematical induction. Show, further, that if $x_1^n + x_2^n$ is divisible by 5, then also $x_1^{n-3} + x_2^{n-3}$ is divisible by 5.

179. The square of the first polynomial cannot contain the same number of positive and negative terms which are products $a_i a_j$, but the second polynomial can.

180. Use mathematical induction; investigate separately the even and odd cases.

181. If $99{,}999 + 111{,}111\sqrt{3} = (A + B\sqrt{3})^2$, then
$$99{,}999 - 111{,}111\sqrt{3} = (A - B\sqrt{3})^2.$$

182. Prove that if $\sqrt[3]{2} = p + q\sqrt{r}$, then $\sqrt[3]{2}$ would have to be rational.

183. The second number is the larger.

184. $x = \dfrac{a_1 + a_2 + \cdots + a_n}{n}$.

185. (a) a_1, a_2, a_4, a_3.

    (b) Prove that if a_{i_α} and a_{i_β} are any two of the given numbers ($\alpha < \beta$) and $a_{i_{\alpha-1}}, a_{i_{\beta+1}}$ are numbers standing in the desired positions before a_{i_α} and after a_{i_β}, then

$$(a_{i_\alpha} - a_{i_\beta})(a_{i_{\alpha-1}} - a_{i_{\beta+1}}) > 0 .$$

Answer: $a_1, a_2, a_4, a_6 \cdots, a_{n-2}, a_n, a_{n-1}, a_{n-3}, a_{n-5}, \cdots, a_5, a_3$ if n is even; $a_1, a_2, a_4, a_6, \cdots, a_{n-1}, a_n, a_{n-2}, a_{n-4}, \cdots, a_5, a_3$, if n is odd.

186. (a) Investigate in the plane the broken line $A_0A_1A_2 \cdots A_n$ such that the projections of $A_0A_1, A_1A_2, \cdots, A_{n-1}A_n$ on the x-axis are, respectively, equal to $a_1, a_2, \cdots, a_n$, and on the y-axis are, respectively, equal to $b_1, b_2, \cdots, b_n$. The equality holds if $\dfrac{a_1}{b_1} = \dfrac{a_2}{b_2} = \cdots = \dfrac{a_n}{b_n}$.

    (b) Use the inequality of problem 186 (a).

187. For even n, the problem can be solved geometrically, in a manner analogous to that used for problem 186 (a). The case in which n is odd can be reduced to the case for even n. The equality holds, for even n, if $a_1 = 1 - a_2 = a_3 = 1 - a_4 = \cdots = a_{n-1} = 1 - a_n$, and for odd n, only if $a_1 = a_2 = \cdots = a_n = \dfrac{1}{2}$.

188. Square both sides of the inequality.

189. For all x, cos sin x exceeds sin cos x.

190. (a) If $\log_2 \pi = a$ and $\log_5 \pi = b$, then $\pi^{(1/a) + (1/b)} = 10$.

    (b) If $\log_2 \pi = a$ and $\log_\pi 2 = b$, then $b = \dfrac{1}{a}$.

191. (a) Use the facts that for every angle x in the first quadrant six $x < x$, cos $x < 1$.

    (b) Use the fact that for every angle x in the first quadrant tan $x > x$.

192. Employ the geometric construction of the tangent concept as the ratio of two line segments on the "trigonometric circle"; also, there is a construction in which the tangent is the doubled area of a certain triangle.

193. arc sin [cos(arc sin x)] + arc cos [sin (arc cos x)] = $\dfrac{\pi}{2}$.

194. Substitute for x, in $\cos 32x + a_{31} \cos 31x + \cdots + a_1 \cos x$, the angle $x + \pi$, and add the resulting expression to the original expression.

195. Calculate, successively,

$$2 \sin a_1 \cdot 45° \, ,$$

$$2 \sin \left(a_1 + \frac{a_1 a_2}{2} \right) \cdot 45°$$

$$2 \sin \left(a_1 + \frac{a_1 a_2}{2} + \frac{a_1 a_2 a_3}{4} \right) \cdot 45° \, ,$$

$$\cdots\cdots\cdots\cdots\cdots\cdots\cdots\cdots\cdots \, ,$$

$$2 \sin \left(a_1 + \frac{a_1 a_2}{2} + \cdots + \frac{a_1 a_2 \cdots a_n}{2^{n-1}} \right) \cdot 45° \, ,$$

using the formula

$$2 \sin \frac{\alpha}{2} = \pm \sqrt{2 - 2 \cos \alpha} \, .$$

196. 1.

197. Use the fact that the given polynomials have the same coefficients for x^{20} as does $(1 + x^2 + x^3)^{1000}$ and $(1 - x^2 - x^3)^{1000}$, respectively.

198. Employ the identity $(a + b)(a - b) = a^2 - b^2$.

199. (a) $C_{1001}^{50} = \dfrac{1001!}{50! \, 951!}$.

(b) $1000 \, C_{1001}^{50} - C_{1001}^{52} = \dfrac{51{,}050 \cdot 1001!}{52! \, 950!}$.

200. Designate the given expression by Π_k.

$$\Pi_k = (\Pi_{k-1} - 2)^2 \, .$$

Answer: $\dfrac{4^{2k-1} - 4^{k-1}}{3}$.

201. (a) 6. (b) $6x$.

202. $-x + 3$.

203. Use the fact that the polynomial $x^4 + x^3 + 2x^2 + x + 1$ divides the binomial $x^{12} - 1$.

Answer: 1.

204. (a) $x^4 - 10x^2 + 1 = 0$.

(b) $x^6 - 6x^4 - 6x^3 + 12x^2 - 36x + 1 = 0$.

205. Transform the expression $(\alpha - \gamma)(\beta - \gamma)(\alpha + \delta)(\beta + \delta)$, using the facts that $\alpha + \beta = -p, \alpha\beta = 1; \gamma + \delta = -q, \gamma\delta = 1$.

206. $Q^2 + q^2 - pP(Q + q) + qP^2 + Qp^2 - 2Qq$.

207. $a = 1,\ a = -2$.

208. $a = 8,\ a = 12$.

209. $b = 1, c = 2, a = 3; b = -1, c = -2, a = -3; b = 2, c = -1\ a = 1;$ $b = 1, c = -2, a = -1$.

210. (a) Never.

(b) Only if $n = 2, a_2 = a_1 + 2$ and $n = 4, a_2 = a_1 - 1, a_3 = a_1 + 1,$ $a_4 = a_1 + 2$.

211. Use the fact that if

$$(x - a_1)^2(x - a_2)^2 \cdots (x - a_n)^2 + 1 = p(x)q(x),$$

then $p(x)$ and $q(x)$, as well as $(x - a_1)^2(x - a_2)^2 \cdots (x - a_n)^2 + 1$, will not vanish (become zero) for any x, and therefore cannot change sign. For the rest, the solution is similar to that of problem 210 (a).

212. Use the fact that $14 - 7 = 7$ cannot be factored in integers.

213. If a polynomial is expressible as a product of two polynomial factors with integral coefficients, then for those values of x for which the polynomial takes on value ± 1 the factors also have value ± 1. Also use the fact that a third-degree polynomial cannot take on the same value for more than three values of x.

214. If p and q are distinct integers, then $P(p) - P(q)$ is divisible by $p - q$.

215. Prove that if $P\left(\dfrac{k}{l}\right) = 0$, then $k - pl = \pm 1$ and $k - ql = \pm 1$.

216. (a) Equate coefficients of like powers of x from both members of the equation

$$(a_0 + a_1 x + a_2 x^2 + \cdots + a_n x^n)(b_0 + b_1 x + b_2 x^2 + \cdots + b_m x^m)$$
$$= c_0 + c_1 x + c_2 x^2 + \cdots + c_{n+m} x^{n+m}$$

and use the resulting relations to show that if the polynomial is factored into the product of two polynomial factors, then all the coefficients of one of these factors must be even (but this is impossible because the leading coefficient—the coefficient of the greatest power of x—of the first polynomial is 1).

(b) Make the substitutions $x = y + 1$ and then, as in part (a), show that if the new polynomial is factored into two polynomial factors, all the coefficients of one of these factors are divisible by the prime number 251.

217. Use the same formula as in problem 216 (a) (see the hint to that problem).

218. Prove that $P\left(\dfrac{p}{q}\right)$, where $\dfrac{p}{q}$ is in lowest terms, cannot be an integer.

219. Let $P(N) = M$; prove that $P(N + kM) - P(N)$ is divisible by M for all k.

220. Represent the polynomial $P(x)$, which takes on integral values for integral x, in the form of a sum $P(x)=b_0P_0(x)+b_1P_1(x)+\cdots+b_nP_n(x)$, where the coefficients $b_0, b_1, \cdots, b_n$ are to be determined, and find these coefficients by successively substituting in the latter equation $x = 0, 1, 2, \cdots, n$.

221. (a) See the hint to problem 220.
      (b) Make the substitution $y = x - k$.
      (c) Investigate the polynomial $Q(x) = P(x^2)$.

222. Use De Moivre's formula.

223. Use the results of problem 222 (b).

224. If $x + \dfrac{1}{x} = 2\cos\alpha$, then $x = \cos\alpha \pm i\sin\alpha$.

225. Use De Moivre's formula.

226. Use the result of problem 225.
Answer: $\cos^2\alpha + \cos^2 2\alpha + \cdots + \cos^2 n\alpha$

$$= \frac{n-1}{2} + \frac{\sin(n+1)\alpha\cos n\alpha}{2\sin\alpha}\ ;$$

$$\sin^2\alpha + \sin^2 2\alpha + \cdots + \sin^2 n\alpha$$

$$= \frac{n+1}{2} - \frac{\sin(n+1)\alpha\cos n\alpha}{2\sin\alpha}\ .$$

227. Use De Moivre's formula and Newton's binomial theorem.
Answer: The given expressions are, respectively, equal to

$$2^n\cos^n\frac{\alpha}{2}\cos\frac{n+2}{2}\alpha, \text{ and } 2^n\cos^n\frac{\alpha}{2}\sin\frac{n+2}{2}\alpha.$$

228. Use the identity $\sin A \sin B = \dfrac{1}{2}\left[\cos(A - B) - \cos(A + B)\right]$ and the result of problem 225.

229. Investigate the roots of the equation $x^{2n+1} - 1 = 0$.

230.   Use the formulas of problem 222 (b).

231.   Use the result of problem 230 (b).

Answer: (a) $\dfrac{n(2n-1)}{3}$.   (b) $\dfrac{2n(n+1)}{3}$.

232.   Use the result of problem 230 (a).

Answer: (a) $\dfrac{\sqrt{2n+1}}{2^n}$ and $\dfrac{\sqrt{n}}{2^{n-1}}$.

   (b) $\dfrac{1}{2^n}$ and $\dfrac{\sqrt{n}}{2^{n-1}}$.

233.   Use the fact that if α is an angle in the first quadrant, then $\sin\alpha < \alpha < \tan\alpha$.

234.   (a, b) Use the formula of problem 225.
   (c)   Use the result of problem (b).

235.   (a)   Use the proposition of problem 234 (a).
   (b)   Use the formula of problem 225.

236.   (a)   Use the proposition of problem 234 (a).
   (b)   See the hint to problem 235 (b).
   (c)   Use the result of problem 232 (a).

237.   Use De Moivre's formula, representing $\sin^{50}\alpha$ in the form of a sum of cosines of angles (multiples of α) with suitable coefficients.

Answer: $5000\, C_{50}^{25} R^{50} = \dfrac{5000\cdot 50!}{(25!)^2} R^{50}$.

238.   Use the fact that $2\cos A = \dfrac{b^2 + c^2 - a^2}{bc}$, rational, and $2\cos nA$ is, for any n, a polynomial of degree n in $2\cos A$; also use the theorem of problem 218.

239.   (a)   In the second formula of problem 222 (b) substitute $\theta = \dfrac{m}{n}180°$, $\cos\theta = \dfrac{1}{p}$, $\sin\theta = \dfrac{\sqrt{p^2-1}}{p}$.

   (b)   If $\theta = \dfrac{m}{n}180°$, then $(1 + i\tan\theta)^n = (1 - i\tan\theta)^n$.

Putting here $\tan\theta = \dfrac{p}{q}$, $p \neq q$, leads to an impossible equality.

240.   Use the fact that if a is not divisible by p, then the numbers $a, 2a, 3a, \cdots, (p-1)a$ all yield different remainders when divided by p.

241.   Use the fact that if $k_1\, k_2, \cdots, k_r$ are integers less than N, and are relatively prime to N, then the integers $k_1a, k_2a, \cdots, k_ra$ yield different remainders when divided by N.

242. Use mathematical induction.

243. Use Euler's theorem (problem 241).

244. Prove by induction that, no matter what the integer N is, there exists a power of 2 whose last N digits are all ones or twos. To prove this, use Euler's theorem (problem 241) and the proposition of problem 242.

245. See the hint to problem 240.

246. Use Wilson's theorem (problem 245). The conditions of the problem are satisfied by the number $x = \left(\dfrac{p-1}{2}\right)!$.

247. (a) Show that the product $(a^2 + b^2)(a_1^2 + b_1^2)$ can be represented as the sum of the squares of two polynomials.

(b) Prove that if the product mp, where p is a prime number, and m is not 1 and is less than p, can be represented as a sum of squares of two integers, then m can be reduced by some factor. That is, a number $n < m$ can be found such that np also can be represented as a sum of squares of two integers. Also, use the proposition of problem 246.

(c) Use the results of parts (a) and (b) and the proposition given in the hint to part (b).

248. Prove that for every odd prime number p it is possible to find two numbers x and y, each less than $\dfrac{p}{q}$, such that x^2 and $-y^2 - 1$ yield the same remainder upon division by p.

249. (a) Show that the product

$$(x_1^2 + x_2^2 + x_3^2 + x_4^2)(y_1^2 + y_2^2 + y_3^2 + y_4^2)$$

can be represented as a sum of squares of four polynomials.

(b) The solution is similar to that of problem 247 (see the hint to that problem). Instead of using the proposition of problem 246 in that solution, we must use that of problem 248.

250. Investigate what remainders can be obtained when the sum of three squares is divided by 8.

251. First, show that if a number is representable as the sum of four squares of integers, then six times its square can be represented as the sum of twelve fourth powers of integers. Further, use the fact that every integer yields a remainder of at most 5 when divided by 6, and apply twice the theorem of problem 249 (b).

252. The idea of the solution is as follows. It is necessary to find some solution in positive rational numbers for the equation

$$x^3 + y^3 + z^3 = a ,$$

where a is a given rational number. We have here one equation and three unknowns. If we assume that two of these unknowns are functionally related in some way, then the equation may be greatly simplified—for example, if we let $y = -z$, then we have simply $x^3 = a$. From the simplified equation it is sometimes possible to express one of the unknowns in terms of a; in this instance, for example, we obtained $x = \sqrt[3]{a}$. Here x is irrational, which will generally be the case with such a procedure. We must try to find a functional dependence between two of the unknowns such that the third unknown comes out of the simplified equation with rational value (without the radical). If this can be done, it remains only to find suitable rational solutions for the two other unknowns, and these are related by the function which was set up between them. To do this, it is convenient to change the unknowns x, y, and z to new unknowns. For the determination of the new variables, it is useful to begin with the identity of problem 162 (b).

253. Let $p_1, p_2, \cdots, p_n$ be n primes. Find an integer not divisible by any of these primes and larger than any of them.

254. (a) Let $p_1, p_2, \cdots, p_n$ be n primes of form $4k - 1$ (or $6k - 1$). Find an integer of form $4n - 1$ (or $6n - 1$) which is not divisible by any of the primes $p_1, p_2, \cdots, p_n$ and is larger than any of them.

  (b) The idea of the solution is somewhat similar to that of problem (a). The proof involves the result of problem 247 (c).

  (c) The proof is quite similar to that of problem (a).

255. Apply the inequality $\sqrt{ab} \leqq \dfrac{a + b}{2}$. (See Section 11.)

256. Use the inequality $\left(\dfrac{a + b}{2}\right)^2 \leqq \dfrac{a^2 + b^2}{2}$. (See Section 11.)

257. Use the inequality of the hint to problem 255.

258. See the hints to problems 255 and 256.

259. See the hint to problem 255.

260. See the hint to problem 255.
Answer: $x = \sqrt[4]{a/b}$.

261. See the hint to problem 255.

262. (a) Apply the expressions for the arithmetic, geometric, and harmonic means to the numbers. (See Section 11.)
(b) This follows from part (a).

263. Use the result of problem 162 (a).

264. Apply the inequality of problem 263.

265. Apply the inequality of problem 263.
Answer: $x = y = z = \dfrac{a}{3}$.

266. Use the inequality of problem 263.

267. Apply the inequality of the hint to problem 255 (*m* times).

268. *First Proof.* Prove that if the proposition holds for $n + 1$ numbers, then it is valid also for n numbers; further, use the result of problem 267.
Second Proof. Use mathematical induction, using the fact that if a_{n+1} is the greatest of the $n + 1$ numbers $a_1, a_2, \cdots, a_{n+1}$, then

$$a_{n+1} \geqq \frac{a_1 + a_2 + \cdots + a_n}{n}.$$

Third Proof. Prove, by mathematical induction, that

$$b_1^n + b_2^n + \cdots + b_n^n \geqq n b_1 b_2 \cdots b_n$$

is equivalent to the theorem of the arithmetic and geometric means. (See Section 11.)

269. Use the theorem of the arithmetic and geometric means (problem 268).

270. See the hint to problem 269.

271. Use the inequality of problem 268.

272. Use problems 268 and 270.

273–275. Use inequality of problem 268.

276. Put the left side of the inequality into the form

$$(1^a 2^{2^{n-1}} 4^{2^{2n-1}} \cdots 2^n)^{1/2n} \, ,$$

277–278. Use the inequality of problem 268.
Answer: To problem 278: If $x = 0$.

279. Use the proposition of problem 269 (a).

280. See the hint to problem 279.
Answer: $b = \dfrac{a}{3}$.

281. (a) Proof is similar to that of the theorem of arithmetic and geometric means (problem 268).

(b) For $n = 2$ the inequality is proven by induction. The proof is analogous to the solution of problem (a).

282. Apply the theorem of arithmetic and geometric means (problem 268).

283. For rational $\dfrac{\alpha}{\beta}$, the problem reduces to the theorem of arithmetic and geometric means (problem 268). If $\dfrac{\alpha}{\beta}$ is an irrational number then a limiting process is required.

284. Use the theorem of the power means (problem 283).
Answers: (a) 12, 24. (b) $18\sqrt{6}, 3\sqrt{6}$.

285. Use mathematical induction.

286. Use the result of the preceding problem.

287. Apply the theorem of symmetric means (problem 286).
Answer: 8, 16.

288. Use the theorem of symmetric means.
Answer: (a) $\sqrt{3}$. (b) $\dfrac{\sqrt{3}}{9}$.

289. *First Proof.* Use the inequality

$$(xa_1 + b_1)^2 + (xa_2 + b_2)^2 + \cdots + (xa_n + b_n)^2 \geq 0,$$

which holds for all real numbers x, and in particular if

$$x = -\frac{a_1b_1 + a_2b_2 + \cdots + a_nb_n}{a_1^2 + a_2^2 + \cdots + a_n^2}.$$

Second proof. Write

$$\bar{a}_i = \frac{a_i}{A},$$

$$\bar{b}_i = \frac{b_i}{B} \qquad (i = 1, 2, \cdots, n),$$

where

$$A = \sqrt{a_1^2 + \cdots + a_n^2},$$
$$B = \sqrt{b_1^2 + \cdots + b_n^2},$$

and use the obvious inequality

$$\bar{a}_i \bar{b}_i \leqq \frac{1}{2} \bar{a}_i^2 + \frac{1}{2} \bar{b}_i^2 \qquad (i = 1, 2, \cdots, n).$$

Prove, first, that

$$\bar{a}_1 \bar{b}_1 + \cdots + \bar{a}_n \bar{b}_n \leqq 1.$$

Third Proof. Use mathematical induction.
Fourth Proof. Prove, first, the equality

$$\left(\sum_{k=1}^{n} a_k^2 \right)\left(\sum_{k=1}^{n} b_k^2 \right) - \left(\sum_{k=1}^{n} a_k b_k \right)^2 = \frac{1}{2} \sum_{k=1}^{n} \sum_{l=1}^{n} (a_k b_l - a_l b_k)^2.$$

290. Use the Cauchy-Buniakowski inequality on the two sets

$$\sqrt{a_1}, \sqrt{a_2}, \cdots, \sqrt{a_n}$$

and

$$\sqrt{\frac{1}{a_1}}, \sqrt{\frac{1}{a_2}}, \cdots, \sqrt{\frac{1}{a_n}}.$$

291. Substitute into the Cauchy-Buniakowski inequality

$$b_1 = b_2 = \cdots = b_n = 1.$$

292. Use the Cauchy-Buniakowski inequality.

293. Substitute into the Cauchy-Buniakowski inequality

$$a_1 = x_1 + y_1, a_2 = x_2 = x_2 + y_2, \cdots, a_n = x_n + y_n;$$
$$b_1 = x_1, b_2 = x_2, \cdots, b_n = x_n,$$

and,

$$a_1 = x_1 + y_1, a_2 = x_2 + y_2, \cdots, a_n = x_n + y_n;$$
$$b_1 = y_1, b_2 = y_2, \cdots, b_n = y_n.$$

294–296. Use the Cauchy-Buniakowski inequality.

297. Compare the given inequality with that of Cauchy-Buniakowski.

298. Apply the Cauchy-Buniakowski inequality twice.

299. If the sequence of numbers a_i is an increasing sequence, $a_1 \leqq a_2 \leqq \cdots \leqq a_n$, then the relationship being examined is maximal when the sequence of b_j is decreasing: $b_1 \geqq b_2 \geqq \cdots \geqq b_n$. Write

$$a_i^2 = \alpha_i a_1^2 + \beta_i a_n^2$$
$$b_i^2 = \alpha_i b_1^2 + \beta_i b_n^2 \quad (i = 2, 3, \cdots, n-1)$$

and use inequalities (2), after problem 288, and (1) of Section 11.

300. Consider the difference

$$n(a_1 b_1 + a_2 b_2 + \cdots + a_n b_n) - (a_1 + a_2 + \cdots + a_n)(b_1 + b_2 + \cdots + b_n) .$$

301. Use the theorem of arithmetic and geometric means (problem 268).

302. See the hint to problem 301.

303. *First Solution.* Use the inequality of problem 302.
Second Solution. Use the inequality of problem 301.

304. The solution is analogous to that of problem 302.

305. Use the inequality of problem 304.

306. Compare the given inequality with that of problem 304.

307. Use the inequality of problem 304 and the theorem of arithmetic and geometric means (problem 263).

308. Use Hölder's inequality (problem 303).

309. (a) Use the formula

$$a^k - b^k = (a - b)(a^{k-1} + a^{k-2}b + a^{k-3}b^2 + \cdots + ab^{k-2} + b^{k-1}) .$$

(b) Apply the result of problem (a).

310. First calculate the greatest coefficient of the series $u_n^{(1)} = b_1 n^{k-1} + \cdots$ [see problem 309 (a)].

311. Use mathematical induction.

312. See the hint to problem 311.

313. Use the formula of problem 311 (a).

314. Use the formula of problem 313.

315. See the hint to problem 314.
Answer: $1^4 + 2^4 + 3^4 + \cdots + n^4 = \dfrac{n(n+1)(2n+1)(3n^2 + 3n - 1)}{30}$.

316. (a) Investigate the series

$$u_n = 1^k + 2^k + 3^k + \cdots + n^k,$$

and use the results of problems 309 (b) and 314.

(b) Apply the result of problem 310 to the series $u_n = 1^k + 2^k + 3^k + \cdots + n^k$, and to the series $u_n^{(1)} = (n + 1)^k$. Using the formula of problem 313, determine the numbers of the $(k-1)$st difference of the kth degree sequence.

Answer: The coefficient of n^{k+1} is equal to $\dfrac{1}{k+1}$; the coefficient of n^k is $\dfrac{1}{2}$.

317. Use the formula of problem 313.

318. Construct an addition triangle for the numbers $C_n^0, C_n^1, C_n^2, \cdots, C_n^n$. *Answer:* C_{2n}^n.

319. Investigate the addition triangle for the numbers of the sequence $1, \dfrac{a}{b}, \dfrac{a^2}{b^2}, \cdots, \dfrac{a^n}{b^n}$.

320. Prove that the difference triangle of the sequence $1, \dfrac{1}{2}, \dfrac{1}{3}, \cdots, \dfrac{1}{n}, \cdots$ has the form

$$\frac{1}{C_0^0} \quad \frac{1}{2C_1^1} \quad \frac{1}{3C_2^2} \quad \frac{1}{4C_3^3} \cdots$$

$$-\frac{1}{2C_1^0} - \frac{1}{3C_2^1} - \frac{1}{4C_3^2} \cdots$$

$$\frac{1}{3C_2^0} \quad \frac{1}{4C_3^1} \cdots$$

$$-\frac{1}{4C_3^0} \cdots$$

A CATALOG OF SELECTED
DOVER BOOKS
IN ALL FIELDS OF INTEREST

A CATALOG OF SELECTED DOVER
BOOKS IN ALL FIELDS OF INTEREST

CONCERNING THE SPIRITUAL IN ART, Wassily Kandinsky. Pioneering work by father of abstract art. Thoughts on color theory, nature of art. Analysis of earlier masters. 12 illustrations. 80pp. of text. 5⅜ x 8½.   0-486-23411-8

CELTIC ART: The Methods of Construction, George Bain. Simple geometric techniques for making Celtic interlacements, spirals, Kells-type initials, animals, humans, etc. Over 500 illustrations. 160pp. 9 x 12. (Available in U.S. only.)   0-486-22923-8

AN ATLAS OF ANATOMY FOR ARTISTS, Fritz Schider. Most thorough reference work on art anatomy in the world. Hundreds of illustrations, including selections from works by Vesalius, Leonardo, Goya, Ingres, Michelangelo, others. 593 illustrations. 192pp. 7⅛ x 10¼.   0-486-20241-0

CELTIC HAND STROKE-BY-STROKE (Irish Half-Uncial from "The Book of Kells"): An Arthur Baker Calligraphy Manual, Arthur Baker. Complete guide to creating each letter of the alphabet in distinctive Celtic manner. Covers hand position, strokes, pens, inks, paper, more. Illustrated. 48pp. 8¼ x 11.   0-486-24336-2

EASY ORIGAMI, John Montroll. Charming collection of 32 projects (hat, cup, pelican, piano, swan, many more) specially designed for the novice origami hobbyist. Clearly illustrated easy-to-follow instructions insure that even beginning papercrafters will achieve successful results. 48pp. 8¼ x 11.   0-486-27298-2

BLOOMINGDALE'S ILLUSTRATED 1886 CATALOG: Fashions, Dry Goods and Housewares, Bloomingdale Brothers. Famed merchants' extremely rare catalog depicting about 1,700 products: clothing, housewares, firearms, dry goods, jewelry, more. Invaluable for dating, identifying vintage items. Also, copyright-free graphics for artists, designers. Co-published with Henry Ford Museum & Greenfield Village. 160pp. 8¼ x 11.   0-486-25780-0

THE ART OF WORLDLY WISDOM, Baltasar Gracian. "Think with the few and speak with the many," "Friends are a second existence," and "Be able to forget" are among this 1637 volume's 300 pithy maxims. A perfect source of mental and spiritual refreshment, it can be opened at random and appreciated either in brief or at length. 128pp. 5⅜ x 8½.   0-486-44034-6

JOHNSON'S DICTIONARY: A Modern Selection, Samuel Johnson (E. L. McAdam and George Milne, eds.). This modern version reduces the original 1755 edition's 2,300 pages of definitions and literary examples to a more manageable length, retaining the verbal pleasure and historical curiosity of the original. 480pp. 5³⁄₁₆ x 8¼.   0-486-44089-3

ADVENTURES OF HUCKLEBERRY FINN, Mark Twain, Illustrated by E. W. Kemble. A work of eternal richness and complexity, a source of ongoing critical debate, and a literary landmark, Twain's 1885 masterpiece about a barefoot boy's journey of self-discovery has enthralled readers around the world. This handsome clothbound reproduction of the first edition features all 174 of the original black-and-white illustrations. 368pp. 5⅜ x 8½.   0-486-44322-1

STICKLEY CRAFTSMAN FURNITURE CATALOGS, Gustav Stickley and L. & J. G. Stickley. Beautiful, functional furniture in two authentic catalogs from 1910. 594 illustrations, including 277 photos, show settles, rockers, armchairs, reclining chairs, bookcases, desks, tables. 183pp. 6½ x 9¼. 0-486-23838-5

AMERICAN LOCOMOTIVES IN HISTORIC PHOTOGRAPHS: 1858 to 1949, Ron Ziel (ed.). A rare collection of 126 meticulously detailed official photographs, called "builder portraits," of American locomotives that majestically chronicle the rise of steam locomotive power in America. Introduction. Detailed captions. xi+ 129pp. 9 x 12. 0-486-27393-8

AMERICA'S LIGHTHOUSES: An Illustrated History, Francis Ross Holland, Jr. Delightfully written, profusely illustrated fact-filled survey of over 200 American lighthouses since 1716. History, anecdotes, technological advances, more. 240pp. 8 x 10¾.
0-486-25576-X

TOWARDS A NEW ARCHITECTURE, Le Corbusier. Pioneering manifesto by founder of "International School." Technical and aesthetic theories, views of industry, economics, relation of form to function, "mass-production split" and much more. Profusely illustrated. 320pp. 6⅛ x 9¼. (Available in U.S. only.) 0-486-25023-7

HOW THE OTHER HALF LIVES, Jacob Riis. Famous journalistic record, exposing poverty and degradation of New York slums around 1900, by major social reformer. 100 striking and influential photographs. 233pp. 10 x 7⅞. 0-486-22012-5

FRUIT KEY AND TWIG KEY TO TREES AND SHRUBS, William M. Harlow. One of the handiest and most widely used identification aids. Fruit key covers 120 deciduous and evergreen species; twig key 160 deciduous species. Easily used. Over 300 photographs. 126pp. 5⅜ x 8½. 0-486-20511-8

COMMON BIRD SONGS, Dr. Donald J. Borror. Songs of 60 most common U.S. birds: robins, sparrows, cardinals, bluejays, finches, more—arranged in order of increasing complexity. Up to 9 variations of songs of each species.
Cassette and manual 0-486-99911-4

ORCHIDS AS HOUSE PLANTS, Rebecca Tyson Northen. Grow cattleyas and many other kinds of orchids—in a window, in a case, or under artificial light. 63 illustrations. 148pp. 5⅜ x 8½. 0-486-23261-1

MONSTER MAZES, Dave Phillips. Masterful mazes at four levels of difficulty. Avoid deadly perils and evil creatures to find magical treasures. Solutions for all 32 exciting illustrated puzzles. 48pp. 8¼ x 11. 0-486-26005-4

MOZART'S DON GIOVANNI (DOVER OPERA LIBRETTO SERIES), Wolfgang Amadeus Mozart. Introduced and translated by Ellen H. Bleiler. Standard Italian libretto, with complete English translation. Convenient and thoroughly portable—an ideal companion for reading along with a recording or the performance itself. Introduction. List of characters. Plot summary. 121pp. 5¼ x 8½. 0-486-24944-1

FRANK LLOYD WRIGHT'S DANA HOUSE, Donald Hoffmann. Pictorial essay of residential masterpiece with over 160 interior and exterior photos, plans, elevations, sketches and studies. 128pp. 9¹/₄ x 10¾. 0-486-29120-0

THE CLARINET AND CLARINET PLAYING, David Pino. Lively, comprehensive work features suggestions about technique, musicianship, and musical interpretation, as well as guidelines for teaching, making your own reeds, and preparing for public performance. Includes an intriguing look at clarinet history. "A godsend," *The Clarinet,* Journal of the International Clarinet Society. Appendixes. 7 illus. 320pp. 5⅜ x 8½. 0-486-40270-3

HOLLYWOOD GLAMOR PORTRAITS, John Kobal (ed.). 145 photos from 1926-49. Harlow, Gable, Bogart, Bacall; 94 stars in all. Full background on photographers, technical aspects. 160pp. 8⅜ x 11¼. 0-486-23352-9

THE RAVEN AND OTHER FAVORITE POEMS, Edgar Allan Poe. Over 40 of the author's most memorable poems: "The Bells," "Ulalume," "Israfel," "To Helen," "The Conqueror Worm," "Eldorado," "Annabel Lee," many more. Alphabetic lists of titles and first lines. 64pp. 5$\frac{3}{16}$ x 8¼. 0-486-26685-0

PERSONAL MEMOIRS OF U. S. GRANT, Ulysses Simpson Grant. Intelligent, deeply moving firsthand account of Civil War campaigns, considered by many the finest military memoirs ever written. Includes letters, historic photographs, maps and more. 528pp. 6⅛ x 9¼. 0-486-28587-1

ANCIENT EGYPTIAN MATERIALS AND INDUSTRIES, A. Lucas and J. Harris. Fascinating, comprehensive, thoroughly documented text describes this ancient civilization's vast resources and the processes that incorporated them in daily life, including the use of animal products, building materials, cosmetics, perfumes and incense, fibers, glazed ware, glass and its manufacture, materials used in the mummification process, and much more. 544pp. 6⅛ x 9¼. (Available in U.S. only.) 0-486-40446-3

RUSSIAN STORIES/RUSSKIE RASSKAZY: A Dual-Language Book, edited by Gleb Struve. Twelve tales by such masters as Chekhov, Tolstoy, Dostoevsky, Pushkin, others. Excellent word-for-word English translations on facing pages, plus teaching and study aids, Russian/English vocabulary, biographical/critical introductions, more. 416pp. 5⅜ x 8½. 0-486-26244-8

PHILADELPHIA THEN AND NOW: 60 Sites Photographed in the Past and Present, Kenneth Finkel and Susan Oyama. Rare photographs of City Hall, Logan Square, Independence Hall, Betsy Ross House, other landmarks juxtaposed with contemporary views. Captures changing face of historic city. Introduction. Captions. 128pp. 8¼ x 11. 0-486-25790-8

NORTH AMERICAN INDIAN LIFE: Customs and Traditions of 23 Tribes, Elsie Clews Parsons (ed.). 27 fictionalized essays by noted anthropologists examine religion, customs, government, additional facets of life among the Winnebago, Crow, Zuni, Eskimo, other tribes. 480pp. 6⅛ x 9¼. 0-486-27377-6

TECHNICAL MANUAL AND DICTIONARY OF CLASSICAL BALLET, Gail Grant. Defines, explains, comments on steps, movements, poses and concepts. 15-page pictorial section. Basic book for student, viewer. 127pp. 5⅜ x 8½. 0-486-21843-0

THE MALE AND FEMALE FIGURE IN MOTION: 60 Classic Photographic Sequences, Eadweard Muybridge. 60 true-action photographs of men and women walking, running, climbing, bending, turning, etc., reproduced from rare 19th-century masterpiece. vi + 121pp. 9 x 12. 0-486-24745-7

ANIMALS: 1,419 Copyright-Free Illustrations of Mammals, Birds, Fish, Insects, etc., Jim Harter (ed.). Clear wood engravings present, in extremely lifelike poses, over 1,000 species of animals. One of the most extensive pictorial sourcebooks of its kind. Captions. Index. 284pp. 9 x 12. 0-486-23766-4

1001 QUESTIONS ANSWERED ABOUT THE SEASHORE, N. J. Berrill and Jacquelyn Berrill. Queries answered about dolphins, sea snails, sponges, starfish, fishes, shore birds, many others. Covers appearance, breeding, growth, feeding, much more. 305pp. 5¼ x 8¼. 0-486-23366-9

ATTRACTING BIRDS TO YOUR YARD, William J. Weber. Easy-to-follow guide offers advice on how to attract the greatest diversity of birds: birdhouses, feeders, water and waterers, much more. 96pp. 5³⁄₁₆ x 8¼. 0-486-28927-3

MEDICINAL AND OTHER USES OF NORTH AMERICAN PLANTS: A Historical Survey with Special Reference to the Eastern Indian Tribes, Charlotte Erichsen-Brown. Chronological historical citations document 500 years of usage of plants, trees, shrubs native to eastern Canada, northeastern U.S. Also complete identifying information. 343 illustrations. 544pp. 6½ x 9¼. 0-486-25951-X

STORYBOOK MAZES, Dave Phillips. 23 stories and mazes on two-page spreads: Wizard of Oz, Treasure Island, Robin Hood, etc. Solutions. 64pp. 8¼ x 11. 0-486-23628-5

AMERICAN NEGRO SONGS: 230 Folk Songs and Spirituals, Religious and Secular, John W. Work. This authoritative study traces the African influences of songs sung and played by black Americans at work, in church, and as entertainment. The author discusses the lyric significance of such songs as "Swing Low, Sweet Chariot," "John Henry," and others and offers the words and music for 230 songs. Bibliography. Index of Song Titles. 272pp. 6½ x 9¼. 0-486-40271-1

MOVIE-STAR PORTRAITS OF THE FORTIES, John Kobal (ed.). 163 glamor, studio photos of 106 stars of the 1940s: Rita Hayworth, Ava Gardner, Marlon Brando, Clark Gable, many more. 176pp. 8⅜ x 11¼. 0-486-23546-7

YEKL and THE IMPORTED BRIDEGROOM AND OTHER STORIES OF YIDDISH NEW YORK, Abraham Cahan. Film Hester Street based on *Yekl* (1896). Novel, other stories among first about Jewish immigrants on N.Y.'s East Side. 240pp. 5⅜ x 8½. 0-486-22427-9

SELECTED POEMS, Walt Whitman. Generous sampling from *Leaves of Grass*. Twenty-four poems include "I Hear America Singing," "Song of the Open Road," "I Sing the Body Electric," "When Lilacs Last in the Dooryard Bloom'd," "O Captain! My Captain!"–all reprinted from an authoritative edition. Lists of titles and first lines. 128pp. 5³⁄₁₆ x 8¼. 0-486-26878-0

SONGS OF EXPERIENCE: Facsimile Reproduction with 26 Plates in Full Color, William Blake. 26 full-color plates from a rare 1826 edition. Includes "The Tyger," "London," "Holy Thursday," and other poems. Printed text of poems. 48pp. 5¼ x 7. 0-486-24636-1

THE BEST TALES OF HOFFMANN, E. T. A. Hoffmann. 10 of Hoffmann's most important stories: "Nutcracker and the King of Mice," "The Golden Flowerpot," etc. 458pp. 5⅜ x 8½. 0-486-21793-0

THE BOOK OF TEA, Kakuzo Okakura. Minor classic of the Orient: entertaining, charming explanation, interpretation of traditional Japanese culture in terms of tea ceremony. 94pp. 5⅜ x 8½. 0-486-20070-1

CATALOG OF DOVER BOOKS

PSYCHOLOGY OF MUSIC, Carl E. Seashore. Classic work discusses music as a medium from psychological viewpoint. Clear treatment of physical acoustics, auditory apparatus, sound perception, development of musical skills, nature of musical feeling, host of other topics. 88 figures. 408pp. 5⅜ x 8½. 0-486-21851-1

LIFE IN ANCIENT EGYPT, Adolf Erman. Fullest, most thorough, detailed older account with much not in more recent books, domestic life, religion, magic, medicine, commerce, much more. Many illustrations reproduce tomb paintings, carvings, hieroglyphs, etc. 597pp. 5⅜ x 8½. 0-486-22632-8

SUNDIALS, Their Theory and Construction, Albert Waugh. Far and away the best, most thorough coverage of ideas, mathematics concerned, types, construction, adjusting anywhere. Simple, nontechnical treatment allows even children to build several of these dials. Over 100 illustrations. 230pp. 5⅜ x 8½. 0-486-22947-5

THEORETICAL HYDRODYNAMICS, L. M. Milne-Thomson. Classic exposition of the mathematical theory of fluid motion, applicable to both hydrodynamics and aerodynamics. Over 600 exercises. 768pp. 6⅛ x 9¼. 0-486-68970-0

OLD-TIME VIGNETTES IN FULL COLOR, Carol Belanger Grafton (ed.). Over 390 charming, often sentimental illustrations, selected from archives of Victorian graphics–pretty women posing, children playing, food, flowers, kittens and puppies, smiling cherubs, birds and butterflies, much more. All copyright-free. 48pp. 9¼ x 12¼. 0-486-27269-9

PERSPECTIVE FOR ARTISTS, Rex Vicat Cole. Depth, perspective of sky and sea, shadows, much more, not usually covered. 391 diagrams, 81 reproductions of drawings and paintings. 279pp. 5⅜ x 8½. 0-486-22487-2

DRAWING THE LIVING FIGURE, Joseph Sheppard. Innovative approach to artistic anatomy focuses on specifics of surface anatomy, rather than muscles and bones. Over 170 drawings of live models in front, back and side views, and in widely varying poses. Accompanying diagrams. 177 illustrations. Introduction. Index. 144pp. 8⅜ x11¼. 0-486-26723-7

GOTHIC AND OLD ENGLISH ALPHABETS: 100 Complete Fonts, Dan X. Solo. Add power, elegance to posters, signs, other graphics with 100 stunning copyright-free alphabets: Blackstone, Dolbey, Germania, 97 more–including many lower-case, numerals, punctuation marks. 104pp. 8⅛ x 11. 0-486-24695-7

THE BOOK OF WOOD CARVING, Charles Marshall Sayers. Finest book for beginners discusses fundamentals and offers 34 designs. "Absolutely first rate . . . well thought out and well executed."–E. J. Tangerman. 118pp. 7¾ x 10⅝. 0-486-23654-4

ILLUSTRATED CATALOG OF CIVIL WAR MILITARY GOODS: Union Army Weapons, Insignia, Uniform Accessories, and Other Equipment, Schuyler, Hartley, and Graham. Rare, profusely illustrated 1846 catalog includes Union Army uniform and dress regulations, arms and ammunition, coats, insignia, flags, swords, rifles, etc. 226 illustrations. 160pp. 9 x 12. 0-486-24939-5

WOMEN'S FASHIONS OF THE EARLY 1900s: An Unabridged Republication of "New York Fashions, 1909," National Cloak & Suit Co. Rare catalog of mail-order fashions documents women's and children's clothing styles shortly after the turn of the century. Captions offer full descriptions, prices. Invaluable resource for fashion, costume historians. Approximately 725 illustrations. 128pp. 8⅜ x 11¼. 0-486-27276-1

HOW TO DO BEADWORK, Mary White. Fundamental book on craft from simple projects to five-bead chains and woven works. 106 illustrations. 142pp. 5⅜ x 8. 0-486-20697-1

THE 1912 AND 1915 GUSTAV STICKLEY FURNITURE CATALOGS, Gustav Stickley. With over 200 detailed illustrations and descriptions, these two catalogs are essential reading and reference materials and identification guides for Stickley furniture. Captions cite materials, dimensions and prices. 112pp. 6½ x 9¼. 0-486-26676-1

EARLY AMERICAN LOCOMOTIVES, John H. White, Jr. Finest locomotive engravings from early 19th century: historical (1804–74), main-line (after 1870), special, foreign, etc. 147 plates. 142pp. 11⅜ x 8¼. 0-486-22772-3

LITTLE BOOK OF EARLY AMERICAN CRAFTS AND TRADES, Peter Stockham (ed.). 1807 children's book explains crafts and trades: baker, hatter, cooper, potter, and many others. 23 copperplate illustrations. 140pp. 4⅝ x 6. 0-486-23336-7

VICTORIAN FASHIONS AND COSTUMES FROM HARPER'S BAZAR, 1867–1898, Stella Blum (ed.). Day costumes, evening wear, sports clothes, shoes, hats, other accessories in over 1,000 detailed engravings. 320pp. 9⅜ x 12¼. 0-486-22990-4

THE LONG ISLAND RAIL ROAD IN EARLY PHOTOGRAPHS, Ron Ziel. Over 220 rare photos, informative text document origin (1844) and development of rail service on Long Island. Vintage views of early trains, locomotives, stations, passengers, crews, much more. Captions. 8⅞ x 11¾. 0-486-26301-0

VOYAGE OF THE LIBERDADE, Joshua Slocum. Great 19th-century mariner's thrilling, first-hand account of the wreck of his ship off South America, the 35-foot boat he built from the wreckage, and its remarkable voyage home. 128pp. 5⅜ x 8½. 0-486-40022-0

TEN BOOKS ON ARCHITECTURE, Vitruvius. The most important book ever written on architecture. Early Roman aesthetics, technology, classical orders, site selection, all other aspects. Morgan translation. 331pp. 5⅜ x 8½. 0-486-20645-9

THE HUMAN FIGURE IN MOTION, Eadweard Muybridge. More than 4,500 stopped-action photos, in action series, showing undraped men, women, children jumping, lying down, throwing, sitting, wrestling, carrying, etc. 390pp. 7⅞ x 10⅝. 0-486-20204-6 Clothbd.

TREES OF THE EASTERN AND CENTRAL UNITED STATES AND CANADA, William M. Harlow. Best one-volume guide to 140 trees. Full descriptions, woodlore, range, etc. Over 600 illustrations. Handy size. 288pp. 4½ x 6⅜. 0-486-20395-6

GROWING AND USING HERBS AND SPICES, Milo Miloradovich. Versatile handbook provides all the information needed for cultivation and use of all the herbs and spices available in North America. 4 illustrations. Index. Glossary. 236pp. 5⅜ x 8½. 0-486-25058-X

BIG BOOK OF MAZES AND LABYRINTHS, Walter Shepherd. 50 mazes and labyrinths in all–classical, solid, ripple, and more–in one great volume. Perfect inexpensive puzzler for clever youngsters. Full solutions. 112pp. 8⅛ x 11. 0-486-22951-3

PIANO TUNING, J. Cree Fischer. Clearest, best book for beginner, amateur. Simple repairs, raising dropped notes, tuning by easy method of flattened fifths. No previous skills needed. 4 illustrations. 201pp. 5⅜ x 8½. 0-486-23267-0

CATALOG OF DOVER BOOKS

HINTS TO SINGERS, Lillian Nordica. Selecting the right teacher, developing confidence, overcoming stage fright, and many other important skills receive thoughtful discussion in this indispensible guide, written by a world-famous diva of four decades' experience. 96pp. 5⅜ x 8½. 0-486-40094-8

THE COMPLETE NONSENSE OF EDWARD LEAR, Edward Lear. All nonsense limericks, zany alphabets, Owl and Pussycat, songs, nonsense botany, etc., illustrated by Lear. Total of 320pp. 5⅜ x 8½. (Available in U.S. only.) 0-486-20167-8

VICTORIAN PARLOUR POETRY: An Annotated Anthology, Michael R. Turner. 117 gems by Longfellow, Tennyson, Browning, many lesser-known poets. "The Village Blacksmith," "Curfew Must Not Ring Tonight," "Only a Baby Small," dozens more, often difficult to find elsewhere. Index of poets, titles, first lines. xxiii + 325pp. 5⅜ x 8¼. 0-486-27044-0

DUBLINERS, James Joyce. Fifteen stories offer vivid, tightly focused observations of the lives of Dublin's poorer classes. At least one, "The Dead," is considered a masterpiece. Reprinted complete and unabridged from standard edition. 160pp. 5³⁄₁₆ x 8¼. 0-486-26870-5

GREAT WEIRD TALES: 14 Stories by Lovecraft, Blackwood, Machen and Others, S. T. Joshi (ed.). 14 spellbinding tales, including "The Sin Eater," by Fiona McLeod, "The Eye Above the Mantel," by Frank Belknap Long, as well as renowned works by R. H. Barlow, Lord Dunsany, Arthur Machen, W. C. Morrow and eight other masters of the genre. 256pp. 5⅜ x 8½. (Available in U.S. only.) 0-486-40436-6

THE BOOK OF THE SACRED MAGIC OF ABRAMELIN THE MAGE, translated by S. MacGregor Mathers. Medieval manuscript of ceremonial magic. Basic document in Aleister Crowley, Golden Dawn groups. 268pp. 5⅜ x 8½. 0-486-23211-5

THE BATTLES THAT CHANGED HISTORY, Fletcher Pratt. Eminent historian profiles 16 crucial conflicts, ancient to modern, that changed the course of civilization. 352pp. 5⅜ x 8½. 0-486-41129-X

NEW RUSSIAN-ENGLISH AND ENGLISH-RUSSIAN DICTIONARY, M. A. O'Brien. This is a remarkably handy Russian dictionary, containing a surprising amount of information, including over 70,000 entries. 366pp. 4½ x 6⅛. 0-486-20208-9

NEW YORK IN THE FORTIES, Andreas Feininger. 162 brilliant photographs by the well-known photographer, formerly with *Life* magazine. Commuters, shoppers, Times Square at night, much else from city at its peak. Captions by John von Hartz. 181pp. 9¼ x 10¾. 0-486-23585-8

INDIAN SIGN LANGUAGE, William Tomkins. Over 525 signs developed by Sioux and other tribes. Written instructions and diagrams. Also 290 pictographs. 111pp. 6⅛ x 9¼. 0-486-22029-X

ANATOMY: A Complete Guide for Artists, Joseph Sheppard. A master of figure drawing shows artists how to render human anatomy convincingly. Over 460 illustrations. 224pp. 8⅜ x 11¼. 0-486-27279-6

MEDIEVAL CALLIGRAPHY: Its History and Technique, Marc Drogin. Spirited history, comprehensive instruction manual covers 13 styles (ca. 4th century through 15th). Excellent photographs; directions for duplicating medieval techniques with modern tools. 224pp. 8⅜ x 11¼. 0-486-26142-5

DRIED FLOWERS: How to Prepare Them, Sarah Whitlock and Martha Rankin. Complete instructions on how to use silica gel, meal and borax, perlite aggregate, sand and borax, glycerine and water to create attractive permanent flower arrangements. 12 illustrations. 32pp. 5⅜ x 8½.   0-486-21802-3

EASY-TO-MAKE BIRD FEEDERS FOR WOODWORKERS, Scott D. Campbell. Detailed, simple-to-use guide for designing, constructing, caring for and using feeders. Text, illustrations for 12 classic and contemporary designs. 96pp. 5⅜ x 8½.
0-486-25847-5

THE COMPLETE BOOK OF BIRDHOUSE CONSTRUCTION FOR WOOD-WORKERS, Scott D. Campbell. Detailed instructions, illustrations, tables. Also data on bird habitat and instinct patterns. Bibliography. 3 tables. 63 illustrations in 15 figures. 48pp. 5¼ x 8½.   0-486-24407-5

SCOTTISH WONDER TALES FROM MYTH AND LEGEND, Donald A. Mackenzie. 16 lively tales tell of giants rumbling down mountainsides, of a magic wand that turns stone pillars into warriors, of gods and goddesses, evil hags, powerful forces and more. 240pp. 5⅜ x 8½.   0-486-29677-6

THE HISTORY OF UNDERCLOTHES, C. Willett Cunnington and Phyllis Cunnington. Fascinating, well-documented survey covering six centuries of English undergarments, enhanced with over 100 illustrations: 12th-century laced-up bodice, footed long drawers (1795), 19th-century bustles, 19th-century corsets for men, Victorian "bust improvers," much more. 272pp. 5⅜ x 8¼.   0-486-27124-2

ARTS AND CRAFTS FURNITURE: The Complete Brooks Catalog of 1912, Brooks Manufacturing Co. Photos and detailed descriptions of more than 150 now very collectible furniture designs from the Arts and Crafts movement depict davenports, settees, buffets, desks, tables, chairs, bedsteads, dressers and more, all built of solid, quarter-sawed oak. Invaluable for students and enthusiasts of antiques, Americana and the decorative arts. 80pp. 6½ x 9¼.   0-486-27471-3

WILBUR AND ORVILLE: A Biography of the Wright Brothers, Fred Howard. Definitive, crisply written study tells the full story of the brothers' lives and work. A vividly written biography, unparalleled in scope and color, that also captures the spirit of an extraordinary era. 560pp. 6⅛ x 9¼.   0-486-40297-5

THE ARTS OF THE SAILOR: Knotting, Splicing and Ropework, Hervey Garrett Smith. Indispensable shipboard reference covers tools, basic knots and useful hitches; handsewing and canvas work, more. Over 100 illustrations. Delightful reading for sea lovers. 256pp. 5⅜ x 8½.   0-486-26440-8

FRANK LLOYD WRIGHT'S FALLINGWATER: The House and Its History, Second, Revised Edition, Donald Hoffmann. A total revision—both in text and illustrations—of the standard document on Fallingwater, the boldest, most personal architectural statement of Wright's mature years, updated with valuable new material from the recently opened Frank Lloyd Wright Archives. "Fascinating"—*The New York Times*. 116 illustrations. 128pp. 9¼ x 10¾.   0-486-27430-6

PHOTOGRAPHIC SKETCHBOOK OF THE CIVIL WAR, Alexander Gardner. 100 photos taken on field during the Civil War. Famous shots of Manassas Harper's Ferry, Lincoln, Richmond, slave pens, etc. 244pp. 10⅝ x 8¼.   0-486-22731-6

FIVE ACRES AND INDEPENDENCE, Maurice G. Kains. Great back-to-the-land classic explains basics of self-sufficient farming. The one book to get. 95 illustrations. 397pp. 5⅜ x 8½.   0-486-20974-1

CATALOG OF DOVER BOOKS

MAGIC AND MYSTERY IN TIBET, Madame Alexandra David-Neel. Experiences among lamas, magicians, sages, sorcerers, Bonpa wizards. A true psychic discovery. 32 illustrations. 321pp. 5⅜ x 8½. (Available in U.S. only.) 0-486-22682-4

THE EGYPTIAN BOOK OF THE DEAD, E. A. Wallis Budge. Complete reproduction of Ani's papyrus, finest ever found. Full hieroglyphic text, interlinear transliteration, word-for-word translation, smooth translation. 533pp. 6½ x 9¼.
0-486-21866-X

HISTORIC COSTUME IN PICTURES, Braun & Schneider. Over 1,450 costumed figures in clearly detailed engravings–from dawn of civilization to end of 19th century. Captions. Many folk costumes. 256pp. 8⅜ x 11¾. 0-486-23150-X

MATHEMATICS FOR THE NONMATHEMATICIAN, Morris Kline. Detailed, college-level treatment of mathematics in cultural and historical context, with numerous exercises. Recommended Reading Lists. Tables. Numerous figures. 641pp. 5⅜ x 8½.
0-486-24823-2

PROBABILISTIC METHODS IN THE THEORY OF STRUCTURES, Isaac Elishakoff. Well-written introduction covers the elements of the theory of probability from two or more random variables, the reliability of such multivariable structures, the theory of random function, Monte Carlo methods of treating problems incapable of exact solution, and more. Examples. 502pp. 5⅜ x 8½. 0-486-40691-1

THE RIME OF THE ANCIENT MARINER, Gustave Doré, S. T. Coleridge. Doré's finest work; 34 plates capture moods, subtleties of poem. Flawless full-size reproductions printed on facing pages with authoritative text of poem. "Beautiful. Simply beautiful."–*Publisher's Weekly.* 77pp. 9¼ x 12. 0-486-22305-1

SCULPTURE: Principles and Practice, Louis Slobodkin. Step-by-step approach to clay, plaster, metals, stone; classical and modern. 253 drawings, photos. 255pp. 8⅜ x 11.
0-486-22960-2

THE INFLUENCE OF SEA POWER UPON HISTORY, 1660–1783, A. T. Mahan. Influential classic of naval history and tactics still used as text in war colleges. First paperback edition. 4 maps. 24 battle plans. 640pp. 5⅜ x 8½. 0-486-25509-3

THE STORY OF THE TITANIC AS TOLD BY ITS SURVIVORS, Jack Winocour (ed.). What it was really like. Panic, despair, shocking inefficiency, and a little heroism. More thrilling than any fictional account. 26 illustrations. 320pp. 5⅜ x 8½.
0-486-20610-6

ONE TWO THREE . . . INFINITY: Facts and Speculations of Science, George Gamow. Great physicist's fascinating, readable overview of contemporary science: number theory, relativity, fourth dimension, entropy, genes, atomic structure, much more. 128 illustrations. Index. 352pp. 5⅜ x 8½. 0-486-25664-2

DALÍ ON MODERN ART: The Cuckolds of Antiquated Modern Art, Salvador Dalí. Influential painter skewers modern art and its practitioners. Outrageous evaluations of Picasso, Cézanne, Turner, more. 15 renderings of paintings discussed. 44 calligraphic decorations by Dalí. 96pp. 5⅜ x 8½. (Available in U.S. only.) 0-486-29220-7

ANTIQUE PLAYING CARDS: A Pictorial History, Henry René D'Allemagne. Over 900 elaborate, decorative images from rare playing cards (14th–20th centuries): Bacchus, death, dancing dogs, hunting scenes, royal coats of arms, players cheating, much more. 96pp. 9¼ x 12¼. 0-486-29265-7

LIGHT AND SHADE: A Classic Approach to Three-Dimensional Drawing, Mrs. Mary P. Merrifield. Handy reference clearly demonstrates principles of light and shade by revealing effects of common daylight, sunshine, and candle or artificial light on geometrical solids. 13 plates. 64pp. 5⅜ x 8½. 0-486-44143-1

ASTROLOGY AND ASTRONOMY: A Pictorial Archive of Signs and Symbols, Ernst and Johanna Lehner. Treasure trove of stories, lore, and myth, accompanied by more than 300 rare illustrations of planets, the Milky Way, signs of the zodiac, comets, meteors, and other astronomical phenomena. 192pp. 8⅜ x 11. 0-486-43981-X

JEWELRY MAKING: Techniques for Metal, Tim McCreight. Easy-to-follow instructions and carefully executed illustrations describe tools and techniques, use of gems and enamels, wire inlay, casting, and other topics. 72 line illustrations and diagrams. 176pp. 8¼ x 10⅞. 0-486-44043-5

MAKING BIRDHOUSES: Easy and Advanced Projects, Gladstone Califf. Easy-to-follow instructions include diagrams for everything from a one-room house for bluebirds to a forty-two-room structure for purple martins. 56 plates; 4 figures. 80pp. 8¾ x 6⅜. 0-486-44183-0

LITTLE BOOK OF LOG CABINS: How to Build and Furnish Them, William S. Wicks. Handy how-to manual, with instructions and illustrations for building cabins in the Adirondack style, fireplaces, stairways, furniture, beamed ceilings, and more. 102 line drawings. 96pp. 8¾ x 6⅜. 0-486-44259-4

THE SEASONS OF AMERICA PAST, Eric Sloane. From "sugaring time" and strawberry picking to Indian summer and fall harvest, a whole year's activities described in charming prose and enhanced with 79 of the author's own illustrations. 160pp. 8¼ x 11. 0-486-44220-9

THE METROPOLIS OF TOMORROW, Hugh Ferriss. Generous, prophetic vision of the metropolis of the future, as perceived in 1929. Powerful illustrations of towering structures, wide avenues, and rooftop parks–all features in many of today's modern cities. 59 illustrations. 144pp. 8¼ x 11. 0-486-43727-2

THE PATH TO ROME, Hilaire Belloc. This 1902 memoir abounds in lively vignettes from a vanished time, recounting a pilgrimage on foot across the Alps and Apennines in order to "see all Europe which the Christian Faith has saved." 77 of the author's original line drawings complement his sparkling prose. 272pp. 5⅜ x 8½. 0-486-44001-X

THE HISTORY OF RASSELAS: Prince of Abissinia, Samuel Johnson. Distinguished English writer attacks eighteenth-century optimism and man's unrealistic estimates of what life has to offer. 112pp. 5⅜ x 8½. 0-486-44094-X

A VOYAGE TO ARCTURUS, David Lindsay. A brilliant flight of pure fancy, where wild creatures crowd the fantastic landscape and demented torturers dominate victims with their bizarre mental powers. 272pp. 5⅜ x 8½. 0-486-44198-9

Paperbound unless otherwise indicated. Available at your book dealer, online at **www.doverpublications.com**, or by writing to Dept. GI, Dover Publications, Inc., 31 East 2nd Street, Mineola, NY 11501. For current price information or for free catalogs (please indicate field of interest), write to Dover Publications or log on to **www.doverpublications.com** and see every Dover book in print. Dover publishes more than 500 books each year on science, elementary and advanced mathematics, biology, music, art, literary history, social sciences, and other areas.